Lecture Notes in Artificial Intelligence 9652

Subseries of Lecture Notes in Computer Science

More information about this series at http://www.springer.com/series/1244

James Bailey · Latifur Khan
Takashi Washio · Gillian Dobbie
Joshua Zhexue Huang · Ruili Wang (Eds.)

Advances in Knowledge Discovery and Data Mining

20th Pacific-Asia Conference, PAKDD 2016
Auckland, New Zealand, April 19–22, 2016
Proceedings, Part II

 Springer

Editors

James Bailey
The University of Melbourne
Melbourne, VIC
Australia

Latifur Khan
The University of Texas at Dallas
Richardson, TX
USA

Takashi Washio
Osaka University
Osaka
Japan

Gillian Dobbie
University of Auckland
Auckland
New Zealand

Joshua Zhexue Huang
Shenzhen University
Shenzhen
China

Ruili Wang
Massey University
Auckland
New Zealand

ISSN 0302-9743   ISSN 1611-3349 (electronic)
Lecture Notes in Artificial Intelligence
ISBN 978-3-319-31749-6   ISBN 978-3-319-31750-2   (eBook)
DOI 10.1007/978-3-319-31750-2

Library of Congress Control Number: 2016934425

LNCS Sublibrary: SL7 – Artificial Intelligence

Printed on acid-free paper

This Springer imprint is published by Springer Nature
The registered company is Springer International Publishing AG Switzerland

PC Chairs' Preface

PAKDD 2016 is the 20th conference of the Pacific Asia Conference series on Knowledge Discovery and Data Mining. For the first time, the conference is being held in New Zealand. The conference provides a forum for researchers and practitioners to present and discuss new research results and practical applications.

There were 307 papers submitted to PAKDD 2016 and they underwent a rigorous double blind review process. Each paper was reviewed by three Program Committee (PC) members and meta-reviewed by one Senior Program Committee (SPC) member who also conducted discussions with the reviewers. The Program Chairs then considered the recommendations from SPC members, looked into each paper and its reviews, to make final paper selections. At the end, 91 papers were selected for the conference program and proceedings, resulting in an acceptance rate below 30 %, among which 39 papers were assigned as long presentation and 52 papers were assigned as regular presentation. The review process was supported by the Microsoft CMT system.

The conference started with a day of five high-quality workshops and five tutorials. During the next three days, the Technical Program included 19 paper presentation sessions covering various subjects of knowledge discovery and data mining, a data mining contest, and three keynote talks by world-renowned experts.

We would like to thank all the Program Committee members and external reviewers for their hard work to provide timely and comprehensive reviews and recommendations, which were crucial to the final paper selection and production of a high-quality Technical Program. We would also like to express our sincere thanks to Huiping Cao and Jinyan Li together with the individual Workshop Chairs for organizing the workshop program; Hisashi Kashima and Leman Akoglu together with the individual tutorial speakers for arranging the tutorial program; Ruili Wang for compiling all the accepted papers and for working with the Springer team to produce these proceedings.

We hope that participants in the conference in Auckland, as well as subsequent readers of the proceedings, will find the technical program of PAKDD 2016 to be both inspiring and rewarding.

February 2016

James Bailey
Latifur Khan
Takashi Washio

General Chairs' Preface

It is our great pleasure to welcome you to the 20th Conference of the Pacific Asia Conference series on Knowledge Discovery and Data Mining. PAKDD has successfully brought together researchers and developers since 1997, with the purpose of identifying challenging problems facing the development of advanced knowledge discovery. The 20th edition of PAKDD continues this tradition.

We are delighted to present three outstanding keynote speakers: Naren Ramakrishnan from Virginia Tech, Mark Sagar from The University of Auckland, and Svetha Venkatesh from Deakin University.

We are grateful to the many authors who submitted their work to the PAKDD technical program. The Program Committee was led by James Bailey, Latifur Khan and Takashi Washio. A report on the paper selection process appears in the PC Chairs' Preface.

We also thank the other Chairs in the organization team: Muhammad Asif Naeem for running the Contest; David Tse Jung Huang for publicizing to attract submissions and managing the website; Ranjini Swaminathan for handling the registration process and Yun Sing Koh and Ranjini Swaminathan for the local arrangements ensuring the conference runs smoothly.

We are grateful to the sponsors of the conference, Auckland Tourism Events and Economic Development, and BECA, for their generous sponsorship and support, and the PAKDD Steering Committee for its guidance and Best Paper Award, Student Travel Award and Early Career Research Award sponsorship. We would also like to express our gratitude to The University of Auckland for hosting and organizing this conference. Last but not least, our sincere thanks go to all the local team members and volunteer helpers for their hard work to make the event possible. We hope you enjoy PAKDD 2016 and your time in Auckland, New Zealand.

Gillian Dobbie
Joshua Zhexue Huang

Organization

Organizing Committee

General Co-chairs

Gillian Dobbie	University of Auckland, New Zealand
Joshua Zhexue Huang	Shenzhen University, China

Program Committee Co-chairs

James Bailey	The University of Melbourne, Australia
Latifur Khan	University of Texas at Dallas, USA
Takashi Washio	Institute of Scientific and Industrial Research, Osaka University, Japan

Workshop Co-chairs

Huiping Cao	New Mexico State University, USA
Jinyan Li	University of Technology Sydney, Australia

Tutorial Co-chairs

Leman Akoglu	Stony Brook University, USA
Hisashi Kashima	Kyoto University, Japan

Local Arrangements Co-chairs

Yun Sing Koh	University of Auckland, New Zealand
Ranjini Swaminathan	University of Auckland, New Zealand

Proceedings Chair

Ruili Wang	Massey University, New Zealand

Contest Chair

Muhammad Asif Naeem	AUT University, New Zealand

Publicity and Website Chair

David Tse Jung Huang	University of Auckland, New Zealand

Registration Chair

Ranjini Swaminathan	University of Auckland, New Zealand

Steering Committee

Chairs

Tu Bao Ho (Chair)   Japan Advanced Institute of Science and Technology, Japan

Ee-Peng Lim (Co-Chair)   Singapore Management University, Singapore

Treasurer

Graham Williams   Togaware, Australia (see also under Life Members)

Members

Tu Bao Ho   Japan Advanced Institute of Science and Technology, Japan (Member since 2005, Co-Chair 2012–2014, Chair 2015–2017, Life Member since 2013)

Ee-Peng Lim   Singapore Management University, Singapore (Member since 2006, Co-Chair 2015–2017)

Jaideep Srivastava   University of Minnesota, USA (Member since 2006)

Zhi-Hua Zhou   Nanjing University, China (Member since 2007)

Takashi Washio   Institute of Scientific and Industrial Research, Osaka University (Member since 2008)

Thanaruk   Thammasat University, Thailand (Member since 2009)
  Theeramunkong

P. Krishna Reddy   International Institute of Information Technology, Hyderabad (IIIT-H), India (Member since 2010)

Joshua Z. Huang   Shenzhen Institutes of Advanced Technology, Chinese Academy of Sciences, China (Member since 2011)

Longbing Cao   Advanced Analytics Institute, University of Technology, Sydney (Member since 2013)

Jian Pei   School of Computing Science, Simon Fraser University (Member since 2013)

Myra Spiliopoulou   Information Systems, Otto-von-Guericke-University Magdeburg (Member since 2013)

Vincent S. Tseng   National Cheng Kung University, Taiwan (Member since 2014)

Life Members

Hiroshi Motoda   AFOSR/AOARD and Osaka University, Japan (Member since 1997, Co-Chair 2001–2003, Chair 2004–2006, Life Member since 2006)

Rao Kotagiri   University of Melbourne, Australia (Member since 1997, Co-Chair 2006–2008, Chair 2009–2011, Life Member since 2007, Treasury Co-Sign since 2006)

Huan Liu   Arizona State University, U.S. (Member since 1998, Treasurer 1998–2000, Life Member since 2012)

Ning Zhong	Maebashi Institute of Technology, Japan (Member since 1999, Life member since 2008)
Masaru Kitsuregawa	Tokyo University, Japan (Member since 2000, Life Member since 2008)
David Cheung	University of Hong Kong, China (Member since 2001, Treasurer 2005–2006, Chair 2006–2008, Life Member since 2009)
Graham Williams	Australian National University, Australia (Member since 2001, Treasurer since 2006, Co-Chair 2009–2011, Chair 2012–2014, Life Member since 2009)
Ming-Syan Chen	National Taiwan University, Taiwan, ROC (Member since 2002, Life Member since 2010)
Kyu-Young Whang	Korea Advanced Institute of Science & Technology, Korea (Member since 2003, Life Member since 2011)
Chengqi Zhang	University of Technology Sydney, Australia (Member since 2004, Life Member since 2012)

Past Members

Hongjun Lu	Hong Kong University of Science and Technology (Member 1997–2005)
Arbee L.P. Chen	National Chengchi University, Taiwan, ROC (Member 2002–2009)
Takao Terano	Tokyo Institute of Technology, Japan (Member 2000–2009)

Program Committee

Senior Program Committee Members

Michael Berthold	University of Konstanz, Germany
Tru Cao	Ho Chi Minh City University of Technology, Vietnam
Ming-Syan Chen	National Taiwan University, Taiwan
Peter Christen	The Australian National University, Australia
Ian Davidson	UC Davis, USA
Guozhu Dong	Wright State University
Bart Goethals	University of Antwerp, Belgium
Xiaohua Hu	Drexel University, USA
Joshua Huang	Shenzhen Institutes of Advanced Technology, Chinese Academy of Sciences, China
George Karypis	University of Minnesota, USA
Ming Li	Nanjing University, China
Jiuyong Li	University of South Australia, Australia
Jinyan Li	University of Technology, Sydney
Chih-Jen Lin	National Taiwan University, Taiwan
Nikos Mamoulis	University of Hong Kong, Hong Kong
Wee Keong Ng	Nanyang Technological University, Singapore

Jian Pei	Simon Fraser University, Canada
Wen-Chih Peng	National Chiao Tung University, Taiwan
Rajeev Raman	University of Leicester, United Kingdom
P. Reddy	International Institute of Information Technology, Hyderabad (IIIT-H), India
Dou Shen	Baidu, China
Kyuseok Shim	Seoul National University, Korea
Myra Spiliopoulou	Otto-von-Guericke-University, Germany
Masashi Sugiyama	The University of Tokyo, Japan
Kai Ming Ting	Federation University, Australia
Hanghang Tong	City University of New York, USA
Vincent S. Tseng	National Cheng Kung University, Taiwan
Koji Tsuda	University of Tokyo, Japan
Wei Wang	University of California at Los Angeles, USA
Haixun Wang	Google, USA
Jianyong Wang	Tsinghua University, China
Xindong Wu	University of Vermont, USA
Xing Xie	Microsoft Research Asia, China
Hui Xiong	Rutgers University, USA
Xifeng Yan	UC Santa Barbara, USA
Jeffrey Yu	The Chinese University of Hong Kong, Hong Kong
Osmar Zaiane	University of Alberta, Canada
Yanchun Zhang	Victoria University, Australia
Min-Ling Zhang	Southeast University, China
Yu Zheng	Microsoft Research Asia, China
Ning Zhong	Maebashi Institute of Technology, Japan
Xiaofang Zhou	The University of Queensland, Australia
Zhi-Hua Zhou	Nanjing University, China

Program Committee Members

Mohammad Al Hasan	Purdue University, USA
Shafiq Alam	University of Auckland, New Zealand
Aijun An	York University, Canada
Gustavo Batista	University of Sao Paulo, Brazil
Chiranjib Bhattachar	Indian Institute of Science, India
Albert Bifet	Universite Paris-Saclay, France
Marut Buranarach	National Electronics and Computer Technology Center, Thailand
Krisztian Buza	Budapest University of Technology and Economics, Hungary
Rui Camacho	Universidade do Porto, Portugal
K. Selcuk Candan	Arizona State University, USA
Jeffrey Chan	RMIT University, Australia
Chia-Hui Chang	National Central University, Taiwan
Muhammad Cheema	Monash University, Australia

Nam Huynh	Japan Advanced Institute of Science and Technology, Japan
Akihiro Inokuchi	Kwansei Gakuin University
Motoharu Iwata	NTT Communication Science Laboratories, Japan
Sanjay Jain	National University of Singapore, Singapore
Toshihiro Kamishima	National Institute of Advanced Industrial Science and Technology, Japan
Murat Kantarcioglu	University of Texas at Dallas, USA
Hung-Yu Kao	National Cheng Kung University, Taiwan
Yoshinobu Kawahara	Osaka University, Japan
Irena Koprinska	University of Sydney, Australia
Walter Kosters	Universiteit Leiden, Netherlands
Marzena Kryszkiewicz	Warsaw University of Technology, Poland
Satoshi Kurihara	Osaka University, Japan
Hady Lauw	Singapore Management University, Singapore
Wang-Chien Lee	Pennsylvania State University, USA
Yue-Shi Lee	Ming Chuan University, Taiwan
Philippe Lenca	Telecom Bretagne, France
Carson K. Leung	University of Manitoba, Canada
Geng Li	Oracle Corporation, USA
Chun-hung Li	Hong Kong Baptist University, Hong Kong
Zhenhui Li	Pennsylvania State University, USA
Yidong Li	Beijing Jiaotong University, China
Xiaoli Li	Institute for Infocomm Research, Singapore
Wu-Jun Li	Nanjing University, China
Xuelong Li	University of London, UK
Hsuan-Tien Lin	National Taiwan University, Taiwan
Jerry Chun-Wei Lin	Harbin Institute of Technology Shenzhen, China
Xu-Ying Liu	Southeast University, China
Wei Liu	University of Technology Sydney, Australia
Qingshan Liu	NLPR Institute of Automation Chinese Academy of Science, China
Hua Lu	Aalborg University, Denmark
Jun Luo	Hua Wei Noahs Ark Lab, Hong Kong
Shuai Ma	Beihang University, China
Marco Maggini	Universita degli Studi di Siena, Italy
Hiroshi Mamitsuka	Kyoto University, Japan
Giuseppe Manco	Universita' della Calabria, Italy
Florent Masseglia	INRIA, France
Mohammad Mehedy Masud	United Arab Emirates University
Tomoko Matsui	Institute of Statistial Mathematics, Japan
Xiaofeng Meng	Renmin University of China, China
Nguyen Le Minh	JAIST, Japan
Pabitra Mitra	Indian Institute of Technology Kharagpur, India
Yang-Sae Moon	Kangwon National University, Korea

Guandong Xu	University of Technology Sydney, Australia
Takehisa Yairi	University of Tokyo, Japan
De-Nian Yang	Academia Sinica, Taiwan
Min Yao	Zhejiang University, China
Mi-Yen Yeh	Academia Sinica, Taiwan
Tetsuya Yoshida	Nara Womens University
Yang Yu	Nanjing University, China
De-Chuan Zhan	Nanjing University, China
Daoqiang Zhang	Nanjing University of Aeronautics and Astronautics, China
Du Zhang	California State University, USA
Bo Zhang	Tsinghua University, China
Junping Zhang	Fudan University, China
Wenjie Zhang	University of New South Wales, Australia
Ying Zhang	University of New South Wales, Australia
Zhongfei Zhang	Binghamton University, USA
Zili Zhang	Deakin University, Australia
Mengjie Zhang	Victoria University of Wellington, New Zealand
Zhao Zhang	Soochow University, China
Xiuzhen Zhang	RMIT University, Australia
Peixiang Zhao	Florida State University, USA
Shuigeng Zhou	Fudan University, China
Bin Zhou	University of Maryland Baltimore County, USA
Feida Zhu	Singapore Management University, Singapore
Xingquan Zhu	Florida Atlantic University, USA
Arthur Zimek	Ludwig-Maximilians-University Munchen, Germany

Sponsors

Contents – Part II

Spatiotemporal and Image Data

Denoising Time Series by Way of a Flexible Model for Phase Space
Reconstruction   3
  Minhazul Islam Sk and Arunava Banerjee

Distributed Sequential Pattern Mining in Large Scale Uncertain Databases . . .   17
  Jiaqi Ge and Yuni Xia

DeepCare: A Deep Dynamic Memory Model for Predictive Medicine   30
  Trang Pham, Truyen Tran, Dinh Phung, and Svetha Venkatesh

Indoor Positioning System for Smart Homes Based on Decision Trees
and Passive RFID   42
  Frédéric Bergeron, Kevin Bouchard, Sébastien Gaboury,
  Sylvain Giroux, and Bruno Bouchard

Deep Feature Extraction from Trajectories for Transportation
Mode Estimation.   54
  Yuki Endo, Hiroyuki Toda, Kyosuke Nishida, and Akihisa Kawanobe

Online Learning for Accurate Real-Time Map Matching   67
  Biwei Liang, Tengjiao Wang, Shun Li, Wei Chen, Hongyan Li,
  and Kai Lei

Multi-hypergraph Incidence Consistent Sparse Coding for Image
Data Clustering.   79
  Xiaodong Feng, Sen Wu, Wenjun Zhou, and Zhiwei Tang

Robust Multi-view Manifold Ranking for Image Retrieval   92
  Jun Wu, Jianbo Yuan, and Jiebo Luo

Image Representation Optimization Based on Locally Aggregated
Descriptors. 104
  Shijiang Chen, Guiguang Ding, Chenxiao Li, and Yuchen Guo

Reusing Extracted Knowledge in Genetic Programming to Solve Complex
Texture Image Classification Problems. 117
  Muhammad Iqbal, Bing Xue, and Mengjie Zhang

Personal Credit Profiling via Latent User Behavior Dimensions
on Social Media   130
   Guangming Guo, Feida Zhu, Enhong Chen, Le Wu, Qi Liu,
   Yingling Liu, and Minghui Qiu

Linear Upper Confidence Bound Algorithm for Contextual Bandit Problem
with Piled Rewards   143
   Kuan-Hao Huang and Hsuan-Tien Lin

Incremental Hierarchical Clustering of Stochastic Pattern-Based
Symbolic Data   156
   Xin Xu, Jiaheng Lu, and Wei Wang

Computing Hierarchical Summary of the Data Streams   168
   Zubair Shah, Abdun Naser Mahmood, and Michael Barlow

Anomaly Detection and Clustering

Unsupervised Parameter Estimation for One-Class Support
Vector Machines   183
   Zahra Ghafoori, Sutharshan Rajasegarar, Sarah M. Erfani,
   Shanika Karunasekera, and Christopher A. Leckie

Frequent Pattern Outlier Detection Without Exhaustive Mining   196
   Arnaud Giacometti and Arnaud Soulet

Ensembles of Interesting Subgroups for Discovering High
Potential Employees   208
   Girish Keshav Palshikar, Kuleshwar Sahu, and Rajiv Srivastava

Dynamic Grouped Mixture Models for Intermittent Multivariate
Sensor Data   221
   Naoya Takeishi, Takehisa Yairi, Naoki Nishimura, Yuta Nakajima,
   and Noboru Takata

Parallel Discord Discovery   233
   Tian Huang, Yongxin Zhu, Yishu Mao, Xinyang Li, Mengyun Liu,
   Yafei Wu, Yajun Ha, and Gillian Dobbie

Dboost: A Fast Algorithm for DBSCAN-Based Clustering on High
Dimensional Data   245
   Yuxiao Zhang, Xiaorong Wang, Bingyang Li, Wei Chen, Tengjiao Wang,
   and Kai Lei

A Precise and Robust Clustering Approach Using Homophilic Degrees
of Graph Kernel   257
   Haolin Yang, Deli Zhao, Lele Cao, and Fuchun Sun

Constraint Based Subspace Clustering for High Dimensional
Uncertain Data   271
   Xianchao Zhang, Lu Gao, and Hong Yu

A Clustering-Based Framework for Incrementally Repairing
Entity Resolution   283
   Qing Wang, Jingyi Gao, and Peter Christen

Adaptive Seeding for Gaussian Mixture Models   296
   Johannes Blömer and Kathrin Bujna

A Greedy Algorithm to Construct L1 Graph with Ranked Dictionary   309
   Shuchu Han and Hong Qin

Novel Models and Algorithms

A Rule Based Open Information Extraction Method Using Cascaded
Finite-State Transducer   325
   Hailun Lin, Yuanzhuo Wang, Peng Zhang, Weiping Wang, Yinliang Yue,
   and Zheng Lin

Active Learning Based Entity Resolution Using Markov Logic.   338
   Jeffrey Fisher, Peter Christen, and Qing Wang

Modeling Adversarial Learning as Nested Stackelberg Games.   350
   Yan Zhou and Murat Kantarcioglu

Fast and Semantic Measurements on Collaborative Tagging Quality   363
   Yuqing Sun, Haiqi Sun, and Reynold Cheng

Matrices, Compression, Learning Curves: Formulation, and the
GROUPNTEACH Algorithms.   376
   Bryan Hooi, Hyun Ah Song, Evangelos Papalexakis, Rakesh Agrawal,
   and Christos Faloutsos

Privacy Aware K-Means Clustering with High Utility   388
   Thanh Dai Nguyen, Sunil Gupta, Santu Rana, and Svetha Venkatesh

Secure k-NN Query on Encrypted Cloud Data with Limited Key-Disclosure
and Offline Data Owner.   401
   Youwen Zhu, Zhikuan Wang, and Yue Zhang

Hashing-Based Distributed Multi-party Blocking for Privacy-Preserving
Record Linkage   415
   Thilina Ranbaduge, Dinusha Vatsalan, Peter Christen,
   and Vassilios Verykios

Text Mining and Recommender Systems

Enabling Hierarchical Dirichlet Processes to Work Better for Short Texts
at Large Scale.   431
    Khai Mai, Sang Mai, Anh Nguyen, Ngo Van Linh, and Khoat Than

Query-Focused Multi-document Summarization Based on Concept
Importance.   443
    Hai-Tao Zheng, Ji-Min Guo, Yong Jiang, and Shu-Tao Xia

Mirror on the Wall: Finding Similar Questions with Deep Structured
Topic Modeling   454
    Arpita Das, Manish Shrivastava, and Manoj Chinnakotla

An Efficient Dynamic Programming Algorithm for STR-IC-STR-IC-LCS
Problem.   466
    Daxin Zhu, Yingjie Wu, and Xiaodong Wang

Efficient Page-Level Data Extraction via Schema Induction
and Verification   478
    Chia-Hui Chang, Tian-Sheng Chen, Ming-Chuan Chen,
    and Jhung-Li Ding

Transfer-Learning Based Model for Reciprocal Recommendation   491
    Chia-Hsin Ting, Hung-Yi Lo, and Shou-De Lin

Enhanced SVD for Collaborative Filtering   503
    Xin Guan, Chang-Tsun Li, and Yu Guan

Social Group Based Video Recommendation Addressing
the Cold-Start Problem   515
    Chunfeng Yang, Yipeng Zhou, Liang Chen, Xiaopeng Zhang,
    and Dah Ming Chiu

FeRoSA: A Faceted Recommendation System for Scientific Articles.   528
    Tanmoy Chakraborty, Amrith Krishna, Mayank Singh, Niloy Ganguly,
    Pawan Goyal, and Animesh Mukherjee

Dual Similarity Regularization for Recommendation   542
    Jing Zheng, Jian Liu, Chuan Shi, Fuzhen Zhuang, Jingzhi Li,
    and Bin Wu

Collaborative Deep Ranking: A Hybrid Pair-Wise Recommendation
Algorithm with Implicit Feedback.   555
    Haochao Ying, Liang Chen, Yuwen Xiong, and Jian Wu

Author Index   569

Contents – Part I

Classification

Joint Classification with Heterogeneous Labels Using Random Walk
with Dynamic Label Propagation 3
  Yongxin Liao, Shenxi Yuan, Jian Chen, Qingyao Wu, and Bin Li

Hybrid Sampling with Bagging for Class Imbalance Learning 14
  Yang Lu, Yiu-ming Cheung, and Yuan Yan Tang

Sparse Adaptive Multi-hyperplane Machine 27
  Khanh Nguyen, Trung Le, Vu Nguyen, and Dinh Phung

Exploring Heterogeneous Product Networks for Discovering Collective
Marketing Hyping Behavior 40
  *Qinzhe Zhang, Qin Zhang, Guodong Long, Peng Zhang,
  and Chengqi Zhang*

Optimal Training and Efficient Model Selection for Parameterized Large
Margin Learning 52
  Yuxun Zhou, Jae Yeon Baek, Dan Li, and Costas J. Spanos

Locally Weighted Ensemble Learning for Regression 65
  Man Yu, Zongxia Xie, Hong Shi, and Qinghua Hu

Reliable Confidence Predictions Using Conformal Prediction 77
  Henrik Linusson, Ulf Johansson, Henrik Boström, and Tuve Löfström

Grade Prediction with Course and Student Specific Models 89
  Agoritsa Polyzou and George Karypis

Flexible Transfer Learning Framework for Bayesian Optimisation 102
  *Tinu Theckel Joy, Santu Rana, Sunil Kumar Gupta,
  and Svetha Venkatesh*

A Simple Unlearning Framework for Online Learning
Under Concept Drifts 115
  Sheng-Chi You and Hsuan-Tien Lin

User-Guided Large Attributed Graph Clustering with Multiple
Sparse Annotations 127
  *Jianping Cao, Senzhang Wang, Fengcai Qiao, Hui Wang, Feiyue Wang,
  and Philip S. Yu*

Early-Stage Event Prediction for Longitudinal Data.................. 139
    Mahtab J. Fard, Sanjay Chawla, and Chandan K. Reddy

Toxicity Prediction in Cancer Using Multiple Instance Learning
in a Multi-task Framework................................... 152
    Cheng Li, Sunil Gupta, Santu Rana, Wei Luo, Svetha Venkatesh,
    David Ashely, and Dinh Phung

Shot Boundary Detection Using Multi-instance Incremental
and Decremental One-Class Support Vector Machine 165
    Hanhe Lin, Jeremiah D. Deng, and Brendon J. Woodford

Will I Win Your Favor? Predicting the Success of Altruistic Requests...... 177
    Hsun-Ping Hsieh, Rui Yan, and Cheng-Te Li

Feature Extraction and Pattern Mining

Unsupervised and Semi-supervised Dimensionality Reduction
with Self-Organizing Incremental Neural Network
and Graph Similarity Constraints............................. 191
    Zhiyang Xiang, Zhu Xiao, Yourong Huang, Dong Wang, Bin Fu,
    and Wenjie Chen

Cross-View Feature Hashing for Image Retrieval 203
    Wei Wu, Bin Li, Ling Chen, and Chengqi Zhang

Towards Automatic Generation of Metafeatures 215
    Fábio Pinto, Carlos Soares, and João Mendes-Moreira

Hash Learning with Convolutional Neural Networks for Semantic
Based Image Retrieval..................................... 227
    Jinma Guo, Shifeng Zhang, and Jianmin Li

Bayesian Group Feature Selection for Support Vector Learning Machines ... 239
    Changde Du, Changying Du, Shandian Zhe, Ali Luo, Qing He,
    and Guoping Long

Active Distance-Based Clustering Using K-Medoids 253
    Amin Aghaee, Mehrdad Ghadiri, and Mahdieh Soleymani Baghshah

Analyzing Similarities of Datasets Using a Pattern Set Kernel............ 265
    A. Ibrahim, P.S. Sastry, and Shivakumar Sastry

Significant Pattern Mining with Confounding Variables............... 277
    Aika Terada, David duVerle, and Koji Tsuda

Building Compact Lexicons for Cross-Domain SMT by Mining
Near-Optimal Pattern Sets 290
Pankaj Singh, Ashish Kulkarni, Himanshu Ojha, Vishwajeet Kumar,
and Ganesh Ramakrishnan

Forest CERN: A New Decision Forest Building Technique 304
Md. Nasim Adnan and Md. Zahidul Islam

Sparse Logistic Regression with Logical Features 316
Yuan Zou and Teemu Roos

A Nonlinear Label Compression and Transformation Method
for Multi-label Classification Using Autoencoders..................... 328
Jörg Wicker, Andrey Tyukin, and Stefan Kramer

Preconditioning an Artificial Neural Network Using Naive Bayes 341
Nayyar A. Zaidi, François Petitjean, and Geoffrey I. Webb

OCEAN: Fast Discovery of High Utility Occupancy Itemsets 354
Bilong Shen, Zhaoduo Wen, Ying Zhao, Dongliang Zhou,
and Weimin Zheng

Graph and Network Data

Leveraging Emotional Consistency for Semi-supervised
Sentiment Classification...................................... 369
Minh Luan Nguyen

The Effect on Accuracy of Tweet Sample Size for Hashtag Segmentation
Dictionary Construction....................................... 382
Laurence A.F. Park and Glenn Stone

Social Identity Link Across Incomplete Social Information Sources
Using Anchor Link Expansion 395
Yuxiang Zhang, Lulu Wang, Xiaoli Li, and Chunjing Xiao

Discovering the Network Backbone from Traffic Activity Data............ 409
Sanjay Chawla, Kiran Garimella, Aristides Gionis, and Dominic Tsang

A Fast and Complete Enumeration of Pseudo-Cliques for Large Graphs..... 423
Hongjie Zhai, Makoto Haraguchi, Yoshiaki Okubo, and Etsuji Tomita

Incorporating Heterogeneous Information for Mashup Discovery
with Consistent Regularization 436
Yao Wan, Liang Chen, Qi Yu, Tingting Liang, and Jian Wu

Link Prediction in Schema-Rich Heterogeneous Information Network 449
Xiaohuan Cao, Yuyan Zheng, Chuan Shi, Jingzhi Li, and Bin Wu

FastStep: Scalable Boolean Matrix Decomposition   461
  Miguel Araujo, Pedro Ribeiro, and Christos Faloutsos

Applications

An Expert-in-the-loop Paradigm for Learning Medical Image Grouping   477
  Xuan Guo, Qi Yu, Rui Li, Cecilia Ovesdotter Alm, Cara Calvelli,
  Pengcheng Shi, and Anne Haake

Predicting Post-operative Visual Acuity for LASIK Surgeries   489
  Manish Gupta, Prashant Gupta, Pravin K. Vaddavalli, and Asra Fatima

LBMF: Log-Bilinear Matrix Factorization for Recommender Systems   502
  Yunhui Guo, Xin Wang, and Congfu Xu

An Empirical Study on Hybrid Recommender System
with Implicit Feedback   514
  Sunhwan Lee, Anca Chandra, and Divyesh Jadav

Who Will Be Affected by Supermarket Health Programs? Tracking
Customer Behavior Changes via Preference Modeling   527
  Ling Luo, Bin Li, Shlomo Berkovsky, Irena Koprinska, and Fang Chen

TrafficWatch: Real-Time Traffic Incident Detection and Monitoring
Using Social Media.   540
  Hoang Nguyen, Wei Liu, Paul Rivera, and Fang Chen

Automated Setting of Bus Schedule Coverage Using Unsupervised
Machine Learning.   552
  Jihed Khiari, Luis Moreira-Matias, Vitor Cerqueira, and Oded Cats

Effective Local Metric Learning for Water Pipe Assessment.   565
  Mojgan Ghanavati, Raymond K. Wong, Fang Chen, Yang Wang,
  and Simon Fong

Classification with Quantification for Air Quality Monitoring   578
  Sanad Al-Maskari, Eve Bélisle, Xue Li, Sébastien Le Digabel,
  Amin Nawahda, and Jiang Zhong

Predicting Unknown Interactions Between Known Drugs and Targets
via Matrix Completion.   591
  Qing Liao, Naiyang Guan, Chengkun Wu, and Qian Zhang

Author Index   605

Spatiotemporal and Image Data

Denoising Time Series by Way of a Flexible Model for Phase Space Reconstruction

Minhazul Islam Sk$^{(\boxtimes)}$ and Arunava Banerjee

University of Florida, Gainesville, FL, USA
smislam@cise.ufl.edu

Abstract. We present a denoising technique in the domain of time series data that presumes a model for the uncorrupted underlying signal rather than a model for noise. Specifically, we show how the non-linear reconstruction of the underlying dynamical system by way of time delay embedding yields a new solution for denoising where the underlying dynamics is assumed to be highly non-linear yet low-dimensional. The model for the underlying data is recovered using a non-parametric Bayesian approach and is therefore very flexible. The proposed technique first clusters the reconstructed phase space through a Dirichlet Process Mixture of Exponential density, an infinite mixture model. Phase Space Reconstruction is accomplished by time delay embedding in the framework of Taken's Embedding Theorem with the underlying dimension being determined by the False Neighborhood method. Next, an Infinite Mixtures of Linear Regression via Dirichlet Process is used to non-linearly map the phase space data points to their respective temporally subsequent points in the phase space. Finally, a convex optimization based approach is used to restructure the dynamics by perturbing the phase space points to create the new denoised time series. We find that this method yields significantly better performance in noise reduction, power spectrum analysis and prediction accuracy of the phase space.

1 Introduction

Noise is a high dimensional dynamical process which limits the extraction of quantitative information from experimental time series data. Successful removal of noise from time series data requires a model either for the noise or for the dynamics of the uncorrupted time series. For example, in wavelet based denoising methods for time series [14,20], the model for the signal assumes that the expected output of a forward/inverse wavelet transform of the uncorrupted time series is sparse in the wavelet coefficients. In other words, it is presupposed that the signal energy is concentrated on a small number of wavelet basis elements; the remaining elements with negligible coefficients are considered noise. Hard-threshold wavelet [25] and Soft-threshold wavelet [4] are two widely known noise reduction methods that subscribe to this model. Principal Component Analysis, on the other hand, assumes a model for the noise: the variance captured by the least important principal components. Therefore, denoising is accomplished by

© Springer International Publishing Switzerland 2016
J. Bailey et al. (Eds.): PAKDD 2016, Part II, LNAI 9652, pp. 3–16, 2016.
DOI: 10.1007/978-3-319-31750-2_1

dropping the bottom principal components and projecting the data onto the remaining components.

In many cases, the time series is produced by a low-dimensional dynamical system. In such cases, the contamination of noise in the time series can disable measurements of the underlying embedding dimension [12], introduce extra Lyapunov Exponents [2], obscure the fractal structure [9] and limit prediction accuracy [5]. Therefore, reduction of noise while maintaining the underlying dynamics generated from the time series is of paramount importance.

A widely used method in time series denoising is Low-pass filtering. Here noise is assumed to constitute all high frequency components without reference to the characteristics of the underlying dynamics. Unfortunately, low pass filtering is not well suited to non-linear chaotic time series [23]. Since the power spectrum of low-dimensional chaos resembles a noisy time series, removal of the higher frequencies distorts the underlying dynamics, thereby, adding fractal dimensions [15].

In this article, we present a phase space reconstruction based approach to time series denoising. The method is founded on Taken's Embedding Theorem [21], according to which a dynamical system can be reconstructed from a sequence of observations of the output of the system (considered, here, the time series). This respects all properties of the dynamical system that do not change under smooth coordinate transformations.

Informally stated, the proposed technique can be described as follows: Consider a time series, $x(1), x(2), x(3).....$ corrupted by noise. We first reconstruct the phase space by taking time delayed observations from the noisy time series (for example, $\langle x(i), x(i+1)\rangle$ forms a phase space trajectory in 2-dimensions). The minimum embedding dimension (i.e., number of lags) of the underlying phase space is determined via the False Neighborhood method [11], as detailed in Sect. 2.1. Next, we cluster the phase space non-parametrically without imposing any constraints on the number of clusters. Finally, we apply a non-linear regression to approximate the temporally subsequent phase space points for each point in each cluster via a Non-parametric Bayesian approach. Henceforth, we refer to our technique by the acronym NPB-NR, standing for non-parametric Bayesian approach to noise reduction in Time Series.

To elaborate, the second step clusters the reconstructed phase space of the time series through an Infinite Mixture of Gaussian distribution via Dirichlet Process [7]. We consider the entire phase space to be generated from a Dirichlet Process mixture (DP) of some underlying density [6]. DP allows the phase space to choose as many clusters as fits its dynamics. The clusters pick out small neighborhoods of the phase space where the subsequent non-linear approximation would be performed. As the latent underlying density of the phase space is unknown, modeling this with an Infinite mixture model allows NPB-NR to correctly find the phase space density. This is because of the guarantee of posterior consistency of the Dirichlet Process Mixtures under Gaussian base density [18]. Therefore, we choose the mixing density to be Gaussian. The posterior consistency acts as a frequentist justification of Bayesian methods—as more data arrives, the posterior density concentrates on the true underlying density of the data.

In the third step, our goal is to non-linearly approximate the dynamics in each cluster formed above. We use a DP mixture of Linear Regression to non-linearly map each point in a cluster to its image (the temporally subsequent point in the phase space). In this Infinite Mixtures of Regression, we model the data in a specific cluster via a mixtures of local densities (Normal density with Linear Transformation of the covariates (βX) as the Mean). Although the mean function is linear for each local density, marginalizing over the local distribution creates a non-linear mean function. In addition, the variance of the responses vary among mixture components in the clusters, thereby varying among covariates. The non-parametric model ensures that the data determines the number of mixture components in specific clusters and the nature of the local regressions. Again, the basis for the infinite mixture model of linear regression is the guarantee of posterior consistency [22].

In the final step, we restructure the dynamics by minimizing the sum of the deviation between each point in the cluster and its pre-image (previous temporal point) and post-image (next temporal point) yielded by the non-linear regression described above. To create a noise removed time series out of the phase space, readjustment of the trajectory is done by maintaining the co-ordinates of the phase space points to be consistent with time delay embedding.

We demonstrate the accuracy of the NPB-NR model across several experimental settings such as, noise reduction percentage and power spectrum analysis on several dynamical systems like Lorenz, Van-der-poll, Buckling Column, GOPY, Rayleigh and Sinusoid attractors, as compared to low pass filtering. We also show the forecasting performance of the NPB-NR method in time series datasets from various domain like the "DOW 30" index stocks, LASER dataset, Computer Generated Series, Astrophysical dataset, Currency Exchange dataset, US Industrial Production Indices dataset, Darwin Sea Level Pressure dataset and Oxygen Isotope dataset against some of its competitors like GARCH, AR, ARMA, ARIMA, PCA, Kernel PCA and Gaussian Process Regression.

2 Mathematical Background

2.1 Time Delay Embedding and False Neighborhood Method

Time Delay Embedding has become a common approach to reconstruct the phase space from an experimental time series. The central idea is that the dynamics is considered to be governed by a solution traveling through a phase space and a smooth function maps points in the phase space to the measurement with some error. Given a time series of measurements, $x(1), x(2), ..., x(N)$, the phase space is represented by vectors in D-dimensional Euclidean space.

$$y(n) = \langle x(n), x(n+T), ..., x(n+(D-1)T)\rangle \tag{1}$$

Here, T is the time delay and D is the embedding dimension. The temporally subsequent point to $y(n)$ in the phase space is $y(n+1)$. The purpose of the embedding is to unfold the phase space to a multivariate space, which is representative of the original dynamics. [21] has shown that under suitable conditions, if the dynamical system has dimension d_A and if the embedding dimension is chosen as $D > 2d_A$, then all the self-crossings in the trajectory due to the projection can be eliminated. The False Neighborhood method [11] accomplishes this task, where it views the dynamics as a compact object in the phase space. If the embedding dimension is too low (the system is not correctly unfolded), many points that lie very close to each other (i.e., neighbors) are far apart in the higher dimensional correctly unfolded space. Identification of these false neighbors allows the technique to determine that the dynamical system has not been correctly unfolded.

For, the time series, $x(n)$, in d^{th} and $(d+1)^{th}$ dimensional embedding, the Euclidean distance between an arbitrary point, $y(n)$ and its closest neighbor $y'(n)$ is, $R_d^2(n) = \sum_{k=0}^{d-1}[x(n+kT) - x'(n+kT)]^2$ and $R_{d+1}^2(n) = \sum_{k=0}^{d}[x(n+kT) - x'(n+kT)]^2$ respectively. If the ratio of these two distances exceeds a threshold R_{tol} (we took this as 15 in this paper), the points are considered to be false neighbors in the dth dimension. The method starts from $d = 1$ and increases it to D, until only $1-2\%$ of the total points appear as false neighbors. Then, we deem the phase space to be completely unfolded in $\mathcal{R}^D$, a D-dimensional Euclidean Space.

2.2   Dirichlet Process and Its Stick-Breaking Representation

A Dirichlet Process [7], $D(\alpha_0, G_0)$ is defined as a probability distribution over a sample space of probability distributions, $G \sim DP(\alpha_0, G_0)$. Here, α_0 is the concentration parameter and G_0 is the base distribution. According to the stick-breaking construction [19] of DP, G, which is a sample from DP, is an atomic distribution with countably infinite atoms drawn from G_0.

$$v_i | \alpha_0, G_0 \sim Beta(1, \alpha_0), \quad \theta_i | \alpha_0, G_0 \sim G_0, \quad M_i = v_i \prod_{l=1}^{i-1}(1 - v_l), \quad G = \sum_{i=1}^{\infty} M_i . \delta_{\theta_i}$$
$$(2)$$

In the DP mixture model [6], DP is used as a non-parametric prior over parameters of an Infinite Mixture model.

$$z_n | \{v_1, v_2, ...\} \sim Categorical\{M_1, M_2, M_3....\}, \quad X_n | z_n, (\theta_i)_{i=1}^{\infty} \sim F(\theta_{z_n}) \quad (3)$$

Here, F is a distribution parametrized by θ_{z_n}. $\{M_1, M_2, M_3, ...\}$ is defined by Eq. 3.

3   NPB-NR Model

3.1   Step One: Clustering of Phase Space

Given a time series $\{x(1), x(2), ..x(N)\}$, let the minimum embedding dimension be D (using the False Neighborhood). Hence, the reconstructed phase space is,

$$\begin{bmatrix} x(1) & x(2) & ... & x(N - (D-1)T) \\ x(1+T) & x(2+T) & ... & x(N - (D-2)T) \\ \vdots & \vdots & \ddots & \vdots \\ x(1+(D-1)T) & x(2+(D-1)T) & ... & x(N) \end{bmatrix} \quad (4)$$

Here, each column represents a point in the phase space. The generative model of the points in the phase space is now assumed as,

$$v_i|\alpha_1, \alpha_2 \sim Beta(\alpha_1, \alpha_2), \quad \{\mu_{i,d}, \lambda_{i,d}\} \sim \mathcal{N}\left(\mu_{i,d}|m_d, (\beta_d, \lambda_{i,d})^{-1}\right) Gamma\left(\lambda_{i,d}|a_d, b_d\right)$$
$$z_n|\{v_1, v_2, \ldots\} \sim Categorical\{M_1, M_2, M_3 \ldots\}, \quad X_d(n)|z_n \sim \mathcal{N}(\mu_{z_n,d}, \lambda_{z_n,d})$$
(5)

Here, $X_d(n)$ is the dth co-ordinate of the nth phase space point. $\{z, v, \mu_{i,d}, \lambda_{i,d}\}$ is the set of latent variables. The distribution, $\{\mu_{i,d}, \lambda_{i,d}\}$, is the base distribution of the DP. $\{M_1, M_2, M_3 \ldots\}$ denotes the Categorical Distribution parameters determined by Eq. 3. In this DP mixture, the sequence, $\{M_1, M_2, M_3 \ldots\}$, creates an infinite vector of mixing proportions and $\{\mu_{z_n,d}, \lambda_{z_n,d}\}$ are the atoms representing the mixture components. This Infinite Mixtures of Gaussians picks clusters for each phase space point and lets the phase space data determine the number of clusters. From this perspective, we can interpret the DP mixture as a flexible mixture model in which the number of components (i.e., the number of cells in the partition) is random and grows as new data is observed.

3.2 Step Two: Non-linear Mapping of Phase Space Points

Due to the discretization of the original continuous phase space, our assumption is that a point in the phase space is constructed by a nonlinear map R whose form we wish to approximate. In this section, we approximate this non-linear map of the subsequent phase space points via the proposed non-linear regression. We assume that a specific cluster has N points. We reorder these points according to their occurrence in the time series. We then pick the corresponding image of these points (which are the temporally subsequent phase space points according to the original time delay embedding). We map each phase space points in the cluster through an Infinite Mixtures of Linear Regression to their respective images. The model is formally defined as:

$$y_1(n) = R_1(x(n)) \quad y_2(n) = R_2(x(n)) \ldots y_D(n) = R_D(x(n)) \tag{6}$$

Here, $R_{1:D}$ are non-linear Regressors which is described by the following set of equations. Here, $X_d(n)$ and $Y_1(n)$ represent the dth co-ordinate of the nth phase space point and the first co-ordinate of its post image respectively. $\{z, v, \mu_{i,d}, \lambda_{x,i,d}, \beta_{i,d}, \lambda_{y,i}\}$ is the set of latent variables and the distributions, $\{\mu_{i,d}, \lambda_{x,i,d}\}$ and $\{\beta_{i,d}, \lambda_{y,i}\}$ are the base distributions of the DP. $\{M_1, M_2, M_3, \ldots\}$ is defined by Eq. 3. Although these set of equations are for R_1, the same model applies for $R_{2:D}$, representing $Y_{2:D}(n)$.

$$v_i|\alpha_1, \alpha_2 \sim Beta(\alpha_1, \alpha_2), \{\beta_{i,d}, \lambda_{y,i}\} \sim \mathcal{N}\left(\beta_{i,d}|m_{y,d}, (\beta_y, \lambda_{y,i})^{-1}\right) Gamma\left(\lambda_{y,i}|a_y, b_y\right)$$
$$z_n|\{v_1, v_2, \ldots\} \sim Categorical\{M_1, M_2, M_3 \ldots\},$$
$$Y_1(n)|X(n), z_n \sim \mathcal{N}\left(\beta_{z_n,0} + \sum_{d=1}^{D} \beta_{z_n,d} X_d(n), \lambda_{y,z_n}^{-1}\right)$$
(7)

The Infinite Mixture model approach to the Linear Regression makes the covariate be associated with the model via a non-linear function, resulting from marginalizing over the other mixtures with respect to a specific mixture. Also, now the variance is different across different mixtures, thereby capturing Heteroscedasticity.

3.3 Step Three: Restructuring of the Dynamics

The idea here is to perturb the trajectory to make the modified phase space more consistent with the dynamics, which is equivalent to reducing the error by perturbing the phase space points from its original position and also the error between the perturbed position and the mapped position. We have to choose a new sequence of phase space points, $\widehat{x(n)}$, such that following objective is minimized.

$$\sum_{n=1}^{N}(\|\widehat{x(n)} - x(n)\|^2 + \|\widehat{x(n)} - R(x_{\widehat{pre-image}})\|^2 + \|R(\widehat{x(n)}) - (x_{\widehat{post-image}})\|^2 \tag{8}$$

R is the non-linear Regressors $(R_{1:D})$ that are used to temporally approximate the phase space (Described in the section above). N is the number of points in the specific cluster. This is done across all the clusters. In addition, to create the new noise removed time series, perturbations of $x_d(n)$'s are done consistently for all subsequent points, such that we can revert back from the phase space to a time series. For example, if the time delay is 1 and the embedding dimension is 2, then, the phase space points are perturbed in such a way that when $x(n) = (t(n), t(n+1))$ is moved to $\widehat{x(n)} = (\widehat{t(n)}, \widehat{t(n+1)})$, we make the first co-ordinate of $\widehat{x(n+1)}$ to be $\widehat{t(n+1)}$. These form a set of equality constraints. What results is a convex program, that is then solved to retrieve the denoised time series.

The entire algorithm is summarized in Table 1.

Table 1. Algorithm: Step-wise description of NPB-NR process.

---

1. Form the phase space dynamics from the Noisy Time Series according to Eq. 4 with the embedding dimension determined by False Neighborhood method described in Sect. 3.1

---

2. Cluster the points in the phase space via Infinite Mixture of Gaussian Densities formally defined in Eq. 5

---

3. For each cluster, map each phase space point via an infinite mixtures of linear regression$(R_{1:D})$ to its temporally subsequent point (post-image) which is formally defined in Eqs. 6 and 7

---

4. Infer the latent parameters for both Infinite Mixture of Gaussian Densities and Infinite Mixture of Linear Regression. $\{z, v, \mu_{i,d}, \lambda_{i,d}\}$ and $\{z, v, \mu_{i,d}, \lambda_{x,i,d}, \beta_{i,d}, \lambda_{y,i}\}$ were inferred through variational inference which we could not include due to lack of space problem. The inference gives us the form of the regressors,$(R_{1:D})$

---

5. Restructure the dynamics via optimizing the Convex function in Eq. 8. The restructuring is done consistently for all the subsequent points, which leads to the reconstruction of the noise removed time series

---

4   Experimental Results

4.1   An Illustrative Description of the NPB-NR Process

First, we present an illustrative pictorial description of the complete NPB-NR process with a real world historical stock price dataset. Our model for the historical time series of the stock price is a low-dimensional dynamical system that was contaminated by noise and passed through a measurement function at the output. Our task was to denoise the stock price to not only recover the underlying original phase space dynamics and create the subsequent noise removed stock price via the NPB-NR process, but also to utilize it to make better future predictions of the stock price. We picked historical daily close out stock price data of IBM from March-1990 to Sept-2015 for this task. The original noisy time series is plotted in Fig. 2. The various stages of NPB-NR are illustrated in the subsequent figures. The underlying dimension of the phase space turned out to be 3 from the False Neighborhood Method. The Reconstructed Phase Space with noise is shown in Fig. 3. The completely clustered phase space and one specific cluster in the phase space by Dirichlet Process Mixture of Gaussian of NPB-NR (step one) is shown in Fig. 4. For a 3-dimensional phase space, as is the case

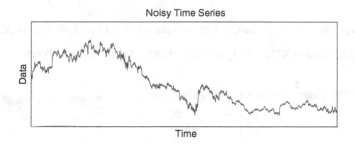

Fig. 1. Plot of the noisy (15 db SNR) and noiseless time series data

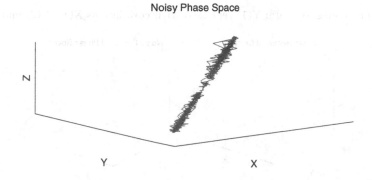

Fig. 2. Depiction of the noiseless and noisy phase space (Reconstructed).

with the IBM stock price data, consider X and Y to be two temporally successive points in one cluster. Therefore, the non-linear regression model (Step Two) in NPB-NR is $Y(1) = R_1(X(1), X(2), X(3)), Y(2) = R_2(X(1), X(2), X(3))$ and $Y(3) = R_3(X(1), X(2), X(3))$. In Fig. 5, we plot $Y(1)$ against $X(1)$, $X(2)$ and $X(3)$ (The first regression-$R(1)$) to depict the non-linearity of the regression model which we have modeled through the Dirichlet Process Mixtures of linear regression (step two). The trajectory adjusted (step three) and consequently the noise removed specific cluster and the complete noise removed phase space are shown in Fig. 6. Finally, the denoised time series is shown in Fig. 7. The error information for prediction for IBM stock data is reported in Table 2.

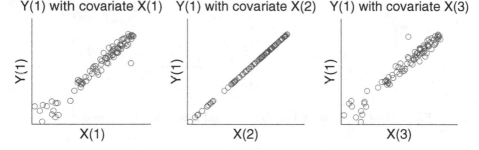

Fig. 3. Depiction of whole clustered phase space (Step One) and one single cluster

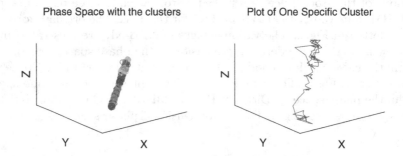

Fig. 4. The regression data: $Y(1)$ regressed with covariate as $X(1)$, $X(2)$ and $X(3)$

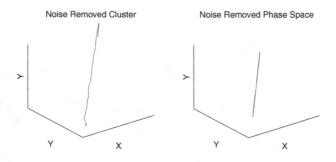

Fig. 5. The single noise removed cluster and whole noise removed phase space

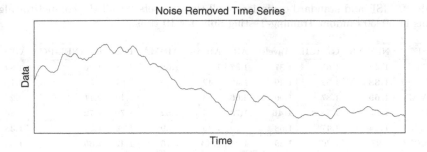

Fig. 6. Plot of the noise removed time series data

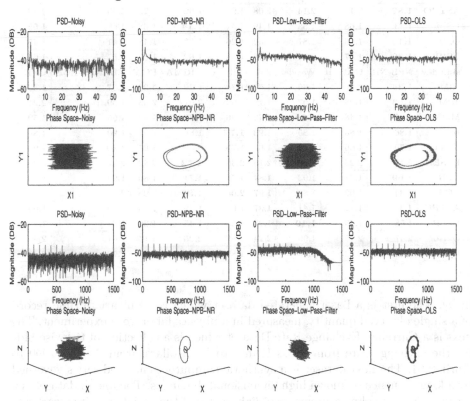

Fig. 7. The Power Spectrum and The Phase Space Plot of Van-der-poll and Sinusoid Attractor

4.2 Prediction Accuracy

NPB-NR was used for time series forecasting. The first dataset was drawn from the stock market. We choose 5 stocks (IBM, JPMorgan, MMM, Home-Depot and Walmart) from March, 2000 to Sept., 2015 with 3239 instances (time points) from "DOW30". The next four datasets came from the Santa Fe competition compiled

Table 2. MSE and standard deviation of all the datasets for all the competitor algorithms in 50–50 random Training-Testing Split for 10 runs.

MSE	NPB-NR	GARCH	Wavelet	AR	ARMA	ARIMA	PCA	KERNEL-PCA	GPR
IBM	**1.43**	1.65	1.37	1.87	1.70	1.68	1.98	1.90	1.84
JPM	**1.38**	1.52	1.49	1.46	1.42	1.39	1.67	1.59	1.73
MMM	**1.69**	1.87	1.96	2.06	1.93	1.83	1.14	2.11	2.23
HD	1.74	1.58	**1.46**	1.73	1.69	1.62	1.79	1.72	1.86
WMT	**1.24**	1.47	1.58	1.39	1.35	1.29	1.49	1.57	1.38
LASER	.97	1.36	1.29	1.31	**.86**	1.15	1.42	1.35	1.34
CER	**.82**	.99	.93	.94	.88	.84	1.18	1.11	1.05
CGS	**1.79**	2.11	1.88	2.03	1.96	1.86	2.38	2.28	2.17
ASTRO	**1.82**	2.19	2.14	2.08	1.91	1.92	2.26	2.33	2.46
DSLP	1.33	1.68	1.53	1.49	1.41	**1.14**	1.68	1.60	1.55
OxIso	**1.19**	1.38	1.87	1.32	1.26	1.53	1.48	1.41	1.45
USIPI	**1.30**	1.57	1.36	1.48	1.43	1.62	1.57	1.57	1.63
stan.dev	NPB-NR	GARCH	Wavelet	AR	ARMA	ARIMA	PCA	KERNEL-PCA	GPR
IBM	**1.34**	1.89	1.86	1.67	1.78	1.39	3.26	2.37	1.42
JPM	**1.63**	1.98	1.97	1.78	1.94	2.01	2.35	1.69	2.21
MMM	1.82	1.48	**1.29**	1.42	1.36	1.82	1.73	1.66	1.59
HD	1.86	1.85	1.85	1.92	**1.77**	1.88	1.93	1.90	1.86
WMT	1.79	1.66	1.82	1.62	1.73	2.31	1.67	**1.61**	1.98
LASER	2.12	2.28	2.39	2.19	2.36	**1.72**	2.42	2.27	2.39
CER	**1.69**	1.78	1.93	1.84	1.74	1.71	1.80	1.91	1.72
CGS	2.34	1.92	1.88	1.97	2.05	**1.87**	1.95	2.17	2.11
ASTRO	**1.13**	1.82	1.41	1.37	1.69	1.79	1.55	1.29	1.62
DSLP	1.58	1.49	**1.19**	1.27	1.35	1.45	1.26	1.42	1.25
OxIso	2.47	2.15	1.99	2.24	1.89	2.25	**1.92**	2.61	2.45
USIPI	**2.23**	2.33	2.49	2.42	1.89	2.29	2.37	2.72	2.25

in [8]. The first is a Laser generated dataset which is a univariate time record of a single observed quantity, measured in a physics laboratory experiment. The next is a Currency Exchange Rate Dataset which is a collection of tickwise bids for the exchange rate from Swiss Francs to US Dollars, from August 1990 to April 1991. The next dataset is a synthetic computer generated series governed by a long sequence of known high dimensional dynamics. The fourth dataset is a set of astrophysical measurements of light curve of the variable white dwarf star PG 1159035 in March, 1989. The next set of datasets are the Darwin sea level pressure dataset from 1882 to 1998, Oxygen Isotope ratio dataset of 2.3 million years and US Industrial Production Indices dataset from Federal Reserve release. All of these are taken from [24]. NPB-NR was compared with the $GARCH$, $AR(\rho)$, $ARMA(p,q)$ and $ARIMA(p,d,q)$ models, where ρ, p, d, q were taken by cross-validations ranging from 1 to 10 fold. We also compared NPB-NR to PCA and kernel PCA [3] with sigma set to 1, and Gaussian Process Based Auto-regression with ρ taken by cross-validations ranging from 1 to 5 fold. We also compared results from Hard Threshold Wavelet denoising using the "wden"

Table 3. Noise reduction percentage of the attractors for the NPB-NR, the low pass filtering method and the hard and soft threshold wavelet method.

	Lorenz	GOPY	Van-Der-Poll	Rossler	Rayleigh
Noise Level–15db SNR					
NPB-NR	40	45	54	29	34
Low Pass Filter	19	27	40	19	31
Wavelet_soft	15	13	29	21	25
Wavelet_hard	17	7	21	18	22
Noise Level–35db SNR					
NPB-NR	51	59	61	40	56
Low Pass Filter	26	31	40	28	39
Wavelet_soft	22	22	36	33	32
Wavelet_hard	23	14	29	24	28
Noise Level–60db SNR					
NPB-NR	63	71	75	79	82
Low Pass Filter	31	35	40	37	41
Wavelet_soft	32	29	41	40	42
Wavelet_hard	29	21	33	32	33
Noise Level–80db SNR					
NPB-NR	72	76	79	81	84
Low Pass Filter	34	39	43	43	44
Wavelet_soft	35	35	46	44	47
Wavelet_hard	34	27	39	36	38
Noise Level–100db SNR					
NPB-NR	80	79	85	85	89
Low Pass Filter	38	43	46	47	46
Wavelet_soft	41	39	50	50	60
Wavelet_hard	36	30	45	40	51

Matlab function. All competitor algorithms were run with a 50-50 training-testing split. We report the Mean Square Error (MSE, L2) of the forecast for all the competitor algorithms in Table 2. Individual time series were reconstructed into a phase space with the dimension determined by the False Neighborhood method, was passed through NPB-NR to find the most consistent dynamics by reducing noise, and subsequently fed into a simple auto-regressor with lag order taken as the embedding dimension of the reconstructed time series. In most datasets, NPB-NR not only yielded better forecasts, but also a smaller standard deviation among its competitors among the 10 runs.

4.3   Noise Reduction Experiment

We evaluated the NPB-NR technique for noise reduction across several well known dynamical systems, namely, Lorenz attractor (chaotic) [13], Van-der-poll attractor [16] and Rossler attractor [17] (periodic), Buckling Column attractor (non strange non chaotic, fixed point), Rayleigh attractor (non strange non chaotic, limit cycle) [1] and GOPY attractor (strange non-chaotic) [10].

Although noise was added to the time series such that the SNR ranged from 15 db to 100 db, it is impossible to calculate numerically or from the Power Spectrum how much noise was actually removed from the noisy time series. Therefore, for both the noise removed and the noisy time series we calculated the fluctuation error:,

$$f_i = \|x_i - x_{i-1} - (dt) \cdot f(x_{i-1}, y_{i-1}, z_{i-1})\|$$

This measures the distance between the observed and the predicted point in the phase space. Here, measurement of the noise reduction percentage is given by,

$$R = 1 - \frac{E_{noise-removed}}{E_{noisy}}, \quad E = \left(\frac{\sum f_i^2}{N}\right)^{\frac{1}{2}}$$

We tabulated the noise reduction percentages of the NPB-NR, the low pass filter, and also wavelet denoising methods in Table 3. For the wavelet method, we used the matlab "wden" function in 'soft' and 'hard' threshold mode. The NPB-NR yielded the highest noise reduction percentage for 15–100 db SNR. Since the faithful reconstruction of the underlying dynamics intrinsically removes the noise, as the noise increases the noise reduction performance of NPB-NR got significantly better as opposed to the other techniques.

4.4   Power Spectrum Experiments

We ran a Power Spectrum experiment for a noise corrupted Van-der-poll attractor (periodic) [16] as well as a time series created by superimposing 6 Sinusoids and subsequently corrupting it with noise. The noise was additive white Gaussian noise with the SNR (Signal-to-Noise ratio) set at 15 db. Var-der-poll is a simple two dimensional attractor with $b = 0.4; x0 = 1; y0 = 1$ and the superimposition of Sinusoids is a simple limit cycle attractor with negative Lyapunov Exponents and no fractal structure. We plot the phase space and the Power Spectrum of the noisy time series generated from these attractors, the noise removed solution with a 6th-order Butterworth low-pass filter (cut-off freq. 30 Hz and 1000 Hz respectively) and the NPB-NR technique. The Power Spectrum and the phase space plot of the Van-der-poll and Sinusoid Attractors is shown in Fig. 1. Note that NPB-NR successfully made the harmonics/peaks more prominent which was originally obscured by the noise. The filtering method was unable to restore the harmonics, although it removed some of the higher frequency components. We also observe that NPB-NR smoothened out the phase space dynamics better than the low pass filter.

5   Conclusion

In this article, we have formulated a Bayesian Non-Parametric Model for noise reduction in time series. The model captures the local non-linear dynamics in the time delay embedded phase space to fit the most appropriate dynamics consistent with the data. We have derived the mean field Variational Inference for the Dirichlet Process Mixture of Linear Regression by maximizing the evidence lower bound to obtain the variational parameters. Finally, we have evaluated the NPB-NR technique on various time series generated from several dynamical systems, stock market data, LASER data, Sea Level Pressure data, etc. The technique yields much better noise reduction percentage, power spectrum analysis, accurate dimension and prediction accuracy. In the experiments, we varied the scale factor from 1 to 5 in increments of .25. This scale factor modulates the number of clusters in the phase space. Developing theoretical insights into how the number of clusters affects the adjustment of the dynamics would be an interesting topic for future research. We also plan to explore which kind of physical systems can be analyzed using Non-parametric Bayesian based noise reduction methods. Finally, considerable effort should be given to analyzing time series generated from higher dimensional systems.

References

1. Abraham, R., Shaw, C.: Dynamics: The Geometry of Behavior. Ariel Press, Santa Cruz (1985)
2. Badii, R., Broggi, G., Derighetti, B., Ravani, M., Ciliberto, S., Politi, A., Rubio, M.: Dimension increase in filtered chaotic signals. Phys. Rev. Lett. **60**, 979–982 (1988)
3. Bishop, C.M.: Pattern Recognition and Machine Learning. Springer, Heidelberg (2006)
4. David, L., Donoho, J.: De-noising by soft-thresholding. IEEE Trans. Inf. Theor. **41**(3), 613–627 (1995)
5. Elshorbagy, A., Panu, U.: Noise reduction in chaotic hydrologic time series: facts and doubts. J. Hydrol. **256**(34), 147–165 (2002)
6. Escobar, D.M., West, M.: Bayesian density estimation and inference using mixtures. J. Am. Stat. Assoc. **90**(430), 577–588 (1995)
7. Ferguson, T.: A bayesian analysis of some nonparametric problems. Ann. Stat. **1**, 209–230 (1973)
8. Gershenfeld, N., Weigend, A.: Time Series Prediction: Forecasting the Future and Under-standing the Past. Addison-Wesley, Reading (1994)
9. Grassberger, P., Schreiber, T., Schaffrath, C.: Non-linear time sequence analysis. Int. J. Bifurcat. Chaos **1**(3), 521–547 (1991)
10. Grebogi, C., Ott, E., Pelikan, S., Yorke, J.A.: Strange attractors that are not chaotic. Physica D **13**(1), 261–268 (1984)
11. Kennel, M.B., Brown, R., Abarbanel, H.D.I.: Determining embedding dimension for phase-space reconstruction using a geometrical construction. Phys. Rev. A **45**(6), 3403–3411 (1992)
12. Kostelich, E.J., Yorke, J.A.: Noise reduction: Finding the simplest dynamical system consistent with the data. Phys. D **41**(2), 183–196 (1990)

13. Lorenz, E.N.: Deterministic nonperiodic flow. J. Atmos. Sci. **20**(2), 130–141 (1963)
14. Mallat, S., Hwang, W.L.: Singularity detection and processing with wavelets. IEEE Trans. Inform. Theor. **38**(2), 617–643 (1992)
15. Mitschke, F., Moller, M., Lange, W.: Measuring filtered chaotic signals. Phys. Rev. A **37**(11), 4518–4521 (1988)
16. Pol, B.V.D.: A theory of the amplitude of free and forced triode vibrations. Radio Rev. **1**, 701–710 (1920)
17. Rossler, O.: An equation for continuous chaos. Phys. Lett. A **57**(5), 397–398 (1976)
18. Ghosal, S., Ghosh, J.K., Ramamoorthi, R.V.: Posterior consistency of dirichlet mixtures in density estimation. Ann. Stat. **27**, 143–158 (1999)
19. Sethuraman, J.: A constructive definition of dirichlet priors. Statistica Sinica **4**, 639–650 (1994)
20. Site, G., Ramakrishnan, A.G.: Wavelet domain nonlinear filtering for evoked potential signal enhancement. Comput. Biomed. Res. **33**(3), 431–446 (2000)
21. Takens, F.: Dynamical systems and turbulence, warwick 1980. In: Rand, D., Young, L.S. (eds.) Detecting strange attractors in turbulence, pp. 366–381. Springer, Heidelberg (1981)
22. Tokdar, S.T.: Posterior consistency of dirichlet location-scale mixture of normals in density estimation and regression. Sankhya: Indian J. Stat. **68**(1), 90–110 (2006)
23. Wang, Z., Lam, J., Liu, X.: Filtering for a class of nonlinear discrete-time stochastic systems with state delays. J. Comput. Appl. Math. **201**(1), 153–163 (2007)
24. West, M.: http://aiweb.techfak.uni-bielefeld.de/content/bworld-robot-control-software/
25. Zhang, L., Bao, P., Pan, Q.: Threshold analysis in wavelet-based denoising. Electron. Lett. **37**(24), 1485–1486 (2001)

Distributed Sequential Pattern Mining in Large Scale Uncertain Databases

Jiaqi Ge$^{(\boxtimes)}$ and Yuni Xia

Department of Computer and Information Science,
Indiana University Purdue University, Indianapolis, IN 46202, USA
{jiaqge,yxia}@cs.iupui.edu

Abstract. While sequential pattern mining (SPM) is an import application in uncertain databases, it is challenging in efficiency and scalability.
In this paper, we develop a dynamic programming (DP) approach to
mine probabilistic frequent sequential patterns in distributed computing
platform Spark. Directly applying the DP method to Spark is impractical because its memory-consuming characteristic may cause heavy JVM
garbage collection overhead in Spark. Therefore, we design a memory-
efficient distributed DP approach and use an extended prefix-tree to
save intermediate results efficiently. The extensive experimental results
in various scales prove that our method is orders of magnitude faster
than straight-forward approaches.

Keywords: Uncertain databases · Sequential pattern mining · Distributed computing

1 Introduction

Sequential pattern mining (SPM) is one of the most important applications in
data mining. It is widely used to analyze customer behaviors in market-basket
databases. For example, online shopping websites usually collect customer pur-
chasing records in databases where sequential patterns are mined to reveal buying habits of consumers. However, in many real applications, events occurring in
a sequence may be uncertain for many reasons. For instance, data collected by
sensors are inherently noisy; in privacy protection applications, artificial noises
are added deliberately; data modeling techniques such as classifications may also
produce indeterministic results [1].

Example 1. Consider an online travel website. To increase sales, a large group of
customers are analyzed in order to discover sequential patterns of user-interested
products. These patterns are useful in intelligenet marketing. For example, by
providing a special hotel offer to a customer who booked a late-night flight, the
website is able to encourage hotel purchases.

Figure 1(a) records user preferences to various travel products. For example,
in the session B, the customer is first attracted by a rental car and then shows

© Springer International Publishing Switzerland 2016
J. Bailey et al. (Eds.): PAKDD 2016, Part II, LNAI 9652, pp. 17–29, 2016.
DOI: 10.1007/978-3-319-31750-2_2

Session	timestamp	Products	Prob.
A	1	flight	0.7
A	2	flight, hotel	0.6
A	3	hotel	0.8
B	1	car	1.0
B	2	hotel	0.7

SID	Sequence
S_A	<1, (flight),0.7>;<2, (flight, hotel), 0.6>; <3, (hotel), 0.8>
S_B	<1, (car), 1.0>;<2, (hotel), 0.7>

(a) uncertain sequence database

world	Sequence	Prob.
1	<(flight)>; <(flight, hotel)>; <(hotel)>	0.336
2	<(flight)>;<(flight, hotel)>	0.084
3	<(flight)>;<(hotel)>	0.224
4	<(flight)>	0.056
5	<(flight, hotel)>;<(hotel)>	0.144
6	<(flight, hotel)>	0.036
7	<(hotel)>	0.096
8	ϕ	0.024

(b) Possible worlds of S_A

Fig. 1. Example application of an uncertain sequence database

interests in a hotel with a probability of 0.7. Here user interests are estimated by posterior probabilities from Naïve Bayesian models which takes features, such as how long a customer stays on a page and so on, into consideration. The website uses a database that represents each visiting session as a single sequence, also shown in Fig. 1(a).

1.1 Problem Statement

The uncertain model applied in this paper is based on possible world semantics with existential *uncertain events*.

Definition 1. *An uncertain event is an event e whose presence in a sequence d is defined by an existential probability $P(e \in d) \in (0, 1]$.*

Definition 2. *An uncertain sequence d is an ordered list of uncertain events. An uncertain sequence database is a collection of uncertain sequences.*

Definition 3. *A sequential pattern $s = \langle s_1, \ldots, s_n \rangle$ is an ordered list of itemsets where an itemset $s_i \in s$ is also called an* element *of s.*

In certain databases, a sequential pattern $s = \langle s_1, \ldots, s_n \rangle$ is *supported* by a sequence $d = \langle e_1, \cdots, e_m \rangle$, denoted by $s \preceq d$, if there exists n integers $1 \leq k_1 < \ldots, k_n \leq m$ that have $s_i \subseteq e_{k_i}$ for $i \in [1, n]$. A sequential pattern is *frequent* if at least τ_s sequences support it, where τ_s is a user-specified threshold. However, in an uncertain database D, the support of a pattern s is uncertain. We define *probabilistic frequent sequential patterns* (p-FSP) as follows:

Definition 4 (Probabilistic Frequent Sequential Pattern). *A sequential pattern s is a probabilistic frequent sequential pattern (p-FSP) if its probability of being frequent is at least τ_p, denoted by $P(sup(s) \geq \tau_s) \geq \tau_p$.*

Here $sup(s)$ is the support of s in D and τ_p is the user-defined minimum confidence in the frequentness of a sequential pattern. We are now able to specify the uncertain SPM problem as: *Given τ_s, τ_p, find all p-FSPs in D.*

In an uncertain database, sequences are often assumed to be mutually independent, which is also known as the *tuple-level independence* [2,7]. Therefore, the overall support $sup(s)$ in D is a sum of uncertain supports in every single sequence. And the probabilistic support of s in an uncertain sequence d_i can be modeled by a Bernoulli random variable $X_i \sim B(1, p_i)$, where $p_i = P(s \preceq d_i)$ is the probability that d_i supports s.

When the size of D grows, we can approximate the distribution of $sup(s)$ by the Gaussian distribution in Eq. (1), according to *the central limit theory.*

$$sup(s) \xrightarrow{|D| \to \infty} \mathcal{N}(\sum_{i=1}^{|D|} p_i, \sum_{i=1}^{|D|} p_i * (1 - p_i)) \tag{1}$$

We are now able to compute the *approximate frequentness probability* of s by:

$$P(sup(s) \geq \tau_s) = 1 - P(sup(s) \leq \tau_s - 1) = 1 - \Phi(\frac{\tau_s - 1 - \mu}{\sigma}) \tag{2}$$

And s is a p-FSP if $P(sup(s) \geq \tau_s) \geq \tau_p$.

Now the key issue of uncertain SPM is the computation of support probabilities. We use possible world semantics to interpret uncertain sequences. Here a possible world is a *certain sequence* instantiated by generating every event according to its existential probability. Each uncertain probability $P(e \in d)$ derives two possible worlds per sequence: One possible world in which event e exists in sequence d, and the other possible world where e does not exist. Therefore, the number of possible worlds increases exponentially in the number of events.

Each possible world w is associated with an existential probability $P(w)$. Figure 1(b) shows all possible worlds derived from sequence S_A in Fig. 1(a). For example, in world 2, the customer in session A is first attracted by a *flight* product and then interested in a *flight-hotel* package. Afterwards, the customer has no interests towards another *hotel* any more.

Uncertain events in a sequence are also assumed to be independent of each other [4,13], since they are often observed independently in real world applications. Therefore, we can compute the existential probability of a possible world w by Eq. (3).

$$P_e(w) = \prod_{e \in d} P(e \in d) * \prod_{e \notin d} (1 - P(e \in d)) \tag{3}$$

For example, the existential probability of world 2 in Fig. 1(b) is $P(w_2) = 0.7 * 0.6 * (1 - 0.8) = 0.084$.

In this paper, we focus on the problem of calculating support probabilities in large scale databases and extracting all p-FSPs. Meanwhile, in order to mine highly scalable databases, we extend the Apriori-like framework of uncertain SPM to Spark [12] which is a distributed computing platform allowing us to load data into a cluster's memory and query it repeatedly.

1.2   Contribution

In this paper, we propose a **D**istributed **S**equential **P**attern (DSP) mining algorithm in large scale uncertain databases. Our main contributions are summarized as follows: (1) We propose a SPM framework for mining large scale uncertain databases in Spark. (2) We present a dynamic programming method to compute support probabilities in linear time. (3) We propose a memory-efficient distributed dynamic programming approach in Spark and design a new data structure to save intermediate results efficiently. (4) Extensive experiments conducted in various scales shows that our algorithm is orders of magnitude faster than both direct extension and existing works.

2   Related Works

Uncertain data mining has been an active area of research recently. Many traditional database and data mining techniques have been extended to be applied to uncertain databases [2]. Specifically, Muzammal and Raman first define the SPM problem in probabilistic database [10]; and Zhao et al. define probabilistic frequent sequential patterns in possible world model and propose their uncertain SPM algorithms [13,14]. Li et al. introduce a dynamic programming approach to mine sequential patterns in a specific spatial-temporal uncertain model [8]. Wan et al. [11] propose a dynamic programming algorithm of mining frequent serial episodes within one uncertain sequence. However, all the above mentioned methods can only be executed in a single machine and may have scalability issues in mining large databases.

Chen et al. extend the classic SPAM algorithm to its MapReduce version SPAMC [5]. Miliaraki et al. propose a gap-constraint frequent sequence mining algorithm in MapReduce [9]. These algorithms are applied in the context of deterministic data, while our work aims to solve large scale uncertain SPM problems. An iterative MapReduce implementation of uncertain SPM in [6] is somehow close to our work; however, it has a quadratic complexity of support probability computation, and the time cost of that in our algorithm is linear.

3   Uncertain SPM Framework in Spark

First of all, we define the following two types of sequential pattern extension.

Definition 5 (Item-extended Pattern). *An item-extended pattern s is a sequential pattern generated by adding a new item i to the first element of another sequential pattern s', denoted by $s = \{i\} \cup s'$.*

Definition 6 (Sequence-extended Pattern). *A sequence-extended pattern s is a sequential pattern generated by adding a new itemset $\{i\}$ to another sequential pattern s' as its first element, denoted by $s = \{i\} + s'$.*

For example, let $s' = \langle (b)(d) \rangle$, then $s_1 = \langle (a,b)(d) \rangle$ is an item-extended pattern of s' and $s_2 = \langle (a)(b)(d) \rangle$ is sequence-extended from s'. The Apriori property of p-FSPs in Lemma 1 allows us to prune a pattern if it is extended from a pattern which is not a p-FSP. [14]

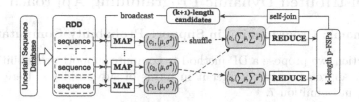

Fig. 2. A framework of DSP in Spark

Lemma 1. *If s is extended from s' and s is a p-FSP, then s' is also a p-FSP.*

Figure 2 shows the Apriori-like uncertain SPM framework of our DSP algorithm in Spark.

An uncertain sequence database $D = \{d_1, \ldots, d_n\}$ is abstracted by an RDD [12] in Spark. These sequences are allocated to a cluster of machines and can be processed in parallel.

Map. A map function is used to compute support probabilities. A set of candidate patterns are broadcasted to all the mappers. For each candidate pattern c, the map function first computes the support probability $p_i = P(c \preceq d_i)$ in the uncertain sequence d_i, then it emits a key-value pair as $\langle c, (\mu_i, \sigma_i^2) \rangle$, if $p_i > 0$. The key field here is the pattern c; the composite value field contains both mean μ_i and variance σ_i^2 of the Bernoulli distributed probabilistic support $X_i \sim \mathcal{B}(1, p_i)$. The key-value pairs are designed to be associative and commutative, so that Spark can aggregate them first in local machines to minimize network usage in shuffling.

Reduce. Pairs with the same key are shuffled to one reducer. In a reduce function, it computes the approximate frequentness probability for each candidate by Eq. (2). All candidates with $P(sup(c) \geq \tau_s) \geq \tau_p$ are saved to a set of k-length p-FSPs, denoted by S_k.

Self-join. we self-join all k-length p-FSPs in S_k to generate a set of $(k+1)$-length candidate patterns in C_{k+1}. Let s_1 and s_2 be two p-FSPs in S_k. Suppose s_1' is the pattern generated by removing the first item i in s_1 and s_2' is the pattern generated by removing the last item of s_2. If $s_1' = s_2'$, we join s_1 and s_2, denoted by $s_1 \bowtie s_2$, to generate a (k+1)-length candidate c according to the following rules: If s_1 is sequence-extended, $c = \{i\} + s_2$; if s_1 is item-extended, $c = \{i\} \cup s_2$. For example, let $s_1 = \langle (a)(bc) \rangle$, $s_2 = \langle (bc)(d) \rangle$ and $s_3 = \langle (c)(de) \rangle$, then $s_1 \bowtie s_2 = \langle (a)(bc)(d) \rangle$; while $s_2 \bowtie s_3 = \langle (bc)(de) \rangle$.

Stop Criterion. If either S_k or C_{k+1} is empty, we terminate the mining process; otherwise, C_{k+1} is broadcasted to all map functions for the next iteration.

4   A Distributed Dynamic Programming Approach

4.1   Dynamic Programming in Support Probability Computation

In this section, we propose a DP method to compute support probabilities. The key to our approach is to consider it in terms of sub-problems. Here we first define $P_{i,j}$ in Definition 7.

Definition 7. $P_{i,j} = P(s_i^n \preceq d_j^m)$ is the probability that s_i^n is supported by d_j^m, where $s_i^n = \langle s_i, \ldots, s_n \rangle$ is a subsequence of a sequential pattern s and $d_j^m = \langle e_j, \ldots, e_m \rangle$ is a subsequence of an uncertain sequence d.

Therefore, $P(s \preceq d) = P_{1,1}$. The idea in our approach is to split the problem of computing $P_{i,j}$ into sub-problems $P_{i,j+1}$ and $P_{i+1,j+1}$. And this can be achieved as follows: in condition of $s_i \subseteq e_j$, $P_{i,j}$ is equal to the probability that s_{i+1}^n is supported by d_{j+1}^m; if $s_i \not\subseteq e_j$, $P_{i,j}$ is equal to the probability that s_i^n is supported by d_{j+1}^m. By splitting the problem in this way we can use the recursion in Lemma 2 to compute $P_{i,j}$ by means of the paradigm of dynamic programming.

Lemma 2

$$P_{i,j} = P(s_i \subseteq e_j) * P_{i+1,j+1} + P(s_i \not\subseteq e_j) * P_{i,j+1} \qquad (4)$$

where $P(s_i \subseteq e_j)$ is the probability that s_i is contained in event e_j. And $P(s_i \subseteq e_j) = P(e_j \in d)$, if $s_i \subseteq e_j$; otherwise, $P(s_i \subseteq e_j) = 0$.

Proof. Referring to *the law of total probability*, we have:

$$P_{i,j} = P(s_i^n \preceq d_j^m | s_i \subseteq e_j) * P(s_i \subseteq e_j) + P(s_i^n \preceq d_j^m | s_i \not\subseteq e_j) * P(s_i \not\subseteq e_j)$$

where $P(s_i^n \preceq d_j^m | s_i \subseteq e_j) = P_{i+1,j+1}$ is the probability that $s' = \langle s_{i+1}, \ldots, s_n \rangle$ is supported by sequence $\langle e_{j+1}, \ldots, e_m \rangle$. And similarly we have $P(s_i^n \preceq d_j^m | s_i \not\subseteq e_j) = P_{i,j+1}$ is the support probability of s_i^n in d_{j+1}^m.   □

This dynamic schema is an adoption of the technique previously used in solving uncertain SPM [10] and frequent episode mining problems [11]. Using this dynamic programming scheme, we can compute the support probability by calculating the cells depicted in Fig. 3. In the matrix, each cell relates to a probability $P_{i,j}$, with i marked on the x-axis and j marked on the y-axis. Referring to Lemma 2, we can compute $P_{i,j}$ from $P_{i,j+1}$ and $P_{i+1,j+1}$ which are cells to the right and lower right of $P_{i,j}$. By definition, if $s = \phi$, then $P(s \preceq d) = 1$; meanwhile, $P(s \preceq d) = 0$ if $s \neq \phi$ and $d = \phi$. Therefore, we iterate the cells from $P_{n+1,m+1}$ to $P_{1,1}$ so that we finally obtain $P(s \preceq d) = P_{1,1}$. The time complexity is $O(n * m)$, as we only need to iterate each cell once.

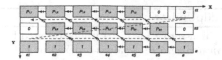

Fig. 3. An example of dynamic programming process

4.2 Distribute Dynamic Programming Schema

A direct extension of the dynamic programming approach in Spark needs to build a $n * m$ matrix for every support probability computation, and this might slow down the entire process because of expensive garbage collection overhead in Spark. Therefore, we refine the original DP schema and design a memory-efficient distributed dynamic programming (ddp) approach here. We first define $P_{s,j}$ as follows.

Definition 8. *Given a sequential pattern s and an uncertain sequence $d = \langle e_1, \ldots, e_m \rangle$, $P_{s,j}$ is defined to be the support probability $P(s \preceq d_j^m)$ where $d_j^m = \langle e_j, \ldots, e_m \rangle$ is a subsequence of d.*

The idea is that we save and reuse computational results in the last iteration. Based on the extension type of sequential pattern s, we have different dynamic programming schemas.

Sequence-Extended. If $s = \{i\} + s'$ is sequence-extended from another pattern s', then we can compute the values of $P_{s,j}$ from $P_{s',j+1}$ and $P(s_1 \subseteq e_j)$ by Eq. (5), according to Lemma 2.

$$P_{s,j} = P(s_1 \nsubseteq e_j) * P_{s,j+1} + P(s_1 \subseteq e_j) * P_{s',j+1} \tag{5}$$

where $s_1 = \{i\}$ is the first element of s.

Item-Extended. Let $s = \{i\} \cup s'$, then s_1' is a strict subset of s_1 and we have

$$P(s_1 \subseteq e_j) = \begin{cases} P(s_1' \subseteq e_j) & \text{if } i \in e_j \\ 0 & \text{otherwise} \end{cases} \tag{6}$$

Referring to Lemma (2), we can compute $P_{s,j}$ by Eq. (7).

$$P_{s,j} = \begin{cases} P_{s,j+1}, & \text{if } i \notin e_j \\ P(s_1 \nsubseteq e_j) * P_{s,j+1} + P(s_1' \subseteq e_j) * P(s_2^n \preceq d_j^m) & \text{otherwise} \end{cases} \tag{7}$$

Note that $s_2^n = s_2'^n = \langle s_2, \ldots, s_n \rangle$, then $P_{s',j}$ can be computed by:

$$\begin{aligned} P_{s',j} &= P(s_1' \nsubseteq e_j) * P_{s',j+1} + P(s_1' \subseteq e_j) * P(s_2'^n \preceq d_j^m) \\ &= P(s_1' \nsubseteq e_j) * P_{s',j+1} + P(s_1' \subseteq e_j) * P(s_2^n \preceq d_j^m) \end{aligned} \tag{8}$$

ALGORITHM 1. ddpUpdate

Input: L_1: a list of n non-zero $P(s_1'' \subseteq e_i)$ values ordered by eid i
      L_2: a list of m non-zero $P_{s',j}$ values ordered by eid j
$L_s \leftarrow \phi,\ p \leftarrow (n-1),\ q \leftarrow (m-1)$
$P_{s',j+1} \leftarrow 0$
while $p \geq 0$ **do**
  $\quad P(s_1 \subseteq e_i) = P(s'' \subseteq e_i) \leftarrow L_1[p];$ // at event e_i
  $\quad P_{s',j} \leftarrow L_2[q]$;   // at event e_j
  $\quad$**while** $j < i \wedge q \geq 0$ **do**
    $\quad\quad$/* find the nearest event e_j with non-zero value $P_{s',j}$ */
    $\quad\quad P_{s',i+1} = P_{s',j} \leftarrow L_2[q]$
    $\quad\quad q \leftarrow q - 1$
  $\quad$**end**
  $\quad q \leftarrow q + 1$
  $\quad$**if** $L_2[q-1] = P_{s',i}$ **then** $P_{s',i} \leftarrow L_2[q-1]$;
  $\quad$**else** $P_{s',i} \leftarrow L_2[q]$;
  $\quad$compute $P_{s,i}$ by Equation (5) or (9) and insert it to the head of L_s
  $\quad p \leftarrow p - 1$
end
return L_s

Therefore, we can derive Eq. (9) by substituting Eq. (8) into Eq. (7).

$$P_{s,j} = \begin{cases} P_{s,j+1} & \text{if } i \notin e_j \\ P(s_1 \nsubseteq e_j) * (P_{s,j+1} - P_{s',j+1}) + P_{s',j} & \text{otherwise} \end{cases} \tag{9}$$

Now we are able to compute $P_{s,j}$ from only the values of $P_{s',j}$, $P_{s',j+1}$ and $P(s_1 \subseteq e_j)$.

Eqs. (5) and (9) constitute our distributed dynamic programming structure for computing support probabilities in parallel. And the time complexity is $O(|d|)$.

4.3 Memory-Efficient Distributed SPM Algorithm

Lemma 3. *It is not necessary to save the value of $P_{s,j}$, if $P(s_1 \subseteq e_j) = 0$.*

Proof. Referring to Eqs. (5) and (9), if $P(s_1 \subseteq e_j) = 0$, then $P_{s,j} = P_{s,k}$ where e_k is the nearest event of e_j which has $k > j$ and $P(s_1 \subseteq e_k) > 0$. ☐

Let s be a k-length sequential pattern generated by joining two $(k-1)$-length patterns as $s = s'' \bowtie s'$. Then, the first element of s and s'' are identical when $k \geq 2$. For example, let $s = \langle(a)(bc)(d)\rangle$ and $s = s'' \bowtie s'$, then $s'' = \langle(a)(bc)\rangle$, $s' = \langle(bc)(d)\rangle$ so that $s_1 = s_1'' = (a)$.

Suppose L_1 is a list of non-zero $P(s_1'' \subseteq e_i)$ values and L_2 is a list of $P_{s',j}$ values with $P(s_1' \subseteq e_j) > 0$. Algorithm 1 computes the values of $P_{s,i}$ from L_1 and L_2. For each value of $P(s_1'' \subseteq e_i) > 0$, we have $P(s \subseteq e_i) = P(s_1'' \subseteq e_i)$ at

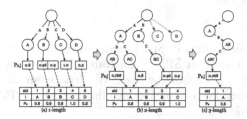

Fig. 4. An example process of updating candidate support probabilities

event e_i because $s_1 = s_1''$. Then we search the nearest event e_k, which satisfies $P_{s',k} > 0$ and $k > i$, to the right of e_i. Thus, we have $P_{s',i+1} = P_{s',k}$ by Lemma 3. If $P(s' \subseteq e_i) > 0$, the value of $P_{s',i}$ must have been saved and we can directly read it from L_2; if $P(s' \subseteq e_i) > 0$, $P_{s',i} = P_{s',i+1}$. Now that we have the values of $P(s \subseteq e_i)$, $P_{s',i+1}$ and $P_{s',i}$, we can compute $P_{s,i}$ by either Eqs. (5) or (9). The time complexity of Algorithm 1 is linear because both L_1 and L_2 are iterated only once.

We extend the data structure *prefix-tree* to save intermediate results in each iteration of uncertain SPM. The root of the prefix tree is the empty pattern ϕ. Each edge in the tree is associated with an item. The key of a node is identified by the path from root to that node. Values are not associated with inner nodes; only leaf nodes point to a list of $P_{s,j}$ values. Each value of $P_{s,j}$ is linked to the event e_j where $P(s_1 \subseteq e_j) > 0$. Here s_1 is the first element of s.

Figure 4 shows an example process of computing support probabilities with the new data structure. In Fig. 4(a), the leaf nodes are 1-length sequential patterns that are linked to events in an uncertain sequence. For example, $s = \langle B \rangle$ is associated with two values ($P_{s,2} = 0.96$, $P_{s,3} = 0.9$) which point to uncertain event e_2 and e_3. The support probability of $\langle B \rangle$ is $P(s \preceq d) = P_{s,2}$ because e_2 is the first event where we have $P(s_1 \subseteq e_2) > 0$.

Another benefit of our algorithm is that we do not need to generate candidates in a centralized node; instead, we broadcast p-FSPs to all mappers and then generate candidates in parallel. In a map function applied to a sequence d, a p-FSP s is not participated in candidate generation if it is not supported in d. In Fig. 4, suppose S_1 is a set of 1-length p-FSPs and $\langle D \rangle \notin S_1$, then we can prune node D from the prefix tree and generate candidates by joining only three 1-length p-FSPs: $\langle A \rangle$, $\langle B \rangle$ and $\langle C \rangle$ for this sequence.

Consider the example candidate $s = \langle (A)(B) \rangle$. We first search leaf nodes associated with $s'' = \langle A \rangle$ and $s' = \langle B \rangle$ in the 1-length pattern tree. Then we retrieve $P(s_1'' \subseteq e_1) = 0.8$, $P_{s',2} = 0.96$ and $P_{s',3} = 0.9$ to compute $P_{s,1} = 0.8 * 0.96 = 0.768$ by Eq. (5). Thereafter, we generate a new leaf node $\langle AB \rangle$ for the 2-length prefix tree.

After expanding the tree for every possible 2-length candidate, we eliminate all 1-length patterns that have not been extended. For instance, node with $\langle C \rangle$ is prune because no 2-length candidate starting with item C are potentially supported in the sequence.

5   Evaluation

We implement our algorithms in Spark and evaluate the performance in large scale datasets. The uncertain SPM algorithm which directly adopts dynamic programming in Sect. 4.1 is denote by *basic*. We denote our distributed uncertain SPM algorithm in Sect. 4.3 by *dsp*. We also implement the IMRSPM algorithm [6] in Spark and name it as *uspm* here.

We employ the IBM market-basket data generator [3] to generate sequence datasets in different scales by varying the parameters: (1) C: number of sequences; (2) T: average number of events per sequence; (3) L: average number of items per event per sequence; (4) I: number of different items.

An existential probability α is added to each event in the synthetic datasets, where $\alpha \in [0.5, 0.9]$ is a parameter to control uncertain levels. We name a synthetic uncertain dataset by its parameters. For example, a dataset C10kT4L10I10k indicates $C = 10k = 10 * 1000$, $T = 4$, $L = 10$ and $I = 10k = 10 * 1000$.

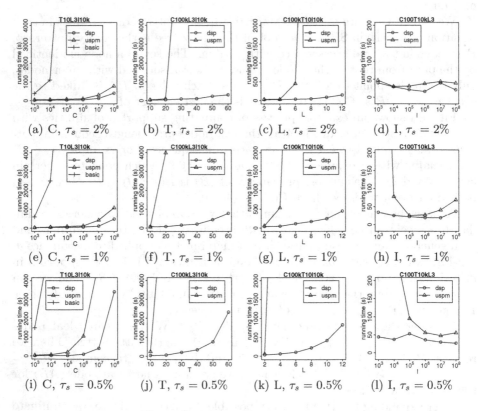

Fig. 5. Scalability of DSP algorithm

In Fig. 5, we evaluate the scalability of our DSP algorithm on various synthetic datasets in a Spark cluster with 100 nodes. Initially, we set uncertain level $\alpha = 0.8$ and frequentness probability threshold $\tau_p = 0.7$. In Fig. 5(a)–(d), we set $\tau_s = 2\% * C$; in Fig. 5(e)–(h), we have $\tau_s = 1\% * C$; and $\tau_s = 0.5\% * C$ in Fig. 5(i)–(l). Under each setting of τ_s, we vary the values of C, T, L and I to evaluate the performance of DSP in different scales:

(1) Figure 5(a), (e), (i) show the running time variations of DSP when C varies from 1000 to 100 000 000, where $T = 10$, $L = 3$ and $I = 10k$.
(2) Figure 5(b), (f), (j) show the running time variations of DSP when T varies from 10 to 60, where $C = 100k$, $L = 4$, $I = 10k$.
(3) Figure 5(c), (g), (k) show the running time variations when L varies from 2 to 12, where $C = 100k$, $T = 4$, $I = 10k$.
(4) Figure 5(d), (h), (l) show the running time variations when I varies from 10 000 to 10 000 000, where $C = 100k$, $T = 4$, $L = 4$.

We observe the following phenomenons in Fig. 5: (1) *dsp* outperforms *basic* and *uspm* under every setting of the parameters. Specifically, *dsp* is orders of magnitude faster than *basic* and *uspm* in datasets with larger values of T or L. When $C > 10k$, *basic* cannot finish because of the garbage collection overhead from Spark. This indicates that directly extending dynamic programming to Spark is not workable and also proves the advantage of our refined schemas. (2) The running time increase with the increment of C, T, L. This is intuitive because increasing these parameters generates larger scale datasets. Comparing to the C scale, both *dsp* and *uspm* are more sensitive to the increment of T and L. *uspm* fails quickly when T or L becomes larger; however, *dsp* still performs well even with large T or L values. (3) The running time first drops and then arises with the increment of I. When the value of I grows, the item occurrences become more sparse, and fewer p-FSPs are mined under the same thresholds; meanwhile, the volume of data shuffled from mapper to reducer via network increases because less key-value pairs are able to be pre-aggregated locally, which slows down the process when I is extremely large.

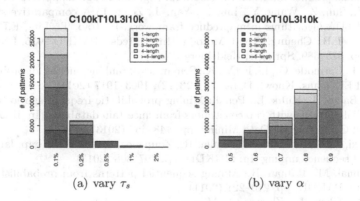

Fig. 6. Number of p-FSPs with different τ_s and α settings

Figure 6(a) shows the number of p-FSPs in the dataset C100kT10L3I10k with uncertain level $\alpha = 0.8$, where we vary the value of τ_s from 0.1 % to 2 %; Fig. 6(b) shows the effect of uncertain level α to the number of p-FSPs in C100kT10L3I10k, where $\tau_s = 0.1$ % and $\alpha \in [0.5, 0.9]$. From Fig. 6, we observe that: (1) With the increment of τ_s, the number of p-FSPs decreases dramatically. This is intuitive because a larger minimal support threshold makes fewer candidates be probabilistically frequent. (2) With the increment of α, the number of p-FSPs increases, which shows the effect of uncertainty in SPM problems. When uncertain level is high (α is small), there are fewer precise information in the data, which makes it more difficult to find p-FSPs under the same thresholds.

6   Conclusions

In this paper, we design a distributed dynamic programming method in Spark to mine sequential patterns in large scale uncertain databases. Our algorithm is proved to be efficient and highly scalable. In the future, we will continue to work on integrating constraints in large scale uncertain SPM problems.

References

1. Agarwal, A., Xie, B., Vovsha, I., Rambow, O., Passonneau, R.: Sentiment analysis of twitter data. In: Proceedings of the Workshop on Languages in Social Media, pp. 30–38 (2011)
2. Aggarwal, C.C., Yu, P.S.: A survey of uncertain data algorithms and applications. IEEE Trans. Knowl. Data Eng. **21**(5), 609–623 (2009)
3. Agrawal, R., Srikant, R.: Fast algorithms for mining association rules in large databases. In: VLDB, pp. 487–499 (1994)
4. Bernecker, T., Kriegel, H.P., Renz, M., Verhein, F., Zuefle, A.: Probabilistic frequent itemset mining in uncertain databases. In: SIGKDD, pp. 119–128. ACM (2009)
5. Chen, C.C., Tseng, C.Y., Chen, M.S.: Highly scalable sequential pattern mining based on mapreduce model on the cloud. In: BigData Congress, pp. 310–317 (2013)
6. Gao, Y., Sun, Z., Wang, Y., Liu, X., Yan, J., Zeng, J.: A comparative study on parallel LDA algorithms in mapreduce framework. In: Cao, T., Lim, E.P., Zhou, Z.H., Ho, T.B., Cheung, David, Motoda, Hiroshi (eds.) PAKDD 2015. LNCS, vol. 9078, pp. 675–689. Springer, Heidelberg (2015)
7. Jestes, J., Cormode, G., Li, F., Yi, K.: Semantics of ranking queries for probabilistic data. IEEE Trans. Knowl. Data Eng. **23**(12), 1903–1917 (2011)
8. Li, Y., Bailey, J., Kulik, L., Pei, J.: Mining probabilistic frequent spatio-temporal sequential patterns with gap constraints from uncertain databases. In: IEEE International Conference on Data Mining, pp. 448–457 (2013)
9. Miliaraki, I., Berberich, K., Gemulla, R., Zoupanos, S.: Mind the gap: large-scale frequent sequence mining. In: SIGKDD, pp. 797–808 (2013)
10. Muzammal, M., Raman, R.: Mining sequential patterns from probabilistic databases. In: PAKDD, pp. 210–221 (2011)
11. Wan, L., Chen, L., Zhang, C.: Mining frequent serial episodes over uncertain sequence data. In: EDBT, pp. 215–226 (2013)

12. Zaharia, M., Chowdhury, M., Das, T., Dave, A., Ma, J., McCauley, M., Franklin, M.J., Shenker, S., Stoica, I.: Resilient distributed datasets: a fault-tolerant abstraction for in-memory cluster computing. In: NSDI 2012 (2012)
13. Zhao, Z., Yan, D., Ng, W.: Mining probabilistically frequent sequential patterns in uncertain databases. In: EDBT, pp. 74–85 (2012)
14. Zhao, Z., Yan, D., Ng, W.: Mining probabilistically frequent sequential patterns in large uncertain databases. IEEE Trans. Knowl. Data Eng. **26**, 1171–1184 (2013)

DeepCare: A Deep Dynamic Memory Model for Predictive Medicine

Trang Pham$^{(\boxtimes)}$, Truyen Tran, Dinh Phung, and Svetha Venkatesh

Center for Pattern Recognition and Data Analytics School
of Information Technology, Deakin University, Geelong, Australia
{phtra,truyen.tran,dinh.phung,svetha.venkatesh}@deakin.edu.au

Abstract. Personalized predictive medicine necessitates modeling of patient illness and care processes, which inherently have long-term temporal dependencies. Healthcare observations, recorded in electronic medical records, are episodic and irregular in time. We introduce Deep-Care, a deep dynamic neural network that reads medical records and predicts future medical outcomes. At the data level, DeepCare models patient health state trajectories with explicit memory of illness. Built on Long Short-Term Memory (LSTM), DeepCare introduces time parameterizations to handle irregular timing by moderating the forgetting and consolidation of illness memory. DeepCare also incorporates medical interventions that change the course of illness and shape future medical risk. Moving up to the health state level, historical and present health states are then aggregated through multiscale temporal pooling, before passing through a neural network that estimates future outcomes. We demonstrate the efficacy of DeepCare for disease progression modeling and readmission prediction in diabetes, a chronic disease with large economic burden. The results show improved modeling and risk prediction accuracy.

1 Introduction

Health care costs are escalating. To deliver cost effective quality care, modern health systems are turning to data to predict risk and adverse events. For example, identifying patients with high risk of readmission can help hospitals to tailor suitable care packages.

Modern electronic medical records (EMRs) offer the base on which to build prognostic systems [11,15,19]. Such inquiry necessitates modeling patient-level temporal healthcare processes. But this is challenging. The records are a mixture of the illness trajectory, and the interventions and complications. Thus medical records vary in length, are inherently episodic and irregular over time. There are long-term dependencies in the data - future illness and care may depend critically on past illness and interventions. Existing methods either ignore long-term dependencies or do not adequately capture variable length [1,15,19]. Neither are they able to model temporal irregularity [14,20,22].

Addressing these open problems, we introduce DeepCare, a deep, dynamic neural network that reads medical records, infers illness states and predicts future

© Springer International Publishing Switzerland 2016
J. Bailey et al. (Eds.): PAKDD 2016, Part II, LNAI 9652, pp. 30–41, 2016.
DOI: 10.1007/978-3-319-31750-2_3

outcomes. DeepCare has several layers. At the bottom, we start by modeling illness-state trajectories and healthcare processes [2,7] based on Long Short-Term Memory (*LSTM*) [5,9]. LSTM is a recurrent neural network equipped with memory cells, which store previous experiences. The current medical risk states are modeled as a combination of *illness memory* and the current medical conditions and are moderated by past and current interventions. The illness memory is partly forgotten or consolidated through a mechanism known as forget gate. The LSTM can handle variable lengths with long dependencies making it an ideal model for diverse sequential domains [6,17,18]. Interestingly, LSTM has never been used in healthcare. This may be because one major difficulty is the handling irregular time and interventions.

We augment LSTM with several new mechanisms to handle the forgetting and consolidation of illness through the memory. First, the forgetting and consolidation mechanisms are time moderated. Second, interventions are modeled as a moderating factor of the current risk states and of the memory carried into the future. The resulting model is sparse and efficient where only observed records are incorporated, regardless of the irregular time spacing. At the second layer of DeepCare, episodic risk states are aggregated through a new time-decayed multiscale pooling strategy. This allows further handling of time-modulated memory. Finally at the top layer, pooled risk states are passed through a neural network for estimating future prognosis. In short, computation steps in DeepCare can be summarized as:

$$P\left(y \mid \boldsymbol{x}_{1:n}\right) = P\left(\text{nnet}_y\left(\text{pool}\left\{\text{LSTM}(\boldsymbol{x}_{1:n})\right\}\right)\right) \tag{1}$$

where $\boldsymbol{x}_{1:n}$ is the input sequence of admission observations, y is the outcome of interest (e.g., readmission), nnet_y denotes estimate of the neural network with respect to outcome y, and P is probabilistic model of outcomes.

We demonstrate our DeepCare in answering a crucial component of the holy grail question "what happens next?". In particular, we predict the next stage of *disease progression* and the risk of *unplanned readmission* for diabetic patients after a discharge from hospital. Our cohort consists of more than 12,000 patients whose data were collected from a large regional hospital in the period of 2002 to 2013. The forecasting of future events may be considerably harder than the classical classification of objects into categories due to inherent uncertainty in unseen interleaved events. We show that DeepCare is well-suited for modeling disease progression, as well as predicting future risk.

To summarize, our main contributions are: (i) Introducing DeepCare, a deep dynamic neural network for medical prognosis. DeepCare models irregular timing and interventions within LSTM – a powerful recurrent neural networks for sequences and (ii) Demonstrating the effectiveness of DeepCare for disease progression modeling and medical risk prediction, and showing that it outperforms baselines.

2 Long Short-Term Memory

This section briefly reviews Long Short-Term Memory (LSTM), a recurrent neural network (RNN) for sequences. A LSTM is a sequence of units that share the same set of parameters. Each LSTM unit has a memory cell that has state $c_t \in \mathbb{R}^K$ at time t. The memory is updated through reading a new input $x_t \in \mathbb{R}^M$ and the previous output $h_{t-1} \in \mathbb{R}^K$. Then an output states h_t is written based on the memory c_t. There are 3 sigmoid gates that control the reading, writing and memory updating: input gate i_t, output gate o_t and forget gates f_t, respectively. The gates and states are computed as follows:

$$i_t = \sigma\left(W_i x_t + U_i h_{t-1} + b_i\right) \tag{2}$$

$$f_t = \sigma\left(W_f x_t + U_f h_{t-1} + b_f\right) \tag{3}$$

$$o_t = \sigma\left(W_o x_t + U_o h_{t-1} + b_o\right) \tag{4}$$

$$c_t = f_t * c_{t-1} + i_t * \tanh\left(W_c x_t + U_c h_{t-1} + b_c\right) \tag{5}$$

$$h_t = o_t * \tanh(c_t) \tag{6}$$

where σ denotes sigmoid function, $*$ denotes element-wise product, and $W_{i,f,o,c}$, $U_{i,f,o,c}$, $b_{i,f,o,c}$ are parameters. The gates have the values in $(0,1)$.

The memory cell plays a crucial role in memorizing past experiences. The key is the additive memory updating in Eq. (5): if $f_t \to 1$ then all the past memory is preserved. Thus memory can potentially grow overtime since new experience is stilled added through the gate i_t. If $f_t \to 0$ then only new experience is updated (memoryless). An important property of additivity is that it helps to avoid a classic problem in standard recurrent neural networks known as vanishing/exploding gradients when t is large (says, greater than 10).

LSTM for Sequence Labeling. The output states h_t can be used to generate labels at time t as follows:

$$P\left(y_t = l \mid x_{1:t}\right) = \text{softmax}\left(v_l^\top h_t\right) \tag{7}$$

for label specific parameters v_l.

LSTM for Sequence Classification. LSTM can be used for classification using a simple mean-pooling strategy over all output states coupled with a differentiable loss function. For example, in the case of binary outcome $y \in \{0,1\}$, we have:

$$P\left(y = 1 \mid x_{1:n}\right) = \text{LR}\left(\text{pool}\left\{\text{LSTM}(x_{1:n})\right\}\right) \tag{8}$$

where LR denotes probability estimate of the logistic regression, and $\text{pool}\{h_{1:n}\} = \frac{1}{n}\sum_{t=1}^{n} h_t$.

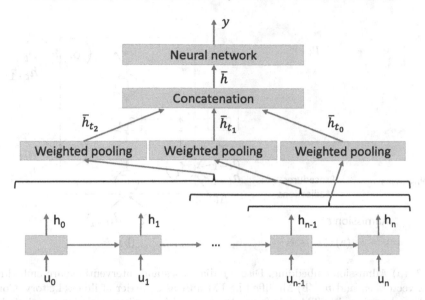

Fig. 1. DeepCare architecture. The bottom layer is Long Short-Term Memory [9] with irregular timing and interventions (see also Fig. 2b)

3   DeepCare: A Deep Dynamic Memory Model

In this section we present our contribution named DeepCare for modeling illness trajectories and predicting future outcomes. As illustrated in Fig. 1, DeepCare is a deep dynamic neural network that has three main layers. The bottom layer is built on LSTM whose memory cells are modified to handle irregular timing and interventions. More specifically, the input is a sequence of admissions. Each admission t contains a set of diagnosis codes (which is then formulated as a feature vector $x_t \in \mathbb{R}^M$), a set of intervention codes (which is then formulated as a feature vector p_t), the admission method m_t and the elapsed time $\Delta t \in \mathbb{R}^+$ between the two admission t and $t - 1$. Denote by $u_0, u_1, \ldots, u_n$ the input sequence, where $u_t = [x_t, p_t, m_t, \Delta t]$, the LSTM computes the corresponding sequence of distributed illness states $h_0, h_1, \ldots, h_n$, where $h_t \in \mathbb{R}^K$. The middle layer aggregates illness states through multiscale weighted pooling $z = \text{pool}\{h_0, h_1, \ldots, h_n\}$, where $z \in \mathbb{R}^{K \times s}$ for s scales.

The top layer is a neural network that takes pooled states and other statistics to estimate the final outcome probability, as summarized in Eq. (1) as $P(y \mid x_{1:n}) = P(\text{nnet}_y(\text{pool}\{\text{LSTM}(x_{1:n})\}))$. The probability $P(y \mid x_{1:n})$ depends on the nature of outputs and the choice of statistical structure. For example, for binary outcome, $P(y = 1 \mid x_{1:n})$ is a logistic function; for multi-class outcome, $P(y \mid x_{1:n})$ is a softmax function; and for continuous outcome, $P(y \mid x_{1:n})$ is Gaussian. In what follows, we describe the first two layers in more detail.

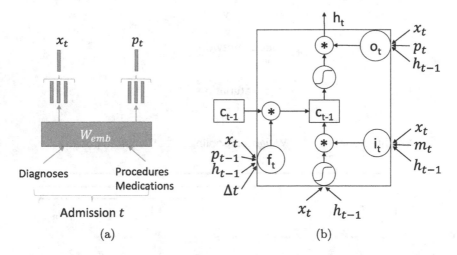

Fig. 2. (a) Admission embedding. Discrete diagnoses and interventions are embedded into 2 vectors x_t and p_t. (b) Modified LSTM unit as a carrier of illness history. Compared to the original LSTM unit (Sect. 2), the modified unit models times, admission methods, diagnoses and intervention

3.1 Admission Embedding

Figure 2a illustrates the admission embedding. There are two main types of information recorded in a typical EMR: (i) diagnoses of current condition; and (ii) interventions. Diagnoses are represented using WHO's ICD (International Classification of Diseases) coding schemes[1]. Interventions include procedures and medications. The procedures are typically coded in CPT (Current Procedural Terminology) or ICHI (International Classification of Health Interventions) schemes. Medication names can be mapped into the ATC (Anatomical Therapeutic Chemical) scheme. These schemes are hierarchical and the vocabularies are of tens of thousands in size. Thus for a problem, a suitable coding level should be used for balancing between specificity and robustness.

Codes are first embedded into a vector space of size M and embedding is learnable. Since each admission typically consists of multiple diagnoses, we average all the present vectors to derive $x_t \in \mathbb{R}^M$. Likewise, we derive the averaged intervention vector $p_t \in \mathbb{R}^M$. Finally, an admission embedding is a $2M$-dim vector $[x_t, p_t]$.

3.2 Moderating Admission Method and Effect of Interventions

There are two main types of admission: planned and unplanned. Unplanned admission refers to transfer from emergency attendance, which typically indicate higher risk. Recall from Eqs. (2, 5) that the input gate i control how much new

---

[1] http://apps.who.int/classifications/icd10/browse/2016/en.

information is updated into memory c. The gate can be modified to reflect the risk level of admission type as follows:

$$i_t = \frac{1}{m_t}\sigma\left(W_i x_t + U_i h_{t-1} + b_i\right) \tag{9}$$

where $m_t = 1$ if emergency admission, $m_t = 2$ if routine admission.

Since interventions are designed to cure diseases or reduce patient's illness, the output gate is moderated by the *current* intervention as follows:

$$o_t = \sigma\left(W_o x_t + U_o h_{t-1} + P_o p_t + b_o\right) \tag{10}$$

Interventions may have long-term impacts than just reducing the current illness. This suggests the illness forgetting is moderated by *previous* intervention

$$f_t = \sigma\left(W_f x_t + U_f h_{t-1} + P_f p_{t-1} + b_f\right) \tag{11}$$

where p_{t-1} is intervention at time step $t-1$.

3.3   Capturing Time Irregularity

We introduce two mechanisms of forgetting the memory by modified the forget gate f_t in Eq. 11:

Time Decay. Recall that the memory cell holds the current illness states, and the illness memory can be carried on to the future time. There are acute conditions that naturally reduce their effect through time. This suggests a simple decay

$$f_t \leftarrow d(\Delta_{t-1:t})f_t \tag{12}$$

where $\Delta_{t-1:t}$ is the time passed between step $t-1$ and step t, and $d\left(\Delta_{t-1:t}\right) \in (0, 1]$ is a decay function, i.e., it is monotonically non-increasing in time. One function we found working well is $d(\Delta_{t-1:t}) = [\log(e + \Delta_{t-1:t})]^{-1}$, where $e \approx 2.718$ is the base of the natural logarithm.

Forgetting Through Parametric Time. Time decay may not capture all conditions, since some conditions can get worse, and others can be chronic. This suggests a more flexible parametric forgetting:

$$f_t = \sigma\left(W_f x_t + U_f h_{t-1} + Q_f q_{\Delta_{t-1:t}} + P_f p_{t-1} + b_f\right) \tag{13}$$

where $q_{\Delta_{t-1:t}}$ is a vector derived from the time difference $\Delta_{t=1:t}$. For example, we may have: $q_{\Delta_{t-1:t}} = (\Delta_{t-1:t}, \Delta_{t-1:t}^2, \Delta_{t-1:t}^3)$ to model the third-degree forgetting dynamics.

3.4 Recency Attention via Multiscale Pooling

Once the illness dynamics have been modeled using the memory LSTM, the next step is to aggregate the illness states to infer about the future prognosis. The simplest way is to use mean-pooling, where $\bar{h} = \text{pool}\{h_{1:n}\} = \frac{1}{n}\sum_{t=1}^{n} h_t$. However, this does not reflect the attention to recency in healthcare. Here we introduce a simple attention scheme that weighs recent events more than old ones: $\bar{h} = \left(\sum_{t=t_0}^{n} w_t h_t\right) / \sum_{t=t_0}^{n} w_t$, where

$$w_t = [m_t + \log(1 + \Delta_{t:n})]^{-1}$$

and $\Delta_{t:n}$ is the elapsed time between the step t and the current step n, measured in months; $m_t = 1$ if emergency admission, $m_t = 2$ if routine admission. The starting time step t_0 is used to control the length of look-back in the pooling, for example, $\Delta_{t_0:n} \leq 12$ for one year look-back. Since diseases progress at different rates for different patients, we employ multiple look-backs: 12 months, 24 months, and all available history. Finally, the three pooled illness states are stacked into a vector: $[\bar{h}_{12}, \bar{h}_{24}, \bar{h}_{all}]$ which is then fed to a neural network for inferring about future prognosis.

3.5 Learning

Learning is carried out through minimizing cross-entropy: $L = -\log P(y \mid x_{1:n})$, where $P(y \mid x_{1:n})$ is given in Eq. (1). For example, in the case of binary classification, $y \in \{0, 1\}$, we use logistic regression to represent $P(y \mid x_{1:n})$, i.e.,

$$P(y = 1 \mid x_{1:n}) = \sigma\left(b_y + \text{nnet}\left(\text{pool}\{\text{LSTM}(x_{1:n})\}\right)\right)$$

where the structure inside the sigmoid is given in Eq. (1). The cross-entropy becomes: $L = -y\log\sigma - (1-y)\log(1-\sigma)$. Despite having a complex structure, DeepCare's loss function is fully differentiable, and thus can be minimized using standard back-propagation. The details are omitted due to space constraint.

4 Experiments

4.1 Data

The dataset is a diabetes cohort of more than 12,000 patients (55.5 % males, median age 73) collected in a 12 year period 2002–2013 from a large regional Australian hospital. Data statistics are summarized in Fig. 3. The diagnoses are coded using ICD-10 scheme. For example, E10 is diabetes Type I, and E11 is diabetes Type II. Procedures are coded using the ACHI (Australian Classification of Health Interventions) scheme, and medications are mapped in ATC codes. We preprocessed by removing (i) admissions with missing key information; and (ii) patients with less than 2 admissions. This leaves 7,191 patients with 53,208 admissions. To reduce the vocabulary, we collapse diagnoses that share the first 2 characters into one diagnosis. Likewise, the first digits in the procedure block are used. In total, there are 243 diagnosis, 773 procedure and 353 medication codes.

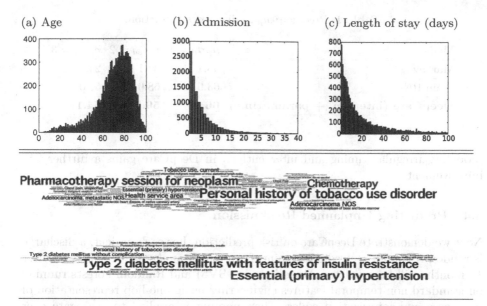

Fig. 3. Top row: Data statistics (y axis: number of patients; x axis: (a) age, (b) number of admissions, (c) number of days); **Bottom row:** Progression from pre-diabetes (upper diag. cloud) to post-diabetes (lower diag. cloud).

4.2 Implementation

The training, validation and test sets are created by randomly picking 2/3, 1/6, 1/6 data points, respectively. We vary the embedding and hidden dimensions from 5 to 50 but the results are rather robust. We report results for $M = 30$ embedding dimensions and $K = 40$ hidden units. Learning is by SGD with mini-batch of 16. Learning rate starts at 0.01. After $n_{waiting}$ epochs, if the model cannot find smaller training cost since the epoch with smallest training cost, the learning rate is divided by 2. At first, $n_{waiting} = 5$, then updated as $n_{waiting} = \min\{15, n_{waiting} + 2\}$ for each halving. Learning is terminated after $n_{epoch} = 200$ or after learning rate smaller than $\epsilon = 0.0001$.

4.3 Modeling Disease Progression

We first verify that the recurrent memory embedded in DeepCare is a realistic model of *disease progression*. We use the bottom layer of DeepCare (Sects. 3.1–3.3) to predict the next n_{pred} diagnoses at each discharge using Eq. (7).

Table 1 reports the Precision@n_{pred}. The Markov model has memoryless disease transition probabilities $P\left(d_t^i \mid d_{t+1}^j\right)$ from disease d^j to d^i at time t. Given an admission with disease subset D_t, the next disease probability is estimated as $Q\left(d^i; t\right) = \frac{1}{|D_t|} \sum_{j \in D_t} P\left(d_t^i \mid d_{t+1}^j\right)$. Using plain RNN improves over memoryless Markov model by 8.8 % with $n_{pred} = 1$ and by 27.7 % with $n_{pred} = 3$.

Table 1. Precision@n_{pred} diagnoses prediction.

Model	$n_{pred} = 1$	$n_{pred} = 2$	$n_{pred} = 3$
Markov	55.1	34.1	24.3
Plain RNN	63.9	58.0	52.0
DeepCare (interven. + param. time)	**66.0**	**59.7**	**54.1**

Modeling irregular timing and interventions in DeepCare gains a further 2% improvement.

4.4   Predicting Unplanned Readmission

Next we demonstrate DeepCare on risk prediction. For each patient, a discharge is randomly chosen as prediction point, from which *unplanned readmission* after 12 months will be predicted. **Baselines** are SVM and Random Forests running on standard non-temporal features engineering using one-hop representation of diagnoses and intervention codes. Then pooling is applied to aggregate over all existing admissions for each patient. Two pooling strategies are tested: *max* and *sum*. Max-pooling is equivalent to the presence-only strategy in [1], and sum-pooling is akin to an uniform convolutional kernel in [20]. This feature engineering strategy is equivalent to zeros-forgetting – any risk factor occurring in the past is memorized.

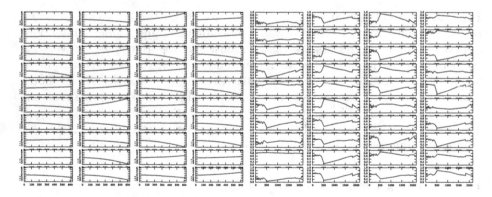

Fig. 4. (Left) 40 channels of forgetting due to time elapsed. (Right) The forget gates of a patient in the course of their illness.

Dynamics of Forgetting. Figure 4(left) plots the contribution of time into the forget gate. The contributions for all 40 states are computed using $Q_f \boldsymbol{q}_{\Delta_t}$ as in Eq. (13). There are two distinct patterns: decay and growing. This suggests that the time-based forgetting has a very small dimensionality, and we will under-parameterize time using decay only as in Eq. (12), and over-parameterize

Table 2. Results of unplanned readmission prediction within 12 months.

	Model	F-score (%)
1	SVM (*max-pooling*)	64.0
2	SVM (*sum-pooling*)	66.7
3	Random Forests (*max-pooling*)	68.3
4	Random Forests (*sum-pooling*)	71.4
5	LSTM (*mean-pooling + logit. regress.*)	75.9
6	DeepCare (*mean-pooling + nnets*)	76.5
7	DeepCare ([*interven. + time decay*]+ *recent.multi.pool. + nnets*)	77.1
8	**DeepCare** ([*interven. + param. time*]+ *recent.multi.pool. + nnets*)	**79.1**

time using full parameterization as in Eq. (13). A right balance is interesting to warrant a further investigation. Figure 4(right) shows the evolution of the forget gates through the course of illness (2000 days) for a patient.

Prediction Results. Table 2 reports the F-scores. The best baseline (non-temporal) is Random Forests with *sum pooling* has a F-score of 71.4 % [Row 4]. Using LSTM with simple mean-pooling and logistic regression already improves over best non-temporal methods by a 4.5 % difference in 12-months prediction [Row 5, ref: Sect. 2]. Moving to deep models by using a neural network as classifier helps with a gain of 5.1 % improvement [Row 6, ref: Eq. (1)]. By carefully modelling the irregular timing, interventions and recency + multiscale pooling, we gain 5.7 % improvement [Row 7, ref: Sects. 3.2, 3.3]. Finally, with parametric time we arrive at 79.1 % F-score, a 7.7 % improvement over the best baselines [Row 8, ref: Sects. 3.2, 3.3].

5 Related Work and Discussion

Electronic medical records (EMRs) are the results of interleaving between the illness processes and care processes. Using EMRs for prediction has attracted a significant interest in recent year [11,19]. However, most existing methods are either based on manual feature engineering [15], simplistic extraction [20], or assuming regular timing as in dynamic Bayesian networks [16]. Irregular timing and interventions have not been adequately modeled. Nursing illness trajectory model was popularized by Strauss and Corbin [2,4], but the model is qualitative but imprecise in time [7]. Thus its predictive power is very limited. Capturing disease progression has been of great interest [10,14], and much effort has been spent on Markov models [8,22]. However, healthcare is inherently non-Markovian due to the long-term dependencies. For example, a routine admission with irrelevant medical information would destroy the illness memory [1], especially for chronic conditions.

Deep learning is currently at the center of a new revolution in making sense of a large volume of data. It has achieved great successes in cognitive domains such as vision and NLP [12]. To date, deep learning approach to healthcare has been an unrealized promise, except for several very recent work [3,13,21], where irregular timing is not property modeled. We observe that there is a considerable similarity between NLP and EMR, where diagnoses and interventions play the role of nouns and modifiers, and an EMR is akin to a sentence. A major difference is the presence of precise timing in EMR, as well as the episodic nature. Our DeepCare contributes along that line.

DeepCare is generic and it can be implemented on existing EMR systems. For that more extensive evaluations on a variety of cohorts, sites and outcomes will be necessary. This offers opportunities for domain adaptations through parameter sharing among multiple cohorts and hospitals.

6   Conclusion

In this paper we have introduced DeepCare, a deep dynamic memory neural network for personalized healthcare. In particular, DeepCare supports prognosis from electronic medical records. DeepCare contributes to the healthcare model literature introducing the concept of *illness memory* into the nursing model of illness trajectories. To achieve precision and predictive power, DeepCare extends the classic Long Short-Term Memory by (i) parameterizing time to enable irregular timing, (ii) incorporating interventions to reflect their targeted influence in the course of illness and disease progression; (iii) using multiscale pooling over time; and finally (iv) augmenting a neural network to infer about future outcomes. We have demonstrated DeepCare on predicting next disease stages and unplanned readmission among diabetic patients. The results are competitive against current state-of-the-arts. DeepCare opens up a new principled approach to predictive medicine.

References

1. Arandjelović, O.: Discovering hospital admission patterns using models learnt from electronic hospital records. Bioinformatics. btv508 (2015)
2. Corbin, J.M., Strauss, A.: A nursing model for chronic illness management based upon the trajectory framework. Res. Theory Nurs. Pract. **5**(3), 155–174 (1991)
3. Futoma, J., Morris, J., Lucas, J.: A comparison of models for predicting early hospital readmissions. J. Biomed. Inform. **56**, 229–238 (2015)
4. Granger, B.B., Moser, D., Germino, B., Harrell, J., Ekman, I.: Caring for patients with chronic heart failure: the trajectory model. Eur. J. Cardiovasc. Nurs. **5**(3), 222–227 (2006)
5. Graves, A.: Generating sequences with recurrent neural networks (2013). arXiv preprint arXiv:1308.0850
6. Graves, A., Liwicki, M., Fernández, S., Bertolami, R., Bunke, H., Schmidhuber, J.: A novel connectionist system for unconstrained handwriting recognition. IEEE Trans. Pattern Anal. Mach. Intell. **31**(5), 855–868 (2009)
7. Henly, S.J., Wyman, J.F., Findorff, M.J.: Health and illness over time: the trajectory perspective in nursing science. Nurs. Res. **60**(3 Suppl), S5 (2011)

8. Henriques, R., Antunes, C., Madeira, S.C.: Generative modeling of repositories of health records for predictive tasks. Data Min. Knowl. Discov. **29**(4), 999–1032 (2015)
9. Hochreiter, S., Schmidhuber, J.: Long short-term memory. Neural Comput. **9**(8), 1735–1780 (1997)
10. Jensen, A.B., Moseley, P.L., Oprea, T.I., Ellesøe, S.G., Eriksson, R., Schmock, H., Jensen, P.B., Jensen, L.J., Brunak, S.: Temporal disease trajectories condensed from population-wide registry data covering 6.2 million patients. Nat. Commun. **5**, 10 (2014)
11. Jensen, P.B., Jensen, L.J., Brunak, S.: Mining electronic health records: towards better research applications and clinical care. Nat. Rev. Genet. **13**(6), 395–405 (2012)
12. LeCun, Y., Bengio, Y., Hinton, G.: Deep learning. Nature **521**(7553), 436–444 (2015)
13. Liang, Z., Zhang, G., Huang, J.X., Hu, Q.V.: Deep learning for healthcare decision making with EMRs. In: 2014 IEEE International Conference on Bioinformatics and Biomedicine (BIBM), pp. 556–559. IEEE (2014)
14. Liu, C., Wang, F., Hu, J., Xiong, H.: Temporal phenotyping from longitudinal electronic health records: a graph based framework. In: Proceedings of the 21th ACM SIGKDD International Conference on Knowledge Discovery and Data Mining, pp. 705–714. ACM (2015)
15. Mathias, J.S., Agrawal, A., Feinglass, J., Cooper, A.J., Baker, D.W., Choudhary, A.: Development of a 5 year life expectancy index in older adults using predictive mining of electronic health record data. J. Am. Med. Inf. Assoc. **20**(e1), e118–e124 (2013)
16. Orphanou, K., Stassopoulou, A., Keravnou, E.: Temporal abstraction and temporal Bayesian networks in clinical domains: a survey. Artif. Intell. Med. **60**(3), 133–149 (2014)
17. Srivastava, N., Mansimov, E., Salakhutdinov, R.: Unsupervised learning of video representations using LSTMS (2015). arXiv preprint arXiv:1502.04681
18. Sutskever, I., Vinyals, O., Le, Q.V.V.: Sequence to sequence learning with neural networks. In: Advances in Neural Information Processing Systems, pp. 3104–3112 (2014)
19. Tran, T., Phung, D., Luo, W., Harvey, R., Berk, M., Venkatesh, S.: An integrated framework for suicide risk prediction. In: KDD 2013 (2013)
20. Tran, T., Luo, W., Phung, D., Gupta, S., Rana, S., Kennedy, R.L., Larkins, A., Venkatesh, S.: A framework for feature extraction from hospital medical data with applications in risk prediction. BMC Bioinform. **15**(1), 6596 (2014)
21. Tran, T., Nguyen, T.D., Phung, D., Venkatesh, S.: Learning vector representation of medical objects via EMR-driven nonnegative restricted Boltzmann machines (eNRBM). J. Biomed. Inform. **54**, 96–105 (2015)
22. Wang, X., Sontag, D., Wang, F.: Unsupervised learning of disease progression models. In: Proceedings of the 20th ACM SIGKDD International Conference on Knowledge Discovery and Data Mining, pp. 85–94. ACM (2014)

Indoor Positioning System for Smart Homes Based on Decision Trees and Passive RFID

Frédéric Bergeron[1(✉)], Kevin Bouchard[1], Sébastien Gaboury[2],
Sylvain Giroux[1], and Bruno Bouchard[2]

[1] Université de Sherbrooke, Sherbrooke, Canada
{frederic.bergeron2,kevin.bouchard,sylvain.giroux}@usherbrooke.ca
[2] Université du Québec à Chicoutimi, Saguenay, Canada
{sebastien.gaboury,bruno.bouchard}@uqac.ca

Abstract. This paper presents a novel Indoor Positioning System (IPS) for objects of daily life equipped with passive RFID tags. The goal is to provide a simple to use, yet accurate, qualitative IPS for housing enhanced with technology (sensors, effectors, etc.). With such a service, the housing, namely called smart home, could enable a wide range of services by being able to better understand the context and the current progression of activities of daily living. The paper shows that classical data mining techniques can be applied to raw data from RFID readers and passive tags. In particular, it explains how we built several datasets using a tagged object in a real smart home infrastructure. Our method was proven very effective as most algorithms result in high accuracy for the majority of the smart home.

Keywords: RFID · Indoor positioning · Smart home · Decision tree · Random forest

1 Introduction

Many occidental countries are faced with the ageing of their population, causing an explosion in health related costs [16]. Indeed, ageing is associated with a higher prevalence of many illnesses like Parkinson's disease, Alzheimer's disease, etc. Persons afflicted by these diseases soon require constant care often leading to institutionalization. In addition to resulting in high costs for the society, institutionalization might significantly decrease social interaction and reduce autonomy which account for a lower quality of life [4]. One way to reduce the impact of these problems is to enable aging in place. Smart homes are being developed by many research teams around the world to address this challenge. One of the primary difficulty faced by researchers lies in the monitoring of progress of ongoing activities of daily living (ADLs), formerly known as the activity recognition problem [6]. Over the years, numerous algorithms have been designed for that purpose [14], but most of them do not distinguish the individual

---

F. Bergeron—We would like to thanks our main sponsor, INTER.

J. Bailey et al. (Eds.): PAKDD 2016, Part II, LNAI 9652, pp. 42–53, 2016.
DOI: 10.1007/978-3-319-31750-2_4

steps of an activity. The main explanation for this limit is that information acquired from sensors data is very noisy and limited.

Information of a spatio-temporal nature could greatly help improve the recognition of ADLs. A smart home contains various technologies that could provide spatio-temporal information. Some research teams have designed Indoor Positioning Systems (IPSs) using microphones to analyse sounds, while others have used low resolution cameras (often considered as very intrusive for the resident) or even communication protocols such as Bluetooth or Wi-Fi to track objects. In this work, we investigated the exploitation of passive Radio Frequency IDentification (RFID) tags because they are inexpensive, small and resistant. They are also the only technology to be small enough the be carried on in an unobtrusive way. On the other hand, they are also very inaccurate and subject to interferences from many sources, causing great difficulties for the design of an IPS. RFID has been used before in the literature for indoor positioning [8]. However, most systems focus primarily on precision, while high accuracy and deployment simplicity are more important.

In this paper, we propose a new Indoor Positioning System for smart home based on passive RFID technology and decision trees. The goal is to enable real-time tracking of the objects used in daily life activities. This new solution palliates the problem of the lack of accuracy of the previous studies by proposing a qualitative solution adapted to the precision of the RFID system. The method is said to be qualitative since it provides a logical position (defined by a qualitative zone) instead of precise coordinates. Our goal was to design an easy-to-deploy solution that would be extensible to the addition of antennas and to the modification of an RFID system. The datasets collected for these tests are available to the scientific community on www.Kevin-Bouchard.com.

2   Related Work

In this section, we present major works that addressed the problem of indoor localization in the last few years. There are many applications for IPSs. For example, in robotics, localization is important for action planning. A number of research has been done on the subject, and yet, good positioning and reliable tracking with passive RFID remains hard to achieve [8]. Many researchers have explored other ways to build an IPS. Some of them are described in the next sub-sections.

2.1   Passive RFID Localization

There are two types of RFID tags: actives and passives. Active tags have an inner energy supply that allows them to emit a strong signal at any time. Passive tags have no inner energy supply. On the contrary, they use the energy contained in an inbound signal to activate themselves and send back a signal with the remnant energy. While having a much longer life, passive tags have a shorter range.

Most RFID systems are based on LANDMARC [13]. This system was the first to use reference tags placed at strategic locations to achieve good precision on positioning. With four readers and one reference tag per square meter, they reached a mean error of one meter. Joho et al. [10] presented a model with a localization error of about 35 cm. They achieved that by using received signal strength information (RSSI), antenna orientation and references tags with a mobile reader that gathers information to compare with the real position.

2.2   Trilateration, Triangulation

In addition to the various declinations of the LANDMARC system, many researchers have worked on the classical algorithms of trilateration (distances) and triangulation (angles). These methods are often ignored in context where incoming data is noisy and inaccurate, but some researchers still obtained very interesting results. In particular, Chen et al. [5] exploited ZigBee (a radio-frequency technology used for communication between objects in a small environment) to perform trilateration. They implemented a fuzzy inference engine composed of only one variable. Their engine correlates the RSSI of a transmitter to localize to the distance separating it from a receiver. Fortin-Simard et al. [8] also worked on trilateration. They exploited multiple filters to preprocess the RSSI as a preliminary step to an elliptical trilateration. Their method achieved a localization of up to 15 cm.

The main problem of these approaches is that they focus on increasing the precision of the localization at the expense of the accuracy over time. As our goal is to obtain an accurate tracking system for activity recognition, these methods are not well-adapted. Also, these systems are usually complex to implement and require a lot of technical expertise.

2.3   Positioning with Data Mining

To palliate the limitation of the literature, learning algorithms could be exploited. Ting et al. [17] studied the possibility of using passive tags to identify qualitative positions. They placed an antenna at each corner of a room and divided the area into nine zones of one square meter. They then recorded 6 readings for each zone and used the average of these readings to form a look-up table. They then used the Euclidian distance to find the nearest neighbor to the RSSI of the detected object whose position is to be determined. Their overall accuracy is 93 % over 90 trials. We extend on this work by using a much more zones we many algorithms.

Decision trees have already been used for the purpose of localization, along with nearest neighbors and Bayesian methods. They are often called fingerprints methods. In a series of papers, Yim [19] makes an extensive use of those methods to build IPSs using wireless local area network (WLAN). In [19], he showed that decision trees have a similar performance to a Bayesian network and a 1-nearest neighbor with WLAN.

3 Methodology

In this section, we explain in detail how we created our datasets to implement our new solution. These datasets are composed of RFID readings for each qualitative zone of the positioning system. But before exploring the details, it is important to describe the intelligent environment that served this project.

This project took place at the DOMUS laboratory of the Universit de Sherbrooke. It is a complete and realistic apartment equipped with more than a hundred sensors distributed in six different rooms: one bedroom, an entry hall, a kitchen, a dining room, a living room and one bathroom. There are twenty RFID antennas disposed to cover the whole surface of the smart home.

The Fig. 1 shows an aerial view of the smart home. As you can see, there are more antennas in the kitchen than in any other room since this is where most of the activities of daily living occur.

3.1 Creating a Target Object

The first step of our project consisted of creating a special object that would allow us to get good RFID readings. This object needed to be something common we could easily find in any house. We chose an empty rigid plastic bottle of water. Then, we had to select the tags to install on the object. To do so, it is important to check tags to ensure they give very similar RSSI when placed side by side. We decided to use four tags to reduce the bad angle of arrival problem. We put those tags on the bottle making sure that each tag was facing a different direction. When collecting data from the bottle, we merged readings for the four tags to keep only the highest reading from each antenna. This gave us an improved special tag that covered each direction, always facing the nearest antennas. A complete reading of our special tag is a vector of twenty negative integers representing the measured signal strength in decibel. Values range from minus thirty to minus seventy, as allowed by the configuration module. We put the tags on an object because for the final application we intent to track only the objects in the housing and not directly the inhabitants.

3.2 Creating Zones

The second step of this project consisted in defining qualitative zones. At first, we had no way to know what kind of precision we would be able to reach with our setup other than estimating. Therefore, we started by dividing the dining room into zones of 100 cm × 100 cm. We evaluated the performance by training a basic decision tree with the SimpleCART algorithm from the Weka tool [9], using cross-validation. Next, we tried with 75 cm × 75 cm, 60 cm × 60 cm, 45 cm × 45 cm, 30 cm × 30 cm and finally 20 cm × 20 cm. We stopped at twenty because it was near the size of the bottle and more precision would not help since the error rate was increasing fast. Additionally, the most precise non qualitative positioning algorithms in the literature report precision around 20 cm [8]. From this preliminary step, we collected many datasets for that single room that were

not used for the final design of our solution, but could be useful in the future
to improve accuracy if the positioning precision proved to be less important for
our final system.

Once we established how precise the system could be, we were able the define
zones for each room. Figure 1 presents a map of all the defined zones. Clearly,
not all of the zones are the same size, because not all rooms require the same
precision for our final goal of recognizing activity of daily living. The figure
also shows that zones located on the edge of each room are bigger because the
precision is less important since they cover places that a person would not usually
go. The precision in the bathroom and in the bedroom is sixty centimeters. The
kitchen and the dining room have a precision of twenty centimeters and the other
rooms have seventy five centimeters. The counter in the bathroom has a higher
precision of thirty centimeters since more complex activities can happen there
(e.g.: brushing teeth, shaving, etc.). This causes the surrounding zones to have
heterogeneous shapes.

3.3 Collecting Datasets

The third step of the project, once we had a special object and all our zones,
was to collect data from them. We collected a single dataset for each room in the
DOMUS' smart home, each one identified by the name of the room. We placed
our special bottle on a bench at the same height as the antennas and collected
exactly fifty readings for each zone. Antennas were calibrated to send a signal
every 750 ms. This allowed us to reduce to a minimum the risk of interference
between antennas as the probability of an antenna to communicate at the same
time as another was very low. The bottle was always placed in the middle of
each zone to ensure the regularity of the readings. For the datasets, we did not
collect any readings close to a zone border.

After several days of data collection, we ended up with a datasets per room.
They contain 47,989 readings from 963 qualitative zones. Moreover, every reading was labeled to its zone, making the datasets perfect for supervised learning.
We decided to make these datasets available for other smart home researchers
on social networks and on our personal websites. The reason we decided to do
so is that it is very difficult to repeat experiments in the context of smart home
research to validate and improve research. Moreover, many researchers in our
field do not have access to RFID technology and they might want to experiment with it before investing a considerable amount of money to equip their
infrastructures.

4 Experiments and Results

An ADL recognition algorithm like the one we planned to design needs a real-
time tracking system. A real-time tracking system needs a fast and accurate
positioning system. This section presents the algorithms that we used and the
accuracy they gave on each dataset. We also briefly describe each family of

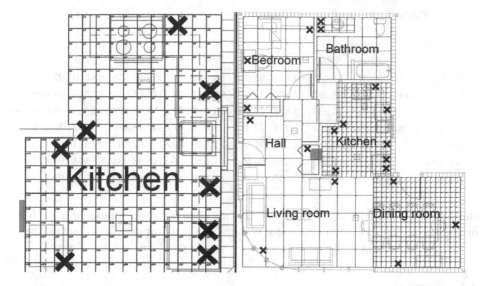

Fig. 1. On the left, a close view of the kitchen map. On the right, the overall map of the DOMUS smart home. The RFID antennas are marked by a X. The grid represents the scaled qualitative zones

Algorithms used. We conclude the section with a simple algorithm that allowed us to merge qualitative zones under certain conditions to improve accuracy.

As mentioned previously, our goal was to determine if we could use known data mining algorithms to create an indoor positioning system using passive RFID. To do so, we tested several algorithms, mostly trees, on each collected dataset. All algorithms are the implementations available in Weka [9] and we used them with default parameters. Table 1 shows an overview of the results we obtained. Accuracy was computed using 10-fold cross-validation, as set by default in Weka. We did this in order to do a maximal number of tests with the widest range of algorithms before searching for the best hyper-parameters. Also, we were looking to develop a simple positioning system that could perform well in almost any given indoor environment without having to spend great effort in fine-tuning the algorithm every time. Thus, using default parameters allowed us to rapidly find the algorithm most suited for positioning in each room. As a side note, it is important to mention that our system supposes the existence of another positioning system able to determine in which room each object is located.

We can see in Table 1 that, as expected, almost all algorithms had a better accuracy for rooms with bigger zones. The size of the zones not only allows for more variation in received signal strength, but also having less target classes makes it easier for most algorithms to discriminate between them. In fact, there are a finite number of RSSI vectors we can obtain in our apartment. The maximum number would be 2040 (40 is the possible variation in decibels and 20

Table 1. Accuracy for different learning algorithms on all datasets

Dataset	Accuracy								
	BNet	NNet	1-NN	Cart	J48	RT	NBT	LMT	RF
Hall	96,375	51,125	94,875	95,625	96,875	95,625	96,375	95,500	97,750
Living Room	97,657	91,257	94,800	92,400	92,057	89,314	94,686	93,314	97,600
Kitchen	84,983	75,639	78,824	73,403	74,966	67,815	83,227	74,908	88,916
Dinning Room	74,931	65,205	73,251	72,492	73,375	71,084	73,578	70,392	76,727
Bedroom	95,697	86,485	93,212	91,515	92,849	89,818	92,485	92,424	96,485
Bathroom	95,111	87,482	90,370	90,266	91,630	88,148	90,963	92,593	96,074
Average	90.792	76.199	87.555	85.950	86.959	83.634	88.552	86.522	92.259

the number of antennas), but in fact the number is much smaller due to the boundary of the environment. Many vectors are just impossible to get as they would be outside, in walls, or because they were out of the antennas' reach.

4.1 Trees

First, we will present some trees we used. We trained some very classical trees, but we also trained more exotic ones to see if they would perform well with our data. We have explained them in the next few paragraphs.

Decision Trees. The first type of trees we present are the decision trees. They are simple yet efficient on our data. J48 is the name given by Weka to its implementation of the C4.5 algorithm [15]. C4.5 uses the information gain metric to find the best attribute to split on in each node until there is only one class left or a minimum number of examples is reached. Then a pruning step is done trying to remove leaves that do not bring any accuracy gain. Next, we tested Simple-Cart [3], Cart in Table 1, which is another decision tree. It differs from C4.5 at the pruning step where it uses the cost complexity pruning. There was not any conclusive gain with SimpleCart over C4.5, which suggests that pruning does not yield significant differences with our datasets.

The last simple tree we present is the Random Tree. It is simply a tree where only a subset of attributes is considered for each split. It is therefore well adapted to datasets consisting of a large number of attributes. The size of this subset is $\log_2(D)+1$, where D is the number of attributes. There is no pruning. It performed slightly worse than the other trees. Indeed, our datasets are composed of only 20 numerical attributes and Random Tree is better suited for contexts with very high dimensionality.

Decision Trees with a Model on Leaves. We just saw three different trees that work under the same basic principles. This sub-section presents two other trees that have a model on their leaves or nodes.

NBTree [11] is a tree with the main characteristic that there is naive Bayes classifiers on the leaves. The author of this model affirms that it can outperform both the normal tree and the naive Bayes classifier, especially on large datasets. His affirmation proved to be true in our experiment. NBTree was better or equivalent to C4.5 on most datasets, while being significantly better in the kitchen. As our final goal is to build a system for ADL recognition, it makes this model a particularity good choice.

Logistic Model Tree [12], LMT in Table 1, is a tree whit a logistic regression function at the leaves. It produces big trees, thus being longer to train than other with tried for similar accuracy. Indeed, a bigger tree implies more leaves and more logistic functions to learn.

Trees Forest. The last tree algorithm that we trained was the Random Forest [2]. Random Forest is simply 100 Random Trees trained separately, exactly the same way as described before. The predicted class is then the modal class between all 100 predicted classes. This forest surpassed the single random tree, thus being by far the most accurate model on our datasets. It even outperformed the Bayesian Network while being shorter to train and faster to use. Nevertheless, it is slower than basic classifiers such as C4.5 and SimpleCART.

4.2 Other Algorithms

In the previous section we discussed the main decision trees algorithms that we decided to test in this work. This section focuses on other algorithm families we tried, the nearest neighbors, the Bayesian Network and the Multilayer Perceptron.

Nearest Neighbors. Alone in his family, the k-nearest neighbors (k-nn) algorithm [1] is a very simple method that consists of finding the k closest known examples to a given unclassified example and predicting the modal class among those examples. There are many distance functions that can be used by k-nn. As we are in a continuous 2-dimentional context, we chose to use the Euclidian distance, which is also the default distance in Weka.

We tried this algorithm with k values ranging from 1 to 5. Table 2 shows the accuracy obtained on the hall dataset. It shows that the best value for k is one. We observed the same results for all datasets, where the accuracy always diminishes as k increase. However, the difference between k equals one and k equals five could be as big as eight percent. Still, in most cases, the difference was not that big, varying of about one percent per k. The main reason behind these results is that a bigger k makes boundaries between classes less distinct. In future work, it would be interesting to verify if the k-nn algorithm performs better when coupled to a heuristic to select a good k.

Bayesian Network. Next, we trained a Bayesian network. It is a probabilistic model presenting itself as a directed acyclic graph that we can use to represent a probability distribution of classes over attributes. It only works with discrete

Table 2. Accuracy for different K

Dataset	Accuracy for K				
	1	2	3	4	5
Hall	94,875	94,625	92,375	91,75	91,000
Living Room	94,800	93,600	92,800	92,514	91,429
Kitchen	78,824	74,706	74,261	72,605	71,319
Dinning Room	73.251	71.318	70.979	70.164	69.787
Bedroom	93,212	92,364	92,424	91,636	91,455
Bathroom	90,370	88,667	88,444	87,482	86,074
Average	87.555	85.880	85.214	84.359	83.511

values, so our data were discretized before the training. Training a Bayesian network can be seen as two separate steps: learning the network structure and learning the probability tables. The one present in Weka offers numerous possibilities in the choice of algorithm for each of those steps. We used the K2 algorithm to learn the network structure. K2 is a hill climbing method that uses a fixed ordering of variables to maximize quality measure of the network structure. In our case, the quality measure was the Bayesian metric from Cooper and Herskovits [7] (see Eq. (1)), a measure that tends to approximate the likelihood of the graph. The graph was initialized as a Naive Bayes Network. This means that the classifier node is connected to all other nodes. In Eq. 1, BS represents the network structure of the database D. $P(B_S)$ is then the prior network structure and r_i is the cardinality of the data. N_{ijk} is the number of cases in D where the variable x_i as the value v_{ik}.

$$Q_{K2}(B_S, D) = P(B_S) \prod_{i=0}^{n} \prod_{j=1}^{q_i} \frac{(r_i - 1)!}{(r_i - 1 + N_{ij})!} \prod_{k=1}^{r_i} N_{ijk}! \qquad (1)$$

To learn the probability tables, we used the SimpleEstimator. It estimates the conditional probabilities directly using the given data. A smoothing constant of 0.5 is used by default when computing the probability tables.

This model proved to be one of the best we have trained, only matched by the Random Forest. However, it requires a very long training, especially on the bigger datasets like the kitchen. Indeed, the complexity of complete inference is NP [18]. Classification could be longer as well because it requires a lot more computation than decision trees which are usually under $O(\log n)$. In future work, it would be interesting to verify if a Bayesian Network is usable to perform real time classification for an indoor positioning system.

Multilayer Perceptron. The Multilayer Perceptron consists of many non-linear nodes linked together by weighted links. In Weka, all nodes are a sigmoid function. Weights and biases in the network are trained using the back-propagation algorithm that allows an efficient transmission of the gradient throughout all nodes. This gradient comes from the loss function that the learning phase tries to minimize. A learning rate of 0.3 is used to monitor the learning, along with a momentum of 0.2. Each training example was seen 500 times by the network. There was only one hidden layer of size $D + C$, where D is the number of attributes and C the number of classes to predict.

Using only the default parameters, the Multilayer Perceptron was very inconsistent over our different datasets. Its accuracy was at most satisfactory and was often worse than all other algorithms. However, we believe that with appropriate fine-tuning it could be one of the best algorithms. Still, it was not in the scope our experiment to fine-tune the learning algorithms and it would have been biased to adjust this algorithm when the other algorithms could also have been fine-tuned. We tested the multilayer perceptron to see if it could be exploited as easily as simple decision trees. As we expected, it requires a bit more effort to use it. We also think that it would perform better with a larger dataset of a higher dimensionality.

4.3 Merging Qualitative Zones

One of the problems we discovered after running the algorithms mentioned above was that they all suffer from the high number of qualitative zones they have to classify. We tried to diminish the impact of this problem by merging automatically certain zones that were hardly distinguishable. In order to achieve this, we designed a simple statistical method. The general idea is presented below:

1. Train a classifier;
2. Record all erroneous classification on test set;
3. For each error, if the predicted zone is neighboring the target zone, merge the zones.

By merging only neighboring zones, we ensured that the accuracy would not drop but only improve since there would be at least one more example that would be correctly classified. Moreover, we thought that those small local merges would not affect our ADL recognition algorithms due to the small number of them. We tested the effects of the merging on CART algorithm and the accuracy improving from 1 to 7 % depending on the room.

5 Discussion

In the previous section, we explained and tested several algorithms. We saw that they all perform relatively well on most datasets. Still, some are faster in making a decision and thus they should be prioritized. Indeed, decision time is an important factor for a tracking system like the one we intend to do. Neural networks could work, but they need much bigger datasets with more attributes to perform at their best. For all those reasons, we consider the Random Forest as the best algorithm to fit our needs. Not only is it one of the best in terms of accuracy, it is also fast to execute. Adding more tree does not really affect the decision speed, as it stays under $\log_2 n$, where n is the number of classes. Another argument in favor of the Random Forest is that we can get more than one prediction. In this paper, we used only the modal prediction. But, for the tracking system, we could use the zones with the most votes and then pick the one that is closest to the previous position.

For the kitchen and the dining room dataset, we did not get as high an accuracy as the other rooms. We could solve this problem by exploiting bigger zones. In fact, we tested datasets composed of zones of 30 cm × 30 cm, 45 cm × 45 cm, 60 cm × 60 cm, 75 cm × 75 cm and 100 cm × 100 cm in the dining room and the accuracy climbed to respectively 78.98 %, 81.78 %, 88.61 %, 95.04 % and 96.07 % with C4.5 from the 73.35 % with the zones of 20 cm × 20 cm. However, we believe that the accuracy will greatly increase when the models will be exploited in a complete real-time tracking system by adding filtering on the incoming positions. Indeed, the reason why other methods such as trilateration work so well is because they use many filters on a high number of readings [8]. The more readings we can get, the better the algorithm will be.

6   Conclusion

In this paper, we presented a novel method for the indoor positioning of objects used in daily life in a smart home. We have shown that traditional data mining algorithms can be effectively exploited for qualitative indoor positioning. Trees are very simple algorithms that can be very accurate on this task. Moreover, they are fast to train and even faster to use, which is important when the goal is to do real-time positioning. Our qualitative approach is also adaptable, allowing us to decide the size of each zone to reflect the precision needed in each room or even for a particular sector of a room. Moreover, like trilateration techniques, data mining algorithms can use the raw received signal strength. However, they do not need a data rate as high as the former because there is no filtering. We also showed that we could merge some neighboring zones to increase the accuracy of our system. Another important contribution of this work is the twenty datasets composed of more than 47 thousands lines that are available to other researchers through our website (www.Kevin-Bouchard.com). Experimenting on realistic datasets is often difficult for smart home researchers due to the time required to perform the data collection and the cost of the infrastructure.

The main drawback of our method is that we need to construct datasets by manually collecting readings for each qualitative zone. While it is an easy task that does not require a particular human expertise, it is also a task that is time consuming. In future work, we aim to address this issue by automating this data collection phase in order to reduce the human involvement and decrease the length of the learning phase. For example, the special object could be placed in each zone with laser meter, the controller program could be on a tablet, a robot could place the object, etc. The next step of this project will be to test this indoor positioning system for real time and continuous tracking of objects within the DOMUS smart home.

References

1. Aha, D.W., Kibler, D., Albert, M.K.: Instance-based learning algorithms. Mach. Learn. **6**(1), 37–66 (1991)
2. Breiman, L.: Random forests. Mach. Learn. **45**(1), 5–32 (2001)
3. Breiman, L., Friedman, J., Stone, C.J., Olshen, R.A.: Classification and Regression Trees. CRC Press, New York (1984)
4. Brown, M., Vandergoot, D.: Quality of life for individuals with traumatic brain injury: comparison with others living in the community. J. Head Trauma Rehabil. **13**(4), 1–23 (1998)
5. Chen, C.-Y., Yang, J.-P., Tseng, G.-J., Wu, Y.-H., Hwang, R.-C., et al.: An indoor positioning technique based on fuzzy logic. In: Proceedings of the International MultiConference of Engineers and Computer Scientists, vol. 2 (2010)
6. Chen, L., Nugent, C.D., Wang, H.: A knowledge-driven approach to activity recognition in smart homes. IEEE Trans. Knowl. Data Eng. **24**(6), 961–974 (2012)
7. Cooper, G.F., Herskovits, E.: A Bayesian method for the induction of probabilistic networks from data. Mach. Learn. **9**(4), 309–347 (1992)
8. Fortin-Simard, D., Bouchard, K., Gaboury, S., Bouchard, B., Bouzouane, A.: Accurate passive RFID localization system for smart homes. In: 2012 IEEE 3rd International Conference on Networked Embedded Systems for Every Application (NESEA), pp. 1–8. IEEE (2012)
9. Hall, M., Frank, E., Holmes, G., Pfahringer, B., Reutemann, P., Witten, I.H.: The WEKA data mining software: an update. ACM SIGKDD Explor. Newsl. **11**(1), 10–18 (2009)
10. Joho, D., Plagemann, C., Burgard, W.: Modeling RFID signal strength and tag detection for localization and mapping. In: IEEE International Conference on Robotics and Automation, 2009. ICRA 2009, pp. 3160–3165. IEEE (2009)
11. Kohavi, R.: Scaling up the accuracy of naive-bayes classifiers: a decision-tree hybrid. In: KDD, pp. 202–207. Citeseer (1996)
12. Landwehr, N., Hall, M., Frank, E.: Logistic model trees. Mach. Learn. **59**(1–2), 161–205 (2005)
13. Ni, L.M., Liu, Y., Lau, Y.C., Patil, A.P.: LANDMARC: indoor location sensing using active RFID. Wirel. Netw. **10**(6), 701–710 (2004)
14. Palmes, P., Pung, H.K., Gu, T., Xue, W., Chen, S.: Object relevance weight pattern mining for activity recognition and segmentation. Pervasive Mobile Comput. **6**(1), 43–57 (2010)
15. Quinlan, J.R.: C4.5: Programs for Machine Learning. Elsevier, New York (2014)
16. Rafalimanana, H., Lai, M.: World Population Ageing 2013. New York: United Nations, Department of Economic and Social Affairs. Population Division. United Nations, New York (2013)
17. Ting, S., Kwok, S.K., Tsang, A.H., Ho, G.T.: The study on using passive RFID tags for indoor positioning. Int. J. Eng. Bus. Manag. **3**(1), 9–15 (2011)
18. Wu, D., Butz, C.J.: On the complexity of probabilistic inference in singly connected Bayesian networks. In: Ślęzak, D., Wang, G., Szczuka, M.S., Düntsch, I., Yao, Y. (eds.) RSFDGrC 2005. LNCS (LNAI), vol. 3641, pp. 581–590. Springer, Heidelberg (2005)
19. Yim, J.: Introducing a decision tree-based indoor positioning technique. Expert Syst. Appl. **34**(2), 1296–1302 (2008)

Deep Feature Extraction from Trajectories for Transportation Mode Estimation

Yuki Endo[✉], Hiroyuki Toda, Kyosuke Nishida, and Akihisa Kawanobe

NTT Service Evolution Laboratories, Yokosuka, Japan
{endo.yuki,toda.hiroyuki,nishida.kyosuke,kawanobe.akihisa}@lab.ntt.co.jp

Abstract. This paper addresses the problem of feature extraction for estimating users' transportation modes from their movement trajectories. Previous studies have adopted supervised learning approaches and used engineers' skills to find effective features for accurate estimation. However, such hand-crafted features cannot always work well because human behaviors are diverse and trajectories include noise due to measurement error. To compensate for the shortcomings of hand-crafted features, we propose a method that automatically extracts additional features using a deep neural network (DNN). In order that a DNN can easily handle input trajectories, our method converts a raw trajectory data structure into an image data structure while maintaining effective spatio-temporal information. A classification model is constructed in a supervised manner using both of the *deep* features and hand-crafted features. We demonstrate the effectiveness of the proposed method through several experiments using two real datasets, such as accuracy comparisons with previous methods and feature visualization.

Keywords: Movement trajectory · Deep learning · Transportation mode

1 Introduction

Estimating users' contexts from their movement trajectories obtained from devices such as mobile phones with GPS is crucial for location-based services (e.g., Google Now[1] and Moves[2]). This paper focuses on a specific aspect of human movement, the transportation mode of individual users when they move. The ability to accurately determine the transportation mode on mobile devices will have a positive impact on many research and industrial fields, such as personalized navigation routing services [7] and geographic information retrieval [17]. According to previous studies [21–23], transportation mode estimation involves two steps: extraction of segments of the same transportation modes and estimation of transportation modes on each segment (see also Fig. 1(a)).

---

[1] http://www.google.com/landing/now/.
[2] https://play.google.com/store/apps/details?id=com.protogeo.moves.

© Springer International Publishing Switzerland 2016
J. Bailey et al. (Eds.): PAKDD 2016, Part II, LNAI 9652, pp. 54–66, 2016.
DOI: 10.1007/978-3-319-31750-2_5

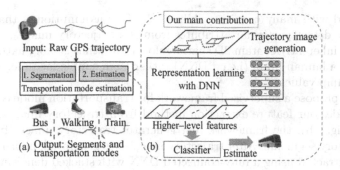

Fig. 1. Transportation mode estimation and our contributions.

In estimating transportation modes, researchers have manually discovered effective features for supervised classification (e.g., movement distance, velocities, acceleration, and heading change rate [21–23]) using their skills. While this heuristic approach is basically important for discriminating between transportation modes, hand-crafted features do not always work well because human behaviors are diverse, and movement trajectories also include various aspects. For example, movement distance and velocity, which are especially fundamental and effective features, depend on users' contexts even when they are using the same transportation mode. Such features are also susceptible to GPS measurement error, which becomes larger especially in urban environments.

To compensate for the above shortcomings, we utilize additional features automatically extracted by *representation learning. Deep learning* [2,6] is a well-known example of this, which learns a deep neural network (DNN) model with multiple intermediate layers and can automatically extracts effective higher-level features for tasks from lower-level features of input data. Recently, this technique fundamentally improved performance in some fields including image recognition [8] and speech recognition [4].

The effectiveness of deep features for a task depends on an input data structure. For example, while raw pixel values are often used as input of a DNN for image data [6,8], spectrograms are calculated from raw signals for audio data so that a DNN can easily handle them [4]. These approaches cannot be directly adapted to the locational information because which has a different data structure (a series of latitude, longitude, and timestamp) from image and audio data. Consequently, how to apply deep learning to locational information has not been properly studied.

We propose a method that extracts features from raw GPS trajectories using deep learning. Our key idea is to represent GPS trajectories as 2D image data structures (called *trajectory images*) and use these trajectory images as input of deep learning. This is based on the knowledge that deep learning works well in the field of image recognition. For example, a DNN can detect local semantic attributes of images, such as skin patterns and tail shapes of animals, as human can understand by looking them. This is because a DNN has a structure that approximates the operation of the neocortex of a human brain, which is associated with many cognitive abilities [1]. Our assumption is that a DNN can suitably detect particular attributes from the trajectory images: movement trajectories inherently contain 2D spatial information that is more naturally perceivable for a human brain (i.e., a DNN) rather than simple latitude, longitude, and timestamp values.

We also propose a supervised framework for transportation mode estimation, which includes our feature extraction method from trajectory images. As illustrated in Fig. 1(b), the framework first generates trajectory images from given GPS trajectory segments. After trajectory images are generated, higher-level features are extracted using a fully-connected DNN with stacked denoising autoencoder (SDA) [19], which is a representative method of deep learning. Intuitively, higher-level features are obtained by appropriately filtering trajectory images for picking up discriminative parts of the images. Finally, transportation modes are estimated using a classifier that is learned from the higher-level features and transportation mode annotations.

Our main contributions are summarized as follows:

- We propose a method for generating informative trajectory images for deep learning from raw GPS trajectories (Sect. 3).
- We propose a supervised framework for trajectory classification including feature extraction from trajectory images using deep learning (Sect. 4).
- Extensive evaluations are provided to confirm the effectiveness of our method using two real datasets (Sect. 5).

2   Related Work

GPS Trajectory Mining. An overview of trajectory data mining is outlined in a survey [20]. In particular, there have been many studies on trajectory mining tasks such as user activity estimation [5,12], transportation mode estimation [10, 13,14,21–23], and movement destination estimation [16]. Several methods [5,12] use not only GPS trajectories as features but also body temperature, heart rate, humidity, and light intensity obtained from other sensors, and construct a model for predicting user activities such as walking, running, cycling, and rowing. While these methods can estimate various user activities, users need to carry many devices. Estimating a user's context with few sensors is ideal to lighten his/her burden. Therefore, using sensor information other than GPS trajectories is out of the scope in this paper.

Liao et al. [10], Patterson et al. [13], and Shah et al. [14] reported on methods for estimating transportation modes, such as walking, bus, and car, using only GPS trajectories as sensor data. However, their methods require external information including a street map. Static information, such as a street map, might not be applied to the task because structures of cities dynamically change over time. We therefore do not target methods that require external information.

For an approach that does not use external information, Zheng et al. [23] proposed a method that can estimate transportation modes using only GPS trajectories. They describe a method for segmenting GPS trajectories by detecting change points of transportation modes on the basis of velocity and acceleration. Transportation modes are then estimated from features of segments using a classifier. Zheng et al. first presented basic features such as moving distance, velocity, and acceleration [21]. They also introduced advanced features including velocity change rate (VCR), stop rate (SR), and heading change rate (HCR), which achieved more accurate estimation [22]. While their method uses hand-crafted features, our method tackles the problem of automatically extracting effective features from trajectory images.

Deep Learning. One of the major goals of deep learning is to obtain effective higher-level features from signal-level input using a DNN. For example, while traditional approaches for image recognition use hand-crafted features such as scale-invariant feature transform (SIFT) [18], a DNN can automatically extract effective features from raw image pixels. In fact, it has been reported supervised learning with deep features can achieve high recognition accuracy [6].

Although a DNN has high expressiveness, learning a DNN model efficiently using conventional approaches is difficult due to a vanishing gradient problem. Specifically, back-propagation used to optimize a DNN does not sufficiently propagate a reconstruction error to deep layers, and the error vanishes midway through an intermediate layer. To solve this problem, greedy layer-wise training was proposed [2,6], and it has allowed the topic of deep learning to gain significant attention. This technique pre-trains parameters of intermediate layers layer-by-layer in an unsupervised manner before fine-tuning for the entire network. This enables error information to be efficiently propagated to deep layers and consequently improved performance in many tasks.

There are several techniques for deep learning such as deep belief nets (DBN) [6], deep Boltzmann machine (DBM) [3], and SDA [19] for pre-training. These and other techniques are outlined in a survey [3] that can be referred to for more information. In this paper, we adopt fully-connected DNN with SDA for transportation mode estimation for the first time and demonstrate its effectiveness.

3   Trajectory Image Generation

There are several difficulties for generating informative trajectory images so that DNNs can discriminate between transportation modes. First, most of the DNNs must fix the dimensions of input vectors. That is, input images must be the same size when the pixel values are directly used as the input vectors. However, different-sized images are obtained by simply clipping an entire GPS segment when a spatial length of one pixel is fixed. The reason is that topographic ranges of the GPS segments differ especially depending on transportation modes (walking is often narrow while a train is broad). Although one straightforward approach to solving this problem is to resize different-sized images to the same size, distance information in a trajectory is lost since each scale differs. Second, DNNs require sufficient as well as informative training data to improve its performance. If images are high resolution (number of pixels is large), detailed movement can be obtained; however, the trajectory pixels (non-zero pixels) in the images become sparse, and such sparse images degrade the generalization capability of a DNN. As a result, many trajectory images are required in order to overcome this sparsity problem. If images are low resolution (number of pixels is small), the sparsity problem is alleviated, but much information of GPS points corresponding to the same pixel is lost.

Based on the above, our trajectory image generation method consists of two steps: (1) determining the *target range* of a segment that is converted into a fixed size image, and (2) determining the *number and value of pixels* of the image. For the first step, we simply clip a certain area from each segment. To do this, we define a rectangle region for clipping by ranges of latitude and longitude. Although information outside the defined region is lost, we verified that this method outperforms the resizing method through our experiments because distance information in a trajectory is preserved. For the second step, we use stay time to determine pixel values; i.e., the longer a user stays in the same pixel (a rectangular region), the higher the pixel value becomes. This manner can maintain temporal information of a segment with a small number of pixels, and thus can alleviate the sparsity problem rather than using large binary images that maintain the details of movements.

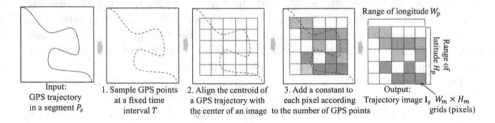

Input: GPS trajectory in a segment P_s | 1. Sample GPS points at a fixed time interval T | 2. Align the centroid of a GPS trajectory with the center of an image | 3. Add a constant to each pixel according to the number of GPS points | Output: Trajectory image $\mathbf{I}_s$ $W_m \times H_m$ grids (pixels)

Fig. 2. Overview of trajectory image generation.

An overview of trajectory image generation is shown in Fig. 2. We first define some terms used in our method. We refer to each data point given a positioning system as a *GPS point*. Given segment s as input, let $P_s = (p^{(i)})_{i=1}^{N_s}$ be a sequence of continuous GPS points, where N_s denotes the number of GPS points in the segment. Let $p^{(i)}$ represent the i-th GPS point and each GPS point be represented as a three-tuple $p^{(i)} = (lat, lng, t)$; latitude lat, longitude lng, and timestamp t. Let W_p and H_p denote ranges of longitude and latitude for pixelizing trajectories, respectively, whereas W_m and H_m denote width and height of images, respectively. Let T is a time interval for sampling GPS points from input GPS trajectories of P_s. $\mathbf{I}_s \in \mathbb{R}^{W_m \times H_m}$ denotes a generated trajectory image that has one-channel value (intensity) per pixel like a grayscale image.

To extract a trajectory image from a GPS trajectory in a segment, we first evenly sample GPS points from P_s. The GPS points in each segment are not always positioned at a fixed time interval due to differences in GPS sensors and signal quality. If sequential GPS points positioned at different time intervals are converted into trajectory images, short time intervals result in a long stay in one pixel even if a user stays in the pixel for a short time. We therefore sample GPS points from P_s at T intervals on the basis of timestamps $p^{(i)}.t$. If the next GPS point is not obtained after just T, we sample the nearest time GPS point. As a result, we obtain a sequence of the sampled GPS points and denote it as P'_s.

In the next step, the target range of a GPS trajectory in a segment is determined. For all segments, we compute the centroid of the sampled GPS points of P_s using $p^{(i)}.lat$ and $p^{(i)}.lng$ and then align the centroid with the center of a trajectory image to unify the basic geographical coordinates. We define a clipped region as a rectangular area measuring W_p and H_p. The rectangular area is divided into $W_m \times H_m$ grids and each grid corresponds to each pixel of the trajectory image. The number of pixels $W_m \times H_m$ is searched for using grid search, and the range of grid search is empirically determined as explained in the evaluation section.

Finally, GPS points of P'_s are then plotted when the GPS points exist in a defined grid. When plotting GPS points, we add a constant $c = 1$ to the corresponding pixel to express the stay time in the pixel. After plotting all GPS points of P'_s on the segment s, trajectory image $\mathbf{I}_s$ is obtained.

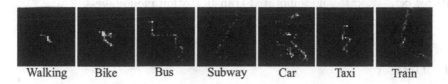

| Walking | Bike | Bus | Subway | Car | Taxi | Train |

Fig. 3. Examples of trajectory images extracted from real GPS trajectories. Brighter color means longer stay time.

Figure 3 shows several examples of trajectory images extracted from real GPS trajectories. The intensity of pixels indicates a user's stay time: the brighter the color, the longer the stay time. These trajectory images show that the images store distance information by clipping in the same range and represent time information through the pixel values. For instance, pixels near the center of the walking images become bright since the moving distance of walking is relatively short and the user stays in the same pixels for a long time. On the other hand, the images of bus and subway include rectilinear lines that are geographically widespread. There are such easy-to-understand features in the images, whereas it is time-consuming and difficult to discover all features and quantify them. We therefore extract effective features from trajectory images using deep learning in the next section.

4   Deep Feature Extraction and Classification

To use trajectory images as input of a fully-connected DNN, we convert trajectory image matrices $\mathbf{I}_s$ into $W_m \times H_m$ dimensional vectors $\mathbf{x}_s$ by simply aligning each pixel value. The number of intermediate layers L of the DNN is determined by grid search as explained in the evaluation section. We use a sigmoid function $s(\cdot)$ as an activation function of each layer. To pre-train parameters (weighting matrices $\mathbf{W}_{(l)}$ and bias terms $\mathbf{b}_{(l)}$ at each intermediate layer l) of DNN using SDA [19], we use a minibatch L-BFGS method because of its effectiveness for classification problems [9]. After pre-training with SDA, supervised fine-tuning adjusts parameters of the entire DNN using annotated labels. For fine-tuning, an output sigmoid layer is added to the DNN, and parameters are updated using a stochastic gradient descent (SGD) method on the basis of the squared error between vectors on the output layer and binary vectors obtained from annotations. By using the learned DNN, higher-level features $\mathbf{x}_{(L+1)}$ are extracted from the deepest L intermediate layer of the DNN:

$$
\mathbf{x}_{(l)} = \begin{cases} s(\mathbf{W}_{(l-1)}\mathbf{x}_{(l-1)} + \mathbf{b}_{(l-1)}) & (l > 1); \\ \mathbf{x}_s & (l = 1). \end{cases} \tag{1}
$$

These image-based higher-level features are concatenated with the hand-crafted features $\mathbf{x}_e$ (movement distance, mean velocity, etc.). We construct a classifier, such as logistic regression and support vector machine, using the concatenated features $[\mathbf{x}_{(L+1)}^T, \mathbf{x}_e^T]^T$ and annotated transportation mode labels.

5   Evaluation

5.1   Dataset

GeoLife (GL). We used a GeoLife dataset [21–23] published by Microsoft Research. The GPS trajectories in the dataset were basically positioned every 1–3 s and 69 users annotated labels of transportation modes. We removed the data of users who have only 10 annotations or fewer and used the data of 54 users for our experiments. Each annotation contains a transportation mode and beginning and end times of the transportation. In the experiments, we labeled each section of GPS trajectories between the beginning and end times with an annotation, and used these sections as a segment of the same transportation mode. Although there are 11 types of annotations, we used only 7 (walking, bus, car, bike, taxi, subway, and train) because the other 4 are in too few trajectories, and 9,043 segments were obtained.

Kanto Trajectories (KT). To verify that our method works in other regions, we used other trajectory data collected in the Kanto area of Japan. The data contains 30 users' trajectories for 20 days obtained from a Nexus7 2012 with a GPS sensor. The trajectories were basically positioned every 3 s. Each trajectories were annotated with a label of the seven transportation modes (walking, bike, car, bus, taxi, motorcycle and train). In this dataset, we additionally segmented each labeled segment at three-minute intervals, and 14,019 segments were obtained. This is because we assume the use of our method for a real-time application, which estimates transportation modes from sequential segments for a relatively-short time window.

5.2   Compared Methods

Feature Extraction Methods. To evaluate our feature extraction method, we prepared the following baseline features and our features:

- Basic Features (BF) [21]: Ten dimensional features such as velocity.
- BF+Advanced Features (AF) [22,23]: Thirteen dimensional features including BF and advanced features (VCR, SR, HCR).
- BoVW (Bag of Visual Words): Image features extracted from trajectory images using Dense-SIFT [18].
- SDNN: deep features extracted using a DNN from vectors simply consisting of a series of latitude, longitude, and movement time at each GPS point.
- IDNN: deep features extracted using a DNN from trajectory images.
- BF+AF+IDNN: Features consisting of hand-crafted ones (BF+AF) and deep ones of trajectory images (IDNN).

For SDNN, the dimensions of input vectors are fixed to be the same number as those of the trajectory images of IDNN. Since one GPS point consists of three dimensional components (i.e., latitude, longitude, and movement time), when three times the number of GPS points in a segment is smaller than the fixed dimensions, the empty element of the vector is set to 0. When that value is larger than the fixed dimensions, the newer GPS points are discarded.

Classification Methods. To build a classifier for estimating transportation modes, supervised learning is done using the extracted features and transportation mode annotations. We compared three classification methods, logistic regression (LR), support vector machine (SVM), and decision tree (DT). The experiment showed that the effectiveness of the classification method differs according to the features. For BF and BF+AF, we used DT in the following experiments since DT obtains the highest accuracy [21–23]. For BoVW, we used SVM. For SDNN, IDNN and BF+AF+IDNN, we used LR.

5.3   Evaluation Method

As an evaluation metric, we use accuracy that is the ratio of segments of correctly estimated labels out of all segments. We used 5-fold cross validation (CV) over users, that is, each dataset was divided into training segments of *80 % users* and test segments of *20 % users*, while previous studies [21–23] mentioned nothing about discriminating users. This is because the training data of the test users are not often obtained in a realistic scenario. The problem setting in our study is more difficult than the previous studies. This is because movement features depend on users due to their habits or environments but their data cannot be trained, and we also handle more transportation modes than the previous studies.

For the GL dataset, we search for model parameters using grid search based on 5-fold-CV with training data (i.e., nested CV):

- For DT, the splitting criterion is selected from the Gini coefficient or entropy, and the maximum ratio of features used for classification is searched for from $\{0.1, 0.2, \ldots, 1.0\}$.
- For SVM, the rbf kernel is used, the trade-off parameter is searched for from $\{0.01, 0.1, 1, 10, 100\}$, and the kernel coefficient is searched for from $\{0.001, 0.01, 0.1, 1, 10\}$.
- For trajectory image generation, the interval of sampling GPS points T is searched for from $\{10, 30, 60, 120\}$ seconds, ranges of longitude W_p and latitude H_p from $\{0.01, 0.05, 0.1, 0.2\}$, and the image size $W_m \times H_m$ from $\{20 \times 20, 25 \times 25, 30 \times 30, 35 \times 35, 40 \times 40, 50 \times 50\}$.
- For the DNN, the number of intermediate layers L is searched for from $\{1, 2, \ldots, 5\}$ (often 3 performed best), the number of each layer's neurons from $\{10, 50, 100, 200\}$ (often 100 performed best). For fine-tuning, the learning rate is set to 0.1 and the number of epochs is searched for from $\{1, 2, \ldots, 15\}$.

For the KT dataset, we empirically set the parameters by referring to the parameters automatically determined for the GL dataset.

5.4   Performance of Feature Extraction

Overall Performance. Table 1 compares the accuracies of transportation mode estimation with our features and the other features. The bold font denotes the condition that yielded the highest accuracy. In the results for both datasets, the accuracy of IDNN is modestly higher than those of BF and BF+AF. This indicates that the features extracted from trajectory images using deep learning work at least similarly to the hand-crafted features, without complicated features designing. IDNN also significantly outperformed BoVW, that is, deep learning is more effective than the common image feature extraction approach. In contrast, SDNN does not work well despite using deep learning. It implies that simply applying deep learning to almost raw trajectory data cannot extract effective features for this task. Finally, it can be seen that the proposed method with the hand-crafted and deep features (i.e., BF+AF+IDNN) achieves the best performance among all the methods. In other word, our deep features make up for the deficiencies of the existing features.

Table 1. Performance comparison of transportation mode estimation.

Features	GL dataset	KT dataset
BF	0.632 ± 0.025	0.771 ± 0.0040
BF+AF	0.648 ± 0.025	0.780 ± 0.0030
BoVW	0.602 ± 0.044	0.760 ± 0.015
SDNN	0.386 ± 0.014	0.474 ± 0.025
IDNN	0.663 ± 0.029	0.797 ± 0.0060
BF+AF+IDNN	$\mathbf{0.679 \pm 0.028}$	$\mathbf{0.832 \pm 0.0047}$

Noise Robustness. We also evaluated our method's robustness against noise. For this purpose, we generated noisy trajectory data from the KT dataset. While the original dataset already contains some noise due to measurement error, the measurement can degenerate even more depending on the performance of a GPS sensor equipped in a mobile device and urban environments. For example, the KT dataset has about a 10 m error on average according to the measurement accuracy reported from a function of Android OS. This value seems to be relatively low because we use devices with a relatively accurate GPS sensor (Nexus7 2012), but all people do not have high-performance devices and some people may also move in noisier environments. In fact, measurement accuracy may be worse than 100 m in actual situations while current positioning systems in smartphones are accurate to within 10 m under ideal conditions [15]. We therefore evaluated noise robustness by adding some noise to the relatively clean trajectories in the KT dataset. The measurement is modeled as random Gaussian noise with zero mean and σ^2 variance [24].

Figure 4(a) shows the accuracy with different noisy levels in the KT dataset. In this experiment, we empirically fixed the DNN parameters for simplifying the experiment. The accuracy of BF+AF decreased with increasing noisy levels, whereas that of IDNN was barely affected by the noise. Although the accuracy of BF+AF+IDNN modestly decreased, it only reached that of IDNN. This is because BF+AF does not work well when the noisy level is high, but our DNN-based method is robust against measurement error.

There are two reasons our method is robust against measurement error. First, noise is reduced in the process of trajectory image generation. For example, if W_p and H_p are 0.01, the images are generated in the range of about 1000 m. When the image size $W_m \times H_m$ is 40×40, one pixel represents $25\,\text{m}^2$. Therefore, noise of tens of meters has an insignificant effect on trajectory image generation. Second, the DNN can automatically detect features from trajectory images even if the data have some noise. In particular, the DNN with SDA learns a model to be able to reconstruct de-noised data from noisy data.

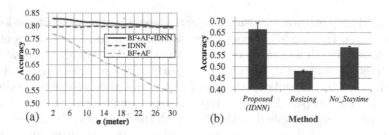

Fig. 4. (a) Accuracy with different noisy levels in KT dataset. (b) Performance comparison in GL dataset with each trajectory image generation method.

5.5   Effectiveness of Image Generation Method

We now discuss the effectiveness of our method at generating trajectory images. Our image generation method does not use the information of the GPS points that are (1) outside of the defined region and (2) not sampled at T intervals, and (3) detailed latitude and longitude values (discretization into pixels).

For the validation of the first point, as shown in Fig. 4(b), we compared the proposed method (*Proposed*), which maintains the scale of trajectories and also stores the stay time in image pixels, with the following two methods. One method (*Resizing*) generates different-sized trajectory images by clipping an entire region in each segment where a spatial range of one pixel is fixed to a small constant. It then resizes the different-sized images to the same size (i.e., $W_m \times H_m$) using the nearest neighbor method [11]. The stay time information is stored in the same way as with *Proposed*. The other method (*No_Staytime*) assigns the same constant value to pixels in which multiple GPS points exist. The scale is maintained in the same way as with *Proposed*. Obviously, *Proposed* performed best among the three methods, which suggests effectiveness of maintaining scale and storing stay time with our method.

Second, we evaluated our method at different sampling intervals from 10 to 120 s. We confirmed that smaller intervals (less than 60 s for the GL dataset) worsened the accuracy via the grid search (explained in Sect. 5.3). The GPS points in each segment are not always positioned at a fixed time interval. Therefore, the sampling method, which generates GPS points at more regular intervals, is effective for accurately maintaining the stay and velocity information of the trajectories in images, and results in accuracy improvement.

Third, as we mentioned in Sect. 5.4, we confirmed the discretization into pixels improved the robustness to spatial noises in GPS trajectories.

We concluded that our image generation method can extract important information of GPS trajectories and convert them into images effectively.

5.6   Feature Visualization

We analyzed deep features by visualizing activity states of neurons on the learned DNN. In Fig. 5, the two left images show visualization results on states of activated neurons of each intermediate layer of the DNN. We can see that each layer acts as filters for extracting characteristic parts of trajectories such as moving range, moving interval, and distribution. The features also become more abstract as layers become deeper. The seven right images visualize the activity states of neurons that strongly respond to the data with the label of each transportation mode. While it is difficult to understand all meanings of them by visualization, we can distinguish between walking, bike, and bus on the basis of moving range. Interestingly, we can see that the activity state of bus includes more dark regions than that of car. This is seemingly because buses are driven on specific roads unlike cars. These results verify that activated neurons differ depending on transportation modes and that deep learning for trajectory images can extract features that effectively distinguish between transportation modes.

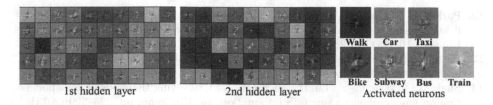

Fig. 5. Visualization results on states of activated neurons of hidden layers.

6 Conclusion

We have proposed a method for extracting features from raw GPS trajectories using deep learning. While we used a fully-connected DNN with SDA, which is a standard method of deep learning, a convolutional neural network (CNN) is known as a closely related approach to deep learning. Although a basic CNN was proposed before deep learning emerged, a recent approach based on CNN significantly improved performance of image recognition [8]. Several advanced learning algorithms for DNNs were also proposed, such as dropout and maxout. Nevertheless, we demonstrated that our framework for transportation mode estimation attained the highest overall performance and significant improvement in noisy environment. It is hoped that our study will become a bridge between the recently advanced approaches of deep learning and trajectory mining.

References

1. Arel, I., Rose, D.C., Karnowski, T.P.: Deep machine learning - a new Frontier in artificial intelligence research. IEEE Comput. Int. Mag. **5**(4), 13–18 (2010)
2. Denglu, Y., Lamblin, P., Popovici, D., Larochelle, H.: Greedy layer-wise training of deep networks. In: NIPS, pp. 153–160 (2006)
3. Bengio, Y.: Learning deep architectures for AI. FTML **2**(1), 1–127 (2009)
4. Dahl, G.E., Yu, D., Deng, L., Acero, A.: Context-dependent pre-trained deep neural networks for large-vocabulary speech recognition. TASLP **20**(1), 30–42 (2012)
5. Ermes, M., Parkka, J., Mantyjarvi, J., Korhonen, I.: Detection of daily activities and sports with wearable sensors in controlled and uncontrolled conditions. IEEE Trans. Inform. Tech. Biomed. **12**(1), 20–26 (2006)
6. Hinton, G.E., Salakhutdinov, R.: Reducing the dimensionality of data with neural networks. Science **313**(5786), 504–507 (2006)
7. Hung, C.-C., Peng, W.C., Lee, W.C.: Clustering and aggregating clues of trajectories for mining trajectory patterns and routes. VLDB J. **24**(2), 169–192 (2015)
8. Krizhevsky, A., Sutskever, I., Hinton, G.: Imagenet classification with deep convolutional neural networks. In: NIPS. pp. 1106–1114 (2012)
9. Le, Q.V., Ngiam, J., Coates, A., Lahiri, A., Prochnow, B., Ng, A.Y.: On optimization methods for deep learning. In: ICML, pp. 265–272 (2011)
10. Liao, L., Fox, D., Kautz, H.: Learning and inferring transportation routines. In: AAAI 2004, pp. 348–353 (2004)
11. Parker, J.A., Kenyon, R.V., Troxel, D.: Comparison of interpolating methods for image resampling. IEEE Trans. Med. Imaging **2**(1), 31–39 (1983)

12. Parkka, J., Ermes, M., Korpippa, P., Mantyjarvi, J., Peltola, J.: Activity classification using realistic data from wearable sensors. IEEE Trans. Inform. Technol. Biomed. **10**(1), 119–128 (2006)
13. Patterson, D., Liao, L., Fox, D., Kautz, H.: Inferring high-level behavior from low-level sensors. In: UbiComp, pp. 73–89 (2003)
14. Shah, R.C., Wan, C.-Y., Lu, H., Nachman, L.: Classifying the mode of transportation on mobile phones using GIS information. In: UbiComp, pp. 225–229 (2014)
15. Shaw, B., Shea, J., Sinha, S., Hogue, A.: Learning to rank for spatiotemporal search. In: WSDM, pp. 717–726 (2013)
16. Song, X., Zhang, Q., Sekimoto, Y., Shibasaki, R.: Prediction of human emergency behavior and their mobility following large-scale disaster. In: KDD, pp. 5–14 (2014)
17. Toda, H., Yasuda, N., Matsuura, Y., Kataoka, R.: Geographic information retrieval to suit immediate surroundings. In: GIS, pp. 452–455 (2009)
18. Vedaldi, A., Fulkerson, B.: VLFeat: an open and portable library of computer vision algorithms. In: MM. pp. 1469–1472 (2010)
19. Vincent, P., Larochelle, H., Lajoie, I., Bengio, Y., Manzagol, P.-A.: Stacked denoising autoencoders: learning useful representations in a deep network with a local denoising criterion. JMLR **11**, 3371–3408 (2010)
20. Zheng, Y.: Trajectory data mining: an overview. ACM TIST **6**(3), 29 (2015)
21. Zheng, Y., Liu, L., Wang, L., Xie, X.: Learning transportation mode from raw GPS data for geographic applications on the web. In: WWW, pp. 247–256 (2008)
22. Zheng, Y., Li, Q., Chen, Y., Xie, X., Ma, W.-Y.: Understanding mobility based on GPS data. In: Ubicomp, pp. 312–321 (2008)
23. Zheng, Y., Chen, Y., Li, Q., Xie, X., Ma, W.-Y.: Understanding transportation modes based on GPS data for web applications. TWEB. **4**(1), 1 (2010)
24. Zheng, Y., Zhou, X. (eds.): Computing with Spatial Trajectories. Springer, New York (2011)

Online Learning for Accurate Real-Time Map Matching

Biwei Liang[1,4], Tengjiao Wang[1,2,4(✉)], Shun Li[5], Wei Chen[2,4],
Hongyan Li[2,3], and Kai Lei[1]

[1] School of Electronics and Computer Engineering (ECE),
Peking University, Shenzhen, 518055, China
1301213711@sz.pku.edu.cn, tjwang@pku.edu.cn
[2] School of Electronics Engineering and Computer Science,
Peking University, Beijing, 100871, China
[3] Key Laboratory of Machine Perception,
Peking University, Ministry of Education, Beijing, 100871, China
[4] Key Laboratory of High Confidence Software Technologies, Peking University,
Ministry of Education, Beijing, 100871, China
[5] School of Information Science and Technology,
University of International Relations, Beijing, 100871, China
ls1977@gmail.com

Abstract. For the reason that deviation exists between GPS traces obtained by
real-time positioning system and actual paths, real-time map matching which
identifies the correct traveling road segment, becomes increasingly important. In
order to effectively improve map matching accuracy, most state-of-art real-time
map matching algorithms use machine learning which calls for time-consuming
human labeling in advance. We propose an accurate real-time map matching
method using online learning called OLMM. It takes into account a small piece
of trajectory data and their matching result to support the subsequent matching
process. We evaluate the effectiveness of the proposed approach using ground
truth data. The results demonstrate that our approach can obtain more accurate
matching results than existing methods without any human labeling beforehand.

Keywords: Real-time map-matching · Online learning · Without human labeling

1 Introduction

In many intelligent transportation system applications, such as traffic sensing [1], traffic
incident detection [2] and travel time prediction [3], real-time map matching is a crucial
step. It is because deviation generally exists between GPS traces obtained by real-time
positioning system and actual paths. Real-time map matching is the process of identifying the correct road segment on which the vehicle is traveling and the vehicle location

This research is supported by the Natural Science Foundation of China (Grant No.
61572043, 61300003).

on it. It receives a trajectory point stream which comes in a high rate and output the matching road segment in sequence in a real-time fashion.

The most obvious real-time map matching algorithm is to simply match each point with the nearest road segment, which is called geometric analysis [7, 8]. Due to measurement noise, however, this algorithm is prone to error [5]. Topological analysis [6] extends geometric analysis by taking into account the relationship between road segments while results are far from expected. Although either geometric analysis or topological analysis is easy to implement and runs fast, its matching accuracy could not satisfy our matching accuracy demands. Because of problems like this, modern map matching algorithms paid great effort on improving the matching accuracy and probabilistic model is widely used in map matching methods. The matching process can be illustrated by Fig. 1(a) These algorithms can achieve a better matching accuracy for the reason that they use human labeled datasets as inputs to machine learning models to get some parameters used in the matching process. Nevertheless, they are not suitable for real-time map matching in different traffic conditions unless taking a lot of time labeling and training different dataset collected from various regions.

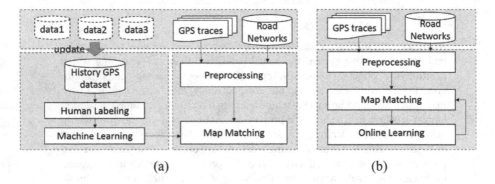

(a)   (b)

Fig. 1. Comparison of algorithms

Our work was motivated by an idea of developing an real-time map matching algorithm that can adapt to the complex and changeable city traffic conditions without time-consuming human labeling in advance. See Fig. 1(b), we use online learning techniques, which takes into account a small piece of trajectory data and their matching result to support the subsequent matching process. We evaluate the effectiveness of the proposed approach using ground truth dataset supplied by Microsoft in [5]. The results demonstrate that our method is superior to the state-of-art real-time map matching algorithms at matching accuracy. To summarize, we make the following contributions:

- We propose an accurate real-time map matching algorithm called OLMM without time-consuming human labeling beforehand.
- Online learning is used in the matching process which makes the algorithm more universality to different travel conditions.

- The proposed approach has a comparable running time with existing real-time map matching algorithms. However, it improves matching accuracy even without human labeling and parameter training process in advance.

The paper is organized as follows: the next Sect. 2 discuss various types of map matching algorithms. It is followed by Sect. 3, in which some preliminary definitions and baseline model is induced. The next Sect. 4 illustrates our algorithm in detail. Conducted experiments are demonstrated in Sect. 5 and finally Sect. 6 concludes our contributions.

2 Related Works

In this section, we review some important existing studies that are related to map-matching. More than thirty map-matching algorithm are surveyed by Quddus et al. in [6]. They were categorized into four groups: geometric, topological, probabilistic and other advanced techniques.

Geometric Analysis: Algorithms using geometric analysis take into account only the shape of the links, and not consider the way how they are connected. The simplest way known as point-to-point matching [7] simply match the trajectory point to the closest node of a road segment. This approach is easy to implement but low in matching accuracy. Another option is point-to-curve matching [8] which matches the current position onto the closest curve in the road network. This approach gives better results while unstable in dense urban networks. The other geometric approach known as curve-to-curve matching compare the vehicle's trajectory against known roads and determining "curve distance" between them. Also, this approach is quite sensitive to outliers and not accuracy enough.

Topological Analysis: A map-matching algorithm which makes use of the geometry of the links as well as the connectivity and contiguity of the links is known as a topological map-matching algorithm [6]. The brief version of the algorithm uses only coordinate information on observed position of the user [9] which makes it very sensitive to outliers. To improve the performance of the algorithm, additional criteria and parameters are introduced including heading and speed [10].

Probabilistic Map-Matching Algorithms: Probabilistic map-matching algorithms define a confidence region around the trajectory point obtained from GPS sensor and use evaluations including heading, connectivity and closeness to determine the optimal match segment [11]. An enhance version of probabilistic map-matching algorithm only construct an error region when the vehicle travels through a junction thus eliminate the introduction of incorrect link and unnecessary computation process [12].

Advanced Techniques: Advanced algorithms usually combine both topological and probabilistic information, while applying various techniques to assign road links to GPS readings [13]. They use more refined concepts such as Kalman Filter [14], Dempster-Shafer's theory [15], Hidden Markov Model [4, 5, 16–18] and etc.

While some map-matching algorithms using HMM have already been proposed, there are some issues that haven't been considered in an real-time algorithm before. Most state-of-art real-time algorithms trained human-labeled dataset in advance in order to obtain parameters used in their algorithms [4]. However, these parameters cannot been applied to process trajectories in different regions and time intervals. Moreover, it would spend a lot of time in labeling the dataset and train the parameters. For these reasons, using dynamic parameters is a feasible way to deal with these problems. We use local sequence learning process and sliding window to obtain dynamic parameters that will be used in our matching process. Detailed information will be illustrated in Sect. 4.

3 Preliminaries

In this section, we will give the preliminaries and baseline model to formally define the problem of real-time map matching.

3.1 Problem Statement

Definition 1 Trajectory. A trajectory $T = \{p_n | n = 1, 2, \ldots, N\}$ is a series of N points collected by a GPS device. Each point p_i is defined by three attributes: the longitude $p_i.lon$, the latitude $p_i.lat$ and the time stamp $p_i.t$.

Definition 2 Road Segment. A road segment is a tuple $R = (p_s, p_e, id, v)$ where $R.p_s$ and $R.p_e$ are the start node and end node of the road segment separately. R.id is the identify number of the road, R.v is the speed limit.

Definition 3 Road Network. Road network is a collection of road segments which express digital map in an undirected graph $G = (V, E)$. Nodes $V \in G$ indicate intersections or end points of road segments while arcs $E_{ij} \in G$ demonstrates road segments from V_i to V_j.

Definition 4 Hidden Markov Model. Hidden Markov Model is a statistical Markov model in which the system being modeled is assumed to be a Markov process with unobserved (hidden) states. It can be represented in a tuple $H = (S, O, \pi, A, B)$ where H.S is a set of hidden states, H.O is a set of observations, H.π is a matrix for initial state probabilities, H.A is a matrix for transition probabilities between hidden states and H.B is a matrix for emission probabilities which describes how likely an observation will appear for a given hidden state.

Figure 2, shows an ideal Hidden Markov Model process. The upper sequence $S_1, S_2, \ldots, S_n$ represents hidden states while lower sequence $O_1, O_2, \ldots, O_n$ represents observations. The arrows in the diagram denote conditional dependencies. And a_{12} represents the transition probability between S_1 and S_2, $b_1(O_1)$ represents the emission probability that S_1 correspond to O_1.

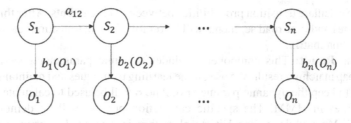

Fig. 2. Ideal Hidden Markov Model process

3.2 Baseline Model

Before illustrating the flow path of our algorithm, some important concepts in Hidden Markov Model should be explained as background knowledge.

As defined above, the Hidden Markov Model is a statistical model that is well known for providing solutions to temporal recognition applications. In our map-matching algorithm, the hidden states of the HMM are the candidate road segments, the observations are trajectory points, the matrix of initial state probabilities describes how likely the first matching point is on its initial candidate road segments. Moreover, transition probabilities depict the transition chance between road segments. Obviously, road segments close to each other or have a same connection node have higher transition probability than other road segment pairs. Finally, emission probability matrix describe the probability that the vehicle is on the candidate road segment. Our goal is to match each location measurement with its proper road segment. Map-matching problem naturally fits the HMM, and practice has proved that it performs well in map matching algorithms.

4 OLMM Map Matching Algorithm

4.1 Algorithm Framework

The framework of our proposed OLMM map-matching algorithm is shown in Fig. 3. It is compose of three major components: candidate preparation, transition analysis and matching result. The calculation of N-connection Matrix in transition analysis phase is the only offline pre-computing part in our algorithm.

- **Candidate Preparation:** Data receiving and range query will be done in this phase. In data receiving process, road network information and trajectory point stream will be input to our framework. Range query retrieve roads whose distance from the trajectory point is less than a threshold as the candidate road segment set of the point. In our algorithm, we set this threshold to be 200 m. There are two reasons for choosing this threshold: first, this threshold could leave out road segments whose transition probability and emission probability is very low; second, remove those road segments could reduce computation time in our matching process.
- **Transition Analysis:** This component constructs n-connection matrix construction and transition probability calculation. We analyze and construct an n-connection

matrix to calculate transition probabilities between road segments. After this component, some candidate road segments will be removed using transition features in our n-connection matrix.

- **Matching Result:** This component includes dynamic parameter generation and output map matching result. We use online learning techniques and maintain a sliding window to keep the dynamic parameter σ. And σ will be used to calculate emission probabilities in HMM. The specific calculation process will be demonstrate in Sect. 4.3. Meanwhile, online Viterbi algorithm is used to do our map matching process and matching results will be output.

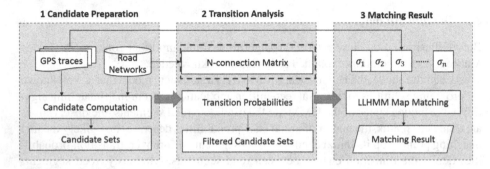

Fig. 3. OLMM algorithm framework

Here is an example: In offline pre-computing part, we input the road network and finish the computation of n-connection matrix. After that, in online computing and matching phases, we first identify a set of candidate road segments for every trajectory point. Then using transition probability from last candidate set and emission probability, we could find the maximum likelihood path over the Markov chain. The process is demonstrated in Fig. 4.

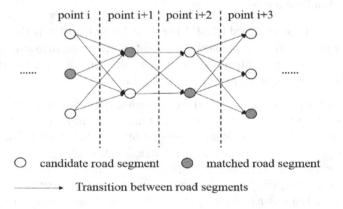

Fig. 4. An example of OLMM algorithm

4.2 Transition Analysis

In transition analysis, we use topological information of the road network to construct an n-connection matrix and calculate transition probabilities between road segments.

Definition 5 n-connection. If two road segments have one common end point, they are 1-connection, else if they can be connected via n-1 road segments, they are n-connection. However, if the possible path violates traffic rules (e.g. a reverse drive is forbidden in a one-way road segment), the transition probability between them is zero. Transition probability decreases while n increases in n-connection road segment adjacency matrix.

Figure 5(a) is a road network modeled by a directed graph G = (V, E), where V are graph nodes describes road segments' end points, E are straight road segments connected by two end points. Figure 5(b) shows the n-connection adjacency matrix for road network in Fig. 5. The header of columns represents the "from" road segment and the header of the rows are "to" road segments. Finally, we got the normalized formula to calculate the transition probability from a given road segment to others.

$$a_{ij} = 0 \left(if\ a[i][j] == 0 \right) \tag{1}$$

$$a_{ij} = \frac{\frac{1}{a[i][j]}}{\sum_k \frac{1}{a[i][k]}} \left(if\ (a[i][j]\ != 0 \right) \tag{2}$$

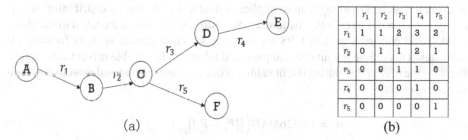

	r_1	r_2	r_3	r_4	r_5
r_1	1	1	2	3	2
r_2	0	1	1	2	1
r_3	0	0	1	1	0
r_4	0	0	0	1	0
r_5	0	0	0	0	1

(a)   (b)

Fig. 5. Road network and n-connection matrix

Here, a_{ij} refers to the transition probability from road segment r_i to road segment r_j while a[i][j] stands for the adjacency coefficient in the matrix. In addition, k refers to all road segments' id that satisfy a[i][k]! = 0.

In candidate preparation step, we find road segments whose distance is less than 200 m to the current GPS point as its candidate road segments. For those whose transition probability is zero from the last candidate set, we can simply remove them from the current candidate set to reduce computation time in our matching process.

4.3 Matching Result

In matching result process, we first use online learning technique to obtain dynamic parameter σ, and then calculate the emission probabilities from the candidate road segment to the current trajectory point.

Emission probability refers to the possibility that the trajectory point is on that road segment. We model the emission probability with Gaussian distribution because this model is proved to be effective in previous works. [5] We denote emission probability that p_i is on road segment r_j as $b_j(p_i)$, and it can be calculated using the following formula:

$$b_j(p_i) = \frac{1}{\sqrt{2\pi}\sigma}\exp(-\frac{|p - p'|_{great-circle}}{2\sigma^2}) \qquad (3)$$

σ is the estimated standard deviation of GPS error, $|p - p'|_{great-circle}$ denotes the ground circle distance between the trajectory point and its projection point on the road segment. In state-of-the-art algorithms, σ is a fixed parameter which will be trained before running the algorithms using human-labeled datasets. However, due to the difference in datasets and training algorithms, the trained parameter σ cannot be used in different matching conditions to obtain a best matching result. Moreover, human labeling and parameter training will take unpredictable time which makes it not universality in real time navigation systems.

Therefore, we use online learning technique to get a dynamic parameter σ during our matching process. Firstly, we assume that the distribution of trajectory points from its correct matched road segments complies with standard Gaussian distribution, that is to say, we set σ = 1 as an initial parameter. Secondly, we maintain the latest m matching results, and choose a median GPS error as our standard deviation σ, as formula (4) indicates. This online learning technique could adapt to changeable urban traffic conditions for the reason that during our matching process, σ will be updated refer to the latest matching result.

$$\sigma_z = 1.4826MAD\left(||P_t - P_i||_{gc}\right) \qquad (4)$$

When we calculate the distance from point p to its projection point on a candidate road segment, we use Haversine Formula. The Haversine formula is an equation important in navigation, giving great-circle distances between two points on a sphere from their longitudes and latitudes [18]. For any two points on a sphere, the great-circle distance is given by

$$d = 2r\arc\left(\sqrt{sin^2(\frac{lat_2 - lat_1}{2}) + \cos\left(lat_1\right)\cos(lat_2)sin^2(\frac{lon_2 - lon_1}{2})}\right) \qquad (5)$$

Our matching process uses online Viterbi Algorithm which was detailed described in [4], so we just leave out this part.

5   Experiments

5.1   Experiments Setup

In our experiment, we conduct our algorithm on a ground truth dataset. The dataset is collected by Microsoft in [5] and is publicly available. It contains 7531 time stamped latitude/longitude pairs sampled at every second on the Seattle road network that consists of 418 k cross points and 875 k road segments. The route is about 80 km (50 miles) long. OLMM map matching algorithm is implemented in C++. Experiments are conducted on PC with Intel® Core™ i5-2450 M CPU and 4G memory under Windows.

Our algorithm is evaluated by two standards: accuracy and running time. Accuracy of the map matching algorithm is quantified by comparing the ground truth route to the route determined by our algorithm. We sum the lengths of correctly matched route and divide this sum by the length of the ground truth route, which is the matching accuracy value we report. This standard can be demonstrate by formula (6).

$$A = \frac{\sum The\ length\ of\ matched\ road\ segments}{The\ length\ of\ the\ trajectory} \tag{6}$$

Running time is the actual program execution time.

5.2   Accuracy Performance

Figure 6(a) shows the impact of sliding window size on the average of σ. In the figure we could infer that dynamic parameter varies along with the sliding window size changes, while fixed parameter stays stable. When the sliding window size is up to fifteen or more, the average number of σ becomes almost the same and we could say that using a sliding window larger than fifteen could get a stable output of matching result. Figure 6(b) illustrates the impact of sliding window size on the matching accuracy of OLMM. In the figure, it is obvious that OLMM could get an almost stable and high matching accuracy around 98.5 % when the sliding window size is up to fifteen or more.

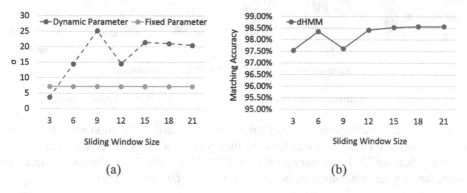

(a)   (b)

Fig. 6. Accuracy performance using different sliding window size (Color figure online)

Table 1 demonstrates the best accuracy performance of different map-matching algorithm using the same dataset. Point-to-point map-matching matches the trajectory point to its nearest road segment, OHMM map-matching [4] uses SVM to train the human labeled dataset and a fixed trained parameter was used to do real-time map matching, OLMM is our algorithm which uses online learning. From the table we can conclude that our matching algorithm outperforms point-to-point map-matching and OHMM map matching in best matching accuracy.

Table 1. Comparison of Accuracy

Algorithm	P2P	OHMM	OLMM
Accuracy	89.79 %	Around 98 %	98.57 %

5.3 Running Time Performance

In Fig. 7, comparison of candidate road segments for OLMM whether using transition analysis is displayed. It is obvious that using transition analysis to filter incorrect road segments greatly reduce the number of candidate road segments for trajectory points while accuracy performance is not affected according to the experiment result. For 7531 trajectory points in our dataset, more than 2200 points only have candidate road segments of 0 ~ 5 using transition analysis filter instead of around 600. This will sharply reduce running time in the matching process.

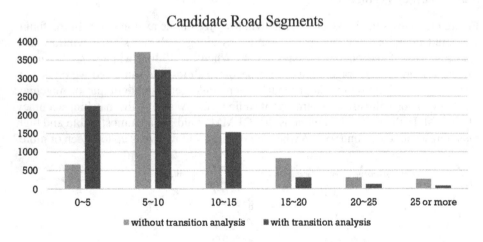

Fig. 7. Comparison of candidate road segment number (Color figure online)

Running time performance of matching algorithms using transition analysis or not are compared in Table 2. It can be seen that transition analysis reduced the whole matching time for 7531 trajectory point from 1275.11 s to 504.844 s. For each trajectory point, the average matching time is cut down from 0.1693 s to 0.0670 s.

Table 2. Comparison of running time

Using transition analysis or not	No	Yes
Running time/points	1275.11s/7531	504.844 s/7531
Average matching time	0.1693 s	0.0670 s

6 Conclusions

In this paper, we describe an accurate real-time map matching algorithm called OLMM. The algorithm employs online learning technique to execute map matching process in a real-time fashion, without any human labeling in advance.

We design and do the experiments to verify the validation of our algorithm. Furthermore, we analyze the performance on ground truth data in two dimensions: accuracy and running time. The experiment results show that our algorithm can be used in real time GPS devices to acquire a more precision matching result than existing real-time map matching algorithms even without time-consuming human labeling in advance.

References

1. Li, Z., Zhu, Y., Zhu, H., et al.: Compressive sensing approach to urban traffic sensing. In: 2011 31st International Conference on IEEE Distributed Computing Systems (ICDCS), pp. 889–898 (2011)
2. Li, M., Zhang, Y., Wang, W.: Analysis of congestion points based on probe car data. In: 12th International IEEE Conference on Intelligent Transportation Systems, ITS C 2009, pp. 1–5. IEEE (2009)
3. De Fabritiis, C., Ragona, R., Valenti, G.: Traffic estimation and prediction based on real time floating car data. In: 11th International IEEE Conference on Intelligent Transportation Systems, ITSC 2008, pp. 197–203. IEEE (2008)
4. Goh, C.Y., Dauwels, J., Mitrovic, N., et al.: Online map matching based on Hidden Markov Model for real-time traffic sensing applications. In: 2012 15th International IEEE Conference on Intelligent Transportation Systems (ITSC). IEEE, pp. 776–781 (2012)
5. Newson, P., Krumm, J.: Hidden Markov map matching through noise and sparseness. In: Proceedings of the 17th ACM SIGSPATIAL International Conference on Advances in Geographic Information Systems, pp. 336–343. ACM (2009)
6. Quddus, M.A., Ochieng, W.Y., Noland, R.B.: Current map-matching algorithms for transport applications: state-of-the art and future research directions. Transp. Res. Part C Emerg. Technol. **15**(5), 312–328 (2007)
7. Ferenc, K., Derek, B.: An introduction to map matching for personal navigation assistants. Geom. Distrib. **122**(7), 1082–1083 (1996)
8. White, C.E., Bernstein, D., Kornhauser, A.L.: Some map matching algorithms for personal navigation assistants. Transp. Res. Part C Emerg. Technol. **8**(1), 91–108 (2000)
9. Greenfeld, J.S.: Matching GPS observations to locations on a digital map. In: Proceedings of Annual Meeting of the Transportation Research Board, Washington D.C., 2002
10. Quddus, M.A., Ochieng, W.Y., Zhao, L., et al.: A general map matching algorithm for transport telematics applications. GPS Solutions **7**(3), 157–167 (2003)
11. Newby, R.M., Annett, D.A., Berry, M.J., et al.: Vehicle location and navigation system (2013). EP, EP1864084

12. Ochieng, W.Y., Quddus, M.A., Noland, R.B.: Map-matching in complex urban road networks. Rev. Bras. De Cartografia **2**, 1–14 (2003)
13. Szwed, P., Pekala, K.: An incremental map-matching algorithm based on Hidden Markov Model. In: Rutkowski, L., Korytkowski, M., Scherer, R., Tadeusiewicz, R., Zadeh, L.A., Zurada, J.M. (eds.) ICAISC 2014, Part II. LNCS, vol. 8468, pp. 579–590. Springer, Heidelberg (2014)
14. El Najjar, M.E., Bonnifait, P.: A road-matching method for precise vehicle localization using belief theory and Kalman filtering. Auton. Robots **19**(2), 173–191 (2005)
15. Nassreddine, G., Abdallah, F., Denoeux, T.: Map matching algorithm using interval analysis and Dempster-Shafer theory. In: 2009 IEEE Intelligent Vehicles Symposium, pp. 494–499 (2009)
16. Krumm, J., Letchner, J., Horvitz, E.: Map matching with travel time constraints. In: SAE World Congress (2007)
17. Thiagarajan, A., Ravindranath, L., Lacurts, K., et al.: VTrack: accurate, energy-aware road traffic delay estimation using mobile phones. Sensys Proc. ACM Conf. Embed. Netw. Sens. Syst. **5554**(1), 85–98 (2009)
18. Najjar, M.E.E., Bonnifait, P.: A road-matching method for precise vehicle localization using belief theory and Kalman filtering. Auton. Robots **19**(2), 173–191 (2005)

Multi-hypergraph Incidence Consistent Sparse Coding for Image Data Clustering

Xiaodong Feng[1(✉)], Sen Wu[2], Wenjun Zhou[3], and Zhiwei Tang[1]

[1] School of Political Science and Public Administration, University of Electronic Science and Technology of China, Chengdu 611731, Sichuan, China
fengxd1988@hotmail.com, tangzw@uestc.edu.cn
[2] Donlinks School of Economics and Management, University of Science and Technology Beijing, Beijing 100083, China
wusen@manage.ustb.edu.cn
[3] Department of Business Analytics and Statistics, University of Tennessee, Knoxville, TN 37996, USA
wzhou7@gmail.com

Abstract. Sparse representation has been a powerful technique for modeling image data and thus enhance the performance of image clustering. Sparse coding, as an unsupervised way to extract sparse representation, learns a dictionary that represents high-level semantics and the new representations on the dictionary. Though existing sparse coding schemes are considering local manifold structure of the data with graph/hypergraph regularization, more from the manifold should be exploited to utilize intrinsic manifold characteristics in the data. In this paper, we firstly propose a Hypergraph Incidence Consistency regularization term by minimizing the reconstruction error of the hypergraph incidence matrix with sparse codes to further regulate the learned sparse codes with hypergraph-based manifold. Moreover, a multi-hypergraph learning framework to automatically select the optimal manifold structure is integrated into the objective of sparse coding learning, resulting in multi-hypergraph incidence Consistent Sparse Coding (MultiCSC). We show that the MultiCSC objective function can be optimized efficiently, and that several existing sparse coding methods are special cases of MultiCSC. Extensive experimental results on image clustering demonstrate the effectiveness of our proposed method.

Keywords: Image representation · Image clustering · Sparse coding · Hypergraph incidence consistency · Multiple hypergraph learning

1 Introduction

Recently, sparse representation has been a powerful technique for image representation and widely applied in image data clustering [1–3]. To learn the sparse representation of signals effectively, it is vital to learn a well-constructed dictionary, i.e. dictionary learning. Dictionary learning can be divided into three categories:

© Springer International Publishing Switzerland 2016
J. Bailey et al. (Eds.): PAKDD 2016, Part II, LNAI 9652, pp. 79–91, 2016.
DOI: 10.1007/978-3-319-31750-2_7

unsupervised [4,5], supervised [6,7], and semi-supervised learning [8], according to whether labels of signals are used while learning the dictionaries.

In real applications, it is worth noting that a huge amount of unlabeled data can be more easily obtained than labeled data, and thus unsupervised dictionary learning is very useful. For example, successful unsupervised image clustering can enable the subsequent supervised learning tasks to recognize image with accuracy that would not be possible without the use of image clustering. In this paper, we focus on developing an effective unsupervised dictionary learning method for image representation and then image clustering, which is useful and potentially powerful for image mining tasks. For unsupervised dictionary learning, sparse coding [4] receives much attention in machine learning and image processing. It aims at learning a dictionary that consists of a set of basis items and the sparse coordinates with respect to the dictionary.

Following the work of original sparse coding, graph regularized sparse coding (GSC) [9] and hypergraph Laplacian sparse coding (HSC) [10] have been developed to encode the geometrical information in the data space, which can preserve the locality and similarity between the image features or instances in the sparse coding space . Even though these manifold regularized sparse coding can achieve much improvement in image processing. in general, it is nontrivial to determine the intrinsic manifold in a systematical way. Usually, cross validation-based parameter selection does not scale up well for a huge number of possible parameters and inevitably overfits to the training set [11]. Therefore, it is crucial to efficiently select the optimal manifold to make the performance of the employed graph or hypergraph regularized sparse coding method robust, or even better. Another issue of GSC or HSC comes from the insufficient exploitation of manifold structure as they only add the Laplacian regularization to the objective of sparse coding.

Therefore, in this paper, we propose a novel sparse coding scheme, named multi-hypergraph incidence Consistent Sparse Coding (MultiCSC). MultiCSC exploits the hypergraph model to regularize the sparse coding, as the hypergraph is able to capture the high-order manifold structure of high dimensional data compared to simple graph. A hypergraph incidence consistency regularization term (HIC) is presented to further leverage the hypergraph-based manifold, based on the assumption that the hypergraph structure can be well reconstructed by sparse codes associated with data instances. Moreover, multi-hypergraph learning term is also integrated to automatically select the optimal manifold structure. Finally, alternative optimization of MultiCSC is presented. The improved performance of image clustering on real image datasets validates the advantage of the proposed unsupervised dictionary learning method.

2   Preliminaries

2.1   Graphs, Hypergraphs and the Laplacian Matrix

A hypergraph $G(X, E)$ consists of a set of vertices $X = [x_1, x_2, \ldots, x_n]$, and a set of hyperedges $E = [e_1, e_2, \ldots, e_{|E|}]$. Each hyperedge is a non-empty subset of the vertices (i.e., $e_j \subseteq X$, $\forall j$). The incidence matrix $H \in \{0,1\}^{n \times |E|}$ of a hypergraph records each hyperedge in a column, so $H_{ij} = 1$ if $x_i \in e_j$, and $H_{ij} = 0$ otherwise ($\forall i = 1, 2, \ldots, n$ and $\forall j = 1, 2, \ldots, |E|$), or H can be defined in a probabilistic way as $H_{ij} \in [0, 1]$ if $x_i \in e_j$.

Moreover, suppose that each hyperedge e has a corresponding weight $w(e)$. The weight matrix $W \in \mathbb{R}^{n \times n}$ of a hypergraph is to measure how close each two vertexes in the manifold space are, defined as the number of shared hyperedges multiplied by their weights. In matrix form, the normalized weight matrix and Laplacian matrix are defined as:

$$W = D_v^{-1/2} H W_e D_e^{-1} H^T D_v^{-1/2}; L = I - W, \tag{1}$$

where D_e, D_v and W_e denote the diagonal matrices of hyperedge degrees, vertex degrees, and hyperedge weight, respectively , and I is the identity matrix.

2.2 Basic Idea of Sparse Coding

Given a data matrix $X = [x_1, x_2, \ldots, x_n] \in \mathbb{R}^{m \times n}$, consisting of n data points measured on m dimensions, the goal of dictionary learning and sparse coding is to decompose the data matrix X into a dictionary matrix $B = [b_1, b_2, \ldots, b_r] \in \mathbb{R}^{m \times r}$ and a sparse coefficient matrix $S = [s_1, s_2, \ldots, s_n] \in \mathbb{R}^{r \times n}$, so that X can be reconstructed using B and S. Since r is typically much smaller than n, the sparse coefficients can represent the data objects with much lower dimensions. Formally, the sparse coding problem can be described as the following optimization problem:

$$\min_{B,S} O^{SC} = \|X - BS\|_F^2 + \beta \sum_{i=1}^{n} \|s_i\|_1 \, ; s.t. \, \|b_j\|_2^2 \leq 1, \forall j = 1, 2.., r$$

where $\| \cdot \|_F$ denotes the matrix Frobenius norm (i.e., $\|A\|_F = \sqrt{\sum_i \sum_j a_{ij}^2}$), $\| \cdot \|_1$ denotes the l_1 norm (i.e., $\|x\|_1 = \sum_{i=1}^{n} |x_i|$), and β is a coefficient that represents the tradeoff between reconstruction error and sparsity.

2.3 Hypergraph Laplacian Sparse Coding

To preserve the locally geometrical structure of the data points in the original space, hypergraph Laplacian sparse coding (HSC) [10] was proposed to regularize the sparse coefficient S with the hypergraph Laplacian term, as follows:

$$\min_{B,S} O^{HSC} = \|X - BS\|_F^2 + \beta \sum_{i=1}^{n} \|s_i\|_1 + \alpha R^{HL}; s.t. \, \|b_i\|_2^2 \leq 1, \forall i = 1, 2.., r$$

The hypergraph Laplacian regularization term R^{HL} is based on the manifold assumption that if two data points x_i and x_j are close in the intrinsic manifold space, then their corresponding sparse representations, s_i and s_j, should also be close to each other. This term is detailed as:

$$R^{HL} = Tr(SLS^T) = \frac{1}{2} \sum_{i=1}^{n} \sum_{j=1}^{n} w_{ij} \|s_i - s_j\|_2^2, \tag{2}$$

where $Tr(\cdot)$ stands for the trace of a square matrix, and L is the Laplacian matrix of the hypergraph constructed from the data objects [10].

3   Multi-hypergraph Incidence Consistent Sparse Coding

3.1   Hypergraph Incidence Consistency

As presented in related works of supervised dictionary learning [6,7], one assumption is that the sparse codes S should be able to well reconstruct supervised information (e.g., class labels), that is, a well classification performance using S. Inspired by this idea, for hypergraph-based unsupervised learning, we conjecture that data points in the same hyperedge are more likely to share the same label. Thus, if we use the incidence matrix H to provide additional information of supervision, this matrix should also be reconstructible with sparse codes S.

Aiming at reconstructing the hyperedge incidence matrix H using sparse coding, we propose to minimize the reconstruction error, as measured by the following hypergraph incidence consistency regularization term (HIC):

$$R^{HIC} = \|H^T - QS\|_F^2 \tag{3}$$

where $Q = [q_1, q_2, \dots, q_r] \in \mathbb{R}^{|E| \times r}$ is the linear transformation matrix, and $|E|$ denotes the number of hyperedges (i.e., the number of columns in H).

The dictionary learned while regulating the incidence matrix H reconstruction error by minimizing Eq. (3) is adaptive to the underlying structure of the dataset, which leads to a good representation of each data point in the set with strict sparsity constraints. HIC generates hypergraph-oriented discriminative sparse codes, and addresses the desirable ability in hypergraph construction regardless of the size of the dictionary. These sparse codes can be utilized directly by clustering algorithms and classifiers, such as in [3]. The hypergraph-oriented discriminative property of sparse code is very important for the performance of unsupervised image clustering tasks.

3.2   Multi-hypergraph Learning

For both hypergraph Laplacian [10] sparse coding and proposed HIC term, it is vital to construct an optimal hypergraph to represent the intrinsic manifold. Instead of using exhaustive search (that does not scale well) or cross validation (that tends to overfit), multi-hypergraph learning techniques have been proposed to approximate the intrinsic manifold for hypergraph Laplacian, such as ensemble manifold regularizer [11] and the multi-hypergraph regularizer in matrix factorization [12,13]. Multiple hypergraph learning works well since the intrinsic manifold of the collected data points is assumed to lie in a convex hull of a set of previously given candidate manifolds, each of which indicates one kind of manifold data structure, defined as follows.

Given the set of t candidate hypergraphs $\Gamma = \{G_1, G_2, \ldots, G_t\}$ with different weighting scheme and neighbor size parameter, the corresponding Laplacian matrices and incidence matrices can respectively be derived as $\Omega = \{L_1, L_2, \ldots, L_t\}$ and $\Phi = \{H_1, H_2, \ldots, H_t\}$.

The ensemble manifold regularizer considers the multi-hypergraph Laplacian as a linear combination of the hypergraph Laplacians of these candidate manifolds, where each candidate graph G_k is associated with a coefficient τ_k, defined as:

$$\mathcal{L} = \sum_{i=1}^{t} \tau_i L_i, \tag{4}$$

where $\boldsymbol{\tau} = [\tau_1, \tau_2, \ldots, \tau_t]$ is the hypergraph weight vector, s.t. $\sum_{k=1}^{t} \tau_k = 1$, $\tau_k \geq 0, \forall k = 1, 2, \ldots, t$.

Assume the candidate hypergraphs are constructed in the same way using k-nearest neighbor selection, so the j-th hyperedge in each hypergraph G_i is associated with the identical data point x_j, which covers the nearest neighbors and x_j itself, and the difference lies in the weighting scheme and neighbor size parameter. In this case, these incidence matrices have additive property as all the j-th columns in H_i denotes the same hyperedge corresponding to data point x_j. Thus, the incidence matrix $\mathcal{H}$ of optimal hypergraph can be also assumed as the linear combination of the candidate hypergraphs' incidence matrices.

$$\mathcal{H} = \sum_{i=1}^{t} \tau_i H_i. \tag{5}$$

In the multi-hypergraph framework, we try to select the optimal manifold by determining the optimal linear combination weights for a group of pre-computed graph candidates. More specifically, by substituting $\mathcal{L}$ in Eq. (4) into L in Eq. (2), the multi-hypergraph Laplacian regularization term can be written as:

$$R^{MultiHL} = Tr\left(S\mathcal{L}S^T\right) = Tr\left(S\left(\sum_{i=1}^{t} \tau_i L_i\right)S^T\right) = \sum_{i=1}^{t} \tau_i Tr\left(SL_iS^T\right). \tag{6}$$

Similarly, replacing H in Eq. (3) by $\mathcal{H}$ in Eq. (5), the multi-hypergraph incidence consistency regularization term (multiHIC) can be written as:

$$R^{MultiHIC} = \left\|\mathcal{H}^T - QS\right\|_F^2 = \left\|\left(\sum_{i=1}^{t} \tau_i H_i\right)^T - QS\right\|_F^2 = \left\|\sum_{i=1}^{t} \tau_i H_i^T - QS\right\|_F^2 \tag{7}$$

To further prevent $\boldsymbol{\tau}$ from overfitting to a single hypergraph, the l_2-norm term (i.e., $\|\boldsymbol{\tau}\|_2^2$) will be added to the objective function for optimization (see

Eq. (8)). By minimizing the multi-hypergraph Laplacian regularization term in Eq. (6) and the multi-hypergraph incidence consistency regularization term in Eq. (7), a larger weight is expected for a hypergraph with better weighting and parameter selection scheme.

3.3 Overall Fourmulation of Multi-hypergraph Incidence Consistent Sparse Coding

Adding the multi-hypergraph Laplacian regularization term in Eq. (6) and the multi-hypergraph incidence consistency regularization term in Eq. (7) to the objective function of sparse coding, we propose a new sparse representation framework called multi-hypergraph incidence consistent sparse coding (MultiCSC). More specifically, the objective becomes

$$
\begin{aligned}
O^{MultiCSC} &= O^{SC} + \alpha R^{MultiHL} + \gamma R^{MultiHIC} + \lambda \|\boldsymbol{\tau}\|_2^2 \\
&= \|X - BS\|_F^2 + \beta \sum_{i=1}^{n} \|\boldsymbol{s}_i\|_1 + \alpha \sum_{i=1}^{t} \tau_i Tr(SL_i S^T) \\
&\quad + \gamma \left\| \sum_{i=1}^{t} \tau_i H_i^T - QS \right\|_F^2 + \lambda \|\boldsymbol{\tau}\|_2^2
\end{aligned}
\tag{8}
$$

where α, γ, and λ are tradeoff parameters. In particular, we call γ the hypergraph consistent tradeoff parameter, and λ is the multi-hypergraph combination parameter. In summary, the MultiCSC problem can be summarized as follows:

$$
\min_{B,Q,S,\tau} O^{MultiCSC} s.t. \|\boldsymbol{b}_i\|_2^2 \leq 1, \|\boldsymbol{q}_i\|_2^2 \leq 1; 0 \leq \tau_i \leq 1; \sum_{i=1}^{t} \tau_i = 1.
\tag{9}
$$

Usually the data instance vector $\boldsymbol{x}_i$ and each row vector in H will be firstly normalized to be a unit norm, so the constraints $\|\boldsymbol{b}_i\|_2^2 \leq 1$ and $\|\boldsymbol{q}_i\|_2^2 \leq 1$ would make B and Q comparable with X and H. As we can see that, HSC [10] is a special case of MultiCSC when $\gamma = 0$ and $\tau = [0, \cdots, 0, 1, 0, \cdots, 0]^T$ ($\tau_k = 1$; $\tau_l = 0, l \neq k$). We define Hypergraph consistent sparse coding (CSC) as a special case of MultiCSC when $\tau = [0, \cdots, 0, 1, 0, \cdots, 0]^T$ ($\tau_k = 1$; $\tau_l = 0, l \neq k$) using only one single hypergraph.

Given sparse codes S of images resulting from optimization Eq. (9), clustering can be implemented under the framework of sparse representation-based image clustering [1–3].

4   Optimization Procedure

As the objective is jointly non-convex with (B, S, Q, $\boldsymbol{\tau}$), direct optimization of MultiCSC in Eq. (9) is infeasible. Fortunately, we can optimize sparse coding along with reconstruction matrix (B, S, Q) and hypergraph combination weights

τ by a two-step iterative algorithm, as suggested in related works using multiple manifold learning [12,13]. At each iteration, (B, S, Q) and τ are alternately optimized while the others are fixed, and then the roles are alternately reversed. These iterations are repeated until convergence is achieved or a maximum number of iterations is reached.

Theorem 1. *When τ is fixed, (B, S, Q) can be optimized as a new Laplacian regularized sparse coding algorithm by the following transformation:*

$$\mathcal{X} = \begin{bmatrix} X \\ \sqrt{\gamma}\mathcal{H} \end{bmatrix} \quad and \quad \mathcal{B} = \begin{bmatrix} B \\ \sqrt{\gamma}Q \end{bmatrix}. \tag{10}$$

Proof. By fixing τ, the objective function in Eq. (9) can be rewritten as:

$$O^{MultiCSC} = \|X - BS\|_F^2 + \alpha Tr(S\mathcal{L}S^T) + \gamma\|\mathcal{H} - QS\|_F^2 + \beta\sum_{i=1}^{n}\|s_i\|_1 + \zeta_{\tau}$$

$$= \left\| \begin{bmatrix} X \\ \sqrt{\gamma}\mathcal{H} \end{bmatrix} - \begin{bmatrix} B \\ \sqrt{\gamma}Q \end{bmatrix}S \right\|_F^2 + \alpha Tr(S\mathcal{L}S^T) + \beta\sum_{i=1}^{n}\|s_i\|_1 + \zeta_{\tau}$$

$$= \|\mathcal{X} - \mathcal{B}S\|_F^2 + \alpha Tr(S\mathcal{L}S^T) + \beta\sum_{i=1}^{n}\|s_i\|_1 + \zeta_{\tau} \tag{11}$$

where $\mathcal{X} \in \mathbb{R}^{(m+|E|)\times n}$ and $\mathcal{B} \in \mathbb{R}^{(m+|E|)\times r}$ are those in Eq. (10), and ζ_{τ} is a constant when τ is fixed.

Therefore, when τ is fixed, (B, S, Q) can be optimized as a new Laplacian regularized sparse coding algorithm with the above transformation. Now, the objective function in Eq. (11) is the same as Eq. (2) of HSC, where $\mathcal{X}$ can be seen as the new data matrix and $\mathcal{B}$ is the new dictionary. As suggested in [4,9,10], the optimization of Eq. (11) can be implemented by alternatingly minimizing over S or $\mathcal{B}$ while holding the other fixed.

Theorem 2. *When fixing (B, S, Q), the optimization problem Eq. (9) can be transformed into a constrained quadratic programming problem.*

Proof. By fixing (B, S, Q) and eliminating irrelevant terms, the objective function in Eq. (9) may be transformed into:

$$O^{MultiCSC} = \zeta_{B,S,Q} + \alpha\sum_{i=1}^{t}\tau_i Tr(SL_iS^T) + \gamma\left\|\sum_{i=1}^{t}\tau_i H_i - QS\right\|_F^2 + \lambda\|\tau\|_2^2 \tag{12}$$

where

$$\zeta_{B,S,Q} = \|X - BS\|_F^2 + \beta\sum_{i=1}^{n}\|s_i\|_1$$

is a constant term since B, S, and Q are all fixed.

Let

$$\boldsymbol{H} = [H_1(:), H_2(:), \ldots, H_t(:)] \in \mathbb{R}^{n|E| \times t}; \boldsymbol{h} = (QS)(:) \in \mathbb{R}^{n|E| \times 1}$$

$$\boldsymbol{l} = \left[Tr(SL_1S^T), Tr(SL_2S^T), \ldots, Tr(SL_tS^T) \right]^T \in \mathbb{R}^t$$

where $A(:)$ denotes the operation of flattening a m-by-n matrix into a column vector of length $m * n$. Then, Eq. (12) can be rewritten as:

$$O^{MultiCSC} = \zeta_{B,S,Q} + \gamma \left\| \boldsymbol{H\tau} - \boldsymbol{h} \right\|_2^2 + \alpha \boldsymbol{\tau}^T \boldsymbol{l} + \lambda \left\| \boldsymbol{\tau} \right\|_2^2, \qquad (13)$$

which is a well-defined constrained quadratic programming problem.

Since the optimization of $\boldsymbol{\tau}$ when fixing (B, S, Q) can be transformed into a constrained quadratic programming problem, we can now efficiently solve it using the quadric optimization solver in the CVX Matlab toolbox [14].

Based on Theorems 1 and 2, the overall process of optimizing MultiCSC can be described in Algorithm 1. In the algorithm, FeatureSign denotes the feature-sign algorithm to solve l_1 regularized convex optimization for S, and LagrangeDual is the Lagrange dual algorithm to optimize $\mathcal{B}$, detailed in GSC [9] and HSC [10]. ConstrQuad is the standard solver for constrained quadratic optimization.

Algorithm 1. The optimization procedure of MultiCSC

Input   : Original data matrix X, Laplacian matrices $\Phi = \{L_1, L_2, \ldots, L_t\}$, incidence matrices $\Omega = \{H_1, H_2, \ldots, H_t\}$, maximum iteration number $maxIter$, $maxIterForS$, and minimum convergence error ε.

Parameters: Nonnegative parameters α, β, γ and λ

Output   : Dictionary with sparse codes (B, S), hypergraph reconstruction coefficient matrix Q, multi-hypergraph combination weight vector τ.

1 $\boldsymbol{\tau}^{(0)} \leftarrow \frac{1}{t} \mathbf{1}_{t \times 1}$; $\mathcal{B}^{(0)} \leftarrow [rand()]_{m \times n}$; $i \leftarrow 0$;

2 **repeat**

　　//Fixing τ, update (B, S, Q)

3　　$X^{(i)} \leftarrow \begin{bmatrix} \mathcal{X} \\ \sqrt{\gamma} \sum_{k=1}^{t} \tau_k^{(i)} H_k \end{bmatrix}$; $\mathcal{L}^{(i)} \leftarrow \sum_{k=1}^{t} \tau_k^{(i)} L_k$;

4　　$S^{(i+1)} \leftarrow \text{FeatureSign}(\mathcal{X}^{(i)}, \mathcal{B}^{(j)}, \mathcal{L}^{(i)})$; $\mathcal{B}^{(i+1)} \leftarrow \text{ConstrQuad}(\mathcal{X}^{(i)}, S^{(i+1)})$;

　　//Fixing (B, S, Q), update τ

5　　$\boldsymbol{\tau}^{(i+1)} \leftarrow \text{LagrangeDual}(\mathcal{B}^{(i+1)}, S^{(i)})$; $i \leftarrow i + 1$;

6 **until** $\|\boldsymbol{\tau}^{(i)} - \boldsymbol{\tau}^{(i+1)}\| \leq \varepsilon$ or $i > maxIter$;

7 $B \leftarrow$ first n rows in $\mathcal{B}^{(i)}$; $Q \leftarrow$ last n rows in $\mathcal{B}^{(i)} / \sqrt{\gamma}$;

8 Output (B, S, Q) and τ. ;

Table 1. A summary of datasets.

Dataset	#Classes	#Objects/Class	#Features	Dataset	#Classes	#Objects/Class	#Features
COIL20	100	72	32×32	MNIST	10	400	784
CMU-PIE	68	21	32×32	USPS	10	900	16×16

5 Experimental Results

In this section, we apply the proposed MultiCSC method to image clustering tasks, which is implemented on real word image datasets.

5.1 Benchmark Datasets

Four popular real image datasets have been used in our experiments as benchmark datasets, summarized in Table 1[1]. For each dataset, we first normalize each image vector into unit norm. Then, as suggested in [9], we use principal component analysis (PCA) to eliminate correlations among features, and take the first 64 principal components as the new transformed features.

5.2 Competing Models and Setup

Our proposed model will be compared with a number of baseline models.

– **Original, KSVD** and **SC**. The "original" method is to cluster image objects in the original data space. K-SVD (KSVD) [15] and sparse coding (SC) [4] are two basic dictionary learning methods without any regularization except sparsity constraints.
– **GSC, HSC**, and **CSC**. GSC [9] and HSC (HSC) [10] respectively add the graph Laplacian or hypergraph Laplacian regularization term to the original sparse coding framework. CSC is our proposed method, which adds the hypergraph incidence consistency regularization term to the HSC framework.
– **MultiGSC, MultiHSC**, and **MultiCSC**. MultiGSC and MultiHSC respectively extends the GSC [9] and (HSC) [10] framework with the multi-laplacian term in Eq. (6). MultiCSC is our proposed method as in Eq. (9).

The dictionary size r of all these models is set to be 128, since several recent works on sparse coding have advocated the use of overcomplete representations for images. For GSC, HSC, and CSC, graph or hypergraph is constructed by 3-nearest neighbor search in Euclidean distance as suggested in [9,10]. For MultiGSC, MultiHSC, and MultiCSC, 12 candidate hypergraphs are constructed in the experiments with different neighbor sizes and weighting schemes (detailed in [12]). All the incidence matrices of hypergraphs are normalized such that the row vector representing each image is unit norm.

We use sparse representation based spectral clustering [2] on sparse code space to group image datasets, and compare the clustering results under

---

[1] We downloaded them from: http://www.cad.zju.edu.cn/home/dengcai/Data.

two standard metrics: accuracy (ACC) and normalized mutual information (NMI) [9]. To ensure stability of results, the clustering are repeated for 50 times, each time with a random set of initial centers in the K-means step in [2], and the average are reported. The regularization parameters of each competed algorithm (β in KSVD and SC; α in GSC and HSC; λ in all multiple-based learning; γ in CSC) are tuned to get the best clustering performance (highest average of NMI in 50 runs).

Table 2. Comparison of clustering results using different methods

Methods	COIL20		CMU-PIE		MNIST		USPS	
	ACC	NMI	ACC	NMI	ACC	NMI	AC	NMI
Original	.6948	.8654	.6240	.7943	.6696	.6739	.7749	.7647
KSVD	.6022	.6912	.8506	.9548	.4931	.4052	.6631	.5602
SC	.6137	.7145	.8188	.9416	.5435	.5069	.6937	.5750
GSC	.7345	.8436	.8491	.9588	.6763	.6640	.8253	.7670
HSC	.7691	.8550	.8484	.9549	.6581	.6548	.8296	.7709
CSC	.8538	.9239	.8800	.9703	.7195	.6863	.8378	.7730
MultiGSC	.7486	.8537	.8597	.9534	.6782	.6668	.8281	.7706
MultiHSC	.7841	.8717	.8610	.9552	.6747	.6720	.8315	.7769
MultiCSC	**.8697**	**.9280**	**.8834**	**.9754**	**.7413**	**.7051**	**.8774**	**.7973**

5.3   Clustering Results on Image Datasets

We implement sparse coding on each dataset, and then do clustering in the sparse code space. The image clustering results are summarized in Table 2, further explained below. According to results in Table 2, methods with manifold regularization outperform those without manifold regularization. In other words, KSVD and SC are outperformed by all other manifold regularized dictionary learning methods. Furthermore, methods using hypergraph consistency regularization (i.e., CSC) perform consistently better than those using simple graph or hypergraph regularization (i.e., GSC and HSC). These two points illustrate the effectiveness of manifold structure in sparse coding and the superiority of proposed hypergraph incidence consistency regularization. Moreover, we specifically compare single graph/hypergraph models with multiple graph/or hypergraph models, included in Table 2 and visualized in Fig. 1. We can see that methods using multiple manifold learning (i.e., MultiGSC, MultiHSC, and MultiCSC) outperform their respective single graph/or hypergraph counterparts (i.e., GSC, HSC, and CSC, respectively). In all datasets, whether using single-graph or multi-graph, sparse coding exploiting hypergraph incidence consistency term beats that only with hypergraph Laplacian regularization. Although only ACC by NCuts is shown in Fig. 1 due to space limitations, we observe similar results using NMI. These observations reveal that both hypergraph incidence consistency regularization and multiple manifold learning make difference for sparse coding, and thus, clustering results on sparse codes learned with MultiCSC are more reliable.

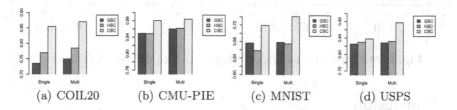

| (a) COIL20 | (b) CMU-PIE | (c) MNIST | (d) USPS |

Fig. 1. Comparison of ACC using single vs. multiple graphs or hypergraphs.

5.4 Robustness on Clustering

In our proposed MultiCSC algorithm, there are two parameters: the hypergraph consistent tradeoff parameter γ, and the multiple combination parameter λ. In this subsection, we test how clustering performance varies with these two tradeoff parameters changing. Figure 2 shows the clustering performance versus the tradeoff parameter λ when γ is fixed as γ^*_{HIC}. Figure 3 shows the clustering performance versus the hypergraph consistent tradeoff parameter γ when λ is fixed as $\lambda^*_{MultiHIC}$. From these figures, we can observe that the performance of proposed MultiCSC keeps steady with varying λ. The performance rises with increasing γ and stabilizes when γ is greater than 10 (except for CMU-PIE when γ is greater than 0.5). These observations prove robustness of the clustering results, and provide references for parameter settings when using our proposed sparse coding methods.

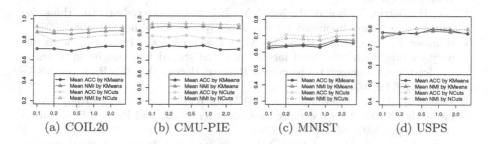

| (a) COIL20 | (b) CMU-PIE | (c) MNIST | (d) USPS |

Fig. 2. ACC and NMI (y-axis) of MultiCSC with different λ (x-axis)

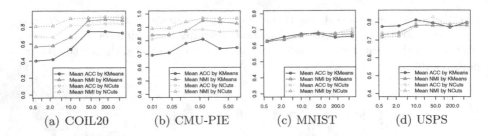

Fig. 3. ACC and NMI (*y*-axis) of MultiCSC with different γ (*x*-axis)

6   Conclusion

This paper presents a novel sparse coding method called MultiCSC, which explicitly considers sufficient exploitation of manifold structure and automatic optimal manifold selection. By introducing a hypergraph incidence consistency term, the hypergraph of input dataset can be well reconstructed using learned sparse codes, and the optimal manifold represented by hypergraph model lies in the linear combination of candidate hypergraphs, where the combination weight is learned in addition to the dictionary and the sparse codes. Experimental results on image clustering have illustrated that our proposed algorithm achieves better discriminating power and significantly enhances the clustering performance.

References

1. Elhamifar, E., Vidal, R.: Sparse subspace clustering. In: 2009 IEEE Conference on Computer Vision and Pattern Recognition (CVPR), pp. 2790–2797. IEEE (2009)
2. Wu, S., Feng, X., Zhou, W.: Spectral clustering of high-dimensional data exploiting sparse representation vectors. Neurocomputing **135**, 229–239 (2014)
3. Wright, J., Ma, Y., Mairal, J., Sapiro, G., Huang, T.S., Yan, S.: Sparse representation for computer vision and pattern recognition. Proccedings of IEEE **98**(6), 1031–1044 (2010)
4. Lee, H., Battle, A., Raina, R., Ng, A.Y.: Efficient sparse coding algorithms. In: Advances in Neural Information Processing Systems 2006, pp. 801–808 (2006)
5. Ramirez, I., Sprechmann, P., Sapiro, G.: Classification and clustering via dictionary learning with structured incoherence and shared features. In: 2010 IEEE Conference on Computer Vision and Pattern Recognition (CVPR), pp. 3501–3508. IEEE (2010)
6. Zhang, Q., Li, B.: Discriminative K-SVD for dictionary learning in face recognition. In: 2010 IEEE Conference on Computer Vision and Pattern Recognition (CVPR), pp. 2691–2698. IEEE (2010)
7. Jiang, Z., Lin, Z., Davis, L.S.: Label consistent k-svd: learning a discriminative dictionary for recognition. IEEE Trans. Pattern Anal. Mach. Intell. **35**(11), 2651–2664 (2013)
8. Liu, X., Song, M., Tao, D., Zhou, X., Chen, C., Bu, J.: Semi-supervised coupled dictionary learning for person re-identification. In: 2014 IEEE Conference on Computer Vision and Pattern Recognition (CVPR), pp. 3550–3557. IEEE (2014)

9. Zheng, M., Bu, J., Chen, C., Wang, C., Zhang, L., Qiu, G., Cai, D.: Graph regularized sparse coding for image representation. IEEE Trans. Image Process. **20**(5), 1327–1336 (2011)
10. Gao, S., Tsang, I., Chia, L.: Laplacian sparse coding, hypergraph laplacian sparse coding, and applications. IEEE Trans. Pattern Anal. Mach. Intell. **35**(1), 92–104 (2013)
11. Geng, B., Tao, D., Xu, C., Yang, Y., Hua, X.: Ensemble manifold regularization. IEEE Trans. Pattern Anal. Mach. Intell. **34**(6), 1227–1233 (2012)
12. Jin, T., Yu, J., You, J., Zeng, K., Li, C., Yu, Z.: Low-rank matrix factorization with multiple Hypergraph regularizer. Pattern Recogn. **48**(3), 1011–1022 (2015)
13. Wang, J.J., Bensmail, H., Gao, X.: Multiple graph regularized nonnegative matrix factorization. Pattern Recogn. **46**(10), 2840–2847 (2013)
14. Grant, M., Boyd, S., Ye, Y.: CVX: Matlab software for disciplined convex programming (2008). http://www.stanford.edu/~boyd/cvx/
15. Aharon, M., Elad, M., Bruckstein, A.: K-svd: an algorithm for designing overcomplete dictionaries for sparse representation. IEEE Trans. Signal Process. **54**(11), 4311–4322 (2006)

Robust Multi-view Manifold Ranking for Image Retrieval

Jun Wu[1,2]([⊠]), Jianbo Yuan[2], and Jiebo Luo[2]

[1] Beijing Key Lab of Traffic Data Analysis and Mining,
Beijing Jiaotong University, Beijing 100044, China
[2] Department of Computer Science,
University of Rochester, Rochester, NY 14627, USA
{jwu,jyuan10,jluo}@cs.rochester.edu

Abstract. Graph-based similarity ranking plays a key role in improving image retrieval performance. Its current trend is to fuse the ranking results from multiple feature sets, including textual feature, visual feature and query log feature, to elevate the retrieval effectiveness. The primary challenge is how to effectively exploit the complementary properties of different features. Another tough issue is the highly noisy features contributed by users, such as textual tags and query logs, which makes the exploration of such complementary properties difficult. This paper proposes a Multi-view Manifold Ranking (M2R) framework, in which multiple graphs built on different features are integrated to simultaneously encode the similarity ranking. To deal with the high noise issue inherent in the user-contributed features, a data cleaning solution based on visual-neighbor voting is embedded into M2R, thus called Robust M2R (RM2R). Experimental results show that the proposed method significantly outperforms the existing approaches, especially when the user-contributed features are highly noisy.

Keywords: Image retrieval · Multi-view learning · Manifold ranking · Data cleaning

1 Introduction

The emergence of Web 2.0 technology along with the prevalence of mobile devices leads to an explosion of images being uploaded and shared online, which makes image retrieval become an important research topic during the past two decades [25]. In a typical image retrieval system, a search task may be launched by either keywords or examples provided by the user, termed as Query-by-Keyword (QBK) and Query-by-Example (QBE) respectively, and then the system ranks the images in the database according to their similarities to the user's query. However, QBK is always limited by the so-called 'intent gap', i.e. the user

---

J. Wu—This work was performed when the first author was an academic visitor at University of Rochester.

© Springer International Publishing Switzerland 2016
J. Bailey et al. (Eds.): PAKDD 2016, Part II, LNAI 9652, pp. 92–103, 2016.
DOI: 10.1007/978-3-319-31750-2_8

may not describe the visual content of his or her target using proper keywords, while QBE often suffers from the well-known 'semantic gap' existing between low-level image pixels captured by machines and high-level semantic concepts perceived by humans.

An effective solution to bridge the gaps is to exploit the users' feedbacks that could be obtained in either *explicit* [28] or *implicit* manner [8], such that the initial image ranking list would be refined. Some recent studies along this direction [5,7,20,22,23,26] are designed for QBK, called image re-ranking, while others serve for QBE including both short-term learning [3,12,14,16,29] and long-term learning [4,9,15,17,21]. In the meantime, a surge of efforts have been made for the graph-based similarity ranking, especially in manifold ranking (MR) [27]. By taking the intrinsic geometrical structure into consideration, MR assigns each data point a relative ranking score, instead of an absolute pairwise similarity as traditional ways. The score is treated as a distance metric defined on the data manifold, which is more meaningful to capture the semantics among data points. In addition, the users' feedbacks are easily exploited by MR in both explicit and implicit ways [18], and previous studies have shown that MR is one of the most successful ranking approaches for the image retrieval with relevance feedback [3,16,19].

Despite its success, the regular MR is limited by two major shortcomings when deployed for image retrieval. First, in many search applications, image data are with multiple views, where each view is actually a feature set. For example, images can be represented by their visual information, surrounding text, and users' query logs. However, the regular MR method cannot effectively integrate multiple views to encode the similarity ranking. Second, the user-contributed views are often completed in a 'crowdsourcing' way, such as textual tags and query logs, so the correctness cannot be always guaranteed. Directly exploiting the highly noisy views in MR may degenerate the retrieval performance. In this paper, we propose a novel method named Robust Multi-view Manifold Ranking (RM2R) to address the aforementioned problems. The main contributions can be summarized as follows. First, we extend the regular MR from single-view to multi-view, aiming to exploit the complementary properties of different feature sets. For convenience of discussion in this paper, we focus on the two-view scenario, assuming that only visual features and query logs are available. Furthermore, we develop a data cleaning solution to make the proposed method more robust to the noisy query logs. An empirical study shows encouraging results in comparison to several exiting approaches. In particular, we observe that our RM2R method is quite robust to noisy query logs, even if the noise level reaches 50 %.

In the following we start with a brief review of some related works. Then we propose our RM2R method and report the experimental results. Finally we conclude this paper.

2   Related Work

Graph-based similarity ranking has been extensively studied in the multimedia retrieval area. Its main idea is to describe the dataset as a graph and then decide the importance of each vertex based on local or global structure drawn from the graph. One canonical graph-based ranking technique is the Manifold Ranking (MR) algorithm [27], and He et al. [3] first applied MR to image retrieval. Its limitations are addressed by latter research efforts. For example, Wang et al. [10] improved the MR accuracy using a k-regular nearest neighbor graph that minimizes the sum of edge weights and balances the edges in the graph as well. Wu et al. [16] proposed a self-immunizing MR algorithm that uses an elastic kNN graph to exploit unlabeled images safely. Xu et al. [19] proposed an efficient MR solution based on scalable graph structure to handle large-scale image datasets.

Multi-view learning concentrates on learning from the data with multiple feature sets. Co-training [1] is probably the most famous representative. It constructs two learners each from one view, and then lets them to provide pseudo-labels for the other learner. Zhou et al. [30] regarded the visual content and surrounding texts of images as two views, and applied co-training to image retrieval. In fact, co-training does not really need the existence of multiple views, and the diversity among the learners is the real essence [13]. A variant of co-training used in image retrieval [29] suggested to train two different rankers based on the same visual feature set, where each ranker identifies unlabeled images with highest confidence to enlarge the training set of the other ranker. Besides co-training, there are many other solutions to fuse multi-view data in literatures, such as multiple kernel learning [26], and multi-graph learning [2,11,23], etc.

Our research is also closely related to the collaborative image retrieval (CIR) that aims to combine short-term learning and long-term learning within a unified framework. For example, Yin et al. [21] exploited the long-term experiences fulfilled by different users to select the optimal online ranker from a set of candidates based on reinforcement learning. Hoi et al. [4] regarded the query log as the 'side information', and then, taking that as constraints, learned a distance metric form a mixture of labeled and unlabeled images. Su et al. [9] suggested to discover the navigation patterns from query logs, and using the patterns to facilitate new searching tasks. Wu et al. [15,17] proposed a hybrid similarity measure that preserves both visual and semantic resemblance by incorporating short-term with long-term feedback experiences.

3   The Proposed RM2R Method

Our RM2R method is developed based on two intuitions. First, a 'good' ranker should be able to exploit the complementary property of different views. Moreover, the ranker should be robust to noisy views. We start from the description of notations.

3.1 Preliminaries

For simplicity, assume that we are handling two-view image dataset $\mathcal{D} = \{X_i = (\mathbf{x}_i^{(1)}, \mathbf{x}_i^{(2)}), i = 1, \cdots, n\}$, where each image instance $X_i = (\mathbf{x}_i^{(1)}, \mathbf{x}_i^{(2)})$ is with two views. Concretely, in the first view $\mathbf{x}_i^{(1)} \in \mathbb{R}^{d_1}$ is a d_1 dimensional visual feature vector of image X_i, and in the second view $\mathbf{x}_i^{(2)} \in \{0, 1\}^{d_2}$ is a d_2 dimensional log feature vector recording the clicks made by different users on image X_i, where '1' means that X_i is clicked in a certain query session, and '0' otherwise.

To discover the geometrical structure, we build a couple of graphs $G^{(z)} = (V^{(z)}, E^{(z)}, \mathbf{W}^{(z)})$ on $\mathcal{D}$, where $z \in \{1, 2\}$ is the graph identify. In details, $V^{(z)}$ is the node set, in which each node corresponds to an image; $E^{(z)}$ and $\mathbf{W}^{(z)} \in \mathbb{R}_+^{n \times n}$ are the edge set and the edge weighting matrix respectively; each $W_{ij}^{(z)}$ represents the weight of edge $E_{ij}^{(z)}$. Typically, the weight is defined by a certain similarity measure, and we apply different similarity measures to $G^{(1)}$ and $G^{(2)}$ due to the different input spaces.

For graph $G^{(1)}$, its nodes are real-valued vectors, so the similarity between $\mathbf{x}_i^{(1)}$ and $\mathbf{x}_j^{(1)}$ is defined by the Gaussian kernel

$$W_{ij}^{(1)} = \exp\left(-\frac{d^2(\mathbf{x}_i^{(1)}, \mathbf{x}_j^{(1)})}{\sigma^2}\right) \tag{1}$$

where $d(\mathbf{a}, \mathbf{b})$ is a distance metric between two vectors $\mathbf{a}$ and $\mathbf{b}$ (suggested by [3], L1 distance is considered), and σ is the bandwidth parameter that can be tuned by local scaling technique, the effectiveness of which has been verified in the clustering [24] and ranking [16] tasks.

For $G^{(2)}$, its nodes are binary vectors, so the similarity between $\mathbf{x}_i^{(2)}$ and $\mathbf{x}_j^{(2)}$ is defined by the Jacquard coefficient

$$W_{ij}^{(2)} = \frac{|\mathbf{x}_i^{(2)} \cap \mathbf{x}_j^{(2)}|}{|\mathbf{x}_i^{(2)} \cup \mathbf{x}_j^{(2)}|} \tag{2}$$

where $|\bullet|$ denotes the size of a set.

To exploit the complementary property between $G^{(1)}$ and $G^{(2)}$, we will extend the regular MR method [27] from single-view to multi-view in the next subsection.

3.2 M2R: Multi-view Manifold Ranking

Let $\mathbf{r} : \mathcal{D} \to \mathbb{R}$ be a ranking function that assigns to each image instance X_i a ranking score r_i. We also define a label vector $\mathbf{y} = [y_1, \cdots, y_n]^T$ to collect the user's online feedbacks, where $y_i = 1$ if X_i is labeled as positive, $y_i = -1$ if X_i is labeled as negative, and $y_i = 0$ otherwise. The cost function associated with $\mathbf{r}$ is defined to be

$$\mathcal{Q}(\mathbf{r}) = \frac{\lambda}{2} \sum_{i,j=1}^{n} W_{ij}^{(1)} \left(\frac{r_i}{\sqrt{D_{ii}^{(1)}}} - \frac{r_j}{\sqrt{D_{jj}^{(1)}}} \right)^2$$

$$+ \frac{1-\lambda}{2} \sum_{i,j=1}^{n} W_{ij}^{(2)} \left(\frac{r_i}{\sqrt{D_{ii}^{(2)}}} - \frac{r_j}{\sqrt{D_{jj}^{(2)}}} \right)^2 + \frac{\mu}{2} \|\mathbf{r} - \mathbf{y}\|^2$$

where $0 < \lambda < 1$ is a parameter to adjust the weight between visual view and log view, $\mu > 0$ is a regularization parameter, and $\mathbf{D}^{(z)}$ ($z \in \{1,2\}$) is a diagonal matrix with $D_{ii}^{(z)} = \sum_{j=1}^{n} W_{ij}^{(z)}$. The first two terms in the cost function are two smoothness constraints, which make the nearby examples in both visual space and log space having close ranking scores. The third term is a fitting constrain, which makes the ranking result fitting to the label assignment.

By differentiating $\mathcal{Q}$ with respect to $\mathbf{r}$, we have

$$\left. \frac{\partial \mathcal{Q}}{\partial \mathbf{r}} \right|_{\mathbf{r}=\mathbf{r}^*} = \lambda(\mathbf{r}^* - \mathbf{S}^{(1)}\mathbf{r}^*) + (1-\lambda)(\mathbf{r}^* - \mathbf{S}^{(2)}\mathbf{r}^*) + \mu(\mathbf{r}^* - \mathbf{y}) = 0$$

where $\mathbf{S}^{(z)}$ is the symmetrical normalization of $\mathbf{D}^{(z)}$, i.e.

$$\mathbf{S}^{(z)} = (\mathbf{D}^{(z)})^{1/2} \mathbf{W}^{(z)} (\mathbf{D}^{(z)})^{1/2}. \tag{3}$$

By regrouping, the equation can be transformed into

$$\mathbf{r}^* - \frac{\lambda}{1+\mu}\mathbf{S}^{(1)}\mathbf{r}^* - \frac{1-\lambda}{1+\mu}\mathbf{S}^{(2)}\mathbf{r}^* - \frac{\mu}{1+\mu}\mathbf{y} = 0.$$

Let $\alpha = \lambda/(1+\mu)$, $\beta = (1-\lambda)/(1+\mu)$ and $\gamma = \mu/(1+\mu)$, and then we have

$$(\mathbf{I} - \alpha\mathbf{S}^{(1)} - \beta\mathbf{S}^{(2)})\mathbf{r}^* = \gamma\mathbf{y}.$$

Note that $\alpha + \beta + \gamma = 1$. Since $(\mathbf{I} - \alpha\mathbf{S}^{(1)} - \beta\mathbf{S}^{(2)})$ is invertible, we have

$$\mathbf{r}^* = \gamma(\mathbf{I} - \alpha\mathbf{S}^{(1)} - \beta\mathbf{S}^{(2)})^{-1}\mathbf{y}. \tag{4}$$

We can directly use the above closed form solution to compute the ranking scores of examples. However, in large scale problems, we prefer to use the iteration solution

$$\mathbf{r}(t+1) = (\alpha\mathbf{S}^{(1)} + \beta\mathbf{S}^{(2)})\mathbf{r}(t) + \gamma\mathbf{y}. \tag{5}$$

It is easy to prove that Eq. (5) *converges to* Eq. (4).

Proof. Suppose the sequence $\{\mathbf{r}(t)\}$ converges to $\mathbf{r}^*$. Substituting $\mathbf{r}^*$ for $\mathbf{r}(t+1)$ and $\mathbf{r}(t)$ in the equation. We have $\mathbf{r}^* = (\alpha\mathbf{S}^{(1)} + \beta\mathbf{S}^{(2)})\mathbf{r}^* + \gamma\mathbf{y}$ that can be transformed into $(\mathbf{I} - \alpha\mathbf{S}^{(1)} - \beta\mathbf{S}^{(2)})\mathbf{r}^* = \gamma\mathbf{y}$. Since $(\mathbf{I} - \alpha\mathbf{S}^{(1)} - \beta\mathbf{S}^{(2)})$ is invertible, we have $\mathbf{r}^* = \gamma(\mathbf{I} - \alpha\mathbf{S}^{(1)} - \beta\mathbf{S}^{(2)})^{-1}\mathbf{y}$.

Given the fact that different users may have different opinions on judging the same image, the inherent noise issue of query logs is inevitable. Hence we will study the Query Log Cleaning (QLC) solution in the next subsection.

3.3 Query Log Cleaning via Neighbor Voting

Inspired by neighbor voting [6], our QLC solution is based on the intuition that if a user clicks a group of visually similar images, his or her clicking information is likely to reflect the objective aspects of visual content. This intuition suggests that, given an image, the confidence of a click made on it can be estimated from how its visual neighbors are judged (clicked or not) in the same session, i.e. accumulating the votes from its visual neighbors.

To facilitate voting, we encode each query session to a unique code (e.g. using the session id), and then all sessions can be viewed as a code book. Then, given an image $X_i = (\mathbf{x}_i^{(1)}, \mathbf{x}_i^{(2)})$, all clicks it received can be represented by a bag-of-code $C_i = \{c_j, j = 1, \cdots, l_i\}$, where each code $c_j \in C_i$ is actually the index of each non-zero element in the log vector $\mathbf{x}_i^{(2)}$ and there are l_i codes in total. Given an image $X_i = (\mathbf{x}_i^{(1)}, C_i)$, we define a voting function $f(\mathbf{x}_i^{(1)}, c_j)$ to measure the relevance of each code $c_j \in C_i$ to the visual object $\mathbf{x}_i^{(1)}$

$$f(\mathbf{x}_i^{(1)}, c_j) = \sum_{m=1}^{K} \delta(\mathbf{x}_m, c_j) \tag{6}$$

where $\mathbf{x}_m$ $(m = 1, \cdots, K)$ denote the K visually nearest neighbors of $\mathbf{x}_i^{(1)}$. The binary function $\delta(\mathbf{x}_m, c_j) = 1$ if the visual neighbor $\mathbf{x}_m$ also has the code c_j, otherwise $\delta(\mathbf{x}_m, c_j) = 0$.

After neighbor voting, each click recorded in query logs is associated with a votes. The higher votes a click received, the more confident it is. Therefore, the clicks with low votes (less than a threshold T) can be removed. We apply the 'Three Standard Deviations' principle to set the threshold, i.e. $T = \mu_v - 1.5\sigma_v$, where μ_v and σ_v are the mean and standard deviation of the voted results, respectively.

3.4 RM2R-Based Image Retrieval

In this section, we make a brief summary of applying RM2R to image retrieval, described in Algorithm 1. Note that the affinity matrixes can be calculated offline, and therefore RM2R can be quite efficient in processing online queries.

4 Experiments

4.1 Experimental Setup

We employ the '10K Images' dataset[1] which is publicly available on the web to make our experiments reproducible. The images are from 100 semantic categories, with 100 images per category. Three kinds of visual features are extracted to represent the images, including a 64-dimensional color histogram,

---

[1] http://www.datatang.com/data/44353. The dataset was firstly used in [15].

Algorithm 1. RM2R-Based Image Retrieval

Input:
  Two-View Image dataset: $\mathcal{D} = \{X_i = (\mathbf{x}_i^{(1)}, \mathbf{x}_i^{(2)}), i = 1, \cdots, n\}$;
  Query example: $\mathbf{q}$;
  Parameters: α, β, γ, K
Output:
  Rank-Score Vector: $\mathbf{r}$
 1: Search the K nearest neighbors for each image based on visual feature set, and
    connect two images with an edge if they are neighbors (i.e. construct a visual KNN
    graph);
 2: Clean query log set based on visual KNN graph by Eq. (6);
 3: Calculate affinity matrix $\mathbf{W}^{(1)}$ based on visual KNN graph by Eq.(1);
 4: Calculate affinity matrix $\mathbf{W}^{(2)}$ based on cleaned query log set by Eq. (2);
 5: Calculate normalized affinity matrixes $\mathbf{S}^{(1)}$ and $\mathbf{S}^{(2)}$ by Eq. (3);
 6: Initialize label vector $\mathbf{y}$ according to $\mathbf{q}$;
 7: **repeat**
 8:    Calculate rank $\mathbf{r}$ based on $\mathbf{S}^{(1)}$, $\mathbf{S}^{(2)}$ and $\mathbf{y}$ by Eq. (4) or Eq. (5);
 9:    Update $\mathbf{y}$ according to user's feedback;
10: **until** user is satisfied with retrieval result

---

an 18-dimensional wavelet-based texture and a 5-dimensional edge direction histogram [15].

The log dataset consists of 1000 query sessions which are simulated based on the ground truth of image dataset. The average number of clicks in each query session is 20. Further, to evaluate the robustness of our method, three noised log datasets are used in experiments and the noise levels are 10 %, 30 % and 50 %, respectively.

Our RM2R method is compared with three existing approaches which are introduced as follows.

– Self-Tuning Manifold Ranking (STMR): the regular MR method [3] with self-tuned bandwidth parameter [24].
– Co-training based Ranking (CoR): a multi-view learning based image ranking method [29].
– Hybrid Similarity learning (HySim): a CIR method combining short-term learning with long-term learning [15].

In addition, M2R is also included in comparisons to study whether the QLC solution used by RM2R is effective or not.

A query set with 200 image examples is randomly selected to evaluate the average performance of all compared methods. For each query, only one round of feedback is performed (the top ten returned images are labeled), since users have no patience to do more. We use the precision-recall-graph (PR-graph), Mean Average Precision (MAP) and the precision at top N retrieval results (P@N) for performance evaluation.

4.2   Performance Evaluation

At first, we conduct model selection for our M2R and RM2R methods, and there are four parameters in total: α, β, γ ($\alpha + \beta + \gamma = 1$) used in Eq. (5) and K used in Eq. (6). For convenience, γ is fixed at 0.01, consistent with the previous experiences [3,19,27]. Then, we have $\beta = 0.99 - \alpha$ and thus only need to tune two parameters α and K. We first investigate the impact of α by only evaluating the performance of M2R. Figure 1(a) shows the performance of M2R over different α values, which is evaluated on the log datasets with different noise levels. We observe that it would be desirable to choose smaller α values when the noise level of log data is low, and larger ones when the noise level is high. To avoid bias, we set this parameter as the value that the best performance of M2R is achieved on the middle noise level (30 %), i.e. $\alpha = 0.5$. Also, the performance derived by this parameter value is not too good or too bad, regardless of the noise level is high or low. Further, we evaluate K for the final results of RM2R on the log dataset with 30 % noise, depicted by Fig. 1(b). Fortunately, our RM2R method is not sensitive to K, and we fixed it at 20.

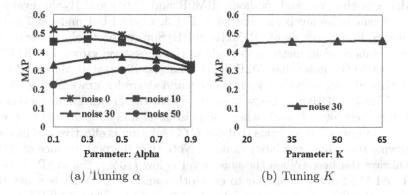

(a) Tuning α (b) Tuning K

Fig. 1. Tuning parameters for our RM2R method.

Next, with the best parameters setting, we compare our RM2R method with STMR, CoR and HySim on the ideal log dataset (without any noise). Figure 2(a) and, (b) print the PR graphs and the P@N curves respectively. By examining all methods, we observe that the methods using both visual features and query logs (RM2R and HySim) outperform the methods only using visual features (STMR, CoR), which indicates that exploiting the multi-view information is beneficial to the similarity ranking. Further, by comparing STMR with CoR, we find that the performance of STMR is much better than CoR, which indirectly verifies the superiority of the graph-based similarity ranking used by our method in comparison to the pairwise similarity ranking.

Furthermore, we compare RM2R with M2R and HySim on the noisy log datasets. To analyze the tolerance of an algorithm to noise, we define a performance lower bound for a method first, which refers to the performance of the method evaluated in the single-view case, i.e. no any query logs are used.

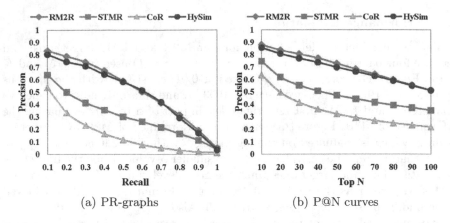

(a) PR-graphs   (b) P@N curves

Fig. 2. The performance of our RM2R method compared with several existing approaches.

The MAPs of the proposed methods (RM2R and M2R) and HySim evaluated under different noise levels are plotted in Fig. 3, where LB-1 and LB-2 are the performance lower bounds of (R)M2R and HySim, respectively. As expected, the performances of all methods degrade as the noise level grows, but RM2R is more robust to the noise than M2R and HySim. As illustrated, the MAP curve of RM2R is always above its lower bound and degrades gracefully, while the curves of M2R and HySim degrade sharply with the growing noise levels and drop below their lower bounds when the noise level hits 50 % and 30 %, respectively. This observation confirms that our QLC solution is effective and essential to achieving the robust similarity ranking. Note that the performance of RM2R is not always the best. When the noise level is low (10 %), the MAP curves of RM2R and M2R are very similar to each other and even RM2R is worse than M2R when there is no noise in log dataset. It is conjectured that QLC is a lossy data cleaning technique, i.e. a very few examples are mistakenly removed, and

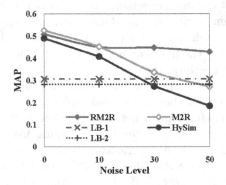

Fig. 3. The robustness of the proposed methods compared with the HySim approach.

Table 1. Comprehensive comparison.

Method\NoiseLevel	0 %	10 %	30 %	50 %
RM2R	0.508	**0.451**	**0.447**	**0.429**
M2R	**0.525**	0.450	0.337	0.272
HySim	0.491	0.408	0.274	0.185
STMR	0.307	0.307	0.307	0.307
CoR	0.157	0.157	0.157	0.157

thus its impact is trivial when the noise level is low. With the growing of the noise level, QLC is increasingly helpful to RM2M.

At last, a comprehensive MAP comparison of all compared methods is summarized in Table 1, where the best performance has been boldfaced. We observe that M2R and HySim perform no better, and often worse, than STMR when the noise level is high, which validates the superiority of our QLC solution again. In summary, when there is no noise in log dataset, the performance of our RM2R method is only marginally better than HySim, but it outperforms STMR and CoR significantly. Considering that our RM2R method is much more robust than M2R and HySim when the log dataset is (highly) noisy, the above presented experiments confirm that our RM2R method is the best among the compared methods.

5 Conclusions

This paper presents a novel RM2R method that aims at the noise-resistant exploitation of the complementary information hidden in multiple views. We first propose a M2R framework to integrate visual feature with query log feature, and then, based on Neighbor Voting, develop a data cleaning solution to noisy query logs. To our best knowledge, not much has been reported on investigating multi-view learning with noisy data in literatures. Experimental study has validated the superiority of the proposed method in comparison to several existing similarity ranking approaches.In this paper, we focus on the two-view data. However, extending to more views will suffer from the inconvenience of tuning a number of parameters. In the future, we will study the self-tuning solution to multi-view manifold ranking.

Acknowledgments. The authors would like to thank the anonymous reviewers for their constructive suggestions. This work was supported in part by the 'Natural Science Foundation of China' (61301185, 61370070 and 61300071), the 'Fundamental Research Funds for the Central Universities' (2015JBM029), and the 'Science Foundation of Beijing Jiaotong University' (2015RC008).

References

1. Blum, A., Mitchell, T.: Combining labeled and unlabeled data with co-training. In: COLT, pp. 92–100 (1998)
2. Deng, C., Ji, R., Tao, D., Gao, X., Li, X.: Weakly supervised multi-graph learning for robust image reranking. IEEE Trans. Multimedia **16**(3), 785–795 (2014)
3. He, J., Li, M., Zhang, H., Tong, H., Zhang, C.S.: Manifold-ranking based image retrieval. In: ACM Multimedia, pp. 9–16 (2004)
4. Hoi, S., Liu, W., Chang, S.F.: Semi-supervised distance metric learning for collaborative image retrieval. In: CVPR, pp. 1–7 (2008)
5. Jain, V., Varma, M.: Learning to re-rank: query-dependent image re-ranking using click data. In: Proceedings of the 20th International Conference on World Wide Web, WWW 2011, pp. 277–286 (2011)
6. Li, X., Snoek, G.C., Worring, M.: Learning social tag relevance by neighbor voting. IEEE Trans. Multimedia **11**(7), 1310–1322 (2009)
7. Pan, Y., Yao, T., Mei, T., Li, H., Ngo, C.W., Rui, Y.: Click-through-based cross-view learning for image search. In: SIGIR, pp. 717–726 (2014)
8. Smith, G., Brien, C., Ashman, H.: Evaluating implicit judgments from image search clickthrough data. J. Am. Soc. Inf. Sci. Technol. **63**(12), 2451–2462 (2012)
9. Su, J., Huang, W., Yu, P., Tseng, V.: Efficient relevance feedback for content-based image retrieval by mining user navigation patterns. IEEE Trans. Knowl. Data Eng. **23**(3), 360–372 (2011)
10. Wang, B., Pan, F., Hu, K.M., Paul, J.C.: Manifold-ranking based retrieval using k-regular nearest neighbor graph. Pattern Recogn. **45**(4), 1569–1577 (2012)
11. Wang, M., Li, H., Tao, D., Lu, K., Wu, X.: Multimodal graph-based reranking for web image search. IEEE Trans. Image Process. **21**(11), 4649–4661 (2012)
12. Wang, T., Dai, G., Ni, B., Xu, D., Siewe, F.: A distance measure between labeled combinatorial maps. Comput. Vis. Image Underst. **116**(12), 1168–1177 (2012)
13. Wang, W., Zhou, Z.-H.: Analyzing co-training style algorithms. In: Kok, J.N., Koronacki, J., Lopez de Mantaras, R., Matwin, S., Mladenič, D., Skowron, A. (eds.) ECML 2007. LNCS (LNAI), vol. 4701, pp. 454–465. Springer, Heidelberg (2007)
14. Wu, J., Lin, Z., Lu, M.Y.: Asymmetric semi-supervised boosting for SVM active learning in CBIR. In: ACM CIVR, pp. 182–188 (2010)
15. Wu, J., Shen, H., Li, Y.D., Xiao, Z.B., Lu, M.Y., Wang, C.L.: Learning a hybrid similarity measure for image retrieval. Pattern Recogn. **46**(11), 2927–2939 (2013)
16. Wu, J., Li, Y., Feng, S., Shen, H.: A self-immunizing manifold ranking for image retrieval. In: Pei, J., Tseng, V.S., Cao, L., Motoda, H., Xu, G. (eds.) PAKDD 2013, Part II. LNCS, vol. 7819, pp. 426–436. Springer, Heidelberg (2013)
17. Wu, J., Lu, M.Y., Wang, C.L.: Collaborative learning between visual content and hidden semantic for image retrieval. In: ICDM, pp. 1133–1138 (2010)
18. Wu, J., Xiao, Z.B., Wang, H.S., Shen, H.: Learning with both unlabeled data and query logs for image search. Comput. Electr. Eng. **40**(3), 964–973 (2014)
19. Xu, B., Bu, J., Chen, C., Wang, C., Cai, D., He, X.: EMR: a efficient manifold ranking model for content-based image retrieval. IEEE Trans. Knowl. Data Eng. **27**(1), 102–114 (2014)
20. Yang, X., Zhang, Y., Yao, T., Ngo, C.W., Mei, T.: Click-boosting multi-modality graph-based reranking for image search. Multimedia Syst. **21**(2), 217–227 (2015)
21. Yin, P.Y., Bhanu, B., Chang, K.C., Dong, A.: Integrating relevance feedback techniques for image retrieval using reinforcement learning. IEEE Trans. Pattern Anal. Mach. Intell. **27**(10), 1536–1551 (2005)

22. Yu, J., Tao, D., Wang, M., Rui, Y.: Learning to rank using user clicks and visual features for image retrieval. IEEE Trans. Cybern. **45**(4), 767–779 (2015)
23. Yu, J., Rui, Y., Chen, B.: Exploiting click constraints and multi-view features for image re-ranking. IEEE Trans. Multimedia **16**(1), 159–168 (2014)
24. Zelnik-Manor, L., Perona, P.: Self-tuning spectral clustering. In: NIPS, pp. 1601–1608 (2004)
25. Zhang, L., Rui, Y.: Image search: from thousands to billions in 20 years. ACM Trans. Multimedia Comput. Commun. Appl. **9**(1s), 36:1–36:20 (2013)
26. Zhang, Y., Yang, X., Mei, T.: Image search reranking with query-dependent click-based relevance feedback. IEEE Trans. Image Process. **23**(10), 4448–4459 (2014)
27. Zhou, D., Weston, J., Gretton, A., Bousquet, O., Scholkopf, B.: Ranking on data manifolds. In: NIPS, pp. 169–176 (2003)
28. Zhou, X.S., Huang, T.S.: Relevance feedback in image retrieval: a comprehensive review. Multimedia Syst. **8**(6), 536–544 (2003)
29. Zhou, Z.H., Chen, K.J., Dai, H.B.: Enhancing relevance feedback in image retrieval using unlabeled data. ACM Trans. Inf. Syst. **24**(2), 219–244 (2006)
30. Zhou, Z.H., Zhan, D.C., Yang, Q.: Semi-supervised learning with very few labeled training examples. In: AAAI, pp. 675–680 (2007)

Image Representation Optimization Based on Locally Aggregated Descriptors

Shijiang Chen[✉], Guiguang Ding, Chenxiao Li, and Yuchen Guo

Tsinghua National Laboratory for Information Science and Technology,
School of Software, Tsinghua University, Beijing 100084, China
chensj13@mails.tsinghua.edu.cn, dinggg@tsignhua.edu.cn,
lichenxiao@gmail.com, yuchen.w.guo@gmail.com

Abstract. Aggregating local descriptors into super vectors achives excellent performance in image classification and retrieval tasks. Vector of locally aggregated descriptors(*VLAD*), which indexes images to compact representations by aggregating the residuals of descriptors and visual words, is a popular super vector encoding method among this kind. This paper will focus on the biggest difficulty of VLAD, the "visual burstiness", reviste the basic assumptions and solutions along this line, then make modifications to two key steps of the initial VLAD process. The main contributions are twofold. Firstly, we start from local coordinate system(LCS) and propose the aggregated version(aggrLCS), which changes the objective and timing of coordinate rotation, for better captures of bursts. Secondly, an adaptive power-law normalization method is adopted to magnify the positive effect of power-law normalization by weighting each dimension respectively. Experiments on image retrieval tasks demonstrate that the proposed modifications show superior performance over the original and several variants of VLAD.

Keywords: Visual burstiness · Power-law · VLAD · Image retrieval · Image representation · Local descriptors

1 Introduction

Effective representation of images is crucial in image recognition and retrieval tasks. One popular method to represent images in vectors is bag of visual words(BoVW)[7,9], after the introduction of BoVW, a lot of notable contributions have been made to this field, including descriptors enhancement [10,11], vector quantization(VQ)[9], sparse coding(SC)[12], locality-constrained linear coding (LLC) [14], fisher vector(FV) [15], vector of locally aggregated descriptors(VLAD) [5,6], and encoding with convolutional neural networks(CNN) [3,4] etc. Among this kind, VLAD is an efficient super vector encoding method that provides successful representations for image retrieval task, especially large scale image search given an instance.

Super vector based encoding methods mainly follow a pipeline of multiple stages, which are "local feature extraction", "codebook learning", "vector quatizing" and "vector pooling".

© Springer International Publishing Switzerland 2016
J. Bailey et al. (Eds.): PAKDD 2016, Part II, LNAI 9652, pp. 104–116, 2016.
DOI: 10.1007/978-3-319-31750-2_9

VLAD is a typical super vector based encoding method, which generally follows the aforementioned pipeline, and it can be regarded as the non probabilistic version of Fisher Vector. VLAD differs from the BoVW method by aggregating the *differences* of each local descriptor and corresponding visual word, while BoVW records the number of descriptors around each centroid. VLAD is demonstrated to be able to generate compact vectors while preserving high retrieval accuracy, however, it suffers from the "visual burstiness" problem, which has been discussed in [1], and local coordinate system(LCS) is proposed to address the problem. "Visual burstiness"[8], caused by visual elements that appear more frequently in an image than statistics, will dominates the similarity measure.

Starting from VLAD and based on LCS for alliviating burstiness in image retrieval task, this paper makes some further modifications to VLAD procedure at two key steps of original VLAD.

Firstly, based on local coordinate system, this paper proposes a novel coordinate rotation algorithm, which ameliorates the objective and timing while performing coordinate rotation independently for each visual word, and benefits the subsequent power-law normalization. Secondly, the classical power-law normalization introduces non-linearity to the final VLAD representation with only one parameter α, which is easy to use but limits the positive effects on alleviating "visual burstiness" for naive parameterization. This paper introduces the adaptive power-law normalization, which weights and restrains bursts on each dimension respectively.

2 Related Works

2.1 VLAD

The vector of locally aggregated descriptors is an encoding and aggregating technique theoretically based on the locality of feature space, and can likewise be regarded as the non probabilistic version of Fisher Vector. Similar to BoVW model, a codebook $D = [d_1, d_2, ..., d_K] \in \mathcal{R}^{d \times K}$ is first learnt from training samples, subsequently the encoding and aggregation of local descriptors are conducted. Typically, $d = 128$ for SIFT descriptor is used to describe regions extracted from an image using an affine invariant detector.

Aggregation. Given a set of local descriptors $X = [x_1, x_2, ..., x_N] \in \mathcal{R}^{d \times N}$ from an imagem, for each visual word $d_j, j \in [1, ..., K]$, a d-dimensional subvector $vsub_j$ is yielded by accumulating the differences between assigned local descriptor (x for convenience) and the corresponding visual word d_j:

$$vsub_j = \sum_{x:NN(x)=d_j} x - d_j. \tag{1}$$

where $NN(x)$ denotes the nearest visual word of x in D. The final VLAD representation is the concetentation of all the d-dimensional subvectors $vsub_j$ and therefore becomes a $d \times K$ dimensional vector $v = [vsub_1, vsub_2, ..., vsub_K]$.

Normalization. The VLAD vector is usually L_2-normalized by $v := \frac{v}{\|v\|_2}$. Since the VLAD representation is encoded by accumulating the differences between local descriptors and their corresponding visual words, leading to that the local descriptors do not contribute equally to the final VLAD representation. Residual normalization (RN) [1] is another normalization method for balancing the contribution of each local descriptor in an image. RN will normalize every residual vector by L_2-normalization prior to accumulation, thus, each d-dimensional subvector $vsub_j$ is normalized to:

$$vsub_j = \sum_{x:NN(x)=d_j} \frac{x - d_j}{\|x - d_j\|_2}. \tag{2}$$

then, each block is L_2-normalized reprectively in the final VLAD representation:

$$v\left(X, D\right) = \left[\frac{vsub_1}{\|vsub_1\|_2}; \frac{vsub_2}{\|vsub_2\|_2}; ...; \frac{vsub_K}{\|vsub_K\|_2} \right]. \tag{3}$$

Experiments show that a codebook of large size is required for good performance while BoVW is adopted, however, VLAD with a vocabulary of size 16 to 256 leads to good enough results. Besides, there are some variants of VLAD. [1] projects all local descriptors of an image on the 64 first principal directions of a offline learned PCA basis prior to accumulation. [2] performs intra-normalization on the final VLAD representation. Several methods brings a substantial performance boosting for negligible additional computational cost.

2.2 LCS

Previous work [6] has shown that performing a PCA of local descriptors boosts VLAD. LCS provides a better handling of burstiness by adapting independently the coordinate system for each visual word. LCS achieves this goal by simply learning a "local" PCA for the partitioned feature space corresponding to each visual word.

To elaborate on the process, for each visual word d_j, which is a cluster center among the training descriptors, a rotation matrix Q_j is offline learned to map from training descriptors to this word. The learned rotation matrices (K in total) are then applied to the normalized residual vectors prior to accumulated to the final VLAD representation. Thus, each aforementioned d-dimensional subvector $vsub_j$ is normalized to:

$$vsub'_j = \sum_{x:NN(x)=d_j} Q_j \frac{x - d_j}{\|x - d_j\|_2}. \tag{4}$$

Under the observation that PCA-based processing of local descriptors improves VLAD representation, we need to dig out the fundamental cause. Since power-law normalization introduces a subsequent non-linearity before generating the final VLAD representation, the basis counts. And the basis produced by

PCA boosts the performance. [1] shows that the during PCA, it is the rotation and its interplay with power-law normalization that improves the results, rather than the dimensionality reduction, which is proved harmful. Although the local coordinate system(LCS) is a simple solution, it benefits the processing of raw local descriptors for avoiding the arbitrariness when power-law normalization is applied during the original VLAD process. This method boosts the performance of VLAD in conjunction with power-law normalization.

3 Aggregated Local Coordinate System

Inspired by [1], this paper proposes a coordinate rotation algorithm based on local aggregated vectors, which improves on upon the local coordinate system by modifying the objective and timing while applying local coordinate rotation, which links up with the subsequent power-law normalization, and thus enhance the improvement of performance by this synergy.

Adequately considered the locality of visual burstiness, LCS proposed to achieve a better handling by adapting independently the coordinate system for each visual word, and related experiments have shown gains in coordination with power-law normalization. However, the encoding function (Eq. 4) does not explain well for timing the coordinate rotation. In order to explicitly describe the problems exist in LCS, this paper mainly made the following observations and will then propose an improved method to address the problem.

Firstly, in Eq. 4, the rotation matrix Q_j corerponding to each visual word is offline learned by performing PCA on the feature space composed by the subset of local descriptors mapped to this word. Residual vectors from the testing set in correlation with the same centroid are then transformed with these rotation matrices. For all that, the rotation matrix Q_j is learned from training descriptors in the original feature space but applied to normalized residual vector space, the inconformity can not be stated reasonably.

Secondly, according to the basic assumptions of LCS, the change of descriptor coordinate system has no impact on similarity measure between resulting vectors in absence of power-law normalization. The PCA-based processing of local descriptors is able to improves VLAD by working with power-law. In other words, PCA does not contribute to the suppression of burstiness directly, but by increasing the effect of power-law normalization. However, the encoding procedure of LCS flaws for performing normalized residual vectors of local descriptors and cluster centers while applying power-law normalization to the representation vector accumulating all the residual vectors.

Aiming at these problems hereinbefore, this paper proposes a coordinate rotation algorithm based on locally aggregated vectors, to aggregate the local descriptors of an image. The encoding procedure in VLAD pipeline can be formulized as:

$$vsub_j'' = Q_j' \cdot \sum_{x:NN(x)=d_j} \frac{x - d_j}{\|x - d_j\|_2}. \tag{5}$$

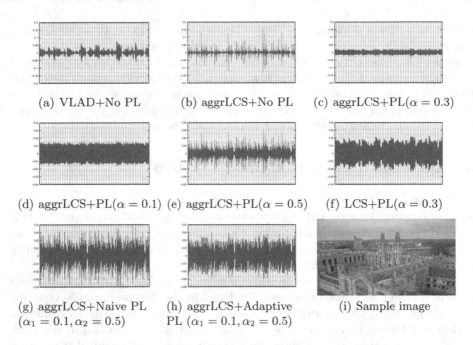

(a) VLAD+No PL   (b) aggrLCS+No PL   (c) aggrLCS+PL($\alpha = 0.3$)

(d) aggrLCS+PL($\alpha = 0.1$) (e) aggrLCS+PL($\alpha = 0.5$)   (f) LCS+PL($\alpha = 0.3$)

(g) aggrLCS+Naive PL   (h) aggrLCS+Adaptive   (i) Sample image
($\alpha_1 = 0.1, \alpha_2 = 0.5$)   PL ($\alpha_1 = 0.1, \alpha_2 = 0.5$)

Fig. 1. Comparison of different strategies of power-law normalization

Equation 5 mainly differs from Eq. 4 in two aspects. First, in the modified formulation, each rotation matrix Q'_j is offline trained with a fixed-size(n) subset of training images by performing PCA on the aggregated local residual vectors $AR_j = [ar_{j,1}, ar_{j,2}, ..., ar_{j,n}]^\mathsf{T}$, the $ar_{j,k}$ is

$$ar_{j,k} = \sum_{x:NN(x)=d_j, x \in X_i} \frac{x - d_j}{\|x - d_j\|_2}. \tag{6}$$

Besides, In Eq. 5, the rotation is performed posterior to accumulating the residuals, which means the rotation for each segment of representation vector occurs after accumulating residuals, in accordance with the subsequent power-law normalization, further enhanced the synergistic effect between them.

As illustrated in Fig. 1, the coordinate rotation works effectively along with power-law normalization. A representative image with "visual burstiness" is singled out and the representation vectors obtained with diverse methods are graphed, thereinto VLAD indicates the traditional VLAD method and the proposed coordinate rotation method in this paper is referred to as aggrLCS. Through comparison and analysis, we can come to the following two conclusions:

First, Fig. 1(a) and (b) show that the coordinate rotation algorithm based on local aggregated vectors cannot be applied independently. As a matter of fact, the fundamental purpose of rotation of local descriptors is to capture the principle directions while acculating a large variety of similar residual vectors, and rotate coordinate system in accord to these principle directions afterwards. The rotation magnifies the impact of "visual burstiness" instead of suppressing it, which brings convinience to the subsequent power-law normalization.

Then, Coordinate rotation is a beneficial processing of local descriptors for restraining burstiness in conjunction with power-law normalization. Although the application of coordinate rotation temporarily blows up the "visual bursti- ness" phenomenon, as shown in Fig. 1(b), while Fig. 1(c) and (e) show that the tunning of power-law normalization parameter α can ultimately provide a final representation with overall balance.

4   Adaptive Power-Law Normalization

Power-law normalization works on a simple principle, when $\alpha < 1$, the normaliza- tion transforms the vector elements by introducing a non-linearity, which adjusts the overall absolute values of vector elements more evenly, and further to some extent, alliviates the similarity measurement problem caused by "visual bursti- ness" in representation vectors component-wise. However, the design concept of power-law is not complicated, and has certain limitations in real world appli- cations. Based on the traditional power-law normalization process, this paper makes some extension hypotheses and proposes a novel adaptive power-law nor- malization method. This paper suspects that there are two major aspects of limitations of traditional power-law normalization, and would like to make some extensional assumptions to address the problem in this section.

First, the traditional power-law normalization is applied to the overall repre- sentation vector with non-linear transformation, without fully taking advantage of the locality of burstiness. Aiming at this thoughtless, this paper makes an extensional assumption, that the positive impact of the power-law normalization can be magnified by applying it to each visual word segment of the representa- tion vector, rather than the overall one, and customizing different adjustment policy on each segment respectively.

In the next place, there is only one parameter in traditional power-law nor- malization, which limits the flexibility of its application. This paper observes that the absolute values of the representation vector components vary greatly, and considers bringing in additional adjustable parameters. As can be inferred from the property of power-law normalization, the closer the value of α is to 0, the stronger effect power-law normalization has on balancing the components of representation vectors, and vice versa. Performing segment-wise power-law nor- malization to respectively with different parameters, i.e. applying minor para- meter α_1 to vector elements with greater absolute values, and a major α_2 to others, will maglify the detraction of visual burstiness.

Based on the above assumptions, this paper proposes a naive 2-phase power- law normalization in the first place, which is applied subsequently after perform- ing rotation to the coordinate systems and obtained the subvector $vsub_j$:

$$vsub_j^i = \begin{cases} \left| vsub_j^i \right|^{\alpha_1} \times sign\left(vsub_j^i\right), & i \in \ top\ 10\,\%\ of\ \mathbf{RankInd}_j, \\ \left| vsub_j^i \right|^{\alpha_2} \times sign\left(vsub_j^i\right), & i \in last\ 90\,\%\ of\ \mathbf{RankInd}_j, \end{cases} \quad (7)$$
$$s.t.\ \ 0 < \alpha_1 < \alpha_2 < 1.$$

Thereinto, $\mathbf{RankInd}_j$ denotes the indices of elements in $\mathbf{vsub}_j$ according to the descending order of absolute values of the elements. Equation 7 achieves 2-phase power-law normalization to $\mathbf{vsub}_j$: for the top 10 % elements with greater absolute values, a smaller adjustment parameter α_1 is utilized to enhance the restraining of power-law normalization to visual burstiness, a bigger parameter α_2 is meanwhile used to descrese tensile strength for the last 90 %, thus, balancing the absolute values of elements in adjusted representation vectors as a whole.

The choice of segmented percentage mentioned in Eq. 7 is heuristical parameters for spotting the segmentation position. Its universal effectiveness on other datasets cannot be guaranteed. On the basis of this, this paper further proposes an adaptive 2-phase power-law normalization algorithm, the improved algorithm segments the representation vector by the particularities of subvectors.

$$vsub_j^i = \begin{cases} \left| vsub_j^i \right|^{\alpha_1} \times sign\left(vsub_j^i\right), & rank(i,j) \leq \mathbf{segPos}_j, \\ \left| vsub_j^i \right|^{\alpha_2} \times sign\left(vsub_j^i\right), & rank(i,j) > \mathbf{segPos}_j, \end{cases} \quad (8)$$
$$s.t.\ \ 0 < \alpha_1 < \alpha_2 < 1.$$

$rank(i, j)$ denotes the rank of the ith vector element in $\mathbf{RankInd}_j$, $\mathbf{segPos}_j$ indicates the segmentation position on the jth subvector, which can be determined adaptively, it is the first index position that meet the following condition:

$$\frac{\left| vsub_j^{RankInd_j^{pos}} \right|}{\left| vsub_j^{RankInd_j^{pos-5}} \right|} \leq 0.8, \qquad s.t.\ pos > 5. \quad (9)$$

The adaptively determined segmentation position divides the rearranged vector elements into two sections: the first section with greater absolute values and large variety scope, which gathered peeks caused by burstiness and should be strongly suppressed, and the tensile strength should be decreased for the second part. There are some heuristical parameters for spotting the segmentation position, experiments have shown that these parameter are only related to the choice of image features, but nothing on the variety of datasets, and then would not influence the adaptivity of segmentation.

Since the rotation of coordinate systems maginifies visual burstiness in advance, it is a common phenomenon that there are a few (around 10 %) elements with large absolute values, and traditional power-law normalization will never deal effectively with these elements. As shown in Fig. 1(c) and (d), a big $\alpha(=0.5)$ leads to insufficient suppression of burstiness while a small one (0.1) destroy the original representation shape. Besides, the utilizing of naive 2-phase power-law normalization can greatly alliviate the inflexibility with the adjustment of only one parameter. In Fig. 1(e) the peeks are suppressed effectively and keep the distribution of original representation vector well at the same time. Furthermore, there are still some peeks that naive 2-phase power-law normalization fails to handle, the comparison of Fig. 1(e) and (f) shows the effectiveness of the proposed adaptive power-law normalization method.

5   Experiments and Discussion

The effectiveness of our proposed method has been verified on two standard and publicly availabe image retrival benchmarks along this line, which are the *Oxford Buildings* [19] and *Holidays* [20]. In this section, we begin with describing the datasets, with the following experiment settings, evaluation procedure, experiment results and discussion in subsequent order.

5.1   Benchmark Datasets

In this section, we will take a brief overview on the datasets.

Oxford Buildings.[1] The Oxford Buildings, also referred as *Oxford5k*, is widely used in content based image retrieval, which consists of 5,062 images of buildings downloaded from Flickr. Along with the images, there are 55 query images corresponding to 11 distinct buildings in Oxford, each query is specified by a rectangular region of interest.

Paris Buildings.[2] The Paris Buildings [21] dataset, often referred to as *Paris6k*, consists of 6,412 images. It is usually used for unsupervised learning of the parameters when evaluating the results on *Oxford5k* and its extension *Oxford105k*.

INRIA Holidays.[3] The INRIA Holidays(*Holidays*) contains 1,491 high resolution images from personal holiday photos of different locations and objects, 500 of which are used as queries. For large scale retrieval evaluation, another 1 million Flickr images (*Flickr1M* [20]) are taken into account.

Flickr 60k. Similar to the *Paris6k* dataset, *Flickr60k* is usually used to determine the parameters while evaluating *Holidays*. It is also provided by INRIA and contains another 60k images downloaded from Flickr.

5.2   Experiment Settings and Evaluation

Dataset Preprocessing. Since images in *Holidays* dataset and *Flickr60k* dataset are with high resolution, in order to avode negative impact on effectiveness for mass amount of local descriptors, we make proper resizing to images

---

[1] http://www.robots.ox.ac.uk/~vgg/data/oxbuildings/.
[2] http://www.robots.ox.ac.uk/~vgg/data/parisbuildings/.
[3] http://lear.inrialpes.fr/~jegou/data.php.

Table 1. Comparison of content based image retrieval methods (mAP performance)

Method			SIFT		RootSIFT	
			Oxford5k	Holidays	Oxford5k	Holidays
BoW [6]	20k	20k	35.4	43.7		
BoW [6]	200k	200k	-	54.0		
VLAD [5]	64	8192	37.2	52.0	38.7	54.3
Fisher [15]	64	8192	41.8	60.5		
VLAD+PL($\alpha = 0.5$)	64	8192	40.7	55.1	43.1	57.6
LCS [1]+PL($\alpha = 0.5$)	64	8192	45.2	59.8	46.3	61.4
AggrLCS+PL($\alpha = 0.5$)	64	8192	47.6	65.4	48.6	67.5
aggrLCS+Whitening	64	8192	48.2	69.1	48.9	70.5
aggrLCS+Naive PL ($\alpha_1=$ 0.1, $\alpha_2 = 0.5$)	64	8192	50.8	69.2	51.7	72.2
aggrLCS+Adaptive PL ($\alpha_1=$ 0.1, $\alpha_2 = 0.5$)	64	8192	50.8	70.0	**52.3**	**72.8**

in these two datasets and ensure there are no more than 786,432 pixels in each image.

Local Descriptors. Normalized RootSIFT [10] descriptors extracted with Hessian-Affine detectors are used as local descriptors of images, and as standardly done in the literature, in Sect. 5, we report both the results of normalized SIFT descriptors and RootSIFT descriptors.

Experiment Parameters. Parmeter tuning exists only in the adaptive power-law normalization, due to experimental experience, parameters are within limits that $\alpha_1 \in [0.1, 0.3]$ and $\alpha_2 \in [0.3, 0.8]$.

Comparison Methods. The comparison methods are the pairwise combination of aggregating methods for local descriptors, which includes BoW, VLAD, Fisher Vector and LCS, and normalization methods for image representation vectors, including L_2-normalization, power-law normalization and whitening [13].

Evaluation Protocol. Following the standard experimental scenario, retrieval performance is measured in terms of mean average precision(mAP) and the distance measurement is Euclidean distance. In particular, the query image itself is left out while calculating the mAP of specific query in *Holidays*.

5.3   Results and Discussion

Overall Performance. Table 1 reports the retrieval mAP on both *Oxford5k* and *Holidays* using SIFT and RootSIFT descriptors. The second and third column denotes the size of visual vocabulary and representation vector respectively. The proposed coordinate rotation method is referred as aggrLCS and Whitening denotes performing whitening process on the subvector of each visual word. Besides, the naive and adaptive 2-phase power-law normalization proposed in this paper are referred as `Naive PL` and `Adaptive PL` respectively.

As shown in Table 1, the proposed coordinate rotation method in conjunction with adoptive power-law normalization provides the best performance which is superior to all the others, thus it is reasonable to infer that the coordination of these two processes can better suppress the negative impact of visual burstiness on image representation. The effectiveness of `aggrLCS` is verified on both the *Oxford5k* and *Holidays* datasets, with 2 % relative performance gain on *Oxford5k* with aspect to LCS and 5 % to initial VLAD, and on *Holidays*, that is +6 % and +10 % respectively. As for the normalization methods, the whitening process and our proposed 2-phase power-law normalization show their superiority, furthermore, the `Adaptive PL` is the best of all.

Besides, the RootSIFT descriptor gives an improvement to SIFT descriptor under most of the local descriptors aggregating methods.

As a further analysis, Fig. 2 visualizes the statistics of standard deviation of each dimension of the final representation vectors produced by different methods on *Oxford5k*, from the point of energy, reflects the suppression of visual burstiness. RootSIFT descriptor is utilized and the size of vocabulary is fixed to 64 for a fair comparison. The `Adaptive PL` clearly shows its flexibility and effectiveness when it come to the preservation of details in the representation vectors and restrain the peeks of absolute values at the same time.

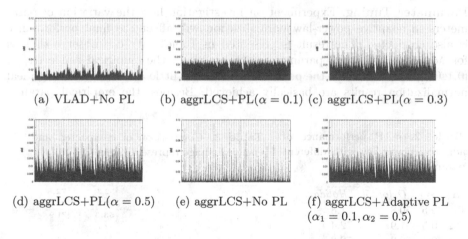

(a) VLAD+No PL   (b) aggrLCS+PL($\alpha = 0.1$) (c) aggrLCS+PL($\alpha = 0.3$)

(d) aggrLCS+PL($\alpha = 0.5$)   (e) aggrLCS+No PL   (f) aggrLCS+Adaptive PL
$(\alpha_1 = 0.1, \alpha_2 = 0.5)$

Fig. 2. Comparison of handling of visual burstiness by standard deviation statistics

Vocabulary Size. We then conduct experiments to show the comparison results when different sizes of vocabulary are used. As a normal practice, RootSIFT descriptor is utilized in all the methods and the adjustment parameter $\alpha = 0.5$ is chosed in traditional power-law normalization and in adaptive ones, we use $\alpha_1 = 0.1$ and $\alpha_2 = 0.5$.

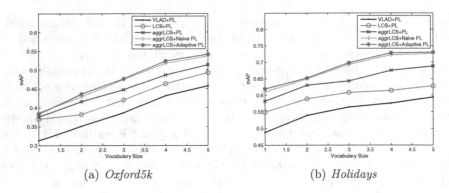

(a) *Oxford5k*   (b) *Holidays*

Fig. 3. Performance with different sizes of visual vocabulary

We can observe the following points from results in Fig. 3. First, the image representation ability boosts with the expansion of visual vocabulary, for all the involved methods. Besides, the trends of the performance present a consistancy, that is, methods utilizing coordinate rotation, such as LCS or aggrLCS, significantly outperform the initial VLAD method, and 2-phase power-law normalization methods shows their superiority. At last, the improvements are out of proportion to the size of vocabulary, indicating the infeasibility of using vocabularies with unlimited scales, which increases computing resource overhead.

Parameter Tuning. Experiment on investigating how the variation of parameters in adaptive power-law normalization will affect the final performance is also conducted. The result is presented in Table 2. We can observe that as for Adaptive PL, when parameters are tuning in the empirical range ($\alpha_1 \in [0.1, 0.3], \alpha_2 \in [0.3, 0.8]$), the performance changes little and comparatively ideal normalization results are basically achieved. Besides, the empirical parame-

Table 2. mAP performance of adaptive power-law with different parameter settings

α_1	α_2	*Oxford5k*	*Holidays*
0.3	0.5	50.8	71.8
0.2	0.5	51.9	72.4
0.1	0.5	52.3	72.8
0.1	0.3	52.2	72.0
0.1	0.4	**52.4**	**72.9**
0.1	0.6	51.8	72.6
0.1	0.7	51.4	72.4
0.1	0.8	50.7	71.6

Table 3. Comparison of compacted and encoded image representations

Method	Vector length	*Oxford5k*	*Holidays*
-	8192	52.3	**72.8**
PCA	1024	52.3	71.8
	512	**53.3**	71.9
	256	52.1	70.8
	128	49.5	68.1
	64	41.0	63.2
	32	32.3	58.0
PCA+PL($\alpha = 0.5$)	1024	42.5	59.2
	512	47.5	69.9
	256	49.3	70.1
	128	47.6	67.4
	64	38.8	62.3
	32	29.0	57.3

ter setting can be adapted to different benchmarks as we can observe with $\alpha_1 = 0.1, \alpha_2 = 0.4$. What's more, the impact of adjusting α_1 is more pronounced, due to its depression effect on the elements with greater absolute values in the subvector corresponding to each visual word.

Projected Representation. Last but not the least, we carried out experiments for applying dimensionality reduction to generated representation vectors. Conclusions can be drawn from Table 3. First of all, compression final representation vectors to reasonable size cause little loss to the performance, the mAP drops only 0.3 % on *Oxford5k* and 2 % on *Holidays* when the 8,192-D representation vectors are compressed to 256-D. Second, there is no distinctive relationship between the reduced dimensionality and the final performance. In addition, performing another power-law normalization after dimensionality reduction won't make any positive contribution to the retrieval performance.

6   Conclusions

This paper introduces two complementary techniques based on VLAD, which are the aggregated coordinate rotation algorithm and adaptive power-law normalization. In comparison with most existing local descriptors aggregating and vector normalization methods, the proposed methods take full advantage of the locality of visual burstiness by performing flexible transformation of the subvectors corresponding to each visual word, to better handle the visual burstiness and enhance the expression of final aggregated vectors. The sufficient experiments have shown the superiority of the proposed methods on standard benchmarks.

Acknowledgments. This research was supported by the National Natural Science Foundation of China (Grant No. 61271394 and 61571269). In the end, the authors would like to sincerely thank the reviewers for their valuable comments and advice.

References

1. Delhumeau, J., Gosselin, P.-H., Jgou, H., Prez, P.: Revisiting the VLAD image representation. In: ACM Multimedia, pp. 653–656 (2013)
2. Arandjelovic, R., Zisserman, A.: All about VLAD. In: CVPR, pp. 1578–1585 (2013)
3. Spyromitros-Xioufis, E., Papadopoulos, S., Ginsca, A.L., Popescu, A., Kompatsiaris, Y.: Improving diversity in image search via supervised relevance scoring. In: ICMR (2015)
4. Ng, J.Y.H., Yang, F., Davis, L.S.: Exploiting local features from deep networks for image retrieval. In: Proceedings of the IEEE Conference on Computer Vision and Pattern Recognition Workshops, pp. 53–61 (2015)
5. Jgou, H., Douze, M., Schmid, C., Prez, P.: Aggregating local descriptors into a compact image representation. In: CVPR, pp. 3304–3311 (2010)
6. Jgou, H., Perronnin, F., Douze, M., Sanchez, J., Perez, P., Schmid, C.: Aggregating local image descriptors into compact codes. In: PAMI, pp. 1704–1716 (2012)
7. Chatfield, K., Lempitsky, V.S., Vedaldi, A., Zisserman, A.: The devil is in the details: an evaluation of recent feature encoding methods. In: BMVC (2011)

8. Jgou, H., Douze, M., Schmid, C.: On the burstiness of visual elements. In: CVPR, pp. 1169–1176 (2009)
9. Sivic, J., Zisserman, A.: Video Google: a text retrieval approach to object matching in videos. In: ICCV, pp. 1470–1477 (2003)
10. Arandjelovi, R., Zisserman, A.: Three things everyone should know to improve object retrieval. In: CVPR, pp. 2911–2918 (2012)
11. Simonyan, K., Vedaldi, A., Zisserman, A.: Descriptor learning using convex optimisation. In: Fitzgibbon, A., Lazebnik, S., Perona, P., Sato, Y., Schmid, C. (eds.) ECCV 2012, Part I. LNCS, vol. 7572, pp. 243–256. Springer, Heidelberg (2012)
12. Yang, J., Yu, K., Gong, Y., Huang, T.: Linear spatial pyramid matching using sparse coding for image classification. In: CVPR, pp. 1794–1801 (2012)
13. Jégou, H., Chum, O.: Negative evidences and co-occurences in image retrieval: the benefit of PCA and whitening. In: Fitzgibbon, A., Lazebnik, S., Perona, P., Sato, Y., Schmid, C. (eds.) ECCV 2012, Part II. LNCS, vol. 7573, pp. 774–787. Springer, Heidelberg (2012)
14. Wang, J., Yang, J., Yu, K., Lv, F., Huang, T., Gong, Y.: Locality-constrained linear coding for image classification. In: CVPR, pp. 3360–3367 (2010)
15. Perronnin, F., Sánchez, J., Mensink, T.: Improving the Fisher Kernel for large-scale image classification. In: Daniilidis, K., Maragos, P., Paragios, N. (eds.) ECCV 2010, Part IV. LNCS, vol. 6314, pp. 143–156. Springer, Heidelberg (2010)
16. Perronnin, F., Liu, Y., Snchez, J., Poirier, H.: Large-scale image retrieval with compressed fisher vectors. In: CVPR, pp. 3384–3391 (2010)
17. Lowe, D.G.: Object recognition from local scale-invariant features. In: ICCV, pp. 1150–1157 (1999)
18. Bay, H., Tuytelaars, T., Van Gool, L.: SURF: Speeded Up Robust Features. In: Leonardis, A., Bischof, H., Pinz, A. (eds.) ECCV 2006, Part I. LNCS, vol. 3951, pp. 404–417. Springer, Heidelberg (2006)
19. Philbin, J., Chum, O., Isard, M., Sivic, J., Zisserman, A.: Object retrieval with large vocabularies and fast spatial matching. In: CVPR, pp. 1–8 (2007)
20. Jgou, H., Douze, M., Schmid, C.: Hamming embedding and weak geometry consistency for large scale image search. In: Proceedings of the 10th European Conference on Computer Vision: Part I, ECCV 2008, pp. 304-317 (2008)
21. Philbin, J., Chum, O., Isard, M., Sivic, J.: Lost in quantization: improving particular object retrieval in large scale image databases. In: CVPR (2008)

Reusing Extracted Knowledge in Genetic Programming to Solve Complex Texture Image Classification Problems

Muhammad Iqbal[✉], Bing Xue, and Mengjie Zhang

Evolutionary Computation Research Group,
School of Engineering and Computer Science, Victoria University of Wellington,
PO Box 600, Wellington 6140, New Zealand
{Muhammad.Iqbal,Bing.Xue,Mengjie.Zhang}@ecs.vuw.ac.nz

Abstract. Transfer learning is a process to transfer knowledge learned in one or more source tasks to a related but more complex, unseen target task, in an effort to facilitate learning in the target task. Genetic programming (GP) is an evolutionary approach to generating computer programs for solving a given problem automatically. Transfer learning in GP has been investigated in complex Boolean and symbolic regression problems, but not much in image classification. In this paper, we propose a novel approach to use transfer learning in GP for image classification problems. Specifically, the proposed novel approach extends an existing state-of-the-art GP method by incorporating the ability to extract useful knowledge from simpler problems of a domain and reuse the extracted knowledge to solve complex problems of the domain. The proposed system has been compared with the baseline system (i.e., GP without using transfer learning) on multi-class texture classification problems from three widely-used texture datasets with different rotations and different levels of noise. The experimental results showed that the ability to reuse the extracted knowledge in the proposed GP method helps achieve better classification accuracy than the baseline GP method.

Keywords: Genetic programming · Transfer learning · Building blocks · Code fragments · Texture image classification

1 Introduction

Human beings have the ability to apply the domain knowledge learned from a simpler problem to more complex problems of the same or a related domain, but currently the vast majority of evolutionary computation techniques lack this ability. This lack of ability to apply the already learned knowledge of a domain results in consuming more resources and time to effectively solve complex problems of the domain.

In order to mimic the human ability of extracting and reusing domain knowledge, transfer learning has been proposed and investigated recently [21]. In transfer learning, the knowledge learned in one or more source tasks is transferred to

© Springer International Publishing Switzerland 2016
J. Bailey et al. (Eds.): PAKDD 2016, Part II, LNAI 9652, pp. 117–129, 2016.
DOI: 10.1007/978-3-319-31750-2_10

a related but more complex, unseen target task, in an effort to facilitate learning in the target task. The reported results in literature showed that transfer learning has helped improve the ability of machine learning in various problem domains [7,10,15,17].

Genetic programming (GP) is an evolutionary approach to generating computer programs for solving a given problem automatically [13]. GP has been successfully applied to solve various classification, symbolic regression, and optimization problems [5]. Transfer learning in GP has been investigated in complex Boolean and symbolic regression problems [3,8,9,11], but not much in image classification.

This paper aims to extend an existing state-of-the-art GP method specifically designed for generating image descriptors to solve multi-class texture classification problems [1], by incorporating transfer learning. Texture classification is an essential task in computer vision that aims at grouping instances that have a similar repetitive pattern into one group. The main goal of this paper is to extract useful knowledge in learning simpler texture classification problems using GP and reuse the extracted knowledge to solve complex texture classification problems. Specifically, this study addresses the following objectives.

– How to identify and extract useful knowledge in GP?
– How to reuse the extracted knowledge in GP?
– Can the reuse of extracted knowledge improve classification accuracy in learning complex texture classification problems?

2   Related Work

GP is an evolutionary approach to generating computer programs for solving a given problem automatically [13]. In GP each individual is a computer program, commonly represented by a tree, that when executed generates the potential solution. For complex problems, the standard monolithic GP may not find a solution due to the large search space leading to an intractable problem. In layered learning, the complex target task is decomposed into subtasks and each subtask is learned in a bottom-up fashion [20]. Gustafson and Hsu [6] implemented layered learning in GP to learn the keep-away soccer game, which is a multi-agent system problem. The main task was decomposed into two subtasks and the final population in the bottom task layer was used as the initial population for the top task layer. The layered learning GP approach evolved better solutions faster than standard GP. Later on transfer learning in GP has been investigated to solve complex Boolean and symbolic regression problems [3,8,9,11].

Iqbal et al. [10] implemented a classifier system using GP tree-like code fragments in order to extract building blocks of knowledge from lower-dimensional problems and reuse them to learn higher-dimensional problems of the domain. The resulting system rapidly solved problems of a scale that existing learning classifier systems and GP techniques could not, e.g. the 135-bit multiplexer problem. Recently, Dinh et al. [4] implemented several methods to transfer a number of good GP individuals or sub-individuals from the source to the target problem.

The reported results show the advantages of using transfer learning in GP to solve symbolic regression problems.

Over the past few decades, GP has received more attention for image related problems [18]. GP has been used to perform texture classification using the raw pixel values directly in [19]. Recently, Al-Sahaf et al. [1] proposed a GP-based method, named GP-criptor, to automatically evolve descriptors for the task of multi-class texture classification. The reported results show that GP-criptor has significantly outperformed two GP-based and nine well-known non-GP methods on two widely-used texture datasets.

Although transfer learning in GP has shown improvements to solve Boolean and regression problems, there is not much work on transfer learning in GP for image classification [12]. In this work, we propose a novel approach to use transfer learning in GP to solve image classification problems.

3   The Proposed Method

In this paper, we extended the GP-criptor method by incorporating the ability of transfer learning to extract and reuse useful knowledge in learning various multi-class texture classification problems. In this section, GP-criptor is briefly introduced first. Then, the proposed method to extract and reuse the potentially useful knowledge, named TLGP-criptor, is explained in detail.

3.1   The Baseline Method

In GP-criptor [1], the content of the dataset is equally divided into a training set and a test set as shown in Fig. 1. During the training, GP-criptor randomly picks two instances from each class in the training set, to evolve a GP program and gener-

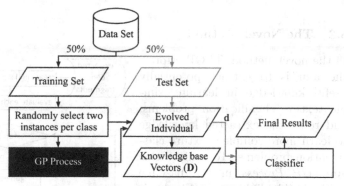

Fig. 1. Evaluation process for GP-criptor [1].

ates knowledge base vectors. During the testing, a feature vector for each instance in the test set generated using the evolved GP program is used by *the-Nearest-Neighbor* (1NN) classifier to predict the class label for the instance being evaluated.

GP-criptor is designed to operate directly on image raw pixel values. Therefore, the pixel values represent the content of the terminal set, as shown in Fig. 2.

The system uses a sliding window of a specific size, and at each position, the pixel values that fall within the window are used as inputs. The function

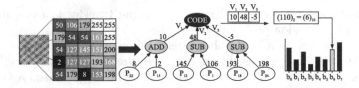

Fig. 2. The process of feature vector generation using GP-criptor [1].

set consists of the *add, sub, mul, div*, and *code* operators. Each of the first four operators has its regular mathematical meaning, i.e., takes two parameters and returns the resulting value after applying the corresponding operator on the inputs. The *div* operator is protected such that it returns zero if the denominator is zero. The *code* operator is a special operator that represents the root of the program tree. The *code* operator uses the input parameters to generate a binary code at each position of the sliding window. The number of children of the *code* operator specifies the length of the generated code.

The core aim of the program evolved by the GP-criptor method is to synthesize a set of equations that are used to convert an image to a feature vector via generating codes using the pixel values. The system scans the pixels of the instance being evaluated using a sliding window. At each pixel, the system computes values for each child of the *code* node. Then, the negative values of the *code* node children are set to 0, whilst positive and zero values are set to 1. The generated code is converted to decimal and the corresponding cell of the feature vector is incremented by 1.

3.2 The Novel Method

In the novel method, TLGP-criptor, the aim is to extract potentially useful knowledge in learning simple texture classification problems; and reuse the extracted knowledge to learn more complex texture classification problems. To achieve this aim, *GP Process* in the baseline method (highlighted in Fig. 1) is modified to enable transfer learning. The design of TLGP-criptor, like any transfer learning based method, presents three challenges. First, how to identify the potentially useful

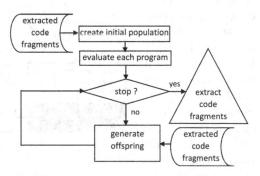

Fig. 3. The proposed approach of using transfer learning in GP.

knowledge to be extracted from simpler problems in a domain. Second, how to extract the identified useful knowledge. Third, how to reuse the extracted knowledge in learning more complex problems of the domain. The flowchart shown in Fig. 3 describes the overall proposed approach. In the following, it is explained in detail how TLGP-criptor manages to cope with these three design challenges.

Identification of Useful Knowledge: It is observed that in GP-criptor the number of attributes in a feature vector is proportional to the number of children of the root node in a GP program; and each child of the root node has an explicit role/importance in feature generation. Therefore, each child of the root node can be considered as a potentially useful chunk of information to be extracted and reused. We will use the term *code fragment* to denote a chunk of information.

Extraction of Useful Knowledge: In TLGP-criptor, children of the root nodes in GP programs in the final population having fitness values greater than or equal to the average fitness value of the whole population are extracted in learning simple problems in a domain.

Reusing the Extracted Knowledge: The extracted code fragments are used to generate initial GP populations as well as to mutate GP programs in learning complex problems of the domain. The reuse of extracted knowledge is explained below.

Program Generation: In GP-criptor, each child of the root node in a GP program in the initial population is a randomly generated subtree. However, in TLGP-criptor it is either a randomly generated subtree or an extracted code fragment with probability of 0.5. This is described in Algorithm 1. It is to be noted that the probability value of 0.5 is empirically chosen, which will be optimized in future.

Algorithm 1. TLGP-criptor_Initial_Program

Data: The number of children n of the root node in a (to be generated) GP
program p.
Result: A newly generated GP program p.
1 initialize GP program p
2 initialize root node's children $p.children$ with length n
3 **for** $i = 1$ *to* n **do**
4   **if** *RandomNumber[0, 1)* < 0.5 **then**
5 | $cf \leftarrow$ randomly select an extracted code fragment
6   **else**
7 | $cf \leftarrow$ randomly generate a new subtree
8   **end**
9     $p.children[i] \leftarrow cf$
10 **end**
11 **return** p

New Mutation Operator: In GP-criptor, the mutation operation randomly selects a subtree in a GP program and replaces it with a randomly generated new subtree. However, in TLGP-criptor, the mutation operation uses the extracted code fragments as described in Algorithm 2. Here, again, the probability value of 0.5 on the second line in Algorithm 2 is empirically chosen, which can be optimized in future.

Algorithm 2. TLGP-criptor_Mutation_Operation

Data: A GP program p to be mutated.
Result: The mutated GP program p.
1 $n \leftarrow$ number of children of the root node in p
2 **if** *RandomNumber[0, 1)* < 0.5 **then**
3      $\quad i \leftarrow$ RandomNumber[1, n]
4      $\quad cf \leftarrow$ randomly select an extracted code fragment
5      $\quad p.children[i] \leftarrow cf$
6 **else**
7      $\quad p \leftarrow$ GP-criptor_Mutation_Operation(p)
8 **end**
9 **return** p

---

4   Experiment Design

4.1   Data Sets

The performance of the proposed method is assessed by using three widely used texture benchmarks in computer vision, i.e., Kylberg [14], Brodatz [2], and Outex [16].

Each instance in the Kylberg dataset is a gray-scale image with size 576×576 pixels that is resized to be 115×115 pixels in order to reduce the computation cost. The instances of the Kylberg texture dataset are split into 28 classes and come in two variations: (1) without rotation; and (2) with rotation. The former group consists of 160 instances in each class. In the rotated version, each image is rotated at successive $30°$ angles, so in this group each class is made up of 1920 instances.

Originally, the Brodatz dataset is made up of 112 classes that each consists of a single gray-scale instance of size 640×640 pixels. In our experiment, only 20 randomly selected classes have been used and the single instance of each class has been re-sampled to 84 sub-images each of size 64×64 pixels. In the rotated version, each image is rotated at successive $30°$ angles, so in this group each class is made up of 1008 instances. The un-rotated Outex dataset, OutexTC00000, consists of 24 classes, each class having 20 gray-scale images with size 128×128 pixels. In the rotated version, OutexTC00010, each image is rotated at 5, 10, 15, 30, 45, 60, and $90°$ angles, so in this group each class is made up of 180 instances.

To investigate the effectiveness of transfer learning in noisy problems, the additive white Gaussian noise was added to each original dataset without rotation. The standard deviation, σ, for the Gaussian distribution was gradually increased from 10^{-4} to 10^1 in 20 intervals with linearly spaced exponents.

4.2   Parameter Settings

This paper uses similar parameter settings as used in the baseline method [1]. The population size is set equal to 100 individuals. The ramped half-and-half method is used to generate the initial GP population. The tournament selection strategy with a tournament of size seven is used to maintain the population diversity. The probabilities of applying crossover, mutation, and reproduction operators are respectively 0.80, 0.19, and 0.01. The tree depth of an evolved program is restricted to be between 2 and 10 levels. Finally, the evolutionary process stops when an ideal individual, i.e., fitness value is 0, is found, or the maximum number of generations, i.e., 30, is reached. In all experiments conducted in this work, the size of the sliding window is set equal to 3×3 pixels. The number of children of the *code* node is set equal to eight.

As only two instances per class of the training pool are randomly selected to form the training set, therefore, the typically used process of 30 independent runs for a stochastic method has been repeated 10 times using different instances in the training set. The mean along with standard deviation statistics on the test set have been calculated and reported.

5   Results

The proposed transfer learning based method, TLGP-criptor, is tested using three widely used texture datasets. Results are compared with the baseline method, GP-criptor.[1] The effectiveness of transfer learning in TLGP-criptor is measured using the mean classification accuracy in various rotated and noisy versions of the original datasets (without rotation and noise). To analyze the results, we applied unpaired t-test on the mean classification accuracy obtained in the test set to determine whether there was any statistically significant difference with a confidence level of 95 %, which is denoted by bold face in the results shown in this section.

In this section, we denote original Kylberg, Brodatz, and Outex datasets with KyWoRo, BrWoRo and OuWoRo; and the rotated versions with KyWiRo, BrWiRo, and OuWiRo, respectively. It is to be noted that the mean classification accuracy in learning the original datasets is the same for both GP-criptor and TLGP-criptor (i.e. 87.26 ± 0.48 in KyWoRo, 94.72 ± 0.56 in BrWoRo, and 94.97 ± 0.47 in OuWoRo) as there is no extracted knowledge to be used by TLGP-criptor. However, after learning the original datasets, potentially useful code fragments have been extracted that are to be used by TLGP-criptor to learn various rotated and noisy versions of the corresponding original dataset.

---

[1] Since GP-criptor is the state-of-the-art method and better than existing methods, we will not compare with other existing methods. This is not the main goal either.

5.1   Overall Results

Table 1 shows the mean classification accuracy obtained in learning
the whole rotated datasets using
GP-criptor and TLGP-criptor. It
is shown that transfer learning
has significantly improved the classification performance of TLGP-
criptor over GP-criptor in learning

Table 1. The accuracies (%) on the test sets
for the complete rotated datasets.

Dataset	GP-criptor	TLGP-criptor
KyWiRo_all	48.81 ± 2.09	**56.01 ± 1.63**
BrWiRo_all	55.27 ± 2.15	**60.73 ± 1.37**
OuWiRo_all	65.10 ± 1.39	**68.95 ± 1.02**

the rotated versions containing all the rotated images in each dataset.

To further investigate the effectiveness of transfer learning in TLGP-criptor,
we conducted three more experiments using the rotated versions of the datasets:
(1) standalone single-rotation, dataset_i, (2) un-rotated plus single-rotation,
dataset_000_i, and (3) un-rotated plus incremental rotations, dataset_upto_i;
where i denotes a rotation angle.

5.2   Kylberg Dataset with Single and Multiple Rotations

The left half of Table 2 shows the mean classification accuracy obtained in learning the rotated version of KyWiRo having only one rotation at a time. It is
observed that the classification performance of GP-criptor in learning the standalone single-rotation version is similar to the original un-rotated version. However, reusing the extracted code fragments from KyWoRo dataset, transfer learning has improved the classification performance of TLGP-criptor in learning the
standalone single-rotation version of KyWiRo. The right half of Table 2 shows
the mean classification accuracy obtained in learning the rotated version containing single-rotation plus the original un-rotated images. As this rotated version is
more complex than the standalone single-rotation version, the classification performance of GP-criptor decreased from ≈87 % to ≈63 %. In this rotated version,

Table 2. The accuracies (%) on the test sets for the single-rotation Kylberg dataset.

Dataset	GP-criptor	TLGP-criptor	Dataset	GP-criptor	TLGP-criptor
KyWiRo_030	87.19 ± 0.60	**88.92 ± 0.47**	KyWiRo_000_030	63.87 ± 1.50	**67.80 ± 1.08**
KyWiRo_060	87.99 ± 0.72	**89.22 ± 0.40**	KyWiRo_000_060	59.41 ± 1.96	**63.69 ± 1.15**
KyWiRo_090	87.73 ± 0.62	**89.10 ± 0.34**	KyWiRo_000_090	60.09 ± 1.23	**63.71 ± 0.97**
KyWiRo_120	87.51 ± 0.73	**89.13 ± 0.35**	KyWiRo_000_120	58.35 ± 1.51	**62.37 ± 1.02**
KyWiRo_150	87.88 ± 0.74	**89.23 ± 0.57**	KyWiRo_000_150	64.47 ± 1.55	**67.49 ± 1.01**
KyWiRo_180	87.93 ± 0.54	**89.24 ± 0.34**	KyWiRo_000_180	75.97 ± 0.86	**78.83 ± 0.85**
KyWiRo_210	87.51 ± 0.63	**88.90 ± 0.51**	KyWiRo_000_210	62.83 ± 1.42	**66.49 ± 0.88**
KyWiRo_240	87.81 ± 0.73	**89.27 ± 0.44**	KyWiRo_000_240	58.70 ± 1.51	**62.62 ± 0.99**
KyWiRo_270	87.78 ± 0.77	**89.06 ± 0.45**	KyWiRo_000_270	59.33 ± 1.77	**63.53 ± 1.00**
KyWiRo_300	87.47 ± 0.62	**89.28 ± 0.31**	KyWiRo_000_300	59.53 ± 1.41	**62.97 ± 1.15**
KyWiRo_330	87.91 ± 0.58	**89.29 ± 0.43**	KyWiRo_000_330	66.89 ± 1.49	**69.89 ± 0.84**

transfer learning again improves the classification performance of TLGP-criptor over the baseline GP-criptor method. As a rotation of an image at 180° has less effect on the image's texture than other rotations, it is observed that the classification accuracy in learning the dataset KyWiRo_000_180 is noticeably greater than other datasets shown in the right half of Table 2. The statistical analysis shows that the improvement is statistically significant.

Table 3 shows the mean classification accuracy obtained in learning the Kylberg dataset with incremental rotations. In this dataset, code fragments are extracted from KyWoRo to learn KyWiRo_upto_030, then code fragments are extracted from KyWiRo_upto_030 to learn KyWiRo_upto_060, and so on. The incremental reuse of extracted knowledge in TLGP-criptor has shown noticeable improvement

Table 3. The accuracies (%) on the test sets for the incrementally rotated Kylberg dataset.

Dataset	GP-criptor	TLGP-criptor
KyWiRo_upto_030	63.87 ± 1.50	**67.80 ± 1.08**
KyWiRo_upto_060	56.64 ± 1.44	**61.40 ± 0.85**
KyWiRo_upto_090	51.87 ± 1.89	**58.62 ± 1.11**
KyWiRo_upto_120	47.92 ± 1.37	**55.74 ± 0.87**
KyWiRo_upto_150	46.59 ± 1.69	**55.29 ± 0.55**
KyWiRo_upto_180	45.07 ± 1.88	**55.68 ± 1.20**
KyWiRo_upto_210	48.34 ± 1.75	**57.24 ± 0.80**
KyWiRo_upto_240	46.92 ± 2.15	**57.77 ± 0.82**
KyWiRo_upto_270	44.85 ± 1.86	**57.03 + 1.20**
KyWiRo_upto_300	47.07 ± 2.24	**57.70 ± 1.28**
KyWiRo_upto_330	48.81 ± 2.09	**62.76 ± 1.27**

in learning Kylberg rotated dataset. The performance of TLGP-criptor is improved from 56.01 ± 1.63 to 62.76 ± 1.27 in learning the whole rotated dataset. The statistical analysis shows that the improvement is extremely statistically significant.

5.3   Brodatz Dataset with Single and Multiple Rotations

The left half of Table 4 shows the mean classification accuracy obtained in learning the rotated version of BrWiRo having only one rotation at a time. It is observed that reusing the extracted code fragments from BrWoRo dataset, transfer learning has improved the classification performance of TLGP-criptor over GP-criptor in learning the standalone single-rotation versions of BrWiRo. The statistical analysis shows that this improvement is significant in all of the rotated versions. The right half of Table 4 shows the mean classification accuracy obtained in learning the rotated version containing single-rotation plus the original un-rotated images. As this rotated version is more complex than the standalone single-rotation version, the classification performance of GP-criptor decreased from ≈96 % to ≈78 %. It is observed that the classification accuracy in learning the dataset BrWiRo_000_180 is noticeably greater than other datasets shown in the right half of Table 4, as expected. In this rotated version, transfer learning has statistically significantly improved the classification performance of TLGP-criptor over the baseline GP-criptor method.

Table 5 shows the mean classification accuracy obtained in learning the rotated version of Brodatz dataset with incremental rotations. The incremental reuse of

Table 4. The accuracies (%) on the test sets for the single-rotation Brodatz dataset.

Dataset	GP-criptor	TLGP-criptor	Dataset	GP-criptor	TLGP-criptor
BrWiRo_030	97.02 ± 0.35	**97.64 ± 0.27**	BrWiRo_000_030	81.39 ± 1.38	**83.75 ± 1.23**
BrWiRo_060	96.06 ± 0.39	**96.59 ± 0.30**	BrWiRo_000_060	74.44 ± 1.09	**76.74 ± 0.80**
BrWiRo_090	95.01 ± 0.42	**96.10 ± 0.33**	BrWiRo_000_090	76.30 ± 1.27	**78.91 ± 0.62**
BrWiRo_120	96.75 ± 0.33	**97.26 ± 0.23**	BrWiRo_000_120	72.17 ± 1.28	**74.43 ± 0.83**
BrWiRo_150	97.05 ± 0.37	**97.57 ± 0.28**	BrWiRo_000_150	78.62 ± 1.46	**81.12 ± 0.85**
BrWiRo_180	94.62 ± 0.48	**95.48 ± 0.35**	BrWiRo_000_180	90.47 ± 0.70	**92.24 ± 0.43**
BrWiRo_210	97.58 ± 0.29	**97.90 ± 0.27**	BrWiRo_000_210	77.87 ± 1.21	**80.86 ± 0.94**
BrWiRo_240	97.39 ± 0.37	**97.78 ± 0.19**	BrWiRo_000_240	74.03 ± 1.28	**76.26 ± 0.86**
BrWiRo_270	94.25 ± 0.47	**95.30 ± 0.27**	BrWiRo_000_270	76.30 ± 0.91	**78.46 ± 0.85**
BrWiRo_300	96.49 ± 0.38	**97.05 ± 0.25**	BrWiRo_000_300	73.80 ± 1.12	**76.75 ± 0.85**
BrWiRo_330	96.88 ± 0.33	**97.30 ± 0.25**	BrWiRo_000_330	82.75 ± 1.24	**84.75 ± 1.07**

extracted knowledge in TLGP-criptor has shown noticeable improvement in learning Brodatz rotated dataset. The performance of TLGP-criptor is improved from 60.73 ± 1.37 to 74.14 ± 0.83 in learning the whole rotated dataset. The statistical analysis shows that the improvement is extremely statistically significant.

5.4   Outex Dataset with Single and Multiple Rotations

The left half of Table 6 shows the mean classification accuracy obtained in learning the rotated version of OuWiRo having only one rotation at a time. It is observed that reusing the extracted code fragments from OuWoRo, transfer learning has improved the classification performance of TLGP-criptor in learning the standalone single-rotation version of OuWiRo. The statistical analysis shows that this

Table 5. The accuracies (%) on the test sets for the incrementally rotated Brodatz dataset.

Dataset	GP-criptor	TLGP-criptor
BrWiRo_upto_030	81.39 ± 1.38	**83.75 ± 1.23**
BrWiRo_upto_060	71.98 ± 1.33	**77.87 ± 0.88**
BrWiRo_upto_090	64.19 ± 1.93	**73.78 ± 1.40**
BrWiRo_upto_120	59.30 ± 2.46	**72.37 ± 0.84**
BrWiRo_upto_150	58.50 ± 1.99	**72.61 ± 1.01**
BrWiRo_upto_180	58.10 ± 2.14	**73.22 ± 1.24**
BrWiRo_upto_210	57.91 ± 1.85	**73.65 ± 1.41**
BrWiRo_upto_240	56.18 ± 1.61	**73.37 ± 0.91**
BrWiRo_upto_270	54.93 ± 2.35	**74.77 ± 0.82**
BrWiRo_upto_300	54.20 ± 2.44	**72.53 ± 1.18**
BrWiRo_upto_330	55.27 ± 2.15	**74.14 ± 0.83**

improvement is significant in all of the rotated versions, except OuWiRo_10 and OuWiRo_15. The right half of Table 6 shows the mean classification accuracy obtained in learning the rotated version containing single-rotation plus the original un-rotated images. It is observed that in this dataset the classification accuracy of GP-criptor noticeably decreased when the rotation was greater or equal to 15°. In this rotated version, transfer learning has statistically

Table 6. The accuracies (%) on the test sets for the single-rotation Outex dataset.

Dataset	GP-criptor	TLGP-criptor	Dataset	GP-criptor	TLGP-criptor
OuWiRo_05	94.97 ± 0.48	**95.52 ± 0.24**	OuWiRo_00_05	94.59 ± 0.44	**95.51 ± 0.31**
OuWiRo_10	95.75 ± 0.43	95.94 ± 0.37	OuWiRo_00_10	92.37 ± 0.53	**93.70 ± 0.42**
OuWiRo_15	95.83 ± 0.43	95.84 ± 0.29	OuWiRo_00_15	88.39 ± 0.94	**89.93 ± 0.64**
OuWiRo_30	96.19 ± 0.57	**96.79 ± 0.31**	OuWiRo_00_30	77.86 ± 1.17	**80.05 ± 0.60**
OuWiRo_45	96.24 ± 0.55	**97.05 ± 0.41**	OuWiRo_00_45	73.78 ± 1.18	**75.77 ± 0.82**
OuWiRo_60	95.05 ± 0.46	**95.46 ± 0.38**	OuWiRo_00_60	72.07 ± 0.99	**74.03 ± 0.83**
OuWiRo_75	95.16 ± 0.38	**95.50 ± 0.37**	OuWiRo_00_75	71.20 ± 1.04	**73.45 ± 0.60**
OuWiRo_90	95.07 ± 0.42	**95.41 ± 0.41**	OuWiRo_00_90	72.37 ± 1.21	**75.24 ± 0.61**

significantly improved the classification performance of TLGP-criptor over the baseline GP-criptor method.

Table 7 shows the mean classification accuracy obtained in learning the rotated version of the Outex dataset with incremental rotations. The incremental reuse of extracted knowledge in TLGP-criptor has shown noticeable improvement in learning the Outex rotated dataset. The performance of TLGP-criptor is improved from 68.95 ± 1.02

Table 7. The accuracies (%) on the test sets for the incrementally rotated Outex dataset.

Dataset	GP-criptor	TLGP-criptor
OuWiRo_upto_05	94.59 ± 0.44	**95.51 ± 0.31**
OuWiRo_upto_10	94.02 ± 0.42	**95.27 ± 0.29**
OuWiRo_upto_15	92.89 ± 0.66	**94.56 ± 0.31**
OuWiRo_upto_30	85.21 ± 0.98	**88.39 ± 0.57**
OuWiRo_upto_45	80.03 ± 0.97	**84.64 ± 0.61**
OuWiRo_upto_60	75.52 ± 1.29	**80.29 ± 0.63**
OuWiRo_upto_75	70.37 ± 1.63	**76.04 ± 0.63**
OuWiRo_upto_90	65.10 ± 1.39	**72.16 ± 0.71**

to 72.16 ± 0.71 in learning the whole rotated dataset. The statistical analysis shows that the improvement is extremely statistically significant.

5.5 Kylberg, Brodatz, and Outex Datasets with Noise

Table 8 shows the mean classification accuracy obtained in learning the unrotated datasets with various noise levels. In the first 10 noise levels, GP-criptor showed robustness to noise and maintained the classification performance to a non-noisy level. After that its performance gradually decreased, as expected. At the last five noise levels, the performance is noticeably decreased, as the severe noise destroyed the textures. Similar to rotated versions, TLGP-criptor showed improvement in classification accuracy in noisy datasets. The statistical analysis shows that the improvement is statistically significant for small amount of noise.

In summary, the proposed technique of transfer learning in GP has shown significant improvement in learning all rotated versions of three texture datasets used in this work. Further, for small amount of noise transfer learning has also shown improvement. It is observed that for severe noise transfer learning does not help.

Table 8. The accuracies (%) on the test sets for the un-rotated noisy datasets.

Noise level	KyWoRo		BrWoRo		OuWoRo	
	GP-criptor	TLGP-criptor	GP-criptor	TLGP-criptor	GP-criptor	TLGP-criptor
0.0001	87.26 ± 0.48	**88.41 ± 0.35**	94.72 ± 0.56	**95.76 ± 0.29**	94.97 ± 0.47	**95.43 ± 0.34**
0.0002	87.26 ± 0.48	**88.41 ± 0.35**	94.72 ± 0.56	**95.76 ± 0.29**	94.97 ± 0.47	**95.43 ± 0.34**
0.0003	87.26 ± 0.48	**88.41 ± 0.35**	94.72 ± 0.56	**95.76 ± 0.29**	94.97 ± 0.47	**95.43 ± 0.34**
0.0006	87.20 ± 0.74	**88.48 ± 0.32**	94.78 ± 0.55	**95.73 ± 0.32**	95.15 ± 0.39	**95.46 ± 0.32**
0.0011	87.18 ± 0.62	**88.38 ± 0.46**	94.64 ± 0.46	**95.92 ± 0.29**	95.19 ± 0.43	**95.45 ± 0.41**
0.0021	87.18 ± 0.54	**88.40 ± 0.44**	94.85 ± 0.55	**95.84 ± 0.37**	95.34 ± 0.35	**95.65 ± 0.29**
0.0038	87.27 ± 0.61	**88.52 ± 0.54**	94.79 ± 0.46	**95.87 ± 0.33**	95.18 ± 0.41	**95.53 ± 0.33**
0.0070	87.48 ± 0.54	**88.52 ± 0.41**	94.89 ± 0.61	**95.98 ± 0.24**	94.74 ± 0.40	**95.25 ± 0.35**
0.0127	87.46 ± 0.54	**88.70 ± 0.41**	94.94 ± 0.35	**95.80 ± 0.28**	93.64 ± 0.47	**94.17 ± 0.41**
0.0234	87.35 ± 0.66	**88.44 ± 0.37**	94.59 ± 0.47	**95.52 ± 0.28**	91.44 ± 0.49	**92.40 ± 0.44**
0.0428	86.92 ± 0.49	**87.94 ± 0.51**	93.30 ± 0.62	**94.31 ± 0.43**	85.96 ± 0.84	**86.84 ± 0.58**
0.0785	83.15 ± 0.77	**84.49 ± 0.61**	91.25 ± 0.64	**92.11 ± 0.48**	**68.04 ± 1.55**	67.08 ± 1.27
0.1438	71.68 ± 1.35	**73.50 ± 0.86**	86.07 ± 1.02	**86.64 ± 0.62**	**35.52 ± 2.02**	35.02 ± 1.13
0.2637	43.04 ± 1.70	**44.70 ± 1.22**	67.66 ± 1.05	**69.35 ± 0.77**	17.18 ± 0.71	17.49 ± 0.60
0.4833	11.74 ± 0.87	**12.75 ± 0.52**	44.94 ± 0.81	**46.62 ± 0.99**	14.66 ± 0.38	14.71 ± 0.41
0.8859	5.68 ± 0.19	5.77 ± 0.19	30.42 ± 0.94	**30.88 ± 0.77**	14.47 ± 0.45	14.56 ± 0.51
1.6238	4.81 ± 0.12	4.87 ± 0.13	19.89 ± 0.59	19.70 ± 0.56	14.41 ± 0.49	14.52 ± 0.44
2.9764	4.79 ± 0.13	4.77 ± 0.11	14.32 ± 0.61	14.32 ± 0.59	14.39 ± 0.52	14.44 ± 0.49
5.4556	4.67 ± 0.13	4.74 ± 0.15	**10.90 ± 0.44**	10.47 ± 0.50	14.35 ± 0.59	14.43 ± 0.51
10.0000	4.75 ± 0.13	4.78 ± 0.14	8.45 ± 0.46	8.28 ± 0.39	14.30 ± 0.37	14.37 ± 0.51

6 Conclusions

The goal of this paper was to investigate the use of transfer learning in GP to solve image classification problems. The said goal is successfully achieved by reusing the extracted knowledge from simpler problems to learn various complex, rotated and noisy problems of three widely-used texture image datasets. Specifically, incremental extraction and reuse of knowledge in learning rotated versions showed noticeable improvement. It is observed that if the noise level is severe then transfer of knowledge has no benefit in all the experiments conducted in this work.

Future work includes the investigation on the sensitivity of newly introduced parameters in the proposed approach. In future, TLGP-criptor will be used in learning two or more datasets simultaneously so that domain independent code fragments can be extracted. The domain independent code fragments will be used to learn other related problem domains.

References

1. Al-Sahaf, H., Zhang, M., Johnston, M., Verma, B.: Image descriptor: a genetic programming approach to multiclass texture classification. In: Proceedings of 2015 IEEE Congress on Evolutionary Computation, pp. 2460–2467. IEEE (2015)
2. Brodatz, P.: Textures: A Photographic Album for Artists and Designers. Dover Publications, New York (1999)

3. Chen, Q., Xue, B., Zhang, M.: Generalisation and domain adaptation in GP with gradient descent for symbolic regression. In: Proceedings of the IEEE Congress on Evolutionary Computation, pp. 1137–1144 (2015)
4. Dinh, T.T.H., Chu, T.H., Nguyen, Q.U.: Transfer learning in genetic programming. In: Proceedings of the IEEE Congress on Evolutionary Computation, pp. 1145–1151 (2015)
5. Gandomi, A.H., Alavi, A.H., Ryan, C.: Handbook of Genetic Programming Applications. Springer, Heidelberg (2015)
6. Gustafson, S.M., Hsu, W.H.: Layered learning in genetic programming for a cooperative Robot Soccer problem. In: Proceedings of the European Conference on Genetic Programming, pp. 291–301 (2001)
7. Gutstein, S., Fuentes, O., Freudenthal, E.: Knowledge transfer in deep convolutional neural nets. Int. J. Artif. Intell. Tools 17(3), 555–567 (2008)
8. Hien, N.T., Hoai, N.X., McKay, B.: A study on genetic programming with layered learning and incremental sampling. In: Proceedings of the IEEE Congress on Evolutionary Computation, pp. 1179–1185 (2011)
9. Hoang, T.H., McKay, R.I.B., Essam, D., Hoai, N.X.: On synergistic interactions between evolution, development and layered learning. IEEE Trans. Evol. Comput. 15(3), 287–312 (2011)
10. Iqbal, M., Browne, W.N., Zhang, M.: Reusing building blocks of extracted knowledge to solve complex, large-scale boolean problems. IEEE Trans. Evol. Comput. 18(4), 465–480 (2014)
11. Jackson, D., Gibbons, A.P.: Layered learning in boolean GP problems. In: Proceedings of the European Conference on Genetic Programming, pp. 148–159 (2007)
12. Jaśkowski, W., Krawiec, K., Wieloch, B.: Cross-task code reuse in genetic programming applied to visual learning. Int. J. Appl. Math. Comput. Sci. 24(1), 183–197 (2014)
13. Koza, J.R.: Genetic Programming: On the Programming of Computers by Means of Natural Selection. The MIT Press, Cambridge (1992)
14. Kylberg, G.: The Kylberg texture dataset v. 1.0. External report (Blue series) 35, Centre for Image Analysis, Swedish University of Agricultural Sciences and Uppsala University, Uppsala, Sweden (2011)
15. Niculescu-Mizil, A., Caruana, R.: Inductive transfer for Bayesian network structure learning. J. Mach. Learn. Res. 27, 167–181 (2012)
16. Ojala, T., Mäenpää, T., Pietikäinen, M., Viertola, J., Kyllonen, J., Huovinen, S.: Outex - new framework for empirical evaluation of texture analysis algorithms. In: Proceedings of the International Conference on Pattern Recognition, pp. 701–706 (2002)
17. Pan, S.J., Yang, Q.: A survey on transfer learning. IEEE Trans. Knowl. Data Eng. 22(10), 1345–1359 (2010)
18. Perez, C.B., Olague, G.: Evolutionary learning of local descriptor operators for object recognition. In: Proceedings of the 11th Annual Conference on Genetic and Evolutionary Computation, pp. 1051–1058. ACM (2009)
19. Song, A., Loveard, T., Ciesielski, V.: Towards genetic programming for texture classification. In: Proceedings of the Australian Joint Conference on Artificial Intelligence, pp. 461–472 (2001)
20. Stone, P., Veloso, M.: Layered learning. In: Proceedings of the European Conference on Machine Learning, pp. 369–381 (2000)
21. Torrey, L., Shavlik, J.: Transfer learning. In: Handbook of Research on Machine Learning Applications and Trends: Algorithms, Methods, and Techniques, chap. 11, pp. 242–264. IGI Global (2010)

Personal Credit Profiling via Latent User Behavior Dimensions on Social Media

Guangming Guo[1,2], Feida Zhu[2], Enhong Chen[1(✉)], Le Wu[1],
Qi Liu[1], Yingling Liu[1], and Minghui Qiu[2]

[1] School of Computer Science and Technology,
University of Science and Technology of China, Hefei 230027, China
[2] School of Information Systems,
Singapore Management University, Singapore, Singapore
guogg@mail.ustc.edu.cn

Abstract. Consumer credit scoring and credit risk management have been the core research problem in financial industry for decades. In this paper, we target at inferring this particular user attribute called credit, i.e., whether a user is of the good credit class or not, from online social data. However, existing credit scoring methods, mainly relying on financial data, face severe challenges when tackling the heterogeneous social data. Moreover, social data only contains extremely weak signals about users' credit label. To that end, we put forward a Latent User Behavior Dimension based Credit Model (LUBD-CM) to capture these small signals for personal credit profiling. LUBD-CM learns users' hidden behavior habits and topic distributions simultaneously, and represents each user at a much finer granularity. Specifically, we take a real-world Sina Weibo dataset as the testbed for personal credit profiling evaluation. Experiments conducted on the dataset demonstrate the effectiveness of our approach: (1) User credit label can be predicted using LUBD-CM with a considerable performance improvement over state-of-the-art baselines; (2) The latent behavior dimensions have very good interpretability in personal credit profiling.

1 Introduction

Accurate assessment of consumers' credit risk has a profound impact on P2P lending's success. Traditional consumer credit scoring literatures have proposed various statistical methods for credit risk management [4]. Advanced methods using data mining approach [24] and machine learning approach [14] have also been proposed in recent years. Mostly, the employed consumer data for credit analysis in these studies is composed of historical loan/payment records, credit reports or demographic information like salary and education. However, according to American Consumer Financial Protection Bureau[1], almost one in ten American consumers has no credit history until 2015, not to mention other less developed countries. Even for users with credit history, online P2P lending

---

[1] http://files.consumerfinance.gov/f/201505_cfpb_data-point-credit-invisibles.pdf.

© Springer International Publishing Switzerland 2016
J. Bailey et al. (Eds.): PAKDD 2016, Part II, LNAI 9652, pp. 130–142, 2016.
DOI: 10.1007/978-3-319-31750-2_11

companies can't access their financial transaction data freely, which is usually dispersed among various institutions and companies. To make it worse, demographic survey data usually costs a lot to collect and validate, which cannot be afforded by these small loan companies.

In the era of social media, the situation is changing. The ever-growing online micro-blogging services have become indispensible for our everyday lives. Most of us rely on social media to share, communicate, discover and network [12]. Meanwhile, tons of User Generated Content (UGC), such as status updates, retweets, replies etc., becomes available on social media. The practice of harnessing this personal UGC on social media for credit profiling, becomes more and more prevalent with the blossoming of online Internet finance startups like Kabbage[2] and ZestFinance[3]. For individuals applying for small loans, the online social data provides great opportunities to investigate their credit risks with unprecedented data scale, coverage, granularity and nearly no cost while preserving their privacy.

However, social data, especially the tweet data, is inherently heterogeneous, dynamic and even noisy. For instance, users on social media frequently invent new words to express their feelings and thoughts. Different from financial data or survey data, tweets are usually informal and fragmented since they are limited to be 140-character-long and diverse in topics [15]. What's more, user credit is a particularly private attribute, even more sensitive than age or gender in most cases. Users seldom generate credit related personal data on the social web. Consequently, social data only contains extremely weak signals about user credit risk. Primary experiment results show that the best prediction accuracy we can achieve is only 57.2 % with thousands of manually defined social features as input. All the above facts pose great challenges for us to leverage the social data for personal credit profiling, i.e., assessing one's credit risk into classes of "good" or "bad" [9] from social data. To our surprise, we find that some kinds of behaviors extracted from tweets, such as posting time of tweets, is informative for credit profiling (Cf.Sect. 2 for details). Unlike tweet content, behavior data is usually precise and formal, and reflects users' behavior habits and characters more comprehensively and directly. This observation is a good example of the old view that characters or habits are also key factors affecting people's credit risk. As far as we know, existing user profiling techniques only treat behavior data as an additional feature source [22], couldn't extract users' habits and characters from it for credit profiling very well.

To achieve the goal of personal credit profiling on social media, we propose the Latent User Behavior Dimension based Credit Model (LUBD-CM), which explicitly models users' behavior data and text data at the same time. Using LUBD-CM, we are able to capture hidden behavior dimensions of users at a much finer granularity, which are especially effective in capturing their habits and characters. Then the credit profiling task can be done using standard $l2$-regularized logistic regression classifier whose input is the latent user behavior dimensions. After the

---

[2] https://www.kabbage.com/.

[3] http://www.zestfinance.com.

classifier learning phase, we can distinguish behavior dimensions that are informative for credit prediction from the classifier easily. By comparing with several state-of-the-art algorithms, we show that LUBD-CM has a much better predictive performance in terms of averaged accuracy, precision, recall and F1-Score, the most common measures in credit scoring. Besides, case studies show that the learnt latent behavior dimensions have excellent abilities in explaining users' credit label, which is very necessary in the practice of credit profiling.

The main contributions of this paper are as follows:

- Our work aims to infer the especially subtle and subjective user attribute – credit. We find that some types of user behaviors on social media are very informative for credit evaluation. To the best of our knowledge, we are the first to formally investigate personal credit profiling problem under the social media data setting.
- We propose a latent variable model LUBD-CM to incorporate as many as 5 different types of user behaviors with text data. In this way, we are able to capture latent user behavior dimensions from the social data at a much finer granularity.
- We conduct comprehensive experiments on a dataset crawled from Sina Weibo[4]. Experimental results demonstrate that LUBD-CM outperforms several state-of-the-art baselines and the learnt latent behavior dimensions are very interpretable for personal credit profiling.

The rest of the paper is organized as follows. In Sect. 2, we introduce the preliminaries and definition of personal credit profiling problem. In Sect. 3, we discuss our approach's framework and present the LUBD-CM model in detail. We present experimental results in Sect. 4. Finally, we review the related work in Sect. 5 and conclude the paper in Sect. 6.

2   Preliminaries and Problem Definition

Dataset Collection and Description. On twitter-style websites like Sina Weibo, one's online tweets are publicly available by nature. Generally speaking, anyone can access others' tweet data even if she is not a friend of the given user. Sina Weibo allows us to access and store one's all tweet data after we are granted with privileges by the given user. With the help of an online P2P lending partner, we obtain more than 200,000 users' Sina Weibo data, whose credit labels are known through the partner's internal data. These users have received at least one credit loan from the P2P lending company. Usually, the credit label is defined by whether the user has defaulted on any loan or not. That is, if the user defaulted on any loan transactions before, he or she is labeled as "bad credit"; if the user has never defaulted, he or she is labeled as "good credit". All users in the Sina Weibo testbed have authorized the company to collect their tweet data, which is a common prerequisite to make loans from the P2P lending companies. As a result, there are no privacy breaches or moral issues to study these users' credit risk based on Sina Weibo data (Table. 1).

---

[4] http://www.weibo.com, the most famous tweet-style platform in China.

Table 1. Some statistics of Sina Weibo Dataset for Credit Profiling.

(a) Some Statistics of the dataset

Description	Value
# of good credit users	3,000
# of bad credit users	3,000
Total size of tweets	904,013
Total number of words	12,301,485
Size of vocabularies	241,197

(b) Summary of behavior types

Behavior Types	# of Possible Values
Retweet or not	2
Posting time(Hours)	24
Posting time(Days)	7
Posting tools	4012
# of emoticons	65

Adequate tweet data is crucial for algorithms' performance, so we set the minimum number of tweets for each user to be 10. Only users with no less than 10 tweets are chosen as experiment data. After removing users with less than 10 tweets, only 3,119 bad credit users are left. Therefore we randomly sample 3,000 good and 3,000 bad credit users from the filtered dataset to construct a balanced dataset for measuring the overall performance of LUBD-CM. For vocabularies, we remove stop words and infrequent words whose document frequency is less than 5. In Table 2(a), we summarize the main statistics of Sina Weibo Dataset. We consider as many as 5 different behavior types, including (1) *whether the tweet is retweeted or not*, (2) *hours of the day when the tweet is posted*, (3) *days of the week when the tweet is posted*, (4) *type of tools used to post the tweet*, (5) *number of emoticons*[5] *in the tweet.* In regards to the 5 behavior types, the number of their possible values ranges from 2 to as large as 4012, and details are listed in Table 2(b). Mostly, the distribution of each behavior type is different from each other, but all somehow follow the power-law distribution.

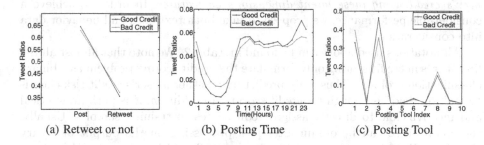

(a) Retweet or not   (b) Posting Time   (c) Posting Tool

Fig. 1. User distribution comparison between good and bad credit users w.r.t. behavior type "retweet or not", "posting time(Hours)" and "posting tools"

Motivating Examples. We demonstrate the motivation of exploiting behavior data by comparing the distribution differences between good and bad credit

---

[5] Icons expressing users' tempers and emotions.

users. Figure 1 examines the distribution difference between good and bad credit users with respect to three behavior types mentioned above. In Fig. 1(a), we can see that good credit users tend to post rather than retweeting compared to bad credit users. Good credit users are more likely to be a creator on the social media to some extent. In Fig. 1(b), there is a clear difference between good and bad credit users in terms of fraction of tweets posted at different hours of the day. Overall, people usually tweet between 9:00 and 24:00. But good credit users show tendency to post more during the daytime than bad credit users, while bad credit users are more likely to tweet during late night, which is an unhealthy lifestyle. It is reasonable that this behavior characteristic of bad credit users increases their risks to have medical emergencies, which may cause them to miss the payments. For the behavior type named "posting tools", we sample 10 representative posting tools. In Fig. 1(c), we can observe obvious differences for these posting tools, indicating that posting tool differences exist between good and bad credit users.

For the above mentioned behavior types, the differences between good and bad credit users all pass significance test at confidence level of 95 %. Similar results can be found with other behavior types. All the above observations validate that behavior data is informative and discriminative for credit prediction. Although the differences between good and bad credit users are very small, a combination of them can lead to a better result. It is worth mentioning that for many other behavior types like "time intervals between posts" or "usage of punctuation", there is no difference between good and bad credit users. Details of these behavior types are omitted due to space limitations.

Problem Definition. The definition of the problem we study can be formalized as follows: *Given a social data composed by $U \times N$ tweets that are generated by U users, our problem is to learn latent user dimensions $\Theta = \{\theta_u\}_{u=1}^{U}$ that can model users' tweet data at a high level, and infer the subjective attribute of each user's credit using these latent dimensions as features.* In order to achieve a considerable performance, we propose to take both text data and behavior data into consideration.

All notations used above can be found in Table 2. We note that almost all the literatures in credit scoring only formulate the credit scoring problem as a binary classification problem. Thus, only predicting whether a user's credit risk class is "good" or "bad" is enough for credit scoring. In addition, it is both impractical and inconvincible to directly assign credit scores to training samples. Usually, the credit score can be obtain after post-processing on the output of binary classifiers. We follow this convention in the study. Although our work aims to separate noises from tweet data for accurate social-data-based credit profiling, we acknowledge that it is very hard to predict users' default risk with a very high accuracy. We view the approach described here as a compliment to existing credit scoring methods. For instance, the latent dimensions we extracted can serve as auxiliary variables when financial data or survey data is also available. And we believe that results will become better when more social data are available from different social media websites.

3   Our Approach

In this section, we first introduce the framework of our approach for social-data-based credit profiling. Figure 2(a) shows our approach's framework, which first takes both behavior and text data into consideration for learning latent user behavior dimensions from social data, and then infer the credit risk label using standard classification algorithms. During the classification phase, all the learnt user behavior dimensions are treated as features. As illustrated in Fig. 2(a), the same behavior habit of posting at late night has different meanings when associated with topics of being drunk and watching football match respectively. Harnessing the behavior patterns inferred from both texts and behaviors of UGC, these latent behavior dimensions can predict whether a user is of good credit or bad credit more effectively. In the following, we will describe our LUBD-CM model that implements the framework of our credit profiling approach, which is also the core component of the framework.

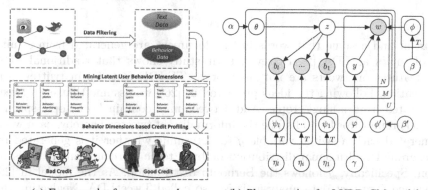

(a) Framework of our approach   (b) Plate notation for LUBD-CM model

Fig. 2. Framework of our credit profiling approach and plate notation for the LUBD-CM model

3.1   The LUBD-CM Model

To learn latent user behavior dimensions from both behavior and text part of social data, we propose a novel multiple behaviors enhanced topic model that integrates users' multiple behaviors simultaneously with textual content of tweets, called LUBD-CM (Latent User Behavior Dimension based Credit Model) for credit prediction. Figure 2(b) shows the plate notation for our proposed LUBD-CM model. The notations and their meanings are summarized in Table 2.

Modeling short text. The tweets, a kind of short text, are informal and heterogeneous. In [11], Hong et al. treat a given user's all tweets as a single pseudo

Table 2. Summary of notations and their meanings presented in Fig. 2(b)

Notations	Notation meanings	Notations	Notation meanings
U	# of users	ϕ	Topic word distribution
M	# of tweets of a user	$\psi_1, ..., \psi_l$	Topic behavior distribution of l types
N	# of words in a tweet	ϕ'	Background word distribution
T	# of topics	θ	User topic distribution
$b_1, ..., b_l$	Behaviors of l types in a tweet	φ	Bernoulli distribution generating y
w	Word in a given tweet	$\alpha, \beta, \beta', \gamma$	Dirichlet priors
z	Topic of a given tweet	$\eta_1, ..., \eta_l$	Dirichlet priors for topic behavior distributions $\psi_1, ..., \psi_l$
y	Switch variable deciding whether or not to sample from ϕ'		

document and assume that words in the document are generated from a mixture of topics as LDA [2]. However, their study shows that traditional LDA topic features on tweets are not superior to TF-IDF features in twitter user classification. Following the ideas presented in [26], we assume that each tweet is generated from a single topic. And a tweet may contain both topic specific and background words[6] to handle the informality of tweets. Thereby, each word is generated with a switch variable y to determine whether it is generated by a background multinomial distribution or by a topic word multinomial distribution. Specifically, y follows the Bernoulli distribution. As the model in [26] is designed for tweet data analysis, it is also called TweetLDA. Neglecting the variables of multiple behaviors $b_1, b_2, ..., b_l$, LUBD-CM can be reduced to Tweet-LDA.

Modeling behavior data. Now we present the techniques used for modeling the behavior data. As behavior data is associated with each tweet, we assume that each behavior is generated after the topic variable z of the tweet is sampled. Each behavior is then sampled from a bag-of-behaviors distribution of the corresponding behavior type, which is also a multinomial distribution. In LUBD-CM, we assume that each tweet has multiple behaviors attached to it, as illustrated in Fig. 2(b). A similar behavior topic model for tweets is proposed by Qiu et al. [20], called B-LDA. Our LUBD-CM model is superior to B-LDA model in that (1) LUBD-CM is able to handle different types of user behaviors simultaneously; (2)With multiple behaviors, LUBD-CM obtains latent behavior dimensions to represent each user at a much finer granularity, which is very crucial for subtle attribute inference, like credit profiling.

---

[6] Background words are like stop words in tweets.

For model inference, we use the most widely adopted collapsed Gibbs sampling method [8] to infer the parameters of LUBD-CM model. Due to space limit, we omit the details of model inference and parameter estimation. Given the presented LUBD-CM model in Fig. 2(b), the generative process for both text and behavior data can be summarized as follows:

Algorithm 1. Generative Process for LUBD-CM

---

for *each topic* $t = 1, ..., T$ **do**

    Sample $\phi_t \sim Dir(\beta)$;

    Sample $\psi_{1,t} \sim Dir(\eta_1)$, ..., Sample $\psi_{l,t} \sim Dir(\eta_l)$;

Sample $\phi' \sim Dir(\beta')$;

Sample $\varphi \sim Dir(\gamma)$;

for *each user* $u = 1, ..., U$ **do**

    Sample topic distribution $\theta_u \sim Dir(\alpha)$;

    for *each tweet* $m = 1, ..., M_u$ *in user* u*'s all* M_u *tweets* **do**

        Sample a topic $z_{u,m}$ from θ_u;

        for *each word* $n = 1, ..., N_{u,m}$ **do**

            Sample $y_{u,m,n}$ from Bernoulli(φ);

            Sample $w_{u,m,n} \sim \phi'$ if $y_{u,m,n} = 0$, otherwise sample $w_{u,m,n} \sim \phi_{z_{u,m}}$;

        for *each behavior of type* l *associated with tweet* m **do**

            sample the behavior $b_{u,m_l} \sim \psi_{z_{u,m_l}}$;

---

4   Experiments

4.1   Experiment Setup

To compare LUBD-CM and the baselines' performance, we run 20 rounds of 10-fold cross validation. The classifier for credit prediction is l_2-regularized logistic regression [7], whose performance is the best in practice. Evaluation metrics include averaged accuracy, precision, recall and F1-Score. Since we are mostly interested in identifying the bad credit ones, the measures of precision, recall and F1-Score are computed based on the bad credit label.

We implement baseline methods including Naive Bayes, LDA, TweetLDA, and LUBD-CM(3). Specifically, Naive Bayes method corresponds to the traditional unigram features based method, which is very effective for user attribute classification [3,21], and LUBD-CM(3) is a variant of LUBD-CM that takes the first 3 types of behavior into account. Similar results can be observed for other cases of combining 3 behavior types. For Naive Bayes methods, no parameters are needed. For LDA, TweetLDA, LUBD-CM(3), and LUBD-CM, we find the optimal values for parameters T, β, β', γ and η using grid search with cross validation. During experiments, α is set to be $50/T$ where T is 150. Both β and β' are set to be 0.01, and γ is set to be 40. In LDA, β is set to 0.1, and in LUBD-CM, $\eta_i = \{0.01, 0.1, 0.1, 1, 1\}$ for $i = \{1, 2, 3, 4, 5\}$.

4.2   Experiment Results

Credit Prediction. Figure 3 shows performance comparison between Naive Bayes, LDA, TweetLDA, LUBD-CM(3), and LUBD-CM w.r.t. averaged accuracy, precision, recall, F1-Score respectively. From Fig. 3, we can clearly see that LUBD-CM consistently outperforms the baselines in credit prediction. We performed a *t*-test on different metrics, and showed that all the differences between LUBD-CM and baselines were statistically significant at confidence level of 95 %. This observation validates that it is superior to consider multiple types of behavior data for inferring user credit. Although the performance improvement of LUBD-CM over baselines is only about 1 %~4 %, this performance improvement can contribute a multitude of revenues to P2P-lending companies in real life. It is worth noting that LDA has comparable performance with LUBD-CM in terms of Accuracy, but its F1-Score value is quite worse. Beside, the precision and recall values of LDA is quite different from other methods. The probable reason may lie in that LDA is usually not suitable for short-text like tweets.

In Fig. 4(a), we show the performance comparison between 5 different LUBD-CM(1)s, which take only one behavior type into consideration. The results show that different behaviors have different impacts on credit prediction and only considering one behavior is not enough for credit profiling. Figure 4(b) shows the performance changes of LUBD-CM as the minimum number of tweets for each user increases. The overall increasing trend demonstrates that the more the data, the better the performance. And we can expect that if more social data per user is available, the performance of LUBD-CM will become even better.

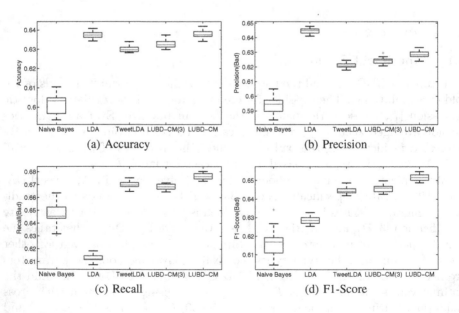

(a) Accuracy

(b) Precision

(c) Recall

(d) F1-Score

Fig. 3. Performance comparison between LUBD-CM and baselines w.r.t. accuracy, precision, recall and F1-Score.

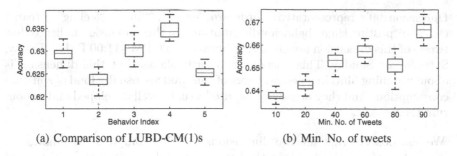

(a) Comparison of LUBD-CM(1)s   (b) Min. No. of tweets

Fig. 4. Performance comparison between LUBD-CM(1)s and LUBD-CM's sensitivity to minimum number of tweets per user.

Case Studies of Latent Behavior Dimensions. After the classifier learning step, each feature, i.e., the latent user behavior dimension, is output with a weight indicating its predictive coefficient within the classifier. Utilizing these weights, we can identify the most predictive behavior dimensions. We find that dimension 29, 51, 6, and 90 are the four most important ones according to the weights associated with them. Among them, dimension 29 is negatively weighted, indicating its contribution to bad credit label, while the rest are positively weighted, indicating their contribution to the good credit label. We analyze the four latent user dimensions in detail as follows:

1. Dimension 29 includes words like "lucky draw", "prize", "money", and "ipad" etc. The probability of this dimension for retweeting and posting is 0.98 and 0.02 respectively, indicating that users of this dimension mostly retweet instead of posting. Users of this dimension often tweet late at night, at time between 3:00 AM~4:00 AM. All these characteristics show that this behavior dimension is about retweeting advertising posts and wining prizes from lucky draws offered by the advertisers. We can infer that users of this dimension are not economically well off, desire for small bonuses and are more likely to miss the payments.
2. Dimension 51 contains words like "highway", "traffic", "jam" etc. The behavior distribution of "posting time" shows that users of this dimension often send tweets between 8:00 AM~9:00 AM, indicating that they are on their way to work and a traffic jam happens. This kind of users often have stable employments, and their credit labels are therefore likely to be good.
3. Dimension 6 includes words like "nation", "society", "government" etc., indicating that users of this dimension often care about affairs related to society and government. From "posting time" behavior, we find that users of this type seldom stay up late in the night sending tweets, and other types of behaviors are all quite normal, indicating that they are ordinary people caring about the public affairs. With no anomaly behavior patterns and paying attention to public affairs, users of this dimension are usually responsible adults and more likely to have good credit.

4. Dimension 90's representative words are "enjoy", "film", "feeling", "tears" etc. The "posting time" behavior distributions on this dimension indicate that tweets of this dimension are often sent between 7:00 PM~11:00 PM on Friday, Saturday or Sunday. This phenomena clearly shows that this dimension is about watching films in cinemas. Users of this kind are usually fond of spiritual consumption. And they are somehow intellectually well developed and seldom ruin their credit.

We also observe that for two dimensions (25, 76) composed by emoticons, their number of emoticons is mostly between 1 and 8 rather than 0. One of them represents the happiness emotions of users and contributes to the good credit label, while the other dimension indicates that users of this kind is very upset and sad and contributes to the bad credit label. This observation also coincides with human intuitions that good credit users shall be more optimistic and happier than bad credit users in real life.

5   Related Work

Social-data-based credit scoring can be viewed as inferring the specific user attribute named credit from social data, which is closely related to user profiling on social media. Rao et al. [21] firstly attempt to classify user attributes including gender, age, region and political affiliation based on features from tweets like unigram and bigram word features and sociolinguistic features. Pennacchiotti and Popescu [19] conducted study for user profiling on twitter with respect to political affiliation, ethnicity, and affinity to a certain brand with more diverse features. Other studies inferring users' attributes including gender [3], age [18], occupation [25] etc., also take advantage of tweet content. Besides, social connections between online users are also explored for user attribute inference in [5,17]. Taking one step further, Li et al. [16] proposed a user co-profiling methodology to model relationship types and user attributes simultaneously. Nonetheless, only text or network data are heavily leveraged in previous user profiling studies. Behavior data on the social web is neglected in most cases, though user behavioral patterns and habits could be very informative for user attribute profiling.

Our work is also related to traditional consumer credit scoring, which also focuses on small loans applied by individual consumers. Abundant research has been devoted to it based on statistical methods [4,9], including discriminant analysis [6], logistic regression [23], decision tree [1], neural networks [13] etc. Recent years have also witnessed the fast development of advanced methods for credit scoring [14,24]. In particular, Harris [10] assesses credit risks using optimal default definition selection algorithm, which selects the best default definition for building models. However, nearly all these works are based on transactional loan/payment records, credit reports or demographic survey data, which is crucially different from social-data-based personal credit scoring.

6   Conclusion

In this paper, we are purposed to harness the social data for personal credit profiling. We found that users' some kinds of behavior data benefits the task greatly, which also coincides with human intuitions. We proposed a joint topic-behavior model LUBD-CM to learn fine-grained latent user behavior dimensions. We conducted extensive experiments on a Sina Weibo dataset. Experimental results validated that our approach using latent dimensions inferred from LUBD-CM outperforms several state-of-the-art baselines with a significant margin. In the future, we plan to investigate more informative behavior types to boost LUBD-CM's performance. In addition, we'd like to improve our model's scalability to make it suitable for dealing with large-scale social data.

Acknowledgement. This research was partially supported by grants from the National Science Foundation for Distinguished Young Scholars of China (Grant No. 61325010), the National High Technology Research and Development Program of China (Grant No. 2014AA015203), the Science and Technology Program for Public Wellbeing (Grant No. 2013GS340302) and the CCF-Tencent Open Research Fund. This work was also partially supported by the Pinnacle Lab for Analytics @ Singapore Management University.

References

1. Arminger, G., Enache, D., Bonne, T.: Analyzing credit risk data: a comparison of logistic discrimination, classification tree analysis, and feedforward networks. Comput. Stat. **12**(2), 293–310 (1997)
2. Blei, D.M., Ng, A.Y., Jordan, M.I.: Latent Dirichlet allocation. J. Mach. Learn. Res. **3**, 993–1022 (2003)
3. Burger, J.D., Henderson, J.C., Kim, G., Zarrella, G.: Discriminating gender on twitter. In: EMNLP, pp. 1301–1309 (2011)
4. Crook, J.N., Edelman, D.B., Thomas, L.C.: Recent developments in consumer credit risk assessment. Eur. J. Oper. Res. **183**(3), 1447–1465 (2007)
5. Dong, Y., Yang, Y., Tang, J., Yang, Y., Chawla, N.V.: Inferring user demographics and social strategies in mobile social networks. In: KDD, pp. 15–24 (2014)
6. Eisenbeis, R.A.: Problems in applying discriminant analysis in credit scoring models. J. Bank. Finance **2**(3), 205–219 (1978)
7. Fan, R.-E., Chang, K.-W., Hsieh, C.-J., Wang, X.-R., Lin, C.-J.: Liblinear: a library for large linear classification. J. Mach. Learn. Res. **9**, 1871–1874 (2008)
8. Griffiths, T.L., Steyvers, M.: Finding scientific topics. Proc. Natl. Acad. Sci. **101**(suppl 1), 5228–5235 (2004)
9. Hand, D.J., Henley, W.E.: Statistical classification methods in consumer credit scoring: a review. J. Royal Stat. Soc. Ser. A (Stat. Soc.) **160**(3), 523–541 (1997)
10. Harris, T.: Default definition selection for credit scoring. Artif. Intell. Res. **2**(4), 49 (2013)
11. Hong, L., Davison, B.D.: Empirical study of topic modeling in twitter. In: Proceedings of the First Workshop on Social Media Analytics, pp. 80–88. ACM (2010)

12. Java, A., Song, X., Finin, T., Tseng, B.: Why we twitter: understanding microblogging usage and communities. In: WebKDD/SNA-KDD, WebKDD/SNA-KDD 2007, pp. 56–65 (2007)
13. Jensen, H.L.: Using neural networks for credit scoring. Manag. Finance 18(6), 15–26 (1992)
14. Kruppa, J., Schwarz, A., Arminger, G., Ziegler, A.: Consumer credit risk: individual probability estimates using machine learning. Expert Syst. Appl. 40(13), 5125–5131 (2013)
15. Kwak, H., Lee, C., Park, H., Moon, S.: What is twitter, a social network or a news media? In: Proceedings of the 19th International Conference on World Wide Web, WWW, pp. 591–600 (2010)
16. Li, R., Wang, C., Chang, K.C.-C.: User profiling in an ego network: co-profiling attributes and relationships. In: Proceedings of the 23rd International Conference on World Wide Web, WWW (2014)
17. Mislove, A., Viswanath, B., Gummadi, P.K., Druschel, P.: You are who you know: inferring user profiles in online social networks. In: WSDM, pp. 251–260 (2010)
18. Nguyen, D., Gravel, R., Trieschnigg, D., Meder, T.: "How old do you think i am?" a study of language and age in twitter. In: ICWSM (2013)
19. Pennacchiotti, M., Popescu, A.-M.: Democrats, republicans and starbucks afficionados: user classification in twitter. In: KDD, pp. 430–438 (2011)
20. Qiu, M., Zhu, F., Jiang, J.: It is not just what we say, but how we say them: Lda-based behavior-topic model. In: SDM, pp. 794–802 (2013)
21. Rao, D., Yarowsky, D., Shreevats, A., Gupta, M.: Classifying latent user attributes in twitter. In: SMUC, pp. 37–44 (2010)
22. Rosenthal, S., McKeown, K.: Age prediction in blogs: a study of style, content, and online behavior in pre- and post-social media generations. In: ACL, pp. 763–772 (2011)
23. Wiginton, J.C.: A note on the comparison of logit and discriminant models of consumer credit behavior. J. Financial Quant. Anal. 15(03), 757–770 (1980)
24. Yap, B.W., Ong, S.H., Husain, N.H.M.: Using data mining to improve assessment of credit worthiness via credit scoring models. Expert Syst. Appl. 38(10), 13274–13283 (2011)
25. Zeng, G., Luo, P., Chen, E., Wang, M.: From social user activities to people affiliation. In: ICDM (2013)
26. Zhao, W.X., Jiang, J., Weng, J., He, J., Lim, E.-P., Yan, H., Li, X.: Comparing twitter and traditional media using topic models. In: ECIR, pp. 338–349 (2011)

Linear Upper Confidence Bound Algorithm for Contextual Bandit Problem with Piled Rewards

Kuan-Hao Huang and Hsuan-Tien Lin[⊠]

Department of Computer Science and Information Engineering,
National Taiwan University, Taipei, Taiwan
{r03922062,htlin}@csie.ntu.edu.tw

Abstract. We study the contextual bandit problem with linear payoff function. In the traditional contextual bandit problem, the algorithm iteratively chooses an action based on the observed context, and immediately receives a reward for the chosen action. Motivated by a practical need in many applications, we study the design of algorithms under the piled-reward setting, where the rewards are received as a pile instead of immediately. We present how the Linear Upper Confidence Bound (LinUCB) algorithm for the traditional problem can be naïvely applied under the piled-reward setting, and prove its regret bound. Then, we extend LinUCB to a novel algorithm, called Linear Upper Confidence Bound with Pseudo Reward (LinUCBPR), which digests the observed contexts to choose actions more strategically before the piled rewards are received. We prove that LinUCBPR can match LinUCB in the regret bound under the piled-reward setting. Experiments on the artificial and real-world datasets demonstrate the strong performance of LinUCBPR in practice.

Keywords: Contextual bandit · Piled rewards · Upper confidence bound

1 Introduction

We study the contextual bandit problem (CBP) [13], which is an interactive process between an algorithm and an environment. In the traditional CBP, the algorithm observes a *context* from the environment in each time step. Then, the algorithm is asked to strategically choose an *action* from the action set based on the *context*, and receives a corresponding feedback, called *reward*, while the reward for other actions are hidden from the algorithm. The goal of the algorithm is to maximize the cumulative reward over all time steps.

Because only the reward of the chosen action is revealed, the algorithm needs to choose different actions to estimate their goodness, called *exploration*. On the other hand, the algorithm also needs to choose the better actions to maximize the reward, called *exploitation*. Balancing between exploration and exploitation is arguably the most important issue for designing algorithms of CBP.

© Springer International Publishing Switzerland 2016
J. Bailey et al. (Eds.): PAKDD 2016, Part II, LNAI 9652, pp. 143–155, 2016.
DOI: 10.1007/978-3-319-31750-2_12

ϵ-Greedy [3] and Linear Upper Confidence Bound (LinUCB) [10] are two representative algorithms for CBP. ϵ-Greedy learns one model per action for exploitation and randomly explores different actions with a small probability ϵ. LinUCB is based on online ridge regression, and takes the concept of *upper-confidence bound* [2,5] to strategically balance between exploration and exploitation. LinUCB enjoys a strong theoretical guarantee [5] and is state-of-the-art in many practical applications [10].

The traditional CBP setting assumes that the algorithm receives the reward immediately after choosing an action. In some practical applications, however, the environment cannot present the reward to the algorithm immediately. This work is motivated from one such application. Consider an online advertisement system operated by a contextual bandit algorithm. For each user visit (time step), the system (algorithm) receives the information of the user (context) from an ad exchange, and chooses an appropriate ad (action) to display to the user. In the application, the click from the user naturally acts as the reward of the action. Nevertheless, to reduce the cost of communication, the ad exchange often does not reveal the individual reward immediately after choosing an action. Instead, the ad exchange stores the individual reward first, and only sends a pile of rewards back to the system until sufficient number of rewards are gathered. We call the scenario as the contextual bandit problem under the piled-reward setting.

A related setting in the literature is the delayed-reward setting, where the reward is assumed to come at several time steps after the algorithm chooses an action. Most existing works on the delayed-reward setting consider constant delays [6] and cannot be easily applied to the piled-reward setting. Several works [8,9,12] propose algorithms for bandit problems with arbitrarily-delayed rewards, but their algorithms are non-contextual. Thus, to the best of our knowledge, no existing work has carefully studied the CBP under the piled-reward setting.

In this paper, we study how LinUCB can be applied under the piled-reward setting. We present a naïve use of LinUCB for the setting and prove its theoretical guarantee in the form of the regret bound. The result helps us understand the difference between the traditional setting and the piled-reward setting. Then, we design a novel algorithm, Linear Upper Confidence Bound with Pseudo Reward (LinUCBPR), which is a variant of LinUCB that allows more strategic use of the context information before the piled rewards are received. We prove that LinUCBPR can match the naïve LinUCB in its regret bound under the piled-reward setting. Experiments on the artificial and real-world datasets demonstrate that LinUCBPR results in strong and stable performance in practice.

This paper is organized as follows. Section 2 formalizes the CBP with the piled-reward setting. Section 3 describes our design of LinUCB and LinUCBPR under the piled-reward setting. The theoretical guarantees of the algorithms are analyzed in Sect. 4. We discuss the experiment results in Sect. 5 and conclude in Sect. 6.

2   Preliminaries

We use bold lower-case symbol like $\mathbf{u}$ to denote a column vector, bold upper-case symbol like $\mathbf{A}$ to denote a matrix, $\mathbf{I}_d$ to denote the $d \times d$ identity matrix, and $[K]$ to denote the set $\{1, 2, ..., K\}$.

We first introduce the CBP under the traditional setting. Let T be the total number of rounds and K be the number of actions. In each round $t \in [T]$, the algorithm observes a context $\mathbf{x}_t \in \mathbb{R}^d$ with $\|\mathbf{x}_t\|_2 \le 1$ from the environment. Upon observing the context $\mathbf{x}_t$, the algorithm chooses an action a_t from K actions based on the context. Right after choosing a_t, the algorithm receives a reward r_{t,a_t} that corresponds to the context $\mathbf{x}_t$ and the chosen action a_t, while other rewards $r_{t,a}$ for $a \ne a_t$ are hidden from the algorithm. The goal of the algorithm is to maximize the cumulative reward after T rounds.

Now, we introduce the CBP under the piled-reward setting. Instead of receiving the reward right after choosing an action (and thus right before observing the next context), the setting assumes that the rewards come as a pile after observing multiple contexts in a round. We shall extend our notation above to the piled-reward setting as follows. In each round t, the algorithm sequentially observes n contexts $\mathbf{x}_{t_1}, \mathbf{x}_{t_2}, ..., \mathbf{x}_{t_n} \in \mathbb{R}^d$ with $\|\mathbf{x}_{t_i}\|_2 \le 1$ from the environment. For simplicity, we assume that n is a fixed number while all the technical results in this paper can be easily extended to the case where n can vary in each round. We use t_i to denote the i-th step in round t. For example, 3_5 means the 5-th step in round 3. Upon observing the context $\mathbf{x}_{t_i}$ in round t, the algorithm chooses an action a_{t_i} from K actions based on the context, and observes the next context $\mathbf{x}_{t_{i+1}}$. In the end of round t, the algorithm receives n rewards $r_{t_1, a_{t_1}}, r_{t_2, a_{t_2}}, ..., r_{t_n, a_{t_n}}$ that correspond to $\mathbf{x}_{t_i}$ and a_{t_i}, while other rewards $r_{t_i, a}$ for $a \ne a_{t_i}$ are hidden from the algorithm. The goal is again to maximize the cumulative reward $\sum_{t=1}^{T} \sum_{i=1}^{n} r_{t_i, a_{t_i}}$ after T rounds.

In other words, the piled-reward setting assumes that the context comes at time steps $\{1_1, 1_2, ..., 1_n, 2_1, 2_2, ..., t_1, t_2, ..., T_{n-1}, T_n\}$, while the rewards come after every n contexts as a pile. Note that the traditional setting is a special case of the piled-reward setting when $n = 1$.

In this paper, we consider the CBP with linear payoff function. We assume that $r_{t_i, a}$ connects with $\mathbf{x}_{t_i}$ linearly through K hidden weight vectors $\mathbf{u}_1, \mathbf{u}_2, ..., \mathbf{u}_K \in \mathbb{R}^d$ with $\|\mathbf{u}_i\|_2 \le 1$. That is, $\mathbb{E}\left[r_{t_i, a} \mid \mathbf{x}_{t_i}\right] = \mathbf{x}_{t_i}^\top \mathbf{u}_a$. Let $a_{t_i}^* = \arg\max_{a \in [K]} \mathbf{x}_{t_i}^\top \mathbf{u}_a$ be the optimal action for $\mathbf{x}_{t_i}$. We define regret of an algorithm to be $\left(\sum_{t=1}^{T} \sum_{i=1}^{n} r_{t_i, a_{t_i}^*} - \sum_{t=1}^{T} \sum_{i=1}^{n} r_{t_i, a_{t_i}}\right)$. The goal of maximizing the cumulative reward is equivalent to minimizing the regret.

Linear Upper Confidence Bound (LinUCB) [5] is a state-of-the-art algorithm for the traditional CBP ($n = 1$). LinUCB maintains K weight vectors $\mathbf{w}_{t_1, 1}, \mathbf{w}_{t_1, 2}, ..., \mathbf{w}_{t_1, K}$ to estimate $\mathbf{u}_1, \mathbf{u}_2, ..., \mathbf{u}_K$ at time step t_1. The K weight vectors are calculated by ridge regression,

$$\mathbf{w}_{t_1, a} = \underset{\mathbf{w} \in \mathbb{R}^d}{\arg\min} \left(\|\mathbf{w}\|^2 + \|\mathbf{X}_{(t-1)_1, a} \mathbf{w} - \mathbf{r}_{(t-1)_1, a}\|^2 \right), \tag{1}$$

where $\mathbf{X}_{(t-1)_1,a}$ is a matrix with rows being the contexts $\mathbf{x}_\tau^\top$ where τ are the time steps before round t and $a_\tau = a$, and $\mathbf{r}_{(t-1)_1,a}$ is a column vector with each element representing the corresponding reward for each context in $\mathbf{X}_{(t-1)_1,a}$. Let $\mathbf{A}_{t_1,a} = (\mathbf{I}_d + \mathbf{X}_{(t-1)_1,a}^\top \mathbf{X}_{(t-1)_1,a})$ and $\mathbf{b}_{t_1,a} = (\mathbf{X}_{(t-1)_1,a}^\top \mathbf{r}_{(t-1)_1,a})$. The solution to (1) is $\mathbf{w}_{t_1,a} = \mathbf{A}_{t_1,a}^{-1} \mathbf{b}_{t_1,a}$.

When a new context $\mathbf{x}_{t_1}$ comes, LinUCB calculates two terms for each action a: the estimated reward $\tilde{r}_{t_1,a} = \mathbf{x}_{t_1}^\top \mathbf{w}_{t_1,a}$ and the uncertainty $c_{t_1,a} = \sqrt{\mathbf{x}_{t_1}^\top \mathbf{A}_{t_1,a}^{-1} \mathbf{x}_{t_1}}$, and chooses the action with the highest score $\tilde{r}_{t_1,a} + \alpha c_{t_1,a}$, where α is a trade-off parameter. After receiving the reward, LinUCB updates the weight vector $\mathbf{w}_{t_1,a_{t_1}}$ immediately, and uses the new weight vector to choose the action for the next context. LinUCB conducts exploration when the chosen action is of high uncertainty. After sufficient (context, action, reward) information is received, $\tilde{r}_{t_1,a}$ shall be close the expected reward, and $c_{t_1,a}$ will be smaller. Then, LinUCB conducts exploitation with the learned weight vectors to choose the action with the highest expected reward.

3   Proposed Algorithm

We first discuss how LinUCB can be naïvely applied under the piled-reward setting. Then, we extend LinUCB to a more general framework that utilizes the additional information within the contexts before the true rewards are received.

Since no rewards are received before the end of the current round t, the naïve LinUCB does not update the model during round t, and only takes the fixed $\mathbf{w}_{t_1,a}$ and $\mathbf{A}_{t_1,a}$ to calculate the estimated reward $\tilde{r}_{t_i,a}$ and the uncertainty $c_{t_i,a}$ for each action a. That is, LinUCB only updates $\mathbf{w}_{t_1,a}$ before the beginning of round t as the solution to (1) with $(\mathbf{X}_{(t-1)_1,a}, \mathbf{r}_{(t-1)_1,a})$ under the traditional setting replaced by $(\mathbf{X}_{(t-1)_n,a}, \mathbf{r}_{(t-1)_n,a})$ under the piled-reward setting. In addition, $\mathbf{A}_{t_1,a}$ can be similarly defined from $\mathbf{X}_{(t-1)_n,a}$ instead.

The naïve LinUCB can be viewed as a baseline upper confidence bound algorithm under the piled-reward setting. There is a possible drawback for the naïve LinUCB. If similar contexts come repeatedly in the same round, because $\mathbf{w}_{t_i,a}$ and $\mathbf{A}_{t_i,a}$ stay unchanged within the round, LinUCB will choose similar actions repeatedly. Then, if the chosen action suffers from low reward, LinUCB suffers from making the low-reward choice repeatedly before the end of the round.

The question is, can we do even better? Our idea is that the contexts $\mathbf{x}_{t_i}$ received during round t can be utilized to update the model *before* the rewards come. That is, at time step t_i, in addition to the *labelled data* (context, action, reward) gathered before time step $(t-1)_n$ that LinUCB uses, the *unlabelled data* (context, action) gathered at time steps $\{t_1, t_2, \ldots, t_{i-1}\}$ can also be included to learn a more decent model. In other words, we hope to design some *semi-supervised* learning scheme within round t to guide the upper-confidence bound algorithm towards more strategic exploration within the round.

Our idea is motivated from the regret analysis. In Sect. 4, we will show that the regret of LinUCB under the piled-reward setting is bounded by the summation of $c_{t_i,a}$ over all time steps. But note that $c_{t_i,a}$ only depends on $\mathbf{x}_{t_i,a}$ and

---

Algorithm 1. LinUCBPR under the piled-reward setting

1: Parameter: $\alpha \in \mathbb{R}^+$
2: Initialize: $\hat{\mathbf{A}}_{11,a} \leftarrow \mathbf{I}_d, \hat{\mathbf{b}}_{11,a} \leftarrow \mathbf{0}_{d \times 1}, \hat{\mathbf{w}}_{11,a} \leftarrow \hat{\mathbf{A}}_{11,a}^{-1} \hat{\mathbf{b}}_{11,a}$
3: **for** $t = 1, 2, 3, ..., T$ **do**
4:     **for** $i = 1, 2, 3, ..., n$ **do**
5:         Observe $\mathbf{x}_{t_i}$ and choose $a_{t_i} = \text{argmax}_{a \in [K]} \mathbf{x}_{t_i}^\top \hat{\mathbf{w}}_{t_i,a} + \alpha \sqrt{\mathbf{x}_{t_i}^\top \hat{\mathbf{A}}_{t_i,a}^{-1} \mathbf{x}_{t_i}}$
6:         Calculate the pseudo reward $p_{t_i,a_{t_i}}$
7:         $\hat{\mathbf{A}}_{t_{i+1},a_{t_i}} \leftarrow \hat{\mathbf{A}}_{t_i,a_{t_i}} + \mathbf{x}_{t_i} \mathbf{x}_{t_i}^\top, \quad \hat{\mathbf{b}}_{t_{i+1},a_{t_i}} \leftarrow \hat{\mathbf{b}}_{t_i,a_{t_i}} + \mathbf{x}_{t_i} p_{t_i,a_{t_i}}$
8:         $\hat{\mathbf{w}}_{t_{i+1},a_{t_i}} \leftarrow \hat{\mathbf{A}}_{t_{i+1},a_{t_i}}^{-1} \hat{\mathbf{b}}_{t_{i+1},a_{t_i}}$
9:     **end for**
10:    Receive rewards $r_{t_1,a_{t_1}}, r_{t_2,a_{t_2}}, ..., r_{t_n,a_{t_n}}$
11:    **for** $a \in [K]$ **do**
12:       $\hat{\mathbf{A}}_{(t+1)_1,a} \leftarrow \hat{\mathbf{A}}_{t_1,a} + \sum_{a_{t_i}=a} \mathbf{x}_{t_i} \mathbf{x}_{t_i}^\top, \quad \hat{\mathbf{b}}_{(t+1)_1,a} \leftarrow \hat{\mathbf{b}}_{t_1,a} + \sum_{a_{t_i}=a} \mathbf{x}_{t_i} r_{t_i,a_{t_i}}$
13:       $\hat{\mathbf{w}}_{(t+1)_1,a} \leftarrow \hat{\mathbf{A}}_{(t+1)_1,a}^{-1} \hat{\mathbf{b}}_{(t+1)_1,a}$
14:    **end for**
15: **end for**

$\mathbf{A}_{t_i,a}$, but not the reward. That is, upon receiving $\mathbf{x}_{t_i}$ and choosing an action a_{t_i}, the term $c_{t_i,a_{t_i}}$ can readily be updated without the true reward. By updating $c_{t_i,a_{t_i}}$ within the round, the algorithm can explore different actions strategically instead of following similar actions when similar contexts come repeatedly in the same round.

This idea can be extended to the following framework. We propose to couple each context $\mathbf{x}_{t_1}, \mathbf{x}_{t_2}, \ldots, \mathbf{x}_{t_{i-1}}$ with a pseudo reward p_{τ,a_τ}, where τ is the time step, before receiving the true reward r_{τ,a_τ}. The pseudo reward can then pretend to be the true reward and allow the algorithm to keep updating the model before the true rewards are received. Note that pseudo rewards have been used to speed up exploration in the traditional CBP [4], and can encourage more strategic exploration in our framework. We name the framework Linear Upper Confidence Bound with Pseudo Reward (LinUCBPR). The framework updates the weight vector and the estimated covariance matrix by

$$\hat{\mathbf{w}}_{t_i,a} = \underset{\mathbf{w} \in \mathbb{R}^d}{\text{argmin}} \left(\|\mathbf{w}\|^2 + \|\mathbf{X}_{(t-1)_n,a}\mathbf{w} - \mathbf{r}_{(t-1)_n,a}\|^2 + \|\hat{\mathbf{X}}_{t_{i-1},a}\mathbf{w} - \mathbf{p}_{t_{i-1},a}\|^2 \right) \quad (2)$$

$$\hat{\mathbf{A}}_{t_i,a} = \mathbf{I}_d + \mathbf{X}_{(t-1)_n,a}^\top \mathbf{X}_{(t-1)_n,a} + \hat{\mathbf{X}}_{t_{i-1},a}^\top \hat{\mathbf{X}}_{t_{i-1},a} \quad (3)$$

where $\hat{\mathbf{X}}_{t_{i-1},a}$ is a matrix with rows being the contexts $\mathbf{x}_\tau^\top$ with $t_1 \leq \tau \leq t_{i-1}$ and $a_\tau = a$, and $\mathbf{p}_{t_{i-1},a}$ is a column vector with each element representing the corresponding pseudo reward for each context in $\hat{\mathbf{X}}_{t_{i-1},a}$.

When receiving the true rewards in the end of round t, we discard the change from pseudo rewards, and use the true rewards to update model again. We show the framework of LinUCBPR in Algorithm 1.

The only remained task is what $p_{\tau,a}$ should be. We will study two variants, one is to use $p_{\tau,a} = \tilde{r}_{\tau,a}$, the estimated reward of actions. We name the variant LinUCBPR with estimated reward (LinUCBPR-ER). Another variant is to be

---

Algorithm 2. BaseLinUCB under the piled-reward setting at round t

---

1: Parameter: $\alpha \in \mathbb{R}^+, \Psi_t \subseteq \{1_1, 1_2, ..., (t-1)_n\}$

2: $\bar{\mathbf{A}}_{t_1} \leftarrow \mathbf{I}_{dK} + \sum_{\tau \in \Psi_t} \bar{\mathbf{x}}_{\tau, a_\tau} \bar{\mathbf{x}}_{\tau, a_\tau}^\top$, $\bar{\mathbf{b}}_{t_1} \leftarrow \mathbf{0}_{dK \times 1} + \sum_{\tau \in \Psi_t} \bar{\mathbf{x}}_{\tau, a_\tau} r_{\tau, a_\tau}$, $\bar{\mathbf{w}}_{t_1} \leftarrow \bar{\mathbf{A}}_{t_1}^{-1} \bar{\mathbf{b}}_{t_1}$

3: **for** $i = 1, 2, 3, ..., n$ **do**

4:     Observe $\mathbf{x}_{t_i}$ and calculate $\bar{\mathbf{x}}_{t_i, 1}, \bar{\mathbf{x}}_{t_i, 2}, ..., \bar{\mathbf{x}}_{t_i, K}$

5:     **for** $a \in [K]$ **do**

6:         $width_{t_i, a} \leftarrow (1 + \alpha) \sqrt{\bar{\mathbf{x}}_{t_i, a}^\top \bar{\mathbf{A}}_{t_1}^{-1} \bar{\mathbf{x}}_{t_i, a}}$

7:         $ucb_{t_i, a} \leftarrow \bar{\mathbf{x}}_{t_i, a}^\top \bar{\mathbf{w}}_{t_1} + width_{t_i, a}$

8:     **end for**

9: **end for**

---

even more aggressive, and set $p_{\tau, a} = \tilde{r}_{\tau, a} - \beta c_{\tau, a}$, a lower-confidence bound of the reward, where β is a trade-off parameter. The lower-confidence bound can be viewed as the underestimated reward, and should allow more exploration within the round, at the cost of more computation. We name the variant LinUCBPR with underestimated reward (LinUCBPR-UR).

4 Theoretical Analysis

In this section, we establish the theoretical guarantee for the regret bound of Lin-UCB and LinUCBPR-ER under the piled-reward setting. Similar to the analysis of LinUCB in the immediate-reward setting [5], there is a difficulty. In particular, the algorithms choose actions based on previous outcomes. Hence, the rewards in each round are not independent random variables. To deal with this problem, we follow the approach of [5]. We modify the algorithm to a base algorithm which assumes the independent rewards, and construct a master algorithm which ensures that the assumption holds.

Note that [5] takes a CBP setting with one context *per action* instead for our setting of the one context *share by actions*. To let the notation be consistent with [5], we simply cast our setting as theirs by following steps. We define a (dK)-dimensional vector $\bar{\mathbf{u}}$ to be the concatenation of $\mathbf{u}_1, \mathbf{u}_2, ..., \mathbf{u}_K$, and define a (dK)-dimensional context $\bar{\mathbf{x}}_{\tau, a}$ per action with $\mathbf{x}_\tau$, where $\bar{\mathbf{x}}_{\tau, a} = \begin{bmatrix} \mathbf{0} & \mathbf{0} & \cdots & \mathbf{0} & \mathbf{x}_\tau^\top & \mathbf{0} & \cdots & \mathbf{0} \end{bmatrix}^\top$ with $\mathbf{x}_\tau$ being the a-th vector within the concatenation. All $\bar{\mathbf{X}}_\tau, \bar{\mathbf{A}}_\tau, \bar{\mathbf{r}}_\tau, \bar{\mathbf{b}}_\tau$, and $\bar{\mathbf{w}}_\tau$ can be similarly defined from $\hat{\mathbf{X}}_{\tau, a}, \hat{\mathbf{A}}_{\tau, a}, \mathbf{r}_{\tau, a}, \hat{\mathbf{b}}_{\tau, a}$, and $\hat{\mathbf{w}}_{\tau, a}$.

4.1 Regret for LinUCB Under the Piled-Reward Setting

Algorithm 2 lists the base algorithm for LinUCB under the piled-reward setting, called BaseLinUCB. We first prove the theoretical guarantee of BaseLinUCB. Let $\bar{c}_{t_i, a} = \sqrt{\bar{\mathbf{x}}_{t_i, a}^\top \bar{\mathbf{A}}_{t_1}^{-1} \bar{\mathbf{x}}_{t_i, a}}$. We can establish the following lemmas.

Lemma 1 (Li et al. [5], Lemma 1). *Suppose the input time step set $\Psi_t \subseteq \{1_1, 1_2, ..., (t-1)_n\}$ given to BaseLinUCB has property that for fixed context $\bar{\mathbf{x}}_{t_i,a}$ with $t_i \in \Psi_t$, the corresponding rewards $r_{t_i,a}$ are independent random variables with means $\bar{\mathbf{x}}_{t_i,a}^{\top} \bar{\mathbf{u}}$. Then, for some $\alpha = \mathcal{O}(\sqrt{\ln(nTK/\delta)})$, we have with probability at least $1 - \delta/(nT)$ that $\left| \bar{\mathbf{x}}_{t_i,a}^{\top} \bar{\mathbf{w}}_{t_i} - \bar{\mathbf{x}}_{t_i,a}^{\top} \bar{\mathbf{u}} \right| \le (1+\alpha)\bar{c}_{t_i,a}$.*

Note that in Lemma 1, the bound is related to the time steps. We want the bound to be related to the rounds, and hence establish Lemmas 2 and 3.

Lemma 2. *Let ψ_t be a subset of $\{t_1, t_2, ..., t_n\}$. Suppose $\Psi_{t+1} = \Psi_t \cup \psi_t$ in BaseLinUCB. Then, the eigenvalues of $\bar{A}_{t_1}$ and $\bar{A}_{(t+1)_1}$ can be arranged so that $\lambda_{t_1,j} \le \lambda_{(t+1)_1,j}$ for all j and $\bar{c}_{t_i,a}^2 \le 10 \sum_{j=1}^{dK} \frac{\lambda_{(t+1)_1,j} - \lambda_{t_1,j}}{\lambda_{t_1,j}}$.*

Proof. The proof can be done by combining Lemmas 2 and 8 in [5].

Lemma 3. *Let $\Phi_{t+1} = \{t \mid t \in [T] \text{ and } \exists j \text{ such that } t_j \in \Psi_{t+1}\}$, and assume $|\Phi_{t+1}| \ge 2$. Then $\sum_{t_i \in \Psi_{t+1}} \bar{c}_{t_i,a} \le 5n\sqrt{dK |\Phi_{t+1}| \ln |\Phi_{t+1}|}$.*

Proof. By Lemma 2 and the technique in the proof of Lemma 3 in [5], we have

$$
\sum_{t_i \in \Psi_{t+1}} \bar{c}_{t_i,a} \le \sum_{t_i \in \Psi_{t+1}} \sqrt{10 \sum_{j=1}^{dK} \frac{\lambda_{(t+1)_1,j} - \lambda_{t_1,j}}{\lambda_{t_1,j}}}
$$

$$
\le \sum_{t \in \Phi_{t+1}} n \sqrt{10 \sum_{j=1}^{dK} \frac{\lambda_{(t+1)_1,j} - \lambda_{t_1,j}}{\lambda_{t_1,j}}} \le 5n\sqrt{dK |\Phi_{t+1}| \ln |\Phi_{t+1}|}.
$$

We construct SupLinUCB on each round similar to [5]. Then, we borrow Lemmas 14 and 15 of [2], and extend Lemma 16 of [2] to the following lemma.

Lemma 4. *For each $s \in [S]$, $\left| \Psi_{T+1}^s \right| \le 5n \cdot 2^s (1+\alpha)\sqrt{dK |\Phi_{t+1}^s| \ln |\Phi_{t+1}^s|}$.*

Based on the lemmas, we can then establish the following theorem for the regret bound of LinUCB under the piled-reward setting.

Theorem 1. *For some $\alpha = \mathcal{O}(\sqrt{\ln(nTK/\delta)})$, with probability $1 - \delta$, the regret of LinUCB under the piled-reward setting is $\mathcal{O}(\sqrt{dn^2 TK \ln^3(nTK/\delta)})$.*

Proof. Let $\Psi^0 = \{1_1, 1_2, ..., T_n\} \setminus \bigcup_{s \in [S]} \Psi_{T+1}^s$. Observing that $s^{-s} \leq 1/\sqrt{T}$, given the previous lemmas and Jensen's inequality, we have

$$
\begin{aligned}
Regret &= \sum_{t=1}^{T} \sum_{i=1}^{n} \left(\mathbb{E}[r_{t_i, a_{t_i}^*}] - \mathbb{E}[r_{t_i, a_{t_i}}] \right) \\
&= \sum_{t_i \in \Psi^0} \left(\mathbb{E}[r_{t_i, a_{t_i}^*}] - \mathbb{E}[r_{t_i, a_{t_i}}] \right) + \sum_{s=1}^{S} \sum_{t_i \in \Psi_{T+1}^s} \left(\mathbb{E}[r_{t_i, a_{t_i}^*}] - \mathbb{E}[r_{t_i, a_{t_i}}] \right) \\
&\leq \frac{2}{\sqrt{T}} |\Psi^0| + \sum_{s=1}^{S} 2^{3-s} |\Psi_{T+1}^s| \\
&\leq \frac{2}{\sqrt{T}} |\Psi^0| + \sum_{s=1}^{S} 40n(1+\alpha)\sqrt{dK |\Phi_{t+1}^s| \ln |\Phi_{t+1}^s|} \\
&\leq 2n\sqrt{T} + 40n(1+\alpha)\sqrt{dK \ln T} \sum_{s=1}^{S} \sqrt{|\Phi_{t+1}^s|} \\
&\leq 2n\sqrt{T} + 40n(1+\alpha)\sqrt{dK \ln T} \sqrt{ST}.
\end{aligned}
$$

The rest of proof is almost identical to the proof of Theorem 6 in [2]. By substituting $\alpha = \mathcal{O}(\sqrt{\ln(nTK/\delta)})$, replacing δ with $\delta/(S+1)S$, substituting $S = \ln(nT)$, and applying Azuma's inequality, we obtain Theorem 1.

Note that if we let $nT = C$ to be a constant, the original regret bound under the traditional setting ($n = 1$) in [5] is $\mathcal{O}(\sqrt{dCK \ln^3(CK/\delta)})$, while the regret bound under the piled-reward setting is $\mathcal{O}(\sqrt{dnCK \ln^3(CK/\delta)})$, which is the original bound multiplied by $\sqrt{n}$.

4.2   Regret for LinUCBPR-ER Under the Piled-Reward Setting

We first prove two lemmas for LinUCBPR-ER.

Lemma 5. *After updating with the context* $\mathbf{x}_{t_i}$ *and the pseudo reward* $p_{t_i, a} = \tilde{r}_{t_i, a}$, *the estimated reward of LinUCBPR-ER is the same. That is,* $\tilde{r}_{t_{i+1}} = \tilde{r}_{t_i}$.

Proof. Because $p_{t_i, a} = \mathbf{x}_{t_i}^\top \hat{\mathbf{w}}_{t_i, a} = \tilde{r}_{t_i, a}$, $\mathbf{x}_{t_i}$ and $p_{t_i, a}$ will not change $\hat{\mathbf{w}}_{t_i, a}$. Thus the reward stays the same.

Lemma 6. *After updating with the context* $\mathbf{x}_{t_i}$ *and the pseudo reward* $p_{t_i, a} = \tilde{r}_{t_i, a}$, *the uncertainty of LinUCBPR-ER for the context is non-increasing. That is, for any* $\mathbf{x}$, $\sqrt{\mathbf{x}^\top \hat{\mathbf{A}}_{t_{i+1}, a}^{-1} \mathbf{x}} \leq \sqrt{\mathbf{x}^\top \hat{\mathbf{A}}_{t_i, a}^{-1} \mathbf{x}}$.

Proof. By Sherman-Morrison formula, we have

$$
\mathbf{x}^\top \hat{\mathbf{A}}_{t_{i+1}, a}^{-1} \mathbf{x} = \mathbf{x}^\top \hat{\mathbf{A}}_{t_i, a}^{-1} \mathbf{x} - \frac{\mathbf{x}^\top \hat{\mathbf{A}}_{t_i, a}^{-1} \mathbf{x}_{t_i} \mathbf{x}_{t_i}^\top \hat{\mathbf{A}}_{t_i, a}^{-1} \mathbf{x}}{1 + \mathbf{x}_{t_i}^\top \hat{\mathbf{A}}_{t_i, a}^{-1} \mathbf{x}_{t_i}} = \mathbf{x}^\top \hat{\mathbf{A}}_{t_i, a}^{-1} \mathbf{x} - \frac{\left(\mathbf{x}^\top \hat{\mathbf{A}}_{t_i, a}^{-1} \mathbf{x}_{t_i} \right)^2}{1 + \mathbf{x}_{t_i}^\top \hat{\mathbf{A}}_{t_i, a}^{-1} \mathbf{x}_{t_i}}.
$$

The second term is greater than or equal to zero, and implies the lemma.

Similarly, we can construct BaseLinUCBPR-ER and SupLinUCBPR-ER. By Lemma 5, we have that for each time step t_i in the round t, the estimated reward $\bar{\mathbf{x}}_{t_i,a}^\top \bar{\mathbf{w}}_{t_i} = \bar{\mathbf{x}}_{t_i,a}^\top \bar{\mathbf{w}}_{t_1}$ does not change. Furthermore, By Lemma 6, we have $\sqrt{\mathbf{x}^\top \hat{\mathbf{A}}_{t_i,a}^{-1} \mathbf{x}} \leq \sqrt{\mathbf{x}^\top \hat{\mathbf{A}}_{t_1,a}^{-1} \mathbf{x}}$. Hence, all lemmas we need also hold for BaseLinUCBPR-ER. Similar to LinUCB, we can then establish the following theorem. The proof is almost identical to Theorem 1.

Theorem 2. *For some* $\alpha = \mathcal{O}(\sqrt{\ln(nTK/\delta)})$, *with probability* $1 - \delta$, *the regret of LinUCBPR-ER under the piled-reward setting is* $\mathcal{O}(\sqrt{dn^2 TK \ln^3(nTK/\delta)})$.

5   Experiments

We apply the proposed algorithms on both artificial and real-world datasets to justify that using pseudo-rewards is useful. In addition, we follow [1], and take the *simple supervised-to-contextual-bandit transformation* [7] on 8 multi-class datasets to evaluate our idea.

Artificial Datasets. For each artificial dataset, we first sample unit vectors $\mathbf{u}_1, \mathbf{u}_2, ..., \mathbf{u}_K$ uniformly from $\mathbb{R}^d$ to simulate the K actions. In each round t, the context $\mathbf{x}_{t_i}$ is sampled from an uniform distribution within $\|\mathbf{x}_{t_i}\| \leq 1$. The reward is generated by $r_{t_i,a_{t_i}} = \mathbf{u}_{a_{t_i}}^\top \mathbf{x}_{t_i} + \epsilon_{t_i}$, where $\epsilon_{t_i} \in [-0.05, 0.05]$ is a uniform random noise. In the experiments, we let $nT = 50000$ to be a constant, and consider parameters $d \in \{10, 30\}$, $K \in \{50, 100\}$, and $n \in \{500, 1000\}$.

Table 1. ACR on artificial datasets (mean ± std)

	$d = 10$				$d = 30$			
	$K = 50$		$K = 100$		$K = 50$		$K = 100$	
	$n = 500$	$n = 1000$	$n = 500$	$n = 1000$	$n = 500$	$n = 1000$	$n = 500$	$n = 1000$
Ideal	0.6607 ±0.0002	0.6607 ±0.0002	0.7061 ±0.0002	0.7061 ±0.0002	0.3930 ±0.0002	0.3930 ±0.0002	0.4252 ±0.0002	0.4252 ±0.0002
ϵ-Greedy	0.6265 ±0.0030	0.6329 ±0.0016	0.6317 ±0.0043	0.6538 ±0.0030	0.3566 ±0.0030	0.3690 ±0.0014	0.3537 ±0.0022	0.3739 ±0.0026
LinUCB	0.6555 ±0.0004	0.6513 ±0.0005	0.6866 ±0.0011	0.6868 ±0.0011	0.3905 ±0.0003	0.3880 ±0.0004	0.4188 ±0.0007	0.4164 ±0.0004
LinUCB PR-ER	**0.6591** ±**0.0001**	**0.6535** ±**0.0002**	**0.7000** ±**0.0012**	**0.7040** ±**0.0003**	**0.3917** ±**0.0002**	**0.3896** ±**0.0002**	**0.4227** ±**0.0004**	**0.4224** ±**0.0004**
LinUCB PR-UR	0.6586 ±0.0001	0.6533 ±0.0001	0.6978 ±0.0011	0.7027 ±0.0003	0.3911 ±0.0002	0.3887 ±0.0002	0.4210 ±0.0002	0.4215 ±0.0003
QPM-D	0.6552 ±0.0003	0.6502 ±0.0004	0.6925 ±0.0010	0.6860 ±0.0013	0.3897 ±0.0003	0.3871 ±0.0003	0.4172 ±0.0007	0.4123 ±0.0010

We compare the performance of ϵ-Greedy, LinUCB, LinUCBPR-ER and LinUCBPR-UR under the piled-reward setting. We also compare Queued Partial Monitoring with Delays (QPM-D) [9], which uses a queue to handle arbitrarily-delay rewards. Furthermore, we consider an "ideal" LinUCB under the traditional setting ($n = 1$) to study the difference between the traditional setting

Table 2. Datasets

Dataset	D	K
shuttle	9	7
poker	10	10
pendigits	16	10
letter	16	26
satimage	36	6
acoustic	50	3
covtype	54	7
usps	256	10

Table 4. t-test at 95 % confidence level (win/tie/loss)

Algorithm	Competitor				
	ϵ-Greedy	LinUCB	LinUCB PR-ER	LinUCB PR-UR	QPM-D
ϵ-Greedy	–	0/0/8	0/0/8	0/0/8	0/0/8
LinUCB	8/0/0	–	0/1/7	2/4/2	1/6/1
LinUCBPR-ER	8/0/0	7/1/0	–	5/3/0	6/2/0
LinUCBPR-UR	8/0/0	2/4/2	0/3/5	–	2/5/1
QPM-D	8/0/0	1/6/1	0/2/6	1/5/2	–

Table 3. ACR on supervised-to-contextual-bandit datasets (mean ± std)

	shuttle	poker	pendigits	letter	satimage	acoustic	covtype	usps
Ideal	0.9373 ±0.0005	0.4866 ±0.0075	0.8929 ±0.0056	0.6271 ±0.0117	0.8344 ±0.0014	0.7216 ±0.0006	0.6987 ±0.0020	0.9358 ±0.0012
ϵ-Greedy	0.8844 ±0.0092	0.4766 ±0.0086	0.8667 ±0.0058	0.4746 ±0.0247	0.8062 ±0.0024	0.6992 ±0.0012	0.6736 ±0.0035	0.9009 ±0.0023
LinUCB	0.9168 ±0.0068	0.4863 ±0.0087	0.8876 ±0.0043	0.5696 ±0.0176	0.8225 ±0.0016	0.7103 ±0.0016	0.6888 ±0.0039	0.9192 ±0.0017
LinUCB PR-ER	**0.9200** ±0.0029	**0.4865** ±0.0046	**0.8901** ±0.0019	**0.6053** ±0.0137	**0.8236** ±0.0022	**0.7112** ±0.0007	0.6915 ±0.0021	**0.9221** ±0.0011
LinUCB PR-UR	0.9170 ±0.0027	0.4846 ±0.0107	0.8872 ±0.0043	0.6017 ±0.0167	0.8189 ±0.0045	0.7099 ±0.0021	0.6913 ±0.0014	0.9179 ±0.0025
QPM-D	0.9166 ±0.0046	0.4860 ±0.0033	0.8844 ±0.0044	0.5585 ±0.0225	0.8221 ±0.0024	0.7101 ±0.0012	**0.6915** ±0.0013	0.9185 ±0.0018

and the piled-reward setting. The parameters of algorithms are selected by grid search, where $\alpha, \beta \in \{0.05, 0.10, ..., 1.00\}$ and $\epsilon \in \{0.025, 0, 05, ..., 0.1\}$. We run the experiment 20 times and show the average cumulative reward (ACR), which is the cumulative reward over the number of time steps, in Table 1. From the table, Ideal LinUCB clearly outperforms others. This verifies that the piled-reward setting introduces difficulty in applying upper-confidence bound algorithms. It also echoes the regret bound in Sect. 4, where LinUCB under the piled-reward setting suffers some penalty when compared with the original bound.

Next, we focus on the influence of the pseudo rewards. LinUCBPR-ER and LinUCBPR-UR are consistently better than LinUCB on all datasets. Figures 1

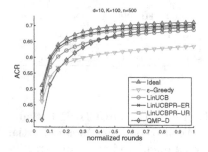

Fig. 1. ACR versus round

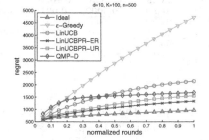

Fig. 2. Regret versus round

and 2 respectively depict the ACR and the regret along normalized rounds, which is t/T, when $d = 10$, $K = 100$, and $n = 500$. Note that LinUCBPR algorithms enjoy an advantage in the early rounds. This is because the exploration is generally more important than the exploitation in the early rounds, and LinUCBPR algorithms encourage more strategic exploration by using pseudo rewards.

We take ϵ-Greedy to compare the effect of conducting exploration within the round based on randomness rather than pseudo rewards. Table 1 suggests LinUCBPR algorithms reach much better performance, and justifies the effectiveness of the strategic exploration. We also compare LinUCBPR algorithms with QPM-D. Table 1 shows that LinUCBPR algorithms are consistently better than QPM-D. The results again justify the superiority of LinUCBPR algorithms.

LinUCBPR-ER and LinUCBPR-UR perform quite comparably across all datasets. The results suggest that we do not need to be more aggressive than LinUCBPR-ER. The simple LinUCBPR-ER, which can be efficiently implemented by updating $\mathbf{A}_{t_i,a}$ only, can readily reach decent performance.

Supervised-to-Contextual-Bandit Datasets. Next, we take 8 public multiclass datasets[1] (Table 2). We randomly split each dataset into two parts: 30 % for parameter tuning and 70 % for testing. For each part, we repeatedly present the examples as an infinite data stream. We let $nT = 10000$ for parameter tuning and $nT = 30000$ for testing. We consider $n = 500$ for all datasets. The parameter setting is the same as the one for artificial datasets.

Table 3 shows the average results of 20 experiments. LinUCBPR-ER is consistently better than others, and LinUCBPR-UR is competitive with others. The results again confirm that LinUCBPR algorithms are useful under the piled-reward setting, and also again confirm that LinUCBPR-ER to be the best algorithm. We further compare these algorithms with a two-sample t-test at 95 % confidence level in Table 4. The results demonstrate the significance of the strong performance of LinUCB-ER.

Real-World Datasets. Finally, We use two real-world datasets R6A and R6B released by Yahoo! to examine our proposed algorithms. The datasets are the only two public datasets for the CBP to the best of our knowledge. They first appear in ICML 2012 workshop on New Challenges for Exploration and Exploitation 3 and also appear in [10]. They are about the news article recommendation.

Note that the action set of the two datasets are dynamic. To deal with this, we let algorithms maintain a weight vector $\mathbf{w}_{t_i,a}$ for each action. The dimensions of the contexts for R6A and R6B are 6 and 136 separately. The rewards for both datasets are in $\{0, 1\}$, which represent the clicks from users. We note that in R6B, there are some examples that do not come with valid contexts. Hence we remove these examples and form a new dataset, R6B-clean. We use *click through rate* (CTR) to evaluate the algorithms, and use the technique described in [11] to achieve an unbiased off-line evaluation.

---

[1] available from http://www.csie.ntu.edu.tw/~cjlin/libsvmtools/datasets/.

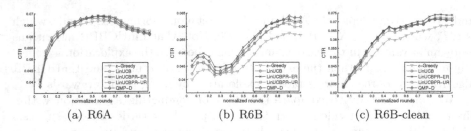

| (a) R6A | (b) R6B | (c) R6B-clean |

Fig. 3. CTR versus round on real-world datasets

We split the datasets into two parts: parameter tuning part and testing part. For R6A, we let $nT = 10000$ for parameter tuning and $nT = 300000$ for testing. For R6B and R6B-clean, we let $nT = 10000$ for parameter tuning and $nT = 100000$ for testing. We consider $n = 500$ for each dataset. The parameter setting is the same as the one for artificial datasets.

Figures 3 shows the experiment results. Unlike the results of artificial datasets, the CTR curve is non-monotonic. This is possibly because the action set is dynamic, and the better actions may disappear from the action set in the middle, which leads to some dropping of CTR.

LinUCBPR algorithms and QPM-D usually perform better than LinUCB and ϵ-Greedy in these datasets. LinUCBPR-ER is stable among the better choices, while LinUCBPR-UR and QPM-D can sometimes be inferior. The results again suggest LinUCBPR-ER to be a promising algorithm for the piled-reward setting.

6   Conclusion

We introduce the contextual bandit problem under the piled-reward setting and show how to apply LinUCB to this setting. We also propose a novel algorithm, LinUCBPR, which uses the pseudo reward to encourage strategic exploration to utilize received contexts that are temporarily without rewards. We prove a regret bound for both LinUCB and the LinUCBPR with estimated reward (-ER), and discuss how the bound compares with the original bound. Empirical results show that LinUCBPR perform better in early time steps, and is competitive in the long term. Most importantly, LinUCBPR-ER yields promising performance on all datasets. The results suggest LinUCBPR-ER to be the best choice in practice.

Acknowledgements. We thank the anonymous reviewers and the members of the NTU CLLab for valuable suggestions. This work is partially supported by the Ministry of Science and Technology of Taiwan (MOST 103-2221-E-002 -148 -MY3) and Asian Office of Aerospace Research and Development (AOARD FA2386-15-1-4012).

References

1. Agarwal, A., Hsu, D., Kale, S., Langford, J., Li, L., Schapire, R.E.: Taming the monster: a fast and simple algorithm for contextual bandits. In: ICML, pp. 1638–1646 (2014)
2. Auer, P.: Using confidence bounds for exploitation-exploration trade-offs. J. Mach. Learn. Res. **3**, 397–422 (2003)
3. Auer, P., Cesa-Bianchi, N., Fischer, P.: Finite-time analysis of the multiarmed bandit problem. Mach. Learn. **47**(2–3), 235–256 (2002)
4. Chou, K.C., Chiang, C.K., Lin, H.T., Lu, C.J.: Pseudo-reward algorithms for contextual bandits with linear payoff functions. In: ACML, pp. 344–359 (2014)
5. Chu, W., Li, L., Reyzin, L., Schapire, R.E.: Contextual bandits with linear payoff functions. In: AISTATS, pp. 208–214 (2011)
6. Dudík, M., Hsu, D., Kale, S., Karampatziakis, N., Langford, J., Reyzin, L., Zhang, T.: Efficient optimal learning for contextual bandits. In: UAI, pp. 169–178 (2011)
7. Dudík, M., Langford, J., Li, L.: Doubly robust policy evaluation and learning. In: ICML, pp. 1097–1104 (2011)
8. Guha, S., Munagala, K., Pal, M.: Multiarmed bandit problems with delayed feedback, arxiv:1011.1161 (2010)
9. Joulani, P., György, A., Szepesvári, C.: Online learning under delayed feedback. In: ICML, pp. 1453–1461 (2013)
10. Li, L., Chu, W., Langford, J., Schapire, R.E.: A contextual-bandit approach to personalized news article recommendation. In: WWW, pp. 661–670 (2010)
11. Li, L., Chu, W., Langford, J., Wang, X.: Unbiased offline evaluation of contextual-bandit-based news article recommendation algorithms. In: WSDM, pp. 297–306 (2011)
12. Mandel, T., Liu, Y.E., Brunskill, E., Popovic, Z.: The queue method: handling delay, heuristics, prior data, and evaluation in bandits. In: AAAI (2015)
13. Wang, C.C., Kulkarni, S.R., Poor, H.V.: Bandit problems with side observations. IEEE Trans. Autom. Control **50**(3), 338–355 (2005)

Incremental Hierarchical Clustering
of Stochastic Pattern-Based Symbolic Data

Xin Xu[1]([✉]), Jiaheng Lu[2], and Wei Wang[3]

[1] Science and Technology on Information System Engineering Laboratory, NRIEE,
Nanjing, China
flora.xin.xu@gmail.com
[2] Department of Computer Science, University of Helsinki, Helsinki, Finland
jiahenglu@gmail.com
[3] State Key Laboratory for Novel Software and Technology, Nanjing University,
Nanjing, China
ww@nju.edu.cn

Abstract. Classic data analysis techniques generally assume that variables have single values only. However, the data complexity during the age of big data has gone beyond the classic framework such that variable values probably take the form of a set of stochastic measurements instead. We refer to the above case as the stochastic pattern-based symbolic data where each measurement set is an instance of an underlying stochastic pattern. In such a case, non existing classic data analysis approaches, such as the crystal item or fuzzy region ones, could apply yet. For this reason, we put forward a novel *I*ncremental *H*ierarchical *C*lustering algorithm for stochastic *P*attern-based *S*ymbolic *D*ata (*IHCPSD*). *IHCPSD* is robust to overlapping and missing measurements and well adapted for incremental learning. Experiments on synthetic and application on real-life emitter parameter data have validated its effectiveness.

Keywords: Symbolic data analysis · Stochastic pattern · Incremental learning · Hierarchical clustering · Emitter parameter analysis

1 Introduction

Classic data analysis techniques are generally designed for variables with single values only. However, when it comes to the age of big data, the data complexity has gone beyond the classic data framework. The values of variables may appear in aggregate form to represent a certain homogeneous behaviours of objects instead, which is referred to as symbolic data. Symbolic data analysis (SDA) approach is of particular interest for huge data set of high complexity when the units are not individual records but some second-order objects.

X. Xu—This work was supported by National Natural Science Foundation of China (No. 61402426, 61373129) and partially supported by Collaborative Innovation Center of Novel Software Technology and Industrialization.

© Springer International Publishing Switzerland 2016
J. Bailey et al. (Eds.): PAKDD 2016, Part II, LNAI 9652, pp. 156–167, 2016.
DOI: 10.1007/978-3-319-31750-2_13

One popular type of symbolic data is the stochastic pattern-based symbolic data, where the variable values are sets of stochastic measurements and each set is an instance of a stochastic process. The second order objects in stochastic pattern-based symbolic data may include the behaviour of some customer of interest in aggregated credit card purchases, the parameter variation patterns of an emitter. In any case, the observed variability for the second-order objects is of utmost interest.

Mean	A1	A2	Type	Measurement Set	A1	A2	Type
o_1	55.65	43	c_1	o_1	\{10.6, 30, 82, 100\}	\{43\}	c_1
o_2	54.1	48	c_1	o_2	\{10.9, 31.5, 79, 95\}	\{48\}	c_1
o_3	54	45	c_1	o_3	\{30, 78\}	\{45\}	c_1
o_4	55.25	44	c_2	o_4	\{21.5, 89 \}	\{44\}	c_2
o_5	55.75	46	c_2	o_5	\{19.5, 92 \}	\{46\}	c_2

(a) Classic data   (b) Stochastic pattern-based symbolic data

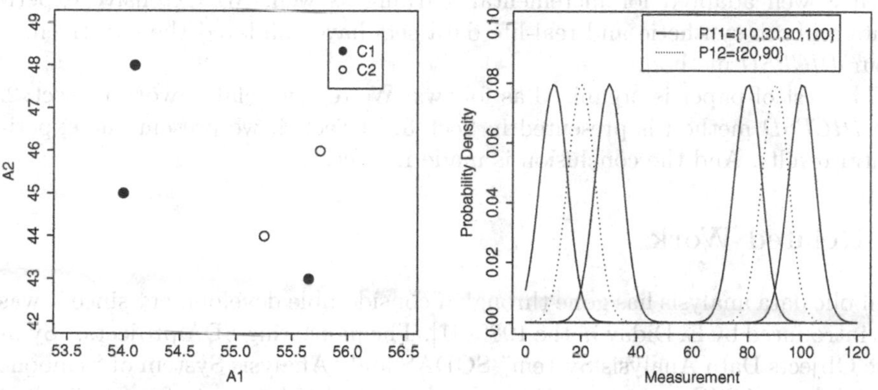

(c) Two classes undistinguishable in the classic data

(d) Two classes distinguishable with the two stochastic patterns

Fig. 1. Comparison of stochastic pattern-based symbolic data and classic data

In classic data analysis, the set of stochastic measurements are generally considered as a multi-valued variable and represented as intervals [13], crystal item sets [14] or fuzzy regions [11,12]. The interval and crystal items are sometimes provided with an associated measure or distribution, such as frequencies, probabilities or weights. However, in practical applications, the situation is even more complicated such that the measurement sets are instances of stochastic patterns whose distributions are heavily overlapping and some measurements may be missing. Therefore, the intervals and item sets become inappropriate. For example, to monitor the physical status of patients, the set of heart rate measurements of individual patients are recorded which follow different stochastic patterns for different individuals. The same goes for radar emitter parameter analysis where the parameter measurement sets from different emitter types comply with the type-specific stochastic patterns.

Figure 1 illustrates a running example of a stochastic pattern-based symbolic data consisted of five objects from two classes. Each object has two attributes ($A1$ and $A2$) and one class label (c_1 or c_2). The objects in the classic data possess a single mean value for each attribute while they have a set of stochastic numeric measurements in the symbolic data instead. The underlying stochastic patterns of attribute $A1$ are denoted as $P11$ and $P12$ respectively. Obviously, the two classes are unable to be discriminated in the classic data but finely distinguishable with the two stochastic patterns.

In this work, we bring forward an incremental hierarchical clustering algorithm for stochastic pattern-based symbolic data ($IHCPSD$) and make the following contributions: (1) We have proposed a novel δ-Jaccard index for evaluation of similarity between a pair of stochastic measurement sets which is robust to overlapping distribution and missing measurements; (2) We have put forward a flexible and effective cluster candidate set model for hierarchical clustering which is well adapted for incremental learning as well; (3) Extensive experiments on both synthetic and real-life data sets have validated the effectiveness of our $IHCPSD$ method.

The rest of paper is organized as follows. We review related work in Sect. 2. Our $IHCPSD$ method is presented in Sect. 3. In Sect. 4, we present the experimental results. And the conclusion is made in Sect. 5.

2   Related Work

Symbolic data analysis has gone through a considerable development since it was first introduced by E. Diday in the 1980s [1]. The pioneering SDA projects, "Symbolic Objects Data Analysis System" SODAS and "Analysis System of Symbolic Official data" ASSO were devoted for a systematic development of symbolic data analysis methodologies. Meanwhile, the first book on SDA, "Analysis of Symbolic Data" [2] was formally published. Though it has been recognized that the values of a symbolic variable could be a symbolic stochastic process as well, not much effort has been made yet [3].

There has been a considerable greater effort in developing methods for interval-valued symbolic data rather than for other types. The benchmark SDA methods, such as the univariate and bivariate descriptive statistics [4], factorial analysis [5], clustering [6], discriminant or unsupervised learning [7], linear regression [8], time series analysis [9], are almost all restricted for interval data. Only a few ones have been adapted for histogram-valued data [9,10].

3   Method

In this section, we formally propose our $IHCPSD$ method which is composed of three major components, pattern similarity evaluation, pattern discovery via

Table 1. Notation

Symbol	Indication
S_i	A set i consisted of stochastic numeric measurements
S_{ip}	The pth numeric measurement in the measurement set S_i
$dist(.,.)$	The distance function for a pair of measurements
C_i	The cluster candidate set for cluster candidate i
C_{ir}	The rth measurement in cluster candidate set C_i
W_i	The set of weights for measurements in cluster candidate set C_i
w_{ir}	The weight of the rth measurement in cluster candidate set C_i
Sup_i	The support of cluster candidate i
$MemSet_i$	The member set of cluster candidate set C_i
$Dis_{Single}(.,.)$	The distance between a pair of cluster candidates with the single linkage
δ	The approximation threshold
$MatchSet(.,.)$	The set of matched pairs between two measurement sets
δ-Jaccard$(.,.)$	The δ-Jaccard index between two stochastic measurement sets
ϵ	The similarity threshold within range $[0,1]$
$minw$	The minimum weight threshold within range $[0,1]$
$minsup$	The minimum support threshold

hierarchical clustering and incremental pattern-based data transformation. The input of *IHCPSD* is the original symbolic data while the output is the set of discovered stochastic patterns and the pattern-based transformed data. Table 1 summarizes the notations in *IHCPSD*.

3.1 Pattern Similarity Evaluation

The Jaccard index is a benchmark statistic for comparing the similarity and diversity of sample sets. However, the traditional Jaccard index only applies for sample sets composed of discrete items or equal-length vectors. This certainly is not our case of stochastic pattern-based symbolic data where the units are consisted of numeric measurements of unequal sizes. To evaluate the pattern similarity between a pair of numeric measurement sets of a certain attribute, S_i and S_j, we propose the δ-Jaccard index based on a specified *approximation threshold* δ and a symmetric distance function $dist()$.

For each measurement $S_{ip} \in S_i$, we find the closest unmatched measurement $S_{jq} \in S_j$ within δ distance away to make a *matched pair* in order. For two measurement sets S_i and S_j, we define their *matched set*, denoted as $MatchSet$ (S_i, S_j), as the set of matched pairs with the shortest distances within δ distance away. Specifically, the first matched pair is the one with the shortest distance below threshold δ, the second pair is the one with the shortest distance among the pairs composed of the remaining unmatched measurements below threshold δ and so on.

Definition 1 *(Match Set). Suppose $S_i = \{S_{ip}\}_p$ and $S_j = \{S_{jq}\}_q$ are two sets of stochastic numeric measurements, given the approximation threshold δ and distance function $dist(S_{ip}, S_{jq}) = \frac{|S_{ip} - S_{jq}|}{max(S_{ip}, S_{jq})}$, the matched set between sets S_i and S_j is the set of matched pairs with the shortest distances below threshold δ, $MatchSet(S_i, S_j) = \{< S_{ip_1}, S_{jq_1} >, < S_{ip_2}, S_{jq_2} >, ..., , < S_{ip_l}, S_{jq_l} >\}$, such that (1) $dist(S_{ip_1}, S_{jq_1}) = min_{S_{ip} \in S_i, S_{jq} \in S_j} dist(S_{ip}, S_{jq})$; (2) assume $\forall 1 \leq k \leq l$, $S_i^k = S_i - \{S_{ip_1}, S_{ip_2}, ..., S_{ip_{k-1}}\}$, $S_j^k = S_j - \{S_{jq_1}, S_{jq_2}, ..., S_{iq_{k-1}}\}$, then $dist(S_{ip_k}, S_{jq_k}) = min_{S_{ip} \in S_i^k, S_{jq} \in S_j^k} dist(S_{ip}, S_{jq})$; and (3) $\nexists S_{ip} \in S_i^l, S_{jq} \in S_j^l$ such that $dist(S_{ip}, S_{jq}) \leq \delta$.*

The δ-Jaccard index is further defined upon the matched set as below:

Definition 2 *(δ-Jaccard index). Suppose $S_i = \{S_{ip}\}$ and $S_j = \{S_{jq}\}$ are two sets of numeric measurements, δ is the specified approximation threshold, the measurement distance function is $dist(S_{ip}, S_{jq}) = \frac{|S_{ip} - S_{jq}|}{max(S_{ip}, S_{jq})}$ and $MatchSet(S_i, S_j)$ is the match set between S_i and S_j, then the δ-Jaccard index between sets S_i and S_j is calculated as*

$$\delta - Jaccard(S_i, S_j) = \frac{|MatchSet(S_i, S_j)|}{|S_i| + |S_j| - |MatchSet(S_i, S_j)|}. \tag{1}$$

The δ-Jaccard index varies between zero and one. Assume the approximation threshold δ is 0.1, for the running example in Fig. 1, the δ-Jaccard index between the measurement sets of attribute $A1$ from object o_1 and o_2 is one. The δ-Jaccard index between object o_1 and o_4 is 0.2.

3.2   Pattern Discovery via Hierarchical Clustering

Based on the δ-Jaccard index, we discover the stochastic patterns by agglomerative hierarchical clustering of cluster candidates. Each cluster candidate is modelled with a *cluster candidate set*. The cluster candidate sets are initialized with the individual measurement sets and then merged iteratively until the δ-Jaccard indexes of all the cluster candidate pairs exceed the specified *similarity threshold* ϵ. During the above agglomerative hierarchical clustering, the measurements, measurement weights, member set and support of the cluster candidate sets would be updated all along the way. The final stochastic patterns would be discovered from the final cluster candidate sets which meet the minimum weight threshold $minw$ and the minimum support threshold $minsup$.

Particularly, the similarity threshold ϵ indicates the minimum $\delta - Jaccard$ index that a pair of cluster candidate sets must obtain for agglomerative merging. The minimum weight threshold $minw$ specifies the minimum weight that the measurements must obtain to be remain in the cluster candidate set. The minimum support threshold $minsup$ specifies the number of measurement sets that the cluster candidate must cover to be identified as a final stochastic pattern.

- **Cluster Candidate Model**

Each cluster candidate is modelled as a weighted *cluster candidate set* $C_i = \{C_{i1}, C_{i2}, ..., C_{i|C_i|}\}$ whose measurement weights are denoted as $W_i = \{w_{i1}, w_{i2}, ..., w_{i|C_i|}\}$, member set as $MemSet_i$ and support as Sup_i. Each weight indicates the probability that the corresponding measurement has a match in the current cluster candidate. The measurement weights of a cluster candidate i all satisfy threshold $minw$, $w_{ik} \geq minw$, $1 \leq k \leq |C_i|$. The member set of a cluster candidate i, $MemSet_i$, is the set of measurement sets that it has merged. The stochastic patterns could be safely discovered as long as the weights of the measurements with missing values satisfy threshold $minw$.

- **Cluster Candidate Initialization**

Each candidate cluster i is initialized with a certain measurement set S_i, such that $C_i = S_i$. The member set of cluster candidate i is initialized as $MemSet_i = \{S_i\}$. Meanwhile, the associated measurement weights and support are all initialized as ones, $w_{i1} = w_{i2} = ... = w_{i|C_i|} = 1$ and $Sup_i = 1$.

For example, with the measurement set S_1 of object o_1 on attribute A_1 in Fig. 1, $S_1 = \{10.6, 30, 82, 100\}$, a cluster candidate set would be initialized as $C_1 = \{10.6(1), 30(1), 82(1), 100(1)\}$, where the associated measurement weights are denoted in brackets, $MemSet_1 = \{S_1\}$ and $Sup_1 = 1$.

- **Hierarchical Clustering**

We make use of the single linkage and the distances between cluster candidates are calculated as the one minus the largest δ-Jaccard index between measurement sets in their member sets.

$$Dis_{Single}(C_u, C_v) = 1 - max_{S_i \in C_u, S_j \in C_v} \delta - Jaccard(S_i, S_j) \qquad (2)$$

During the hierarchical clustering, the pair of cluster candidates $< C_i, C_j >$ with the largest $\delta - Jaccard$ index above threshold ϵ would be merged into a new cluster candidate set $C_{i'}$ with the associated measurement weights $W_{i'}$, support $Sup_{i'}$ and member set $MemSet_{i'}$ iteratively.

The new cluster candidate set $C_{i'}$ is constructed as follows. Firstly, the match set $MatchSet(C_i, C_j)$ between the pair of cluster candidate sets C_i and C_j is inferred. Then, for each matched pair $< C_{ip_k}, C_{jq_k} > \in MatchSet(C_i, C_j)$, $1 \leq k \leq |MatchSet(C_i, C_j)|$, a new measurement $C_{i'r}$ and its associated weight $w_{i'r}$ would be generated according to Eqs. 3 and 4 respectively:

$$C_{i'r} = \frac{C_{ip_k} \times Sup_i + C_{jq_k} \times Sup_j}{Sup_i + Sup_j} \qquad (3)$$

$$w_{i'r} = \frac{w_{ip_k} \times Sup_i + w_{jq_k} \times Sup_j}{Sup_i + Sup_j} \tag{4}$$

On the other hand, for each unmatched measurement within $C_i + C_j - MatchSet(C_i, C_j)$, either $C_{ip'}$ in cluster candidate C_i or $C_{jq'}$ in cluster candidate C_j, the corresponding measurement and associated weight in the new cluster candidate i' would be generated as shown in Eqs. 5 and 6:

$$C_{i'r'} = \begin{cases} C_{ip'} & C_{ip'} \text{ is unmatched} \\ C_{jq'} & C_{jq'} \text{ is unmatched} \end{cases} \tag{5}$$

$$w_{i'r'} = \begin{cases} \frac{w_{ip'} \times Sup_i}{Sup_i + Sup_j} & C_{ip'} \text{ is unmatched} \\ \frac{w_{jq'} \times Sup_j}{Sup_i + Sup_j} & C_{jq'} \text{ is unmatched} \end{cases} \tag{6}$$

Meanwhile, the support and member set of the new cluster candidate i' would be calculated as well as shown in Eqs. 7 and 8:

$$Sup_{i'} = Sup_i + Sup_j \tag{7}$$

$$MemSet_{i'} = MemSet_i \cup MemSet_j \tag{8}$$

Whenever the weight of a measurement is below threshold $minw$, the measurement will be removed from the cluster candidate set. The hierarchical clustering process proceeds iteratively until none existing cluster candidate pairs satisfies threshold ϵ. Finally, the set of measurements in cluster candidate sets that satisfy threshold $minsup$ would be identified as a stochastic pattern.

- **Example**

  For the running example in Fig. 1, given approximation threshold $\delta = 0.1$, similarity threshold $\epsilon = 0.5$, minimum weight threshold $minw = 0.5$ and the minimum support threshold $minsup = 2$, the agglomerative hierarchical clustering proceeds as shown in Fig. 2. There are initially five cluster candidates 1–5 represented by cluster candidate sets C_1, C_2, ..., and C_5. Then, cluster candidates 1 and 2 merge into a new cluster candidate $1'$. So do cluster candidates 4 and 5 which merge into cluster candidate $4'$. Later, cluster candidates $1'$ and 3 merge into $1''$. The final two stochastic patterns above threshold $minsup$ are $C_{1''} = \{10.75(0.67), 30(1), 79.67(1), 97.5(0.67)\}$ and $C_{4'} = \{20.5(1), 90.5(1)\}$, which are rather close to the underlying true patterns.

Dis_{Single}	C_2	C_3	C_4	C_5
C_1	0.0	0.5	0.8	0.8
C_2	-	0.5	0.8	0.8
C_3	-	-	1.0	1.0
C_4	-	-	-	0.0

(a) Initial

Dis_{Single}	C_3	$C_{4'}$
$C_{1'}$	0.5	0.8
C_3	-	1.0

(b) Updated

Dis_{Single}	$C_{4'}$
$C_{1''}$	0.8

(c) Final

Fig. 2. Distance matrices during agglomerative hierarchical clustering

3.3   Incremental Pattern-Based Data Transformation

Our *IHCPSD* method adapts well for incremental pattern-based data transformation on the symbolic data stream. We store the latest symbolic data in G data blocks of equal size $BlockSize$, denoted as D_1, D_2, ..., and D_G and the old data blocks would be discarded. Given the specified approximation threshold δ, the similarity threshold ϵ, the minimum weight threshold $minw$ and the minimum support threshold $minsup$, we discover the set of stochastic patterns from each data block D_g for each attribute, $1 \leq g \leq G$ respectively, denoted as Ω_g. Then, a global merging would be conducted. As long as any two stochastic patterns from different data blocks satisfy the similarity threshold ϵ, we merge them iteratively just as we do in the hierarchical clustering. The set of final stochastic patterns after merging is denoted as Ω.

Meanwhile, we collect the latest $ClsSize$ number of symbolic data records from each class to construct the latest symbolic training data $Train$. The class discriminating power of each discovered stochastic pattern in Ω is then evaluated by the chi-square test on $Train$ and the top M discriminating stochastic patterns are selected for symbolic data transformation.

With each of the M stochastic patterns, the original stochastic measurement sets of each object would be transformed into a δ-Jaccard index. In this way, the original symbolic training data $Train$ would be transformed into a data set composed of M new attributes and the classical machine learning approaches could be applied.

4   Results

We evaluated our *IHCPSD* method on a series of synthetic data sets and one real-life emitter parameter data. Experiments were conducted on a Dell PC running Microsoft Windows XP with a Pentium dual-core CPU of 2.6 GHz and a 4G RAM.

In the synthetic data sets, three overlapping stochastic patterns of length three, five and eight were embedded, where $P_1 = \{30, 60, 90\}$, $P_2 = \{80, 100, 120, 140, 160\}$ and $P_3 = \{90, 150, 180, 200, 220, 240, 260, 280\}$. The measurements for each parameter value p comply with a normal distribution $Norm(p, sd)$, where $sd = c * p$ and coefficient c varied between 0.01 and 0.3. To evaluate the robustness of *IHCPSD* to missing measurements, a missing probability $mprob$ was specified as 20 %. We made use of a data generator with a random variable R for missing measurement simulation. The value of variable R is a random number following a uniform distribution in the range of $[0, 1]$. In case variable R is below $mprob$, the measurement would be missed, otherwise the measurement would be simulated according to the above normal distribution. The number of simulated symbolic data records was varied between $10k$ and $100k$.

The real-life emitter parameter data set was consisted of $12k$ stochastic pattern-based symbolic data records. Each record was consisted of an emitter id, a set of stochastic measurements of PRI (pulse repetition interval) parameter, a numeric RF (radio frequency) parameter value and a label of emitter type.

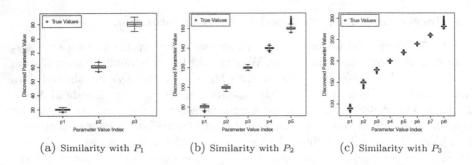

(a) Similarity with P_1　　　(b) Similarity with P_2　　　(c) Similarity with P_3

Fig. 3. Evaluation of pattern discovery on synthetic data

The real-life data had three different types of airborne radar emitters, denoted as "A", "B" and "C" respectively.

4.1 Evaluation of Pattern Discovery on Synthetic Data

We evaluated the effectiveness of *IHCPSD* on the synthetic data sets. In the default setting, we fixed the approximation threshold δ as 0.1, the similarity threshold ϵ as 0.5, the minimum weight threshold $minw$ as 0.5, the minimum support threshold $minsup$ as 0.1 of the data block size, the data block size $BlockSize$ as $10k$ and the number of data blocks G as ten.

We evaluated the similarity between the discovered stochastic patterns and the underlying true patterns. We found that over 90% discovered stochastic patterns achieved a δ-Jaccard index larger than 0.8 with one of the three true patterns. This indicates that the discovered stochastic patterns comply with the underlying true parameter patterns fairly well. The boxplots of these discovered coherent stochastic patterns w.r.t. the corresponding true parameter patterns were given in Fig. 3(a), (b) and (c).

4.2 Evaluation of Pattern Discovery on Real-Life Data

There were three PRI modulation patterns, single, dual and pulse group in the real-life emitter parameter data. In the single PRI modulation, the PRI value across different time slots was unique while for the dual PRI modulation, two different PRI values were observed across the time slots. For the pulse group modulation, on the other hand, pulses were organized in groups, each pulse group with a fixed number of PRI values across different time slots. As can be seen, the PRI modulation patterns across time slots were typical stochastic patterns.

We compared the discovered stochastic PRI patterns against the ground truth modulation patterns provided from domain experts. During experiments, we applied the default parameter setting as the synthetic data.

In Fig. 4, we report the top five stochastic PRI patterns discovered for emitter type "A", denoted as A_1, A_2, ..., A_5 respectively. The estimated PRI measurements in the discovered stochastic patterns were represented as solid points and

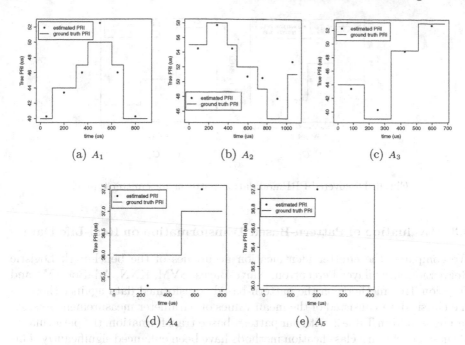

Fig. 4. Discovered PRI modulation patterns for emitter type A

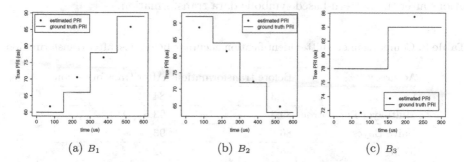

Fig. 5. Discovered PRI modulation patterns for emitter type B

the true PRI modulation pattern values in time slots as horizontal lines. As can be seen, A_1, A_2 and A_3 were in association with the pulse group modulation, A_4 was in association with the dual modulation, while A_5 was in association with the single modulation. Similar observation could be found for discovered stochastic PRI patterns for emitter type "B" and "C", as shown in Figs. 5 and 6.

Comparatively, neither the crystal item-based nor the fuzzy region based pattern mining methods were able to discriminate the delicate emitter parameter patterns. Due to the heavy overlapping between parameter patterns, the traditional discretization strategy and pre-clustering strategy for fuzzy region discovery simply could not work.

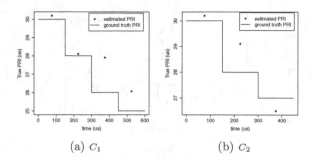

(a) C_1 (b) C_2

Fig. 6. Discovered PRI modulation patterns for emitter type C

4.3   Evaluation of Pattern-Based Transformation on Real-Life Data

We compared the emitter identification accuracies of the benchmark Logistic
Regression, Multilayer Perceptron, Naive Bayes, SVM, KNN, AdaBoostM1 and
Decision Tree methods on the pattern-based transformed data against those on
the classic data consisted of the mean values of parameter measurement sets. As
can be seen from Table 2, with our pattern-based transformation, the performance
of these benchmark classification methods have been enhanced significantly. This
is because we have made a good use of the underlying parameter pattern distributions after the pattern-based symbolic data transformation.

Table 2. Comparison of emitter identification accuracy before v.s. after transformation

Accuracy (%)	Before transformation	After transformation
Logistic	77.3	84
Multilayer Perceptron	89.3	93.5
Naive Bayes	86	93
SVM	86	92
KNN	87.3	92.5
AdaBoostM1	90	96
Decision Tree	75.3	82.5

5   Conclusion

In this paper, we have proposed a novel incremental hierarchical clustering algorithm for stochastic pattern-based symbolic data ($IHCPSD$) to discover stochastic
patterns from numeric set-valued symbolic data. To our knowledge, our $IHCPSD$
method is the first one designed for stochastic pattern-based symbolic data. It's

robust to pattern overlapping and missing values and adaptable for incremental learning. Experimental results on both synthetic and real-life data indicate that *IHCPSD* is effective in identifying delicate stochastic patterns and valuable for enhancing classification performance on symbolic data.

References

1. Diday, E.: Introduction à lapproche symbolique en analyse des données. RAIRO Rech. Opérationnelle **23**(2), 193–236 (1989)
2. Bock, H.-H., Diday, E. (eds.): Analysis of Symbolic Data: Exploratory Methods for Extracting Statistical Information from Complex Data. Springer, Heidelberg (2000)
3. Noirhomme-Fraiture, M., Brito, P.: Far beyond the classical data models: symbolic data analysis. Stat. Anal. Data Min. ASA Data Sci. J. **4**(2), 157–170 (2011)
4. Billard, L.: Sample covariance functions for complex quantitative data. In: Proceedings of the IASC, Joint Meeting of 4th World Conference of the IASC and 6th Conference of the Asian Regional Section of the IASC on Computational Statistics & Data Analysis, Yokohama, Japan (2008)
5. Lauro, C., Verde, R., Irpino, A.: Generalized canonical analysis. In: Diday, E., Noirhomme-Fraiture, M. (eds.) Symbolic Data Analysis and the Sodas Software, pp. 313–330. Wiley, Chichester (2008)
6. De Carvalho, F.A.T., de Souza, R.: Unsupervised pattern recognition models for mixed feature-type symbolic data. Pattern Recogn. Lett. **31**(5), 430–443 (2010)
7. Rasson, J.P., Pircon, J.-Y., Lallemand, P., Adans, S.: Unsupervised divisive classification. In: Diday, E., Noirhomme-Fraiture, M. (eds.) Symbolic Data Analysis and the Sodas Software, pp. 149–156. Wiley, Chichester (2008)
8. Neto, E.A.L., De Carvalho, F.A.T.: Constrained linear regression models for symbolic interval-valued variables. Comput. Stat. Data Anal. **54**(2), 333–347 (2010)
9. Arroyo, J., González-Rivera, G., Maté, C.: Forecasting with interval and histogram data. Some financial applications. In: Ullah, A., Giles, D., Balakrishnan, N., Schucany, W., Schilling, E. (eds.) Handbook of Empirical Economics and Finance. Chapman and Hall/CRC, New York (2010)
10. González-Rivera, G., Arroyo, J.: Time series modeling of histogram-valued data: the daily histogram time series of SP&500 intradaily returns. Int. J. Forecast. **28**(1), 20–33 (2012)
11. Singh, S.K., Wayal, G., Sharma, N.: A review: data mining with fuzzy association rule mining. Int. J. Eng. Res. Technol. (IJERT) **1**(5) (2012)
12. Prabha, K.S., Lawrance, R.: Mining fuzzy frequent itemset using compact frequent pattern (CFP) tree algorithm. In: International Conference on Computing and Control Engineering (ICCCE) (2012)
13. Lin, C.-M., Chen, Y.-M., Hsueh, C.-S.: A self-organizing interval type-2 fuzzy neural network for radar emitter identification. Int. J. Fuzzy Syst. **16**(1), 20–30 (2014)
14. Hahsler, M., Buchta, C., Gruen, B.: arules: Mining Association Rules and Frequent Itemsets. R package version 1.0-10 (2011). http://CRAN.R-project.org/

Computing Hierarchical Summary
of the Data Streams

Zubair Shah[(⊠)], Abdun Naser Mahmood, and Michael Barlow

University of New South Wales, Canberra, Australia
zubair.shah@student.adfa.edu.au

Abstract. Data stream processing is an important function in many
online applications such as network traffic analysis, web applications,
and financial data analysis. Computing summaries of data stream is
challenging since streaming data is generally unbounded, and cannot
be permanently stored or accessed more than once. In this paper, we
have proposed two counter based hierarchical (CHS) ϵ–approximation
algorithms to create hierarchical summaries of one dimensional data.
CHS maintains a data structure, where each entry contains the incoming data item and an associated counter to store its frequency. Since
every item in streaming data cannot be stored, CHS only maintains
frequent items (known as hierarchical heavy hitters) at various levels
of generalization hierarchy by exploiting the natural hierarchy of the
data. The algorithm guarantees accuracy of count within an ϵ bound.
Furthermore, using aperiodic (CHS-A) and periodic (CHS-P) compression strategy the proposed technique offers improved space complexities
of $O(\frac{\eta}{\epsilon})$ and $O(\frac{\eta}{\epsilon} \log \epsilon N)$, respectively. We provide theoretical proofs
for both space and time requirements of CHS algorithm. We have also
experimentally compared the proposed algorithm with the existing
benchmark techniques. Experimental results show that the proposed
algorithm requires fewer updates per element of data, and uses a moderate amount of bounded memory. Moreover, precision-recall analysis
demonstrates that CHS algorithm provides a high quality output compared to existing benchmark techniques. For the experimental validation,
we have used both synthetic data derived from an open source generator, and real benchmark data sets from an international Internet Service
Provider.

Keywords: Hierarchical heavy hitters · Network traffic summarization ·
Heavy hitters · Data streams

1 Introduction

Data stream processing is an emerging domain of applications where data are
modeled not as persistent relations but rather as transient data streams. Examples of such applications include network monitoring, financial applications,
telecommunications data management, sensor networks, web applications, and

© Springer International Publishing Switzerland 2016
J. Bailey et al. (Eds.): PAKDD 2016, Part II, LNAI 9652, pp. 168–179, 2016.
DOI: 10.1007/978-3-319-31750-2_14

so on. The data from these sources are often continuous, rapid, time-varying, possibly unpredictable and unbounded in nature. Such applications cannot afford storing or revisiting the data and often require fast and real-time response.

Data from many domains contain hierarchical attributes, such as Time (Year, Month, Hour, Minute, Second), Geographic Locations (Continent, Country, State, City), IP addresses (192.*, 192.168.*, 192.168.1.*, 192.168.1.1). Analyzing such data at multiple aggregation levels simultaneously– when data is arriving in a stream– is much more challenging (and meaningful) than analyzing flat data. This is known as the Hierarchical Heavy Hitters (HHH) problem. Formally, we compute a ϕ-HHH summary as follows: for a given threshold ϕ, we report only those nodes (prefixes of source or destination IPs) as heavy hitters with frequency exceeding ϕN, after removing the frequency of all its descendant nodes (descendant prefixes of source or destination IPs) that are also heavy hitters. In other words, a reported ϕ-HHH element does not contain the frequency of any other descendant ϕ-HHH element, but may contain the frequencies of non-HHH elements.

A HHH-summary of network traffic is of particular interest to a network monitoring team because it may reveal important patterns in the underlying data. For example, the network traffic may be composed of numerous peer-to-peer traffic connections (such as torrents or online gaming) that can be identified as one generalized traffic connection (composed of many lower level packets having a common pattern) in the HHH-summary.

The exact computation of HHH is not possible without space linearly proportional to at least the number of input elements [1,2], therefore the paradigm of approximation is adopted in a resource constrained environment such as data streams. Consequently, researchers have recently focused on efficient computation (by minimizing memory requirement) of HHH with an acceptable ϵ precision, where ϵ is a given error tolerance between 0 and 1. The algorithms for computing HHH are compared in terms of space usage and update cost that is often bounded using the error ϵ induced by the proposed solution in the estimation.

1.1 Our Contribution

The contributions of this paper are as follows; we have proposed a Counter-based Hierarchical (CHS) ϵ–approximation algorithm that only requires memory bounded in terms of a user supplied parameter $\epsilon < 1$. We provide theoretical proof of accuracy, update cost, and memory requirement. The proposed aperiodic (CHS-A) and periodic (CHS-P) compression strategies offer improved space complexities of $O(\frac{\eta}{\epsilon})$ and $O(\frac{\eta}{\epsilon} \log \epsilon N)$, respectively. Apart from the space complexities and theoretical update costs, experimental results demonstrate that the proposed algorithm requires fewer updates per element of data, and uses a moderate amount of bounded memory. Moreover, precision-recall analysis show that CHS algorithm provides a high quality (e.g., fewer false positives) output compared to existing benchmark techniques on widely compared datasets.

2   Related Work

Computing one-dimensional HHH using the Space Saving algorithm [3] is proposed in [4] using space $O(\frac{\eta^2}{\epsilon})$ which is improved by [5] that require space $O(\frac{\eta}{\epsilon})$. Similarly, another algorithm is proposed in [6] for computing two-dimensional HHH that requires space $O(\frac{\eta^{3/2}}{\epsilon})$. Also, in [7] an algorithm is proposed for finding HHH which is well-suited to commodity hardware such as Ternary Content-Addressable Memories (TCAM). More recently, the technique proposed in [7] is extended in [8] by combining the information provided by the TCAM packet counters with an on-demand sampling mechanism. The sampling mechanism is implemented by a collector device, running on a separate machine. Also, resources allocation for TCAM counters using HHH detection is proposed in [9].

Recently, it has been shown in [10] that the massive graphs can be modeled in a form of hierarchy to extract bicliques using HHH technique. The approach used in [10], however, does not follow the streaming model (i.e., requires multiple passes over the graph) and is just an application of the HHH technique used in graph mining. Also, computing HHH has been used in detection of distributed denial of service attacks [11] and providing customized information to a user based on identifying trends in data [12].

Finally, the propose CHS algorithm is similar to the work from Cormode et al's [13] Full Ancestry (FA) algorithm and Partial Ancestry (PA) algorithm. The space required by their algorithms is $O(\frac{\eta}{\epsilon} \log(\epsilon N))$. Our proposed algorithms has improved on these bounds.

3   Problem Definition

Let $S = \{R_1, R_2, R_3 \cdots\}$ be a continuous stream of records, where each record is characterized by a hierarchical attribute. Let the number of levels or the height of the hierarchy of attribute is denoted by η, which is numbered 0 to $\eta - 1$. As an example, consider a record from network IP data with an IP address 1.2.3.4/32, where the notation x.x.x.x/b is often used to represent IP addresses, and b indicates the number of bits important to distinguish a given IP address. For example, if b = 32, then all the $4 \times 8 = 32$ bits are significant. If b = 24, then only the first 24 bits (the 24 MSB bits) are significant and the remaining 8-bits are disregarded.

Definition 1. *Hierarchical Heavy Hitters: Let ϕ be the support and N is the current count of records processed so far, F_H is the set of all HHH, $F_{H_l} \subseteq F_H$ is set of HHH at level l of the lattice where $0 \leq l \leq \lambda - 1$. Let the symbol $\prec$ denote the descendant–ancestor relation, such that $R \prec R_a$ means that record R_a is the generalization of record R, and $(R \prec R_a) \wedge (R = R_a) \Rightarrow R \preccurlyeq R_a$ then:*

- *F_{H_0} the HHH at level 0 are simply the heavy hitters of S, and*
- *F_{H_l} are HHH at level l of lattice, such that,*

$$(f(R_a) = \sum_{(R \preccurlyeq R_a) \wedge (R \notin F_{H_{l-1}})} f_R) \wedge (f(R_a) \geq \phi N)$$

- $F_H = \bigcup_{l=0}^{\lambda-1} F_{H_l}$

Computation of HHH is a recursive process. The frequency of HHH at any node must contain the frequencies of all its descendant nodes except the nodes that are heavy hitters themselves. At each level it is sufficient to take care of its immediate descendants because the immediate descendants include the frequency of their immediate descendants and so on. This definition require space linearly proportional to the size of input for computing HHH with exact frequencies. In the data stream model with constraint on the resources, HHH detection problem is as follows;

Definition 2. $\epsilon-$*approximate HHH detection problem: Given a data stream S, and user define parameters ϵ and $\phi \in [\epsilon, 1]$, then $\epsilon-$approximate HHH identification problem is to output a set F_H of prefixes and approximate bounds on the frequency of each prefix, such that the following conditions are satisfied;*
1. *Accuracy:* $\hat{f}(R_a) \leq f(R_a) + \epsilon N$
2. *Coverage:* $\forall R_a \notin F_H \Rightarrow (f(R_a) < \phi N)$

The error at a higher level of the hierarchy is larger than the error at a lower level of the hierarchy. Therefore, the precision guarantee at an upper level is not ϵN. However, by an appropriate rescaling of ϵ one can compute the frequencies of the ancestors accurately, nonetheless, with the price of higher space requirements. It had been shown by Hershberger et al. [14] that this factor is unavoidable for computations of HHH, therefore our focus is on providing a methodology to compute the count of HHH from the data structure we maintain.

4   Proposed HHH Algorithms

In this Section, we provide details of $\epsilon-$approximation CHS algorithm for computing HHH.

4.1   Counter-Based Hierarchical Summarization (CHS) Algorithm

In this section, we explain counter-based algorithm for computing HHH. The central idea behind the CHS algorithm is to maintain η number of data structures; one at each level l of the generalization hierarchy of data. The CHS algorithm inserts records from stream into a data structure (i.e., DataStructures[ ] $\mathcal{D}$ where $\mathcal{D}[l]$ is distinguished by an array index $l = 0, 1, \cdots \eta - 1$) at the lowest level of the hierarchy. To keep the space required by the data structure at each level bounded, the algorithm uses a compression strategy, which can either be periodic (every t secs or n records) or after every incoming record R_i. For the periodic compression strategy we use Lossy Counting (LC) algorithm [15], and for the aperiodic compression strategy we use Space Saving (SS) algorithm [3]. The deleted records during compression from lower level data structure are then generalized and inserted at the higher level data structure. When a user query is issued, the algorithm generates the output by starting from the lowest level of the hierarchy and progressing towards

---

Algorithm 1. CHS: An Algorithm for Computing HHH

```
 1: procedure PROCESSSTREAM(R_i, f(R_i))
 2:     D[ ] ; rc++; compression = 'P' or 'A';
 3:     if D[0].contains(R_i) then
 4:         f_{D[0]}(R_i) += f(R_i);
 5:     else
 6:         if compression = 'P' then
 7:             Call CompressP(R_i, f(R_i));
 8:         else
 9:             Call CompressA(R_i, f(R_i));
10:         end if
11:     end if
12: end procedure
13: procedure COMPRESSP(R_i, f(R_i))
14:     cw = ⌈rc/w⌉;
15:     D[0].add(R_i, f(R_i), cw - 1);
16:     if rc%w != 0 then
17:         Return;
18:     end if
19:     for l = 0 to η - 1 do
20:         for each R_j ∈ D[l] do
21:             if (f_{D[l]}(R_j) + cw - 1) < cw then
22:                 D[l].remove(R_j)
23:                 R_a = ancestor(R_j, l + 1)
24:                 if D[l+1].contains(R_a) then
25:                     f_{D[l+1]}(R_a)+ = f_{D[l]}(R_j);
26:                     Δ_{R_a} = max(Δ_{R_j}, Δ_{R_a});
27:                 else
28:                     D[l+1].add(R_a, f_{D[l]}(R_j), Δ_{R_a});
29:                 end if
30:             end if
31:         end for
32:     end for
33: end procedure
```

---

the top. At each level, the output contains all the records whose count exceeds ϕN, after discounting the count of their descendant HHH.

The CHS algorithm (see Algorithm 1) has two simple operations, the first one is *ProcessStream()* which processes each incoming record from stream S, and the second one is a compression operation (*CompressP()* periodic compression or *CompressA()* aperiodic compression). First, each incoming record R_i from stream S is compared with $D[0]$; if R_i exist in $D[0]$, its count is incremented. Otherwise, R_i is passed to one of the compress sub routines, where it is processed for insertion into the data structure $D[l]$. Each entry in $D[l]$ is a tuple $<identifier, f_{D[l]}(R_i), Δ_{R_i} >$ where *identifier* is created from a string representation of R_i, $f_{D[l]}(R_i)$ is the true frequency of the record R_i, and $Δ_{R_i}$ is the error relative to insertion of R_i. The estimated frequency for a given record is computed using $\hat{f}_{D[l]}(R_i) = f_{D[l]}(R_i) + Δ_{R_i}$.

CompressP(): With this option, first, CHS-P inserts $\frac{1}{\epsilon}$ records into the lowest level of $D[l]$, and then perform a compression. During the compression infrequent records are deleted from each level, generalized and inserted into the next higher level of $D[l]$. CHS-P first computes $cw = ⌈\frac{rc}{w}⌉$ (where rc is the number of records counted thus far, and $w = ⌈*\frac{1}{\epsilon}⌉$), and then inserts each incoming

---

Algorithm 2. CHS (Continue)

34: **procedure** COMPRESSA($R_i, f(R_i)$)
35:     $l = 0$;
36:     $T = \{R_i, f(R_i)\}$;
37:     **while** T *is not empty* **do**
38:         $\{R_x, f_T(R_x)\} = T$.remove();
39:         **if** $\mathcal{D}[l]$.contains(R_x) **then**
40:             $f_{\mathcal{D}[l]}(R_x) += f_T(R_x)$;
41:         **else**
42:             **if** sizeOf($\mathcal{D}[l]$)$< \frac{1}{\epsilon}$ **then**
43:                 $\mathcal{D}[l]$.add($R_x, f_T(R_x), 0$)
44:             **else**
45:                 Let R_s be the record with smallest count in $\mathcal{D}[l]$;
46:                 $\Delta_{R_x} = f_{\mathcal{D}[l]}(R_s)$;
47:                 $\mathcal{D}[l]$.remove(R_s);
48:                 $\mathcal{D}[l]$.add($R_x, f_T(R_x) + f_{\mathcal{D}[l]}(R_s), \Delta_{R_x}$)
49:                 $T = \{ancestor(R_s, l+1), f_{\mathcal{D}[l]}(R_s) - \Delta_{R_s}\}$;
50:             **end if**
51:         **end if**
52:         $l + +$;
53:     **end while**
54: **end procedure**
55: **procedure** OUTPUTHHH($\mathcal{D}, \phi$)
56:     **for** $l = 0$ to $\eta - 1$ **do**
57:         **for each** $R_i \in \mathcal{D}[l]$ **do**
58:             **if** $f_{\mathcal{D}[l]}(R_i) \geq \phi N$ **then**
59:                 $F_H = F_H \bigcup \{R_i\}$;
60:             **else**
61:                 $R_a = ancestor(R_i, l+1)$
62:                 **if** $\mathcal{D}[l+1]$.contains(R_a) **then**
63:                     $f_{\mathcal{D}[l+1]}(R_a) += f_{\mathcal{D}[l]}(R_i)$;
64:                 **else**
65:                     $\mathcal{D}[l+1]$.add($R_a, f_{\mathcal{D}[l]}(R_i), \Delta_{R_i}$);
66:                 **end if**
67:             **end if**
68:         **end for**
69:     **end for**
70: **end procedure**

---

record into $\mathcal{D}[0]$. Second, after every $\frac{1}{\epsilon}$ new insertions into $\mathcal{D}[0]$, CHS-P scans $\mathcal{D}[l]$ sequentially from $l = 0$ to $l = \eta - 1$. At any level l of $\mathcal{D}[l]$, if the count of a record R_i is $(f_{\mathcal{D}[l]}(R_i) + \Delta_{R_i}) < cw$, CHS-P deletes R_i from $\mathcal{D}[l]$, generalizes R_i to the next level and inserts it into $\mathcal{D}[l + 1]$. This step, which delete any record whose count is less than cw, is repeated for all the levels of $\mathcal{D}[l]$. Thus, these compression steps keep the data structures $\mathcal{D}[l]$ at each level l under $O(\frac{1}{\epsilon} \log \epsilon N)$.

CompressA(): With this option, CHS-A inserts each incoming record into lowest level of $\mathcal{D}[l]$, and then if this insertion causes a record to be deleted from lower level of $\mathcal{D}[l]$, it is generalized and inserted into next higher level of $\mathcal{D}[l]$. CHS-A inserts incoming record R_i to a container T (that contains only one record and its frequency), and iterate through $\mathcal{D}[l]$ sequentially from $l = 0$ to $l = \eta - 1$, such that T is not empty. CHS-A removes record R_x from T, and first checks if R_x exists in $\mathcal{D}[l]$, if it exist CHS-A just updates its frequency using $f_{\mathcal{D}[l]}(R_x) += f_T(R_x)$ and returns. Second, if the record does not exist in $\mathcal{D}[l]$, CHS-A checks if the number of records in $\mathcal{D}[l]$ (denoted by sizeOf) is less than

$\frac{1}{\epsilon}$, and then inserts the R_x into $\mathcal{D}[l]$ with its frequency $f_T(R_x)$ and Δ_{R_x}. Otherwise, CHS-A finds a record R_s with smallest count in $\mathcal{D}[l]$, replaces it with R_x, and sets the error of Δ_{R_x} equal to $f(R_s)$. Finally, CHS-A generalizes the deleted record R_s to its upper level, and inserts it into T with frequency $f_{\mathcal{D}[l]}(R_s) - \Delta_{R_s}$ for processing in the next iteration that follows the same steps explained above. Thus, just like $CompressP()$, the compression strategy $CompressA()$ keeps the data structures $\mathcal{D}[l]$ at each level under $O(\frac{1}{\epsilon})$.

Finally, at any time to output HHH for a given threshold ϕ, the CHS algorithm scans $\mathcal{D}[l]$ sequentially from $l = 0$ to $l = \eta - 1$. The algorithm outputs the record if $\hat{f}_{\mathcal{D}[l]}(R_i) \geq \phi N$. Otherwise, if $\hat{f}_{\mathcal{D}[l]}(R_i) < \phi N$, then the record is generalized one level up and inserted into the data structure $\mathcal{D}[l+1]$. In this way it extracts all the HHH of the stream at any level of the hierarchy whose frequencies have exceeded the threshold ϕ.

Next, we provide some theoretical observations of CHS algorithm including proof of accuracy and coverage properties of the algorithm.

Theorem 1. *The CHS algorithm satisfies the accuracy and coverage properties from Definition 2.*

Proof. Accuracy: From the theoretical proofs of the compression strategies (LC [15], SS [3]), we are guaranteed to find records from $\mathcal{D}[l]$ with estimated frequency, $\hat{f}_{\mathcal{D}[l]}(R_i) \leq f_{\mathcal{D}[l]}(R_i) + \epsilon N$. The estimated frequency contains an additive error ϵN and requires memory $O(\frac{\eta}{\epsilon} \log \epsilon N)$ or $O(\frac{\eta}{\epsilon})$ (for CHS-P and CHS-A, see *Lemma 2*).

Coverage: Next we prove the coverage requirements satisfied by CHS. The algorithm scans $\mathcal{D}[l]$ sequentially from $l = 0$ to $l = \eta - 1$, and for each l it performs either of the following two operations, (1) if $\hat{f}_{\mathcal{D}[l]}(R_i) \geq \phi N : R_i \in \mathcal{D}[l]$ then it outputs this record, or (2) if $\hat{f}_{\mathcal{D}}(R_i) < \phi N : R_i \in \mathcal{D}[l]$ then it generalizes this record to its upper level $\mathcal{D}[l+1]$. Since operation 1 and 2 are mutually exclusive, this follows that the records generalized by operation 2 do not contain the count from records output (i.e., HHH) by operation 1, i.e., the frequency of HHH is computed as follows:

$$\hat{f}_{\mathcal{D}[j]}(R_a) = \sum_{0 \leq l > j} \hat{f}_{\mathcal{D}[l]}(R_i) \quad st.(R_i \prec R_a) \wedge (R_i \notin F_H) \Rightarrow (R_i \prec R_a) \wedge (R_i \notin F_H)$$

$$\wedge (\hat{f}_{\mathcal{D}[l]}(R_i) < \phi N) \Rightarrow (R_i \prec R_a) \wedge (R_i \notin F_H) \wedge (f_{\mathcal{D}[l]}(R_i) < (\phi - \epsilon)N)$$

Which completes the proof of the coverage requirements of the CHS algorithm.

Lemma 1. *The CHS algorithm performs $O(\eta)$ updates per packet in the worst case.*

Proof. The number of updates in $\mathcal{D}[l]$ depends on the number of compressed records in the previous level data structure $\mathcal{D}[l-1]$. Consider that a record comes into the lowest level of the data structure (i.e., $l = 0$) and gets deleted from here (e.g., because it is not frequent with respect to the current window). This generalized record requires another insertion at the upper level of the data structure (i.e., $l = 1$). In the worst case a record may get deleted at every level of $\mathcal{D}[l]$, except the root, so there can be a maximum of $l - 1$ deletions and η updates. Hence, the worst case update cost by a single record can be $O(\eta)$.

Lemma 2. *The CHS-P algorithm uses $O(\frac{\eta}{\epsilon} \log \epsilon N)$ memory, while CHS-A uses $O(\frac{\eta}{\epsilon})$ memory.*

Proof. The proof of *Lemma 2* is simple; since η is the total number of levels in $\mathcal{D}[l]$, and each level requires $O(\frac{1}{\epsilon} \log \epsilon N)$ or $O(\frac{1}{\epsilon})$ space depending on CSH-P and CHS-A. Therefore, the space required by CHS-P is $O(\frac{\eta}{\epsilon} \log \epsilon N)$, and the space required by CHS-A is $O(\frac{\eta}{\epsilon})$.

5   Implementation and Evaluation

We have compared the CHS-P and CHS-A algorithms with existing FA (termed Full) and PA (termed `Partial`) algorithms for a range of parameters (i.e., ϵ, ϕ, and varying stream length N). We have implemented our proposed algorithms using the Java programming language, and compared them with the open source version of PA [13] and FA [13] algorithms[1]. The data structures used are based on hashing techniques, which require one hashing function to lookup a particular record in the data structure.

Datasets: In our experiments, we have used real Internet traffic datasets [16] that are openly available from the WAND Network Research Group from the University of Waikato, New Zealand. Each of these datasets contain 30 min trace of network traffic in tcpdump[2] format. We used Wireshark[3] to read the tcpdump format data, and to extract the source IPs from each network traffic traces.

Evaluation Criteria: To measure the **efficiency** and **effectiveness** of our proposed algorithms, we have considered different factors including space usage, update efficiency, and **quality** of the approximate output compared to the exact answer. The *space usage* is compared using the maximum number of tuples maintained by the algorithms in their respective data structures. The *update efficiency* is compared using the total number of updates performed during insertion and compression by each algorithm after processing a stream of around 8 Millions records. The *quality of output* is evaluated using a number of measures, for example, we have used Precision, Recall and Dice Coefficient which is the harmonic mean of precision and recall, to evaluate our algorithms. Moreover, we have used **hierarchical measures** such as Optimistic Genealogy Measure (OGM) [17] to compare the output quality of the algorithms.

---

[1] http://www-ai.cs.uni-dortmund.de/SOFTWARE/HHHPlugin/index.html.

[2] http://www.tcpdump.org/manpages/tcpdump.1.html, Accessed: 23/02/2015.

[3] https://www.wireshark.org/, Accessed: 23/02/2015.

Table 1. Updates and Memory comparisons of Full, Partial, CHS-P and CHS-A algorithms. The best value is marked as bold.

	Avg. Updates		Memory (No. tuples)	
	$\epsilon = 0.01$	$\epsilon = 0.001$	$\epsilon = 0.01$	$\epsilon = 0.001$
Full	6.03	5.64	413	2062
Partial	2.01	1.87	**200**	**998**
CHS-P	1.29	1.15	222	1015
CHS-A	**1.26**	**1.13**	400	2053

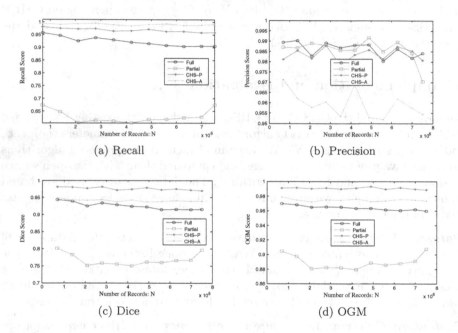

(a) Recall

(b) Precision

(c) Dice

(d) OGM

Fig. 1. Comparisons of Full, Partial, CHS-P and CHS-A algorithms in terms of Recall, Precision, Dice and OGM scores for $\epsilon = 0.01$ and $\phi = 0.02$.

5.1   Update Efficiency

Table 1 compares the number of updates from all four algorithms for $\epsilon = 0.01$ and $\epsilon = 0.001$. It can be seen that for higher values of ϵ (e.g., $\epsilon = 0.01$) all the algorithms performs more updates than for lower values of ϵ (e.g., $\epsilon = 0.001$). The reason is that, for lower values of ϵ, the algorithms have higher space (because space is an inverse function of ϵ), and therefore they delete less number of records compared to higher values of ϵ. The update efficiency of both CHS-P and CHS-A algorithms are better than the update efficiency of both Full and Partial. Particularly, CHS-A is the best algorithm in terms of update cost as it requires the lowest number of updates. The update cost of CHS-P is little higher than CHS-A but it is still better than both Full and Partial.

The Full algorithm requires the highest number of updates of all the four algorithms due to the fact that it has to perform $O(\eta)$ updates for each incoming record, while the remaining three algorithms perform only an $O(1)$ operation for each incoming record. Both Full, and Partial also perform updates during the periodic compression operation. CHS-P and CHS-A algorithms perform an $O(1)$ update for each deleted record into the upper level. We observe from Table 1 that, Partial roughly performs 1.5 times more updates than CHS-P and CHS-A algorithms, and Full roughly perform 5 times more updates than the CHS-P and CHS-A algorithms. This demonstrates that the CHS algorithms are much more efficient than Full and Partial in terms of update performance.

5.2 Output Quality

One of the most important issues with approximate algorithms, such as in the case of HHH, is that these techniques are not guaranteed to find the exact set of HHH (hence they are ϵ-approximate). It is expected that the output may miss some HHH and potentially might include some extra (i.e., non-HHH). Thus to evaluate the proposed algorithms, we have compared them using both flat (e.g., Recall, Precision and Dice coefficient) and hierarchical (e.g., OGM) measures using the exact answer as a frame of reference. All the four measures were plotted against the stream length, in order to observe the performance of the algorithms for varying values of N. We have compared our algorithms for a range of ϵ and ϕ values for different measures, but due to space limitation, we have reported results for $\epsilon = 0.01$, and $\phi = 0.02$ which reflects the overall behavior across the broad range of these parameters.

Figure 1a compares the recall scores of various algorithms. The results clearly show that in terms of not missing true HHH records, both CHS-P and CHS-A perform better than the Full and Partial. Figure 1b compares the precision scores of various algorithms. It can be seen that, both Full and Partial have higher precision than CHS-P and CHS-A algorithms. The high precision of Full and Partial suggest that they almost do not output any non-HHH record, however, their very low recall values shows that they miss a lot of true HHH records. On the other hand, on average both CHS-P and CHS-A may not miss any true HHH records but might output about less than 5 % additional non-HHH records. Moreover, we used other measures (described next) to show the tradeoff between precision and recall.

Figure 1c and d compare the output quality of the CHS-P and CHS-A algorithms against the Full and Partial algorithms in terms of Dice and OGM scores. It can be seen that both CHS-P and CHS-A algorithms outperform both the Full and Partial for the two measures in majority of the cases. In general, the Dice scores of all the algorithms are lower than their corresponding OGM scores. This is because the Dice score shows the exact matching of the actual answer and approximate answer, whereas the OGM measure, in addition to exact matching also take into account the ancestor-descendant relations between an approximate answer and the exact answer. In the case of approximate HHH solutions, if an algorithm does not output an exact HHH, it is highly likely that the algorithm will output its ancestor HHH, because the deleted descendant HHH are generalized and combined at ancestor HHH. Therefore, overall the OGM score of the algorithms are higher than their corresponding Dice score.

In general, our proposed algorithms outperform `Partial` and is consistently better than `Full` for a range of experiments. Thus, we emphasize that even in terms of trade off between precision and recall our proposed algorithms are consistently better than both the `Full` and `Partial` algorithms.

5.3 Space Usage

Table 1 provides the comparison of space usage for `Full`, `Partial`, CHS-P and CHS-A algorithms for $\epsilon = 0.01$ and $\epsilon = 0.001$. Notice that the memory usage of the algorithms is independent of ϕ, i.e., it only depends on ϵ, hence Table 1 does not specify any ϕ values. It can be seen that the memory usage of the `Full` algorithm is highest as it inserts and stores every ancestor of each incoming record. Notice that, the memory usage of CHS-P is almost identical to `Partial`, and the memory usage of CHS-A is identical to `Full`. The memory usage of CHS-P and CHS-A is higher than the memory usage of `Partial`, however, this higher memory usage is a tradeoff for better quality results compared to `Partial` and `Full`. `Partial` uses the lowest memory, but its better memory usage comes at the cost of missing a larger fraction of true HHH records and significantly decreased output quality. We emphasize that the space usage of the CHS algorithm is not a major disadvantage in the situations where high quality output is required as it consumes just a little more memory than `Partial` but produces significantly better results.

We conclude that, `Full` uses largest memory and has highest update cost but it is better than `Partial` in terms of output quality. The `Partial` has low output quality but it is better than `Full` in terms of memory and update performance. On the other hand, CHS is better than both `Full` and `Partial` in terms of update cost and output quality, and uses moderate amount of memory like `Partial`.

6 Conclusion

This paper presented efficient techniques to compute hierarchical heavy hitters from data streams. The theoretical analysis proved that the worst case update cost and space requirements for the CHS algorithm are better than the existing algorithms. The experimental results demonstrated that the output produced by the CHS algorithm for one dimensional data is identical to the exact answer sizes in majority of the cases. The Precision-Recall analysis also suggested that the output produced by the CHS algorithm is of high quality and very close to the exact answers. Finally, accuracy and coverage of the algorithms are theoretically proven to have a bound which can be controlled using user supplied parameter ϵ. This provides the flexibility to make *trade-off* between the space used and the precision required in the estimation.

References

1. Estan, C., Varghese, G.: New directions in traffic measurement and accounting. SIGCOMM Comput. Commun. Rev. **32**(4), 323–336 (2002)
2. Charikar, M., Chen, K., Farach-Colton, M.: Finding frequent items in data streams. In: Widmayer, P., Triguero, F., Morales, R., Hennessy, M., Eidenbenz, S., Conejo, R. (eds.) ICALP 2002. LNCS, vol. 2380, pp. 693–703. Springer, Heidelberg (2002)
3. Metwally, A., Agrawal, D.P., El Abbadi, A.: Efficient computation of frequent and top-k elements in data streams. In: Eiter, T., Libkin, L. (eds.) ICDT 2005. LNCS, vol. 3363, pp. 398–412. Springer, Heidelberg (2005)
4. Lin, Y., Liu, H.: Separator: sifting hierarchical heavy hitters accurately from data streams. In: Alhajj, R., Gao, H., Li, X., Li, J., Zaïane, O.R. (eds.) ADMA 2007. LNCS (LNAI), vol. 4632, pp. 170–182. Springer, Heidelberg (2007)
5. Mitzenmacher, M., Steinke, T., Thaler, J.: Hierarchical heavy hitters with the space saving algorithm, arXiv 1102
6. Truong, P., Guillemin, F.: Identification of heavyweight address prefix pairs in IP traffic. In: 21st International Teletraffic Congress, 2009. ITC 21 2009, pp. 1–8. IEEE (2009)
7. Jose, L., Yu, M., Rexford, J.: Online measurement of large traffic aggregates on commodity switches. In: Proceedings of the USENIX HotICE Workshop (2011)
8. da Cruz, M.A., Correa, S., Cardoso, K.V., et al.: Accurate online detection of bidimensional hierarchical heavy hitters in software-defined networks. In: 2013 IEEE Latin-America Conference on Communications (LATINCOM), pp. 1–6. IEEE (2013)
9. Moshref, M., Yu, M., Govindan, R., Vahdat, A.: Dream: dynamic resource allocation for software-defined measurement. In: ACM SIGCOMM 2014, pp. 419–430. ACM (2014)
10. Hernández, C., Navarro, A.G., Marín, M.: Managing massive graphs, universidad de chile (2014). http://users.dcc.uchile.cl/~gnavarro/algoritmos/tesiscecilia.pdf, Ph.D. thesis, Citeseer (2009)
11. Kalliola, A., Aura, T., Šćepanović, S.: Denial-of-service mitigation for internet services. In: Bernsmed, K., Fischer-Hübner, S. (eds.) NordSec 2014. LNCS, vol. 8788, pp. 213–228. Springer, Heidelberg (2014)
12. Leeder, M.A.: Providing customized information to a user based on identifying a trend, US Patent 8,649,779, 11 February 2014
13. Cormode, G., Korn, F., Muthukrishnan, S., Srivastava, D.: Finding hierarchical heavy hitters in streaming data. ACM Trans. Knowl. Discov. Data (TKDD) **1**(4), 1–48 (2008)
14. Hershberger, J., Shrivastava, N., Suri, S., Tóth, C.D.: Space complexity of hierarchical heavy hitters in multi-dimensional data streams. In: Proceedings of Principles of database systems, pp. 338–347. ACM (2005)
15. Manku, G.S., Motwani, R.: Approximate frequency counts over data streams. In: Proceedings of Very Large Data Bases, VLDB Endowment, pp. 346–357 (2002)
16. Micheel, J., Graham, I., Brownlee, N.: The auckland data set: an access link observed. In: Proceedings of Access Networks and Systems, pp. 19–30 (2001)
17. Ganesan, P., Garcia-Molina, H., Widom, J.: Exploiting hierarchical domain structure to compute similarity. ACM Trans. Inf. Syst. (TOIS) **21**(1), 64–93 (2003)

Anomaly Detection and Clustering

Unsupervised Parameter Estimation
for One-Class Support Vector Machines

Zahra Ghafoori[1(✉)], Sutharshan Rajasegarar[2], Sarah M. Erfani[1],
Shanika Karunasekera[1], and Christopher A. Leckie[1]

[1] Department of Computing and Information Systems, The University of Melbourne,
Melbourne, Australia
{ghafooriz,sarah.erfani,karus,caleckie}@unimelb.edu.edu
[2] School of Information Technology, Deakin University, Geelong, Australia
sutharshan.rajasegarar@deakin.edu.au

Abstract. Although the hyper-plane based One-Class Support Vector
Machine (OCSVM) and the hyper-spherical based Support Vector Data
Description (SVDD) algorithms have been shown to be very effective
in detecting outliers, their performance on noisy and unlabeled training data has not been widely studied. Moreover, only a few heuristic
approaches have been proposed to set the different parameters of these
methods in an unsupervised manner. In this paper, we propose two unsupervised methods for estimating the optimal parameter settings to train
OCSVM and SVDD models, based on analysing the structure of the
data. We show that our heuristic is substantially faster than existing
parameter estimation approaches while its accuracy is comparable with
supervised parameter learning methods, such as grid-search with cross-
validation on labeled data. In addition, our proposed approaches can
be used to prepare a labeled data set for a OCSVM or a SVDD from
unlabeled data.

Keywords: One-Class Support Vector Machine · Support vector data
description · Outlier detection · Parameter estimation

1 Introduction

Abnormal patterns in a data set, which are inconsistent with the majority of the
data, are commonly referred to as outliers or anomalies. In many applications,
such as fraud detection, environmental monitoring, and medical diagnosis, one
of the main tasks is to detect such instances or to remove them [10]. The two
major underlying assumptions of many existing outlier detection methods are
the rarity of outliers and the distinctive differences between them and the normal
data [1].

In general, outlier detection algorithms can be categorized as supervised,
semi-supervised or unsupervised learning methods [1,7]. The former case
assumes that both negative and positive labels are available to train a binary

© Springer International Publishing Switzerland 2016
J. Bailey et al. (Eds.): PAKDD 2016, Part II, LNAI 9652, pp. 183–195, 2016.
DOI: 10.1007/978-3-319-31750-2_15

classifier, while the latter one does not make any assumption regarding the availability of a labeled data set [7]. In comparison with these two approaches, semi-supervised methods assume that only the normal examples are available during training, which makes it possible to build a model of normality that rejects anomalous instances [1,7]. For unsupervised and semi-supervised methods, if it is assumed that the majority of the training data is normal, the methods are also categorised as one-class classification. In this paper, we mainly focus on one-class classification, and interested readers are referred to [1,7] for more comprehensive surveys.

The OCSVM [14] and SVDD [19] algorithms are two widely used one-class classification methods for outlier detection [4,6,9,11,15]. It has been shown that the OCSVM and SVDD algorithms handle small fractions of outliers in the training set [14,19], but if a considerable proportion of such examples exist, both algorithms may end up producing models that are skewed towards outliers [10]. Unfortunately, the availability of a (nearly) clean training data to avoid this problem is not guaranteed in many real applications. Moreover, contributing "good" examples of outliers, i.e., ones that do not lie on normal regions and are far from the normal data points, is sometimes necessary to boost the performance of the OCSVM and SVDD algorithms [19], but we may have no prior knowledge about such examples. Finally, both algorithms have some data dependent parameters whose value can substantially affect the accuracy of the method, and estimating these parameters in an efficient and unsupervised way is an open research problem. Usually, the feature space is searched via grid-search and cross-validation, which are computationally expensive and require labeled examples from both the normal and outlier classes.

This paper addresses the aforementioned problems in the following ways: (i) we propose two fully unsupervised methods to analyse the structure of the data and make a near-optimal estimation of the parameter settings in an efficient way in comparison with existing methods, (ii) we show how our methods can be used to restrict the domain of search in grid-search and improve its efficiency, (iii) we show the application of our proposed methods in pre-processing an unclean data set, comprising a considerable fraction of outliers, and building a labeled data set comprising the normal data and good examples of outliers.

2   Background and Related Work

In this section, a brief explanation of the OCSVM and SVDD algorithms is presented, followed by a review of the related works that have been proposed to find optimal parameter settings for OCSVM or SVDD training.

2.1   One-Class Support Vector Machines

The OCSVM or ν-SVM [14] algorithm is a semi-parametric one-class classification method that finds a boundary around dense areas comprising the normal data [7]. In OCSVM, a training set of $x_i \in R^d (i = 1, 2, ..., l)$ feature vectors are projected to a potentially higher dimensional space using a feature map φ. Then, the algorithm finds a hyper-plane that separates the projected examples from the origin with the maximum possible margin. The primal quadratic problem that the OCSVM classifier solves is as follows:

$$\min_{\omega,\xi,\rho} \frac{1}{2}\|\omega\|^2 + \frac{1}{\nu l}\sum_{i=1}^{l}\xi_i, \ s.t. \ (\omega.\varphi(x_i)) \geq \rho - \xi_i; \ \xi_i \geq 0 \ \forall i, \tag{1}$$

where $\omega \in R^d$ and $0 < \nu \leq 1$. In addition, $\xi_i \geq 0$ are slack variables that relax the problem constraints and allow some examples to fall outside the model boundary. Any given solution for this optimization problem has three separate sets of examples: examples that fall inside the boundary (non-support vectors), examples that lie on the boundary (border support vectors), and examples that fall outside the boundary (outliers or bounded support vectors). One of the important properties of OCSVM is that the user-defined parameter ν is an upper bound on the fraction of outliers and a lower bound on the fraction of support vectors. Using a kernel function as φ, like a Gaussian kernel ($k(x,y) = e^{-\gamma\|x-y\|^2}$) with the kernel parameter γ, it is possible to apply the kernel trick and separate normal data points and outliers that are not linearly separable in the input space. After the training phase, the label of any unseen data x is simply predicted using the decision function $f(x) = sign((\omega.\varphi(x)) - \rho)$.

SVDD [19] has a similar optimization function (Eq. 2), but instead of a hyper-plane, it minimizes the radius R of a hyper-sphere that encompasses almost all normal samples:

$$\min_{R,a} R^2 + C\sum_{i=1}^{l}\xi_i, \ s.t. \ \|x_i - a\|^2 \leq R^2 + \xi_i; \ \xi_i \geq 0 \ \forall i, \tag{2}$$

where a is the center of the hyper-sphere and C is a user-defined regularization parameter that has a similar effect as the ν parameter in a OCSVM.

Assuming two training examples x and y, if an applied kernel only depends on $x - y$, i.e., the kernel is stationary, the SVDD and OCSVM algorithms result in equal solutions [14]. As we use the RBF kernel throughout this paper, which is stationary based on the proposed definition, and the ν and C parameters can be defined based on each other using the $\nu = \frac{1}{Cl}$ formula [18], hereafter, we assume that the discussions made for a OCSVM are also valid for a SVDD.

2.2 Estimating Parameter Settings for the OCSVM Algorithm

The choice of the values for γ (the kernel width parameter) and ν (the regularisation parameter) has a major influence on the accuracy of a model generated by the OCSVM or SVDD algorithms [12,18]. To illustrate the effect of the parameters, we have designed an experiment with a toy problem named Half Kernel, that includes 4,000 normal samples and 5% outliers, which were added to the normal data at random using a uniform distribution. Figure 1 shows the different models that have been generated using the OCSVM algorithm with different

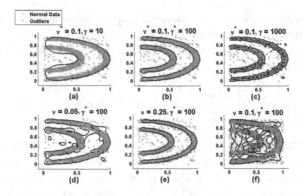

Fig. 1. Sensitivity of a OCSVM to the γ and ν parameters (Half Kernel data set).

values of the γ and ν parameters. The best model, i.e., Fig. 1(b), has been built using the optimal parameter settings (ν^*, γ^*). Figure 1(a), (c) show that choosing γ values less than the optimal value results in building overly general and simple models with a high false positive rate (FPR) with respect to the target class, whereas larger values are prone to building poor models with a high false negative rate (FNR). The ν parameter affects the results in the same manner as shown in Fig. 1(d), (e). To illustrate how the optimal value of this parameter depends on the data, in Fig. 1(f) we have added another 5 % anomalies to the Half Kernel data set and used the same setting as in Fig. 1(b) to train a OCSVM model. This figure shows how dramatically the model may deviate towards outliers, if the ν parameter is not set correctly. Consequently, the optimal ν and γ parameter values depend on the given data set, and thus we require a method to select these parameter values from the data. We now summarize two families of approaches to this problem, namely, supervised and unsupervised learning approaches.

Supervised Learning of the Parameters: If ground truth labels are available, it is possible to optimize the choice of values for the ν and γ parameters via n-fold cross-validation. Zhuang et al. [21] suggested that the most reliable approach to search the parameter space in this manner is to use grid-search, and due to its considerable computational requirements, an efficient parameter search algorithm should be used. To this end, they have first applied a coarse-grained search over the entire parameter space and then performed two further fine-grained searches to reduce the complexity by restricting the search space.

Unsupervised Learning of the Parameters: A major drawback of supervised learning in this context is the need for ground truth labels. Consequently, several heuristic unsupervised approaches have been proposed to estimate the γ and ν parameters when the training set is not clean and no ground truth labels are available to find an optimal parameter setting. Emmott et al. [5] have assumed prior knowledge about the value of the ν parameter. Then, the value of the γ parameter has been increased until a predefined proportion of the data has been rejected. Since there is usually more than one pair of parameters that result in approximately this predefined proportion of outliers, this approach may not be successful in finding an optimal parameter setting. Since the proportion of outliers might be unknown, Rätsch et al. [12] proposed a heuristic to find an appropriate ν value for a OCSVM. Their main assumptions are that outliers are far enough from normal samples and the γ parameter is known, and the idea is to increase ν over the range $(0, 1)$ to find a value that maximizes the separation distance between the normal class and the rejected samples. The distance is defined by the following equation:

$$D_\nu = \frac{1}{N^+} \sum_{f(x) \geq \rho} f(x) - \frac{1}{N^-} \sum_{f(x) < \rho} f(x), \tag{3}$$

where N^+ and N^- are the number of samples in the target and outlier classes, respectively. Rätsch et al. have reported that if there is no clear separation between negative and positive samples, the proposed heuristic may come up with extreme solutions, i.e., 0 or 1. Moreover, the choice of the γ parameter and its effect on finding a good value for the ν parameter has not been discussed in their work. Liu et al. [10] have estimated the γ parameter as $\frac{1}{\gamma} = 2 \sum_{i,j=1}^{l} \|x_i - x_j\|^2 / l^2$, which can be used in combination with Rätsch's method to estimate both the γ and ν parameters in a fully unsupervised manner. Hereafter, we call this method Duplex Max-margin Model Selection (DMMS) as it is based on the max-margin principle and maximizes the separation between the two classes.

Tax et al. [18] have proposed a heuristic to estimate the γ and ν parameters for a SVDD in a fully unsupervised manner. Their proposed heuristic optimizes the estimated FP and FN rates by solving the following minimization problem:

$$\Lambda(\gamma, \nu) = \frac{|SV|}{N} + \lambda |SV_b| \left(\frac{1}{\gamma^{0.5} s_{max}} \right)^d, \tag{4}$$

where SV and s_{max} represent the set of support vectors and the maximum distance in the training set, respectively. SV_b indicates the set of border support vectors (i.e., those with $0 < \alpha_i < C$), and λ is a regularizer. The first term ($\frac{|SV|}{N}$) is an estimate of the error on the target class, and the second term controls the error on the outlier class. Since the RBF kernel has been used for this heuristic, it is also possible to use the same parameter setting to train a OCSVM. Hereafter, we refer to this heuristic as Duplex Error-minimisation Model Selection (DEMS) as its objective is to minimize FPR and TPR.

There is another work by Liu et al. [10] that uses an unsupervised self-guided soft labeling mechanism to train a one-class classifier, different from the OCSVM and SVDD methods, by applying the soft labels directly in the optimization problem of the studied one-class classification algorithm, which is very different from the aim of this paper and so is not discussed here.

In addition to estimating the parameters, Suvorov et al. [17] and Tax et al. [19] have proposed to use samples from the outlier class directly in the optimization function of a OCSVM or a SVDD to boost the accuracy. However,

Tax et al. [19] have shown that choosing "poor" outlier examples, i.e., outliers that fall inside or very close to the target class, reduces the accuracy of the trained model to be similar to a random classifier. They have also discussed that if only examples from the target class are available, generating synthetic outliers in low density regions can help tighten the data description and enhance accuracy, but an automatic method to generate such examples has not been proposed.

We summarize the shortcomings of the existing methods as follows:

1. Even a moderately high resolution grid-search may incur a substantial number of iterations and high time-complexity. Moreover, the granularity of the search can have a major effect on the final result.
2. In many applications, examples from the outlier class are not available for use in finding an optimal parameter setting via cross-validation. Moreover, it is not assured that a nearly clean data set of normal samples is available during training. These problems have not been studied by the existing approaches.
3. Even if negative examples are available as well as positive ones, based on the work presented by Tax et al. [19], it is not guaranteed that their contribution improves the accuracy of the trained model, unless they are far enough from the target class. None of the existing methods resolves this problem.
4. All the existing unsupervised parameter estimation methods (except DEMS) assume prior knowledge of either γ or ν, but both parameters may be unknown in many applications. This makes it impossible to optimise one parameter based on knowing the other one.
5. The DEMS method is a fully unsupervised method, but it requires a mechanism like grid-search to search over the parameter space and suffers from the time-complexity problem in point 1 above. Moreover, this approach has been examined on only a limited number of data sets.

3   Problem Statement

We are given an unlabeled data set DS comprising unlabeled data points $x_i \in R^d (i = 1, 2, ..., l)$ from the normal and outlier classes. Like [16], we assume that outliers are uniformly distributed in the feature space. Our aim is to find a compact region in the search space for the parameters γ and ν that contains the optimal parameter settings γ^* and ν^* for a OCSVM with RBF kernel. Once such a compact region has been found, we can either directly estimate the optimal settings, or efficiently apply a grid-search method within this compact region. In this way, we can estimate the optimal parameter settings for training a OCSVM (or SVDD) without requiring ground truth labels or resorting to exhaustive grid-search, even when the fraction of outliers in DS is high.

In addition, we aim to develop a method of pre-processing the data set DS that can: (1) filter "border-line" data points that can affect the accuracy of the learned OCSVM model; and (2) add synthetic outliers in low density regions of nearly clean data sets to enhance the accuracy of the trained OCSVM by generating labeled data sets.

In the following section we propose two unsupervised methods to automatically estimate optimal parameter values γ^* and ν^*, which address the shortcomings that were identified in Sect. 2 for the existing methods to this problem.

4   Methodology

We divide the problem of finding optimal parameter settings into two steps: (1) estimating the γ parameter, and (2) estimating the ν parameter.

Estimation of γ: Recall from Sect. 2 that the γ parameter is the bandwidth parameter of the RBF kernel, which acts as a scaling factor to smooth the learned density estimate to reflect the true data density. Lihi et al. [20] proposed a method that estimates a local scaling factor for each sample x_i in an affinity matrix $A \in R^{l \times l}$ ($A_{ij} = exp(-\gamma_i\gamma_j d^2(x_i, x_j))$, where $d(.,.)$ is some distance metric). They used the distance between x_i and its Kth nearest neighbor to obtain an estimate of γ_i, and showed that setting $K = 7$ results in good estimates even for high-dimensional data sets.

Inspired by Lihi et al. [20], we introduce a density measure s_K^i for each data point x_i in the training set DS:

$$s_K^i = \frac{1}{K} \sum_{k \in KNN_i} \|x_i - x_k\|; \quad \forall i = 1..l, \tag{5}$$

where KNN_i is the set of the K nearest neighbors of x_i inclusive. The density measure s_K^i reflects the density of points around point x_i. Our challenge is to find the value $s_K^{i^*}$ of a point x_{i^*} that corresponds to the "limit" of the density of normal points, and can thus be used to estimate the γ parameter. To do this, we define an ordered set $S_K = \{s_K^i, \forall i = 1..l | S_k^m \geq S_K^{m-1}, \forall m = 1..l\}$, which can be used to fit a function $FS(m)$ to visualize the densities of points in DS. It can be shown that for data sets that follow the definition of DS in Sect. 3, the function $FS(m)$ is similar to Fig. 2. We propose that the knee-point in this monotonically increasing function, which is shown in Fig. 2 by a circle, carries important information that can be used to set the γ parameter. The knee-point actually represents a sudden change in the densities where we have the normal points near to the data boundaries followed by the outliers.

For a given monotonically increasing function $f(x)$, a knee-point is a point with maximum curvature. The curvature at each point x of the function $f(x)$ is defined below as $C_f(x)$ [13], hence a knee-point can be formulated as Eq. 6.

$$x_{C_f}^{max} = \arg \max_x C_f(x), \text{ where } C_f(x) = \frac{f''(x)}{(1 + f'(x)^2)^{0.5}}. \tag{6}$$

Using this definition, the knee-point of the function $FS(m)$ is $m_{C_{FS}}^{max}$. Thus we set $\gamma = \frac{1}{FS(m_{C_{FS}}^{max})}$ in our heuristic. Later in Sect. 5, we show that this heuristic works reasonably well for a variety of data sets with different inherent structures.

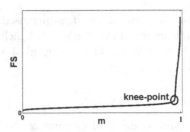

Fig. 2. Illustration of the knee-point of a monotonically increasing function $FS(m)$.

We now propose two variants of our heuristic that provide a means of estimating the ν parameter.

Quick Model Selection (QMS): The information gained via the knee-point of $FS(m)$ can be utilized to estimate the ν parameter as well. As the knee-point is an indicator of a sudden change in $FS(m)$, it shows that densities s_K^i greater than $m_{C_{FS}}^{max}$ are very rare in the training set. As a result, we argue that samples x_i with this property are good representatives of outliers, which leads us to set $\nu = \frac{|S_K| - m_{C_{FS}}^{max}}{|S_K|}$. This is similar to using the K Nearest Neighbor (K-NN) method to detect outliers in a data set, with the key difference that we use this unsupervised method just once in the pre-processing step to estimate an optimal parameter setting for a OCSVM, and after training the OCSVM, unlike the K-NN method, there is no need to compute distances for the test instances.

Another important property of QMS is that it can be used to automatically select good examples of outliers for training purposes. To this end, we introduce a shrinking factor η in the range $(0, 1]$ that can be used to safely divide samples into three groups:

- Normal $(s_K^i < \eta \times FS(m_{C_{FS}}^{max}))$
- Outlier $(s_K^i > (2 - \eta) \times FS(m_{C_{FS}}^{max}))$
- Border-line $(\eta \times FS(m_{C_{FS}}^{max}) \leq s_K^i \leq (2 - \eta) \times FS(m_{C_{FS}}^{max}))$

Now, we can remove the border-line samples from the data set and label the outlier examples as they are sufficiently far from the normal samples. This process is shown in Fig. 3 for a Banana data set comprising 10,000 normal instances and 20 % anomalies, which were generated using a uniform distribution. This example also illustrates the robustness of our method to the percentage of outliers in the training set.

Revised DMMS (RDMMS): We also propose to use our heuristic to estimate the γ parameter for the DMMS method, which was explained in Sect. 2. We further modify the distance metric D_ν (Eq. 3) as Eq. 7 below, for we have found it to be a more practical metric in our experiments:

$$D_\nu = median_{f(x) \geq \rho} f(x) - median_{f(x) < \rho} f(x). \tag{7}$$

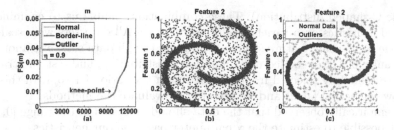

Fig. 3. Pre-processing a Banana data set using QMS; (a) dividing samples into Normal, Border-line, and Outlier sets, (b) the original data set, (c) pruned labeled data set including good examples of outliers.

Our proposed heuristic approaches can address the challenge of the time-complexity of grid-search, by providing an initial guess for the optimal parameter setting and substantially reducing the search space. In the next section, we evaluate our heuristics on a variety of data sets in terms of their accuracy and run-time, and compare them with supervised grid-search and several existing unsupervised approaches.

5 Experimental Evaluation

We evaluated our proposed methods in comparison to the DEMS and DMMS methods. Similar to [20], we set $K = 7$ in our proposed approaches. Since Tax et al. [18] have reported that the DEMS method is not sensitive to the value of the λ parameter, its default value ($\lambda = 1$) was used in our experiments. We also implemented a supervised grid-search method, including two phases of coarse-grained and fine-grained search based on the proposed method by Hsu et al. [8], to make sure that the time-complexity is kept low. This method applies a 10-fold cross validation to find optimal parameter settings.

Several real and synthetic data sets were used to evaluate the accuracy and time-complexity of the methods. To evaluate accuracy, we used the Receiver Operating Characteristic (ROC) curve and the corresponding Area Under the Curve (AUC) as it is insensitive to class balance. The reported AUC values were averaged over 200 runs. The experiments were conducted on a machine with an Intel Core i7CPU at 3.40 GHz and 16 GB RAM. MATLAB LIBSVM toolbox (version 3.20) [2] was used to implement the OCSVM method.

5.1 Data Sets

We ran our experiments on 7 real data sets from the UCI Machine Learning Repository, namely Contraceptive Method Choice (CMC), Cardiotocography (Cardio), Breast Cancer Wisconsin (Cancer), Ozone Level Detection (Ozone), Forest CoverType (Forest), Shuttle, and Vowel. We also generated 3 synthetic data sets: a Banana, a C-shape and a Smile (including a mixture of two Gaussians and one C-shaped distribution). These combinations enable us to examine our proposed heuristics on a variety of data structures. All data sets were scaled in the range [0, 1] using feature scaling technique. For all data sets, 5 % anomalies in the range [0, 1] were added using a uniform distribution, and test and training sets were randomly selected from the data with the ratio of 1 to 4. In this way we know the actual labels and we are able to evaluate the methods.

We empirically observed that the data should be scaled in the range [0, 1] to make it possible to estimate the γ parameter based on our heuristics.

5.2   Results and Discussion

Table 1 reports the accuracy and run-time of the examined methods. As the traditional supervised grid-search (S-Grid-S) method has considerable computational requirements, we set an upper-bound equal to 10,000 data points on the size of the whole data set in all the experiments reported in this table.

Based on the reported results in Table 1, our proposed methods outperform the existing unsupervised parameter estimation methods (i.e., DEMS and DMMS). In comparison with the S-Grid-S method, on 3 real data sets (Cancer, CMC, and Vowel) our methods result in considerably higher accuracy and lower time complexity, while the S-Grid-S method outperforms our QMS method only on Forest and Ozone, and for the rest of the data sets their accuracy is almost the same. To identify the statistical significance of results between the two approaches with highest AUC, i.e., QMS and S-Grid-S, we conducted a $t-$test with a level of significance of $\alpha = 0.05$. The returned $p = 0.63$ for the accuracy measure fails to reject the null hypothesis with a level of significance, i.e., the difference between the AUC of the two approaches is not statistically significant. Moreover, the returned $p = 0.034$ for the training time indicates that the time-complexity of our method is significantly lower than the S-Grid-S method. Since $l \gg d$ in our experiments, the time-complexity of S-Grid-S using traditional matrix inverse is $O(l^3)$ [3], while the most expensive part of our QMS method, i.e., finding the K nearest neighbors, requires $O(l^2)$. Note that S-Grid-S requires a labeled data set to find the optimal parameter settings, but our proposed methods are completely unsupervised, i.e., learning of the parameter settings is performed without having access to the labels (the labels have been used only for testing purposes).

To compare the scalability of the two methods with the best accuracy, i.e., S-Grid-S and QMS, we have conducted another experiment with the Forest data set. The number of data points has been increased between 10 K and 500 K, and we have given both methods at most 6 h to find the optimal parameter settings. As shown in Fig. 4, our QMS method successfully finds the parameter settings in this time limit, but the running time of the S-Grid-S method exceeds the limit even for a data set of 20 K samples.

Table 1. Accuracy and time-complexity of our proposed unsupervised parameter estimation methods (QMS and RDMMS) in comparison with existing supervised (S-Grid-S) and unsupervised (DEMS and DMMS) methods.

Data set	#Features	AUC					CPU_Time (in seconds)				
		DEMS	DMMS	S-Grid-S	QMS	RDMMS	DEMS	DMMS	S-Grid-S	QMS	RDMMS
Cancer	10	0.020	0.691	0.706	**0.813**	0.781	2.50	0.42	34.06	0.04	0.64
Cardio	22	0.632	0.876	**0.964**	0.958	0.959	31.46	6.65	319.71	0.27	7.13
CMC	9	0.522	0.567	0.766	**0.836**	0.805	9.57	2.08	112.40	0.13	2.45
Forest	54	0.591	0.883	**0.985**	0.958	0.958	351.93	155.33	10176.60	7.10	193.41
Ozone	72	0.697	0.877	**0.981**	0.942	0.941	83.74	14.69	689.87	0.53	21.81
Shuttle	9	0.683	0.596	**0.999**	0.995	0.997	236.24	71.17	1555.03	4.30	87.93
Vowel	10	0.784	0.867	0.927	**0.957**	**0.951**	3.77	0.70	49.45	0.06	0.92
Banana	2	0.850	0.570	**0.900**	0.896	0.849	200.04	59.05	2034.83	4.21	145.19
C-shape	2	0.894	0.554	**0.901**	0.898	0.840	196.41	58.34	2277.96	4.15	122.85
Smile	20	0.696	0.596	0.981	**0.991**	**0.992**	387.90	122.02	7966.55	5.29	148.92
Average		0.697	0.708	0.920	0.924	0.907	150.36	49.05	2521.65	**2.61**	73.13

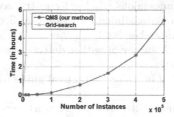

Fig. 4. Scalability of QMS in comparison with the supervised grid-search method on the Forest data set (it took more than 24 h for the S-Grid-S method to find optimal parameter settings when the size of the data set is 20 K).

6   Conclusion

We proposed two parameter estimation algorithms, namely QMS and RDMMS, for the OCSVM and SVDD algorithms, which estimate optimal parameter settings without any need for ground truth labels or exhaustive grid-search over the parameter space. Our experimental evaluation showed that our methods outperformed existing heuristic approaches that found the parameter settings in an unsupervised manner. Moreover, our QMS method had comparable accuracy to the supervised grid-search method, while it was in average more than 900 times faster than the supervised-grid search method on the examined real and synthetic data sets. The QMS method also outperformed all the existing methods in terms of time-complexity. In future work, we aim to use this heuristic in training OCSVMs for concept-drifting data streams, where we need to train a new model using the recent data.

References

1. Chandola, V., Banerjee, A., Kumar, V.: Anomaly detection: a survey. ACM Comput. Surveys (CSUR) **41**(3), 15 (2009)
2. Chang, C.C., Lin, C.J.: LIBSVM: a library for support vector machines. ACM Trans. Intell. Syst. Technol. (TIST) **2**(3), 1–27 (2011)
3. Chapelle, O.: Training a support vector machine in the primal. Neural Comput. **19**(5), 1155–1178 (2007)
4. Chen, Y., Zhou, X.S., Huang, T.S.: One-class SVM for learning in image retrieval. In: Proceedings of the International Conference on Image Processing, vol. 1, pp. 34–37, IEEE, October 2001
5. Emmott, A.F., Das, S., Dietterich, T., Fern, A., Wong, W.K.: Systematic construction of anomaly detection benchmarks from real data. In: Proceedings of the ACM SIGKDD Workshop on Outlier Detection and Description, pp. 16–21. ACM (2013)
6. Heller, K., Svore, K., Keromytis, A.D., Stolfo, S.: One class support vector machines for detecting anomalous windows registry accesses. In: Proceedings of the ICDM Workshop on Data Mining for Computer Security (DMSEC), pp. 2–9, Melbourne, FL, USA, November 2003
7. Hodge, V.J., Austin, J.: A survey of outlier detection methodologies. Artif. Intell. Rev. **22**(2), 85–126 (2004)
8. Hsu, C.W., Chang, C.C., Lin, C.J.: A practical guide to support vector classification (2003)
9. Hu, W., Liao, Y., Vemuri, V.R.: Robust support vector machines for anomaly detection in computer security. In: Proceedings of the 2003 International Conference on Machine Learning and Applications (ICMLA), pp. 168–174, LA, CA, USA, June 2003
10. Liu, W., Hua, G., Smith, J.R.: Unsupervised one-class learning for automatic outlier removal. In: Proceedings of IEEE Conference on Computer Vision and Pattern Recognition (CVPR), pp. 3826–3833. IEEE (2014)
11. Mukkamala, S., Janoski, G., Sung, A.: Intrusion detection using neural networks and support vector machines. In: Proceedings of the International Joint Conference on Neural Networks (IJCNN), vol. 2, pp. 1702–1707, IEEE, May 2002
12. Rätsch, G., Mika, S., Scholkopf, B., Müller, K.R.: Constructing boosting algorithms from SVMs: an application to one-class classification. IEEE Trans. Pattern Anal. Mach. Intell. **24**(9), 1184–1199 (2002)
13. Satopää, V., Albrecht, J., Irwin, D., Raghavan, B.: Finding a "kneedle" in a haystack: detecting knee points in system behavior. In: Proceedings of the 31st International Conference on Distributed Computing Systems Workshops (ICDCSW), pp. 166–171. IEEE (2011)
14. Schölkopf, B., Platt, J.C., Shawe-Taylor, J., Smola, A.J., Williamson, R.C.: Estimating the support of a high-dimensional distribution. Neural Comput. **13**(7), 1443–1471 (2001)
15. Shin, H.J., Eom, D.H., Kim, S.S.: One-class support vector machines: an application in machine fault detection and classification. Comput. Indus. Eng. **48**(2), 395–408 (2005)
16. Subramaniam, S., Palpanas, T., Papadopoulos, D., Kalogeraki, V., Gunopulos, D.: Online outlier detection in sensor data using non-parametric models. In: Proceedings of the 32nd International Conference on Very Large Data Bases, pp. 187–198. VLDB Endowment (2006)

17. Suvorov, M., Ivliev, S., Markarian, G., Kolev, D., Zvikhachevskiy, D., Angelov, P.: OSA: one-class recursive SVM algorithm with negative samples for fault detection. In: Mladenov, V., Koprinkova-Hristova, P., Palm, G., Villa, A.E.P., Appollini, B., Kasabov, N. (eds.) ICANN 2013. LNCS, vol. 8131, pp. 194–207. Springer, Heidelberg (2013)
18. Tax, D.M., Duin, R.P.: Outliers and data descriptions. In: Proceedings of the 7th Annual Conference of the Advanced School for Computing and Imaging, pp. 234–241 (2001)
19. Tax, D.M., Duin, R.P.: Support vector data description. Mach. Learn. 54(1), 45–66 (2004)
20. Zelnik-Manor, L., Perona, P.: Self-tuning spectral clustering. In: Proceedings of Advances in Neural Information Processing Systems (NIPS), pp. 1601–1608 (2004)
21. Zhuang, L., Dai, H.: Parameter estimation of one-class SVM on imbalance text classification. In: Lamontagne, L., Marchand, M. (eds.) Canadian AI 2006. LNCS (LNAI), vol. 4013, pp. 538–549. Springer, Heidelberg (2006)

Frequent Pattern Outlier Detection Without Exhaustive Mining

Arnaud Giacometti$^{(\boxtimes)}$ and Arnaud Soulet

Université François Rabelais Tours, LI EA 6300,
3 Place Jean Jaurès, 41029 Blois, France
{arnaud.giacometti,arnaud.soulet}@univ-tours.fr

Abstract. Outlier detection consists in detecting anomalous observations from data. During the past decade, pattern-based outlier detection methods have proposed to mine all frequent patterns in order to compute the outlier factor of each transaction. This approach remains too expensive despite recent progress in pattern mining field. In this paper, we provide exact and approximate methods for calculating the frequent pattern outlier factor (FPOF) without extracting any pattern or by extracting a small sample. We propose an algorithm that returns the exact FPOF without mining any pattern. Surprisingly, it works in polynomial time on the size of the dataset. We also present an approximate method where the end-user controls the maximum error on the estimated FPOF. Experiments show the interest of both methods for very large datasets where exhaustive mining fails to provide the exact solution. The accuracy of our approximate method outperforms the baseline approach for a same budget in time or number of patterns.

1 Introduction

Outlier detection consists in detecting anomalous observations from data [1]. The outlier detection problem has important applications, such as detection of credit card fraud or network intrusions. Recently, outlier detection methods were proposed for categorical data using the concept of frequent patterns [2–4]. The key idea of such approaches is to consider the number of frequent patterns supported by each data observation. A data observation is unlikely to be an outlier if it supports many frequent patterns since frequent patterns correspond to the "common features" of the dataset. Frequent pattern outlier detection methods first extract all frequent itemsets from the data and then assign an outlier score to each data observation based on the frequent itemsets it contains. These outlier detection methods follow the schema of pattern-based two-step methods.

Pattern-based two-step methods [5] aim at exhaustively mining all patterns (first step) in order to build models (second step) like pattern sets (e.g., classifier [6]) or pattern-based measures (e.g., FPOF [2] or CPCQ index [7]). The completeness of pattern mining is often considered as a crucial advantage for constructing accurate models or measures. However, the completeness of the first

© Springer International Publishing Switzerland 2016
J. Bailey et al. (Eds.): PAKDD 2016, Part II, LNAI 9652, pp. 196–207, 2016.
DOI: 10.1007/978-3-319-31750-2_16

step requires to adjust thresholds which is recognized as being very difficult. If the thresholds are too low, the extraction becomes unfeasible. If the thresholds are too high, some essential patterns are missed. Finally, completeness leads to huge pattern volumes without guaranteeing not missing important patterns. For a smaller budget (in time or number of patterns), we claim that non-exhaustive methods can produce collections of patterns better adapted to the task of the second step. Interestingly, a non-exhaustive method can even guarantee a certain quality on the second step.

This paper revisits the calculation of the Frequent Pattern Outlier Factor (FPOF) based on non-exhaustive methods i.e., by mining a small number of patterns (or zero!). We first propose a method for calculating the exact FPOF of each transaction. Surprisingly, our method is non-enumerative in the sense that no pattern is generated (a fortiori, this is also a non-exhaustive method). For this, we reformulate the FPOF by operating directly on transaction pairs. This method calculates the FPOF in polynomial time on the number of transactions and items of the dataset. Experiments show that this method arrives to calculate the exact FPOF where the usual approach fails. We also propose a non-exhaustive approximate method that exploits a pattern sample instead of the complete collection of frequent patterns. Using Bennett's inequality, this method selects the sample size so as to guarantee a maximum error for a given confidence. Experiments show its efficiency with reasonable error.

The outline of this paper is as follows. Section 2 reviews some related work. Section 3 introduces the basic definitions about the FPOF and states the problem of its exact and approximate calculation. In Sect. 4, we propose our exact non-enumerative method for calculating FPOF. We introduce our approximate method based on sampling in Sect. 5. Section 6 provides experimental results.

2   Related Work

In this paper, we focus on the outlier detection methods based on frequent patterns [2–4]. A broader view of outlier detection is provided by surveys including [1]. Pattern-based methods benefit from the progress of pattern mining made over the past two decades. Such methods have a double interest. On the one hand, they are well suited to handle categorical data unlike most other methods dedicated to numerical data. In addition, they also remain efficient for high-dimensional spaces. The first approach [2] introduced the frequent pattern outlier factor that exploits the complete collection of frequent itemsets (while [3] uses an opposite approach by considering non-frequent itemsets). More recently, [4] replaces the collection of frequent itemsets by the condensed representation of Non-Derivable Itemsets (NDI) which is more compact and less expensive to mine. We would go further by showing that the frequent pattern outlier factor proposed in [2] can be calculated without extracting any pattern or by extracting a small sample.

Recently, there has been a resurgence in pattern mining for non-exhaustive methods through pattern sampling [8,9]. Pattern sampling aims at accessing the pattern space $\mathcal{L}$ by an efficient sampling procedure simulating a distribution $\pi : \mathcal{L} \to [0,1]$ that is defined with respect to some interestingness measure m: $\pi(.) = m(.)/Z$ where Z is a normalizing constant (formal framework and algorithms are detailed in [9]). In this way, the user has a fast and direct access to the entire pattern language and with no parameter (except possibly the sample size). Pattern sampling has been introduced to facilitate interactive data exploration [10]. In this paper, we investigate the use of pattern sampling for assigning an outlier score to each transaction. With a lower (pattern or time) budget than that of an exhaustive method, we obtain a higher quality with a bounded error.

3 Frequent Pattern Based Outlier Detection

3.1 Basic Definitions

Let $\mathcal{I}$ be a set of distinct literals called *items*, an itemset (or a pattern) is a subset of $\mathcal{I}$. The language of itemsets corresponds to $\mathcal{L} = 2^{\mathcal{I}}$. A transactional dataset is a multi-set of itemsets of $\mathcal{L}$. Each itemset, usually called *transaction*, is a data observation. For instance, Table 1 gives three transactional datasets with 4 or 5 transactions t_i described by until 4 items A, B, C and D.

Table 1. Three toy datasets with slight variations

$\mathcal{D}$			$\mathcal{D}'$			$\mathcal{D}''$		
Trans.	Items		Trans.	Items		Trans.	Items	
t_1	A B		t_1	A B		t_1	A B	D
t_2	A B		t_2	A B		t_2	A B	D
t_3	A B		t_3	A B		t_3	A B	D
t_4		C	t_4		C	t_4		C
			t_5	**A B**				

Pattern discovery takes advantage of interestingness measures to evaluate the relevancy of a pattern. The *support* of a pattern X in the dataset $\mathcal{D}$ is the proportion of transactions covered by X [11]: $supp(X, \mathcal{D}) = |\{t \in \mathcal{D} : X \subseteq t\}|/|\mathcal{D}|$. A pattern is said to be *frequent* when its support exceeds a user-specified minimal threshold. The set of all frequent patterns for σ as minimal threshold in $\mathcal{D}$ is denoted by $\mathcal{F}_\sigma(\mathcal{D})$: $\mathcal{F}_\sigma(\mathcal{D}) = \{X \in \mathcal{L} : supp(X, \mathcal{D}) \geq \sigma\}$.

In the following, we manipulate pattern multisets which are collections of patterns admitting several occurrences of the same pattern. The representativeness of a pattern multiset $\mathcal{P}$, denoted by $Supp(\mathcal{P}, \mathcal{D})$, is the sum of the support of each pattern in $\mathcal{P}$: $Supp(\mathcal{P}, \mathcal{D}) = \sum_{X \in \mathcal{P}} supp(X, \mathcal{D})$. The range of $Supp(\mathcal{P}, \mathcal{D})$ is $[0, |\mathcal{P}|]$. Given a cardinality, high representativeness means the multiset contains very common patterns of the dataset. For comparing the content of two pattern multisets, we use the semi-join operator, denoted by $\mathcal{P}_2 \triangleright \mathcal{P}_1$, that returns all the patterns of $\mathcal{P}_2$ occurring in $\mathcal{P}_1$: $\mathcal{P}_2 \triangleright \mathcal{P}_1 = \{X \in \mathcal{P}_2 : X \in \mathcal{P}_1\}$. For instance, $\{A, AB, A, D\} \triangleright \{C, A, B\} = \{A, A\}$.

3.2   Frequent Pattern Outlier Factor

Intuitively, a transaction is more representative when it contains many patterns which are very frequent within the dataset. In contrast, an outlier contains only few patterns and these patterns are not very frequent. The frequent pattern outlier factor [2] formalizes this intuition:

Definition 1 (FPOF). *The frequent pattern outlier factor of a transaction t in $\mathcal{D}$ is defined as follows:*

$$fpof(t, \mathcal{D}) = \frac{Supp(2^t, \mathcal{D})}{\max_{u \in \mathcal{D}} Supp(2^u, \mathcal{D})}$$

The range of *fpof* is $[0, 1]$ where 1 means that the transaction is the most representative transaction of the dataset while a value near 0 means that the transaction is an outlier. Other normalizations (denominator) are possible like $Supp(\mathcal{L}, \mathcal{D})$ or $\sum_{t \in \mathcal{D}} Supp(2^t, \mathcal{D})$. Whatever the normalization method, two transactions remain ordered in the same way (so it does not affect the Kendall's tau that is used for evaluating our method). Under a certain Markov model, the score $fpof(t, \mathcal{D})$ is also the proportion of time that an analyst would dedicate to study the transaction t considering the collection of frequent itemsets [12].

In the first dataset provided by Table 1, t_1 is covered by $\emptyset$ ($supp(\emptyset, \mathcal{D}) = 1$) and, A, B and AB whose support equals to 0.75 ($Supp(\{\emptyset, A, B, AB\}, \mathcal{D}) = 3.25$) while t_4 is only covered by $\emptyset$ and C ($Supp(\{\emptyset, C\}, \mathcal{D}) = 1.25$). Consequently, $fpof(t_1, \mathcal{D}_1) = 3.25/3.25$ and $fpof(t_4, \mathcal{D}_1) = 1.25/3.25$. In this example, t_4 appears to be an outlier. It is easy to see that increasing the frequency of the patterns covering the first transactions (e.g., dataset $\mathcal{D}'$) decreases the FPOF of t_4. Similarly, increasing the number of patterns covering the first transactions also decreases the FPOF factor of t_4 (e.g., dataset $\mathcal{D}''$).

3.3   Problem Formulation

The outlier detection problem consists in computing the FPOF for each transaction:

Problem 1 (Exact Problem). Given a dataset $\mathcal{D}$, compute the frequent pattern outlier factor of each transaction $t \in \mathcal{D}$.

In practice, the exact calculation of frequent pattern outlier factor is performed by mining all patterns appearing at least once in the dataset (i.e., with $\sigma = 1/|\mathcal{D}|$). Of course, this expensive task is not possible for very large datasets. Instead, FPOF is approximated with a collection of frequent patterns i.e., with a higher minimal support threshold:

Definition 2 (σ-Exhaustive FPOF). *Given a minimal support threshold σ, the σ-exhaustive FPOF of a transaction t in $\mathcal{D}$ is defined as follows:*

$$fpof_\sigma(t, \mathcal{D}) = \frac{Supp(\mathcal{F}_\sigma(\mathcal{D}) \triangleright 2^t, \mathcal{D})}{\max_{u \in \mathcal{D}} Supp(\mathcal{F}_\sigma(\mathcal{D}) \triangleright 2^u, \mathcal{D})}$$

The approximation becomes accurate with low support thresholds. Figure 1 (left) plots the Kendall's tau of $fpof_\sigma$ in comparison with $fpof$ for some benchmarks[1]. When the support threshold becomes very low, the number of patterns and the extraction time explode, see Fig. 1 (right). Furthermore, the approximation error is not estimated. With a smaller budget, we claim that it is possible to approximate more precisely FPOF while having a bound on the error.

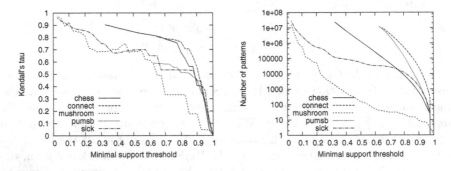

Fig. 1. Kendall's tau and number of patterns with minimal support threshold

The Kendall's tau varies significantly depending on the dataset for a same minimal support threshold. This threshold is not easy to set for obtaining a good compromise between efficiency and quality. Therefore, it seems interesting that the user sets the maximum error that he/she tolerates on the result rather than a threshold related to the method:

Problem 2 (Approximate Problem). Given a dataset $\mathcal{D}$, two reals δ and ϵ, find a function approximating the frequent pattern outlier factor such that for each transaction $t \in \mathcal{D}$, $|fpof(t, \mathcal{D}) - \widetilde{fpof}(t, \mathcal{D})| \leq \epsilon$ with confidence $1 - \delta$.

This problem aims at assigning an approximate FPOF to each transaction with the probability $1 - \delta$ that the error is less than ϵ. Problem 1 is a particular case of Problem 2 by fixing $\epsilon = 0$ and $\delta = 0$.

4   Exact Non-enumerative Method

This section addresses Problem 1. To calculate the FPOF of a transaction t, Definition 1 formulates the problem in terms of frequent patterns appearing in t. The idea is to reformulate this factor by considering what each transaction u brings to the transaction t. For instance, in dataset $\mathcal{D}$, the FPOF of the first transaction relies on $Supp(\{\emptyset, A, B, AB\}, \mathcal{D})$ which is equal to $|\{\emptyset, A, B, AB, \quad \emptyset, A, B, AB, \quad \emptyset, A, B, AB, \quad \emptyset\}|/4$. Each subset

---

[1] It is the proportion of pairs of transactions which would be ranked similarly with the approximate FPOF and with the true FPOF (see Sect. 6 for a formal definition).

Algorithm 1. Exact Non-Enumerative Method

Input: A dataset $\mathcal{D}$
Output: FPOF computed for each transaction
1: Initialize $fpof[t] \leftarrow 0$ for $t \in \mathcal{D}$
2: **for all** $(t, u) \in \mathcal{D} \times \mathcal{D}$ **do**
3:    $fpof[t] \leftarrow fpof[t] + 2^{|t \cap u|}$
4: **end for**
5: Normalize $fpof[t] \leftarrow fpof[t]/Z$ for $t \in \mathcal{D}$ where $Z = \max_{u \in \mathcal{D}} fpof[u]$
6: **return** $fpof$

$\{\emptyset, A, B, AB\}$ and $\{\emptyset\}$ result from the intersection of patterns covering t_1 with those covering another transaction $u \in \mathcal{D}$. Thereby, $Supp(\{\emptyset, A, B, AB\}, \mathcal{D}) = |\{\bigcup_{u \in \mathcal{D}} 2^{t_1} \cap 2^u\}|/|\mathcal{D}| = |\{\bigcup_{u \in \mathcal{D}} 2^{t_1 \cap u}\}|/|\mathcal{D}|$. This observation is at the core of the following property:

Property 1 (Reformulation). Given a dataset $\mathcal{D}$, the frequent pattern outlier factor can be reformulated as follows for all transaction $t \in \mathcal{D}$:

$$fpof(t, \mathcal{D}) = \frac{\sum_{u \in \mathcal{D}} 2^{|t \cap u|}}{\max_{v \in \mathcal{D}} \sum_{u \in \mathcal{D}} 2^{|v \cap u|}}$$

Proof. Let $\mathcal{D}$ be a dataset. Given $u \in \mathcal{D}$, we define $\kappa(X, u) = 1$ if $X \subseteq u$ and 0 otherwise. For each transaction $t \in \mathcal{D}$, we obtain:

$$Supp(2^t, \mathcal{D}) = \frac{1}{|\mathcal{D}|} \sum_{X \subseteq t} \left(\sum_{\{u \in \mathcal{D}: X \subseteq u\}} 1 \right) = \frac{1}{|\mathcal{D}|} \sum_{X \subseteq t} \sum_{u \in \mathcal{D}} \kappa(X, u) = \frac{1}{|\mathcal{D}|} \sum_{u \in \mathcal{D}} \sum_{X \subseteq t} \kappa(X, u)$$

$$= \frac{1}{|\mathcal{D}|} \sum_{u \in \mathcal{D}} \left(\sum_{X \subseteq t \wedge X \subseteq u} 1 \right) = \frac{1}{|\mathcal{D}|} \sum_{u \in \mathcal{D}} \left(\sum_{X \subseteq t \cap u} 1 \right) = \frac{1}{|\mathcal{D}|} \sum_{u \in \mathcal{D}} 2^{|t \cap u|}$$

Injecting this equation into Definition 1 proves that Property 1 is right.   □

From a conceptual point of view, it is interesting to note that ultimately, the FPOF of a transaction is just the sum of its similarity with each of transactions (where similarity between t and u is $2^{|t \cap u|}$). This measure is therefore very close to traditional methods relying on pair-wise distance among data observations.

Interestingly, FPOF formula of Property 1 can be simply calculated with a double loop (see Algorithm 1). Therefore, the first strong result of this paper is to prove that Problem 1 can be solved in polynomial time, contrary to what was envisaged in the literature:

Property 2 (Complexity). The frequent pattern outlier factor of all transactions can be calculated in time $O(|\mathcal{D}|^2 \times |\mathcal{I}|)$.

Due to lack of space, we omit certain proofs. To the best of our knowledge, our proposal is the first method to calculate FPOF in polynomial time. Nevertheless, for large datasets with a lot of transactions, this complexity remains high. Then it makes sense to consider an approximate solution obtained much faster.

5 ε-Approximate Sampling Method

This section addresses Problem 2 by using pattern sampling. First, we propose a method for approximating FPOF from a pattern sample drawn according to frequency. Then we show how to choose the sample size to control the error.

5.1 Pattern Sampling for FPOF

In Sect. 3.3, we showed that the use of the most frequent patterns is insufficient to approximate accurately FPOF. The most frequent patterns do not measure the singularity of each transaction that also relies on more specific patterns (whose frequency varies from small to average). Conversely, do not considering frequent patterns would also be a mistake because they contribute significantly to FPOF. A reasonable approach is to select patterns randomly with a probability proportional to their weight in the calculation of FPOF. Typically, in the dataset $\mathcal{D}$ of Table 1, the itemset AB is 3 times more important than itemset C in the calculation of FPOF due to their frequency.

In recent years pattern sampling techniques have been proposed to randomly draw patterns in proportion to their frequency [9]. Such approaches are ideal to bring us a well-adapted collection of patterns. Of course, it remains the non-trivial task of approximating FPOF starting from this collection:

Definition 3 (k-Sampling FPOF). *Given an integer $k > 0$, a k-sampling frequent pattern outlier factor of a transaction t in $\mathcal{D}$ is defined as follows:*

$$fpof_k(t, \mathcal{D}) = \frac{|\mathcal{S}_k(\mathcal{D}) \triangleright 2^t|}{\max_{u \in \mathcal{D}} |\mathcal{S}_k(\mathcal{D}) \triangleright 2^u|}$$

where $\mathcal{S}_k(\mathcal{D})$ is a sample of k patterns drawn from $\mathcal{D}$ according to support: $\mathcal{S}_k(\mathcal{D}) \sim supp(\mathcal{L}, \mathcal{D})$.

It is important to note that $|.|$ is used here instead of $Supp(., \mathcal{D})$ as in Definition 1. As the sampling technique already takes into account the frequency when it draws patterns, it is not necessary to involve the support here. Indeed, the draw is with replacement for the correct approximation of FPOF (without this replacement the most frequent patterns would be disadvantaged). It induces that the same pattern can have multiple occurrences within the sample $\mathcal{S}_k(\mathcal{D})$.

For the same sample size k and for the same transaction t, it is possible to calculate different values of a k-sampling FPOF due to $\mathcal{S}_k(\mathcal{D})$. But, the higher the threshold k, the less the difference between values stemming from two samples is high. Furthermore, the greater the sample size k, the better the approximation:

Property 3 (Convergence). Given a dataset $\mathcal{D}$, a k-sampling FPOF converges to the FPOF for all transaction $t \in \mathcal{D}$.

Proof. $\mathcal{S}_k(\mathcal{D}) \sim supp(\mathcal{L}, \mathcal{D})$ means that there exists a constant $\alpha > 0$ such that $\forall X \in \mathcal{L}$, $\lim_{k \to \infty} |\mathcal{S}_k(\mathcal{D}) \triangleright \{X\}| = \alpha supp(X, \mathcal{D})$. Then, for each transaction t, we obtain that: $\lim_{k \to \infty} |\mathcal{S}_k(\mathcal{D}) \triangleright 2^t| = \alpha \sum_{X \in 2^t} supp(X, \mathcal{D}) = \alpha Supp(2^t, \mathcal{D})$. By injecting this result into Definition 3, we conclude that Property 3 is right. □

Algorithm 2. ϵ-Approximate Sampling Method

Input: A dataset $\mathcal{D}$, a confidence $1 - \delta$, a bound ϵ
Output: A k-sampling frequent pattern outlier factor of all transactions in $\mathcal{D}$ with an
    error bounded by ϵ for a confidence $1 - \delta$
1: $\tilde{\epsilon} \leftarrow 1$; $\mathcal{S} \leftarrow \emptyset$
2: **while** $\tilde{\epsilon} > \epsilon$ **do**
3:     $\mathcal{S} \leftarrow \mathcal{S} \cup \{X\}$ where $X \sim supp(\mathcal{L}, \mathcal{D})$ // add a pattern in the sample
4:     $m \leftarrow \arg\max_{t \in \mathcal{D}} cov_{\mathcal{S}}(t)$ // select the most covered transaction
        // estimate the maximal error on $cov_{\mathcal{S}}$
5:     $e_t \leftarrow \sqrt{2\overline{\sigma_t}\ln(1/\delta)/|\mathcal{S}|} + \ln(1/\delta)/(3|\mathcal{S}|)$ for each $t \in \mathcal{D}$
        // estimate the maximal error on FPOF
6:     $\tilde{\epsilon} \leftarrow \max_{t \in \mathcal{D}}\{\min\{1; (cov_{\mathcal{S}}(t) + e_t)/(cov_{\mathcal{S}}(m) - e_m)\} - cov_{\mathcal{S}}(t)/cov_{\mathcal{S}}(m)\}$
7:     $\tilde{\epsilon} \leftarrow \max_{t \in \mathcal{D}}\{cov_{\mathcal{S}}(t)/cov_{\mathcal{S}}(m) - \max\{0; (cov_{\mathcal{S}_k}(t) - e_t)/(cov_{\mathcal{S}}(m) + e_m)\}; \tilde{\epsilon}\}$
8: **od**
9: **return** $\langle cov_{\mathcal{S}}(t)/\max_{u \in \mathcal{D}} cov_{\mathcal{S}}(u)\rangle_{t \in \mathcal{D}}$

Beyond convergence, the interest of this approach is the speed of convergence far superior to that of the σ-exhaustive frequent pattern outlier factor as shown in the experimental study (see Sect. 6). This speed is accompanied by a good efficiency due to a reasonable complexity of pattern sampling:

Property 4 (Complexity). A k-sampling FPOF of all transactions can be calculated in time $O(k \times |\mathcal{I}| \times |\mathcal{D}|)$.

Given a number of patterns k, a k-sampling FPOF is therefore effective to calculate. However, the choice of the sample size k is both difficult and essential in order to achieve a desired approximation as suggested by Problem 2. The next section presents an iterative method for fixing this sample size.

5.2 Bounding the Error

This section shows how to determine the right sample size k for computing a k-sampling FPOF satisfying user specified parameters (a maximum error with a given confidence). The idea is to draw a sample and to bound the maximum error of FPOF using a statistical result known as Bennett's inequality. If this error is less than that allowed by the user, the algorithm returns a sampling FPOF based on the current sample. Otherwise, it increases the sample size by drawing more patterns and so on.

We use Bennett's inequality to estimate the current error because it is true irrespective of the probability distribution. After k independent observations of real-valued random variable r with range $[0, 1]$, Bennett's inequality ensures that, with confidence $1 - \delta$, the true mean of r is at least $\overline{r} - \epsilon$ where $\overline{r}$ and $\overline{\sigma}$ are respectively the observed mean and variance of the samples and

$$\epsilon = \sqrt{\frac{2\overline{\sigma}\ln(1/\delta)}{k}} + \frac{\ln(1/\delta)}{3k}$$

In our case, the random variable is the average number of patterns within a sample $\mathcal{S}_k \sim supp(\mathcal{L}, \mathcal{D})$ that cover the transaction t. It is denoted by $cov_{\mathcal{S}_k}(t)$ and defined as follows: $cov_{\mathcal{S}_k}(t) = |\mathcal{S}_k \triangleright 2^t|/k$. It is easy to see that a k-sampling FPOF factor can be rewritten using $cov_{\mathcal{S}_k}$: $fpof_k(t, \mathcal{D}) = cov_{\mathcal{S}_k}(t)/\max_{u \in \mathcal{D}} cov_{\mathcal{S}_k}(u)$. Using Bennett's inequality and the above definition enables us to bound FPOF:

Property 5 (FPOF Bound). Given a dataset $\mathcal{D}$ and confidence $1 - \delta$, the FPOF of a transaction t is bounded as follows:

$$\max\left\{0, \frac{cov_{\mathcal{S}_k}(t) - \epsilon_t}{cov_{\mathcal{S}_k}(u) + \epsilon_u}\right\} \leq fpof(t, \mathcal{D}) \leq \min\left\{\frac{cov_{\mathcal{S}_k}(t) + \epsilon_t}{cov_{\mathcal{S}_k}(u) - \epsilon_u}, 1\right\}$$

where $\mathcal{S}_k \sim supp(\mathcal{L}, \mathcal{D})$, $u = \arg\max_{v \in \mathcal{D}} cov_{\mathcal{S}_k}(v)$ and $\epsilon_v = \sqrt{2\overline{\sigma_v}\ln(1/\delta)/k} + \ln(1/\delta)/(3k)$.

Algorithm 2 returns the approximate FPOF of all transactions by guaranteeing a bounded error of ϵ with confidence $1 - \delta$. Basically, the main loop is iterated until that the maximal error $\tilde{\epsilon}$ is inferior to the expected bound ϵ. Lines 4–7 calculate the maximal error $\tilde{\epsilon}$ using Property 5. If the maximal error is less than ϵ, Line 9 returns the k-sampling FPOF with the current sample $\mathcal{S}$. Otherwise, one more pattern is drawn (Line 3).

As desired by Problem 2, Algorithm 2 approximates the FPOF:

Property 6 (Correctness). Given a dataset $\mathcal{D}$, a confidence $1 - \delta$, a bound ϵ, Algorithm 2 returns a k-sampling frequent pattern outlier factor of a transaction t in $\mathcal{D}$ approximating FPOF with an error bounded by ϵ with confidence $1 - \delta$.

6   Experimental Study

This experimental study aims to compare the speed of the non-enumerative method with that of the exact exhaustive method (i.e., $1/|\mathcal{D}|$-exhaustive) and to estimate the error quality of the ϵ-approximate sampling method faced to the σ-exhaustive method. Due to the lack of space, we do not provide new experiments showing the interest of FPOF for detecting outliers as this aspect is already detailed in literature [2–4]. Experiments are conducted on datasets coming from the UCI Machine Learning repository (archive.ics.uci.edu/ml) and the FIMI repository (fimi.ua.ac.be). Table 2 gives the main features of datasets in first columns. All experiments are performed on a 2.5 GHz Xeon processor with the Linux operating system and 2 GB of RAM memory. Each reported evaluation measure is the arithmetic mean of 10 repeated measurements (interval confidence are narrow enough to be omitted).

6.1   Exact Non-enumerative Method

Table 2 reports the running time required for calculating the exact FPOF using the $1/|\mathcal{D}|$-exhaustive and the non-exhaustive methods (respectively the 4th and the 5th column). Note that the exact exhaustive method (as baseline) benefits from LCM which is one of the most recognized frequent itemset mining algorithm. The exact non-enumerative method is effective and rivals the exact exhaustive one. Its main advantage is to calculate the exact FPOF with datasets where the exact exhaustive method fails (e.g., pumsb where the execution was aborted after 5 h).

Table 2. Time comparison of the different methods

$\mathcal{D}$	$|\mathcal{D}|$	$|\mathcal{I}|$	Exact methods		Approximate method
			Exh. time (s)	Non-enum. time (s)	0.1-Approx. time (s)
chess	3,196	75	439.5	1.1	0.3
connect	67,557	129	748.5	577.7	176.6
mushroom	8,124	119	0.4	5.9	2.0
pumsb	49,096	7,117	time out	1,970.5	175.0
retail	88,162	16,470	8.7	5,969.9	1.3
sick	2,800	58	0.8	0.5	0.5

6.2   Approximate Sampling Method

This section compares our sampling method with the traditional heuristic based on frequent patterns as baseline. For this purpose, we use the Kendall's tau for comparing the ranking stemming from an approximate method f with that stemming from the FPOF (calculated with an exact method):

$$\tau(f, \mathcal{D}) = \frac{|\{(t, u) \in \mathcal{D}^2 : sgn(f(t, \mathcal{D}) - f(u, \mathcal{D})) = sgn(fpof(t, \mathcal{D}) - fpof(u, \mathcal{D}))\}|}{|\mathcal{D}|^2}$$

and in the same way, we also compute the average error per transaction: $\varepsilon(f, \mathcal{D}) = \sum_{t \in \mathcal{D}} |f(t, \mathcal{D}) - fpof(t, \mathcal{D})| / |\mathcal{D}|$.

The left chart of Fig. 2 plots the difference between the Kendall's tau of k-sampling FPOF and that of σ-exhaustive FPOF (when the curve is above 0, it means that the ranking of k-sampling FPOF is better than the ranking of σ-exhaustive FPOF). The right chart reports the average error of σ-sampling FPOF divided by that of k-sampling FPOF (when the curve is above 1, it means that the average error of k-sampling FPOF is smaller than that of σ-exhaustive FPOF). For each point, the minimal support threshold σ is used as parameter of the σ-exhaustive FPOF method. At the same time, the sample size k is fixed with the number of patterns mined with the minimal support threshold σ: $k = |\mathcal{F}_\sigma(\mathcal{D})|$. The k-sampling FPOF is clearly more accurate than the σ-exhaustive FPOF for a same pattern budget. For some datasets (e.g., chess or sick), the difference is always greater than 0. For other datasets, as soon as the number of patterns in the sample increases, the difference becomes positive. The average

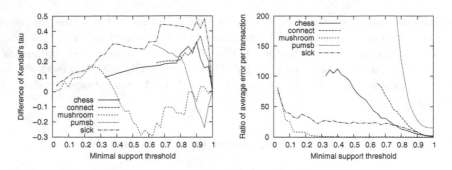

Fig. 2. Error of σ-exhaustive FPOF and k-sampling FPOF with σ

error ratio clearly shows that our method gives a better approximation of FPOF (especially, when the number of patterns is high).

Figure 3 plots the number of patterns and the average error per transaction of the ϵ-approximate sampling method with bound ϵ (for $\delta = 0.1$). As expected, the smaller the error bound ϵ, the greater the number of patterns in the sample. Therefore, the longer the execution time. It is interesting to note that the approximate sampling method (with $\epsilon = 0.1$) is regularly faster than the exact methods (see Table 2). Finally, the real average error per transaction of the approximate sampling method is always much lower than the requested bound ϵ (e.g., $\epsilon = 0.1$ actually gives an error less than 0.01). This difference results from the Bennett's inequality that makes no assumption about the distribution.

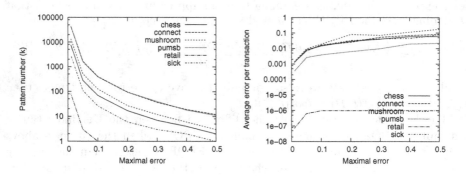

Fig. 3. Number of patterns and average error per transaction with maximum error ϵ

7   Conclusion

We revisited the FPOF calculation in extracting the least possible patterns or no pattern. Despite this constraint, the proposed exact method has a complexity better suited to certain datasets. Our approximate method using a sampling technique provides additional guarantees on the result with a maximum bound on the error. The experiments have shown the interest of these two approaches in terms of speed and accuracy compared to the usual exhaustive approach where all frequent patterns are mined.

Our proposal therefore combines the proven power of pattern-based methods by adding a guarantee on the quality of results without sacrificing speed thanks to sampling techniques. We think it can be generalized to other measures involving patterns or pattern-based models. We would also like to adapt this approach to build anytime algorithms. In the case of FPOF, it consists in extending the pattern sample indefinitely until the end-user wants to stop the process. Then, the algorithm returns the FPOF achieved with the current sample while estimating its error.

Acknowledgements. This work has been partially supported by the Prefute project, PEPS 2015, CNRS.

References

1. Hawkins, D.M.: Identification of Outliers, vol. 11. Springer, The Netherlands (1980)
2. He, Z., Xu, X., Huang, Z.J., Deng, S.: FP-outlier: frequent pattern based outlier detection. Comput. Sci. Inf. Syst. **2**(1), 103–118 (2005)
3. Otey, M.E., Ghoting, A., Parthasarathy, S.: Fast distributed outlier detection in mixed-attribute data sets. Data Min. Knowl. Discovery **12**(2–3), 203–228 (2006)
4. Koufakou, A., Secretan, J., Georgiopoulos, M.: Non-derivable itemsets for fast outlier detection in large high-dimensional categorical data. Knowl. Inf. Syst. **29**(3), 697–725 (2011)
5. Knobbe, A., Crémilleux, B., Fürnkranz, J., Scholz, M.: From local patterns to global models: the lego approach to data mining. In: From Local Patterns to Global Models: Proceedings of the ECML PKDD 2008 Workshop, pp. 1–16 (2008)
6. Liu, B., Hsu, W., Ma, Y.: Integrating classification and association rule mining. In: International Conference on Knowledge Discovery and Data mining (1998)
7. Liu, Q., Dong, G.: CPCQ: contrast pattern based clustering quality index for categorical data. Pattern Recogn. **45**(4), 1739–1748 (2012)
8. Chaoji, V., Hasan, M.A., Salem, S., Besson, J., Zaki, M.J.: ORIGAMI: a novel and effective approach for mining representative orthogonal graph patterns. Stat. Anal. Data Min. **1**(2), 67–84 (2008)
9. Boley, M., Lucchese, C., Paurat, D., Gärtner, T.: Direct local pattern sampling by efficient two-step random procedures. In: ACM SIGKDD International Conference on Knowledge Discovery and Data Mining, pp. 582–590 (2011)
10. van Leeuwen, M.: Interactive data exploration using pattern mining. In: Jurisica, I., Holzinger, A. (eds.) Knowledge Discovery and Data Mining. LNCS, vol. 8401, pp. 169–182. Springer, Heidelberg (2014)
11. Agrawal, R., Srikant, R., et al.: Fast algorithms for mining association rules. In: International conference on Very Large Data Bases, vol. 1215, pp. 487–499 (1994)
12. Giacometti, A., Li, D.H., Soulet, A.: Balancing the analysis of frequent patterns. In: Tseng, V.S., Ho, T.B., Zhou, Z.-H., Chen, A.L.P., Kao, H.-Y. (eds.) PAKDD 2014, Part I. LNCS, vol. 8443, pp. 53–64. Springer, Heidelberg (2014)

Ensembles of Interesting Subgroups
for Discovering High Potential Employees

Girish Keshav Palshikar[✉], Kuleshwar Sahu, and Rajiv Srivastava

TCS Innovation Labs - TRDDC, Tata Consultancy Services Limited,
54B Hadapsar Industrial Estate, Pune 411013, India
{gk.palshikar,kuleshwar.sahu,rajiv.srivastava}@tcs.com

Abstract. We propose a new method for building a classifier ensemble, based on *subgroup discovery* techniques in data mining. We apply subgroup discovery techniques to a labeled training dataset to discover interesting subsets, characterized by a conjuctive logical expression (rule), where such subset has an unusually high dominance of one class. Treating these rules as base classifiers, we propose several simple ensemble methods to construct a single classifier. Another novel aspect of the paper is that it applies these ensemble methods, along with standard anomaly detection and classification, to automatically identify *high potential (HIPO)* employees - an important problem in management. HIPO employees are critical for future-proofing the organization in the face of attrition, economic uncertainties and business challenges. Current HR processes for HIPO identification are manual and suffer from subjectivity, bias and disagreements. Proposed data-driven analytics algorithms address some of these issues. We show that the new ensemble methods perform better than other methods, including other ensemble methods on a real-life case-study dataset of a large multinational IT services company.

1 Introduction

As an important alternative to learning a single classifier (i.e., a single decision surface), ensemble-based classification approaches [31], such as bagging [6], boosting [12] and random forests [7], build several base classifiers (often using sampling) and combine their predictions to determine the final class label for a given query point. In this paper, we propose a new method for building a classifier ensemble, based on the well-known *subgroup discovery* problem in data mining. Many real-life datasets can be viewed as containing information about a set of entities. Further, one or more continuous or discrete valued columns in such datasets can be interpreted as a performance or evaluation measure for each entity. Given such a dataset, it is then of interest to automatically discover *interesting subsets* of entities (also called *subgroups*), such that each subset (a) is characterised by a common (shared) pattern or description; and (b) has unusual characteristics in terms of the performance attribute, as a set, when compared to the remaining set of entities. The problem of automatically discovering interesting subsets is well-known in data mining as *subgroup discovery*.

© Springer International Publishing Switzerland 2016
J. Bailey et al. (Eds.): PAKDD 2016, Part II, LNAI 9652, pp. 208–220, 2016.
DOI: 10.1007/978-3-319-31750-2_17

Much work in subgroup discovery [2,3,13,15,17,18,20,21,26,27,30] is focused on the case when the domain of possible values for the performance measure column is finite and discrete. In this paper, we use descriptors for subsets which have the form $Cond \rightarrow Class$, where $Cond$ is a conjunction of attribute-value pairs of the form $A_i = v_j$.

The idea behind the proposed approach is as follows. We use the binary class label as the performance measure for each entity, which means that each discovered interesting subset (i.e., the associated logical descriptor or *rule*) can be viewed as a relatively high-precision (but possibly low-recall) base classifier. These base classifiers are typically neither disjoint nor exhaustive i.e., an entity may satisfy 0, 1 or more of these rules. We empirically demonstrate that an ensemble of such base classifiers can give us an improved accuracy. Most ensemble techniques create base classifiers based on random selections whereas as we create base classifiers in a non-random manner.

The paper also discusses a new application in the domain of human resources management: identifying *high potential (HIPO)* employees. HIPO employees are critical for "future-proofing" the organization in the face of continual attrition of highly skilled and experienced employees, ever-present economic uncertainties and new business challenges [4,5,10,25,29]. Current HIPO identification processes are mostly manual (supervisor recommendations, shortlisting, interviews and evaluation) and consequently may suffer from subjectivity, bias, disagreements and incompleteness. We present analytics algorithms for identifying HIPO employees using organizational databases, to supplement the manually identified ones. We present a real-life case-study involving a large multinational IT services company.

The contributions of this paper are as follows. We propose a new method for building a classifier ensemble, based on the *subgroup discovery* problem in data mining. We illustrate this new approach on a new application in human resource management domain: identifying high potential employees, where we demonstrate that the proposed method outperforms other classifier ensemble methods. The paper is organized as follows. Section 2 outlines the related work. Section 3 describes the real-life case-study dataset. Section 4 applies anomaly detection and standard classification approaches. Section 5 formalizes the new approach and demonstrates its results on the case-study dataset. Section 6 discusses conclusions and further work.

2   Related Work

There is rich work on classifier ensemble methods. *Bagging* creates m labeled datasets (called *bootstrap samples*) by sampling with replacement the original training dataset D, fit m *base classifier* models to these m datasets and then take majority voting among these m models to predict the class label for a given query point. Bagging is useful to improve the accuracy, when the base classifiers are "unstable". *Boosting* is an iterative procedure that adaptively changes distribution of training samples in each round to focus base classifiers on "hard" examples. Initially each example has equal weight. A sample is drawn (with replacement) according to the sampling distribution. A classifier is induced from the dataset of selected samples. Weights are reduced (increased) for examples that are correctly (wrongly) classified. A voting-based ensemble is created using the base classifiers. The *Random Forest (RF)* algorithm uses decision trees as base classifiers. RF combines the predictions made by multiple decision trees using majority voting. There are several ways to construct each decision tree; e.g., one way is to use only a subset of randomly selected features when building the base decision tree and ignore the rest of the features (i.e., use a vertical partition of the data).

Initial approaches to subgroup discovery were based on a heuristic search framework [13]. More recently, several subgroup discovery algorithms adapt well-known classification rule learning algorithms to the task of subgroup discovery. For example, CN2-SD [21] adapts the CN2 classification rule induction algorithm to the task of subgroup discovery, by inducing rules of the form $Cond \rightarrow Class$. They use a weighted relative accuracy (WRA) measure to prune the search space of possible rules. Roughly, WRA combines the size of the subgroup and its accuracy (difference between true positives and expected true positives under the assumption of independence between Cond and Class). They also propose several interestingness measures for evaluating induced rules. Some recent work has adopted well-known unsupervised learning algorithms to the task of subgroup discovery. [17] adapts the apriori association rule mining algorithm to the task of subgroup discovery. The SD-Map algorithm [2] adopts the FP-tree method for association rule mining to the task of minimum-support based subgroup discovery. Some sampling based approaches to subgroup discovery have also been proposed [26, 27].

There has been insufficient research within the HR community into the comparison and effectiveness of HR processes used within various organizations for identifying and managing HIPO (e.g., maturity of the assessment procedures, objectivity in the identification process) [8]. Gaps are likely between the best practices reported in the literature for HIPO identification and the actual HR practices. In particular, the data and reasoning used for identifying HIPO have not been subject of much research. Even the various possible definitions of *high potential* have not been rigorously compared and analyzed in enough details [23]. Research in HR has focused on identifying and using appropriate attributes for HIPO identification [1, 10, 11, 23, 28]. Commonly used (subjective) variables for identifying high potentials include communication skills, results orientation, flexibility/adaptability, strategic thinking, decision-making skills, learning agility, teamwork, vision etc. Performance appraisals and past results are among the two of the more quantitative data sources used for HIPO identification [9, 22, 23]. However, these data elements are largely evaluated manually and subjectively, without recourse to more quantitative indicators from within organizational databases. Employees' (often unfavourable) psychological reactions to the HIPO programs have been explored [14, 16].

3   Case-Study: Identifying High Potential Employees

We studied the HIPO identification process within a particular business unit of a large multi-national IT organization. Most employees work in software development, maintenance and support related tasks, while some are involved in tasks like consulting, business development etc. Each employee works on a *project* and handles specific tasks within the project; e.g., requirements analysis, software design, coding or testing. The employee is also assigned a *role* that is closely related to the tasks that the employee carries out in the project. About 146 roles were used for these employees. Some example roles are: Solution Architect, Test Engineer, Developer, Design Lead etc. The projects are typically done for a particular customer, who is in a particular business domain such as Insurance, Banking or Telecom. An employee has to log the task-wise efforts (i.e., time) spent by her for each project every day. An employee may undergo certain internal training programs and courses, which are also tracked. The projects vary in their complexity, which is often indicated by the total efforts spent on the project, the team size (larger team size usually implies a more complex project) and duration (longer lasting projects tend to be more complex). We have assumed that an employee works on only one project at a time. It is also possible that an employee may not be assigned to any customer project (e.g., when the person is doing internal work, training, on long leave etc.). We used 51 variables for each employee; e.g., experience, efforts (rolewise, technology-wise etc.), trainings etc. These variables are "cumulative" over time; e.g., the variable leadership_course_count gives the total number of leadership related courses that the employee has completed so far (i.e., since joining the organization).

We had two datasets available: datasets D1 and D2 are summary information about the employees as on 01-April-2011 and 01-April-2012 respectively. For uniformity, we have selected a subset of employees in this unit, consisting of those that belong to 3 linearly ordered grades G3, G4 and G5 (G5 being the higher grade). The grades include mostly software engineers and increase with experience and denote higher responsibilities; employees typically spend 3 to 5 years in a grade. The grade G5 usually correspond to leadership responsibilities, often based on technical knowledge, domain knowledge, project and team management skills or business development skills. Table 1 shows the basic demographic summary of both these datasets. The org. experience denotes the time spent by the employee in the organization and the total experience equals org. experience + the experience the employee had prior to joining the organization.

As ground truth, we had available to us two lists L1 and L2 of HIPO employees for this unit, created on 01-Apr-2011 and 01-Apr-2012 respectively. Both these lists were created using the manual HIPO identification process as practiced within the organization. Table 2 shows summaries of both these HIPO lists. Note that the % of employees in HIPO lists grows along the grades i.e., the higher grades tend to have a higher percentage of employees designated as HIPO. This is because of two reasons. First, the grade sizes give a roughly pyramid shape to the organization, with the senior grades being smaller than the junior grades. Secondly, the progression to higher grades is based on performance and through

Table 1. Summary of datasets D1 and D2.

Dataset	Grade	#employees	Total experience		Org. experience		Age		% male	% post-graduate
			AVG	STDEV	AVG	STDEV	AVG	STDEV		
D1	G3	1076	9.09	2.28	4.93	2.80	33.28	3.61	86.99	17.29
	G4	543	11.77	3.13	7.03	4.01	36.11	4.27	90.61	23.02
	G5	209	14.94	3.68	9.58	5.49	39.38	4.75	89.95	31.58
D2	G3	1564	9.31	2.10	4.67	2.94	33.50	4.30	88.11	34.02
	G4	673	12.44	2.90	7.04	4.18	36.71	4.26	90.49	32.54
	G5	261	14.90	3.94	9.36	5.61	39.62	4.69	87.36	40.61

Table 2. Summary of the lists L1 and L2 of HIPO employees.

List	Grade	#employees	% HIPO in grade	Total experience		Org. experience		Age		% male	% post-graduate
				AVG	STDEV	AVG	STDEV	AVG	STDEV		
L1	G3	120	11.15	8.90	1.15	6.18	1.75	31.90	1.84	90.00	15.00
	G4	130	23.94	11.46	1.56	8.48	2.66	34.10	2.14	90.00	22.31
	G5	59	28.23	14.50	2.14	11.25	3.81	37.44	2.63	93.22	27.12
L2	G3	214	13.68	8.90	1.24	6.53	1.52	32.09	2.40	89.25	32.71
	G4	171	25.41	11.99	2.01	8.94	2.58	34.87	2.56	91.23	35.67
	G5	92	35.25	14.59	2.41	11.22	3.77	37.99	2.95	93.48	41.30

a controlled promotion process. Only employees with proven track record and demonstrated high performance (and not just experience) tend to get promoted into the higher grades.

4   Solution Approaches

4.1   Anomaly Detection

An obvious way to detect HIPO employees in an unsupervised manner is to use anomaly detection techniques. The experimental results (Table 3) show that the anomaly detection techniques are not that effective in identifying HIPO employees. The values in brackets are the parameters used. The Mahalanobis algorithm computes the Mahalanobis distance of each point from the mean vector and lists the top K as potentially anomalies. The Knorr algorithm [19] classifies a point P as an anomaly if the number of points within a given distance R from it is less than the given number M. For each point P, the RRS algorithm [24] computes the distance of the k_0-th nearest neighbour (for a given k_0) and identifies top K points having maximum value for this distance as anomalies. A possible reason why HIPO employees need not correspond to *anomalies* is that HIPO employees may not have unusually large values on all attributes; more often they have unusual *combinations* of values for different attributes - a reason that motivated us to explore subgroup discovery techniques.

Table 3. Anomaly detection for HIPO identification (dataset $D2$).

Anomaly detection algorithm	Grade	#Anomalies (#Predicted HIPO)	#Actual HIPO	Precision P	Recall R	Accuracy F
Mahalanobis ($K = 469$)	G3	469	78	0.166	0.364	0.228
Knorr ($R = 5.2$, $M = 12$)		470	87	0.185	0.405	0.254
RRS ($k_0 = 18, K = 469$)		469	112	0.239	0.523	**0.328**
Mahalanobis ($K = 336$)	G4	336	79	0.235	0.462	0.312
Knorr ($R = 5.2$, $M = 12$)		337	91	0.270	0.532	0.358
RRS ($k_0 = 1, K = 336$)		336	128	0.381	0.748	**0.505**
Mahalanobis ($K = 130$)	G5	130	40	0.308	0.435	0.360
Knorr ($R = 6.2$, $M = 20$)		130	38	0.292	0.413	0.342
RRS ($k_0 = 1, K = 130$)		130	56	0.431	0.609	**0.504**

4.2 Classification

Since we have labeled data, the next obvious step is to try classification techniques. The training data consists of employees, each described by the 51 variables. There are two classes of employees, HIPO or NON-HIPO i.e., there is a binary class label HIPO_Flag for each employee (+1 if the employee is in HIPO list and −1 otherwise). We used several built-in classification algorithms in the well-known WEKA tool to learn the HIPO classification model on dataset D1 (a separate model for each grade) and applied it to predict the HIPO employees in D2. The prediction results are shows in Table 4. Naive Bayes models are overall the best; Decision Tree models have consistently the best precision. The use of boosting with different classifiers sometimes showed a slight improvement, but more often reduced the accuracy; and the best performance was always without boosting. Interestingly, when we applied various feature selection methods to work with a reduced subset of features, generally, the accuracy did not improve.

5 Ensemble of Interesting Subsets

Subgroup discovery is well-known in data mining, where the problem is to discover logically characterized subsets of the given dataset, such that each subset is "interesting" in some way. We use the class label of each point in the training dataset as a measure and adapt a subgroup discovery algorithm to discover interesting subsets of employees in dataset D1. Each *green flag* or *positive* rule has the form $Cond \rightarrow Class = +1$ and corresponds to a subset of employees which has an "unusually high" % of HIPO employees (i.e., employees with label +1). Similarly, each *red flag* or *negative* rule has the form $Cond \rightarrow Class = -1$ and corresponds to a subset of employees which has an "unusually high" % of non-HIPO employees (i.e., employees with label −1). Each discovered rule has high precision and often low recall. The discovered rules need not be disjoint nor exhaustive i.e., an employee may satisfy 0, 1 or more rules. An employee may even satisfy multiple green flag rules as well as multiple red flag rules.

Table 4. Classifiers trained on dataset D1 for HIPO identification in dataset D2.

Classification Model	Grade	#Predicted HIPO	#Actual HIPO	Precision P	Recall R	Accuracy F
Decision Tree (J48)	G3	87	33	0.379	0.155	0.220
J48 + boosting		78	34	0.466	0.159	0.237
J48 + bagging		40	21	0.525	0.098	0.165
SVM		428	114	0.266	0.535	0.356
SVM + boosting		485	104	0.214	0.486	0.298
SVM + bagging		231	106	0.320	0.495	**0.389**
Naive Bayes		643	158	0.246	0.742	0.369
Naive Bayes + boosting		557	142	0.255	0.664	0.368
Naive Bayes + bagging		496	136	0.274	0.636	0.383
Decision Tree (J48)	G4	136	61	0.449	0.357	0.397
J48 + boosting		124	70	0.565	0.409	0.475
J48 + bagging		139	66	0.475	0.386	0.426
SVM		268	114	0.425	0.667	0.519
SVM + boosting		256	83	0.324	0.485	0.389
SVM + bagging		88	43	0.489	0.251	0.332
Naive Bayes		349	137	0.393	0.801	0.527
Naive Bayes + boosting		228	95	0.417	0.556	0.476
Naive Bayes + bagging		213	131	0.419	0.766	**0.541**
Decision Tree (J48)	G5	82	48	0.585	0.522	0.552
J48 + boosting		58	41	0.707	0.446	0.547
J48 + bagging		66	39	0.591	0.424	0.494
SVM		90	39	0.433	0.424	0.429
SVM + boosting		81	46	0.568	0.500	0.532
SVM + bagging		87	50	0.575	0.543	0.559
Naive Bayes		142	76	0.535	0.826	**0.650**
Naive Bayes + boosting		93	51	0.548	0.554	0.551
Naive Bayes + bagging		130	71	0.546	0.772	0.640

Before applying the subgroup discovery algorithm, we discretized each variable, by dividing its range into 5 intervals of equal width, denoted $I_1, \ldots, I_5$. For example, variable V44 has min and max values of 0 and 28 respectively, for grade G5 in dataset D1. Then the corresponding 5 equal length intervals are $I1 = [0, 5.6], I2 = [5.6, 11.2]$ and so forth. To control the quality of the green flag rules, we specify values for parameters *minimum precision* p_0, *minimum subset size* n_0. We retain only those green flag rules whose precision is at least p_0 and which apply to a minimum of n_0 employees. In addition, we define a parameter k_0, which we use to further prune the rules in G as follows: order the rules in G (retained using given values of n_0, p_0) in descending order of their precision, retain only the top k_0 rules and discard the rest. If $k_0 = 0$ then we skip this step. We define corresponding parameters p_1, n_1 and k_1 for red flag rules. Table 5 shows some green flag rules for grade G4, with $n_0 = 15$ and $p_0 = 0.300$. The second rule in Table 5 is true for 20 employees in grade G5 in dataset D1, out of which 12 are in the HIPO list L1, yielding the precision of 0.600. Following are the number of green flag rules generated on dataset D1 for different grades for different values of p_0 ($n_0 = 15$).
G3:: 0.20:32, 0.25:22, 0.30:3

Table 5. Top 10 highest precision green flag rules for grade G5 on dataset D1.

Subgroup	#Employees in subgroup	Precision P
$V39 \in I4 = [60.0, 80.0]$	20	0.600
$V44 \in I2 = [5.6, 11.2]$	20	0.600
$V22 \in I3 = [1054.4, 1581.61]$	30	0.567
$V21 \in I4 = [692.41, 923.21]$	18	0.556
$V42 \in I2 = [3.0, 6.0]$	27	0.556
$V45 \in I3 = [6.0, 9.0]$	20	0.550
$V41 \in I4 = [60.16, 80.1]$	22	0.545
$V15 \in I4 = [1488.0, 1984.0]$	15	0.533
$V28 \in I5 = [2992.0, 3740.0]$	15	0.533
$V44 \in I4 = [60.0, 80.0]$	19	0.526

G4:: 0.30:665, 0.35:203, 0.40:57, 0.45:9, 0.50:4
G5:: 0.30:304, 0.35:170, 0.40:60, 0.45:20, 0.50:14

Let G (R) denote the set of green (red) flag rules generated by analyzing the dataset D1 using particular values for the parameters n_0, p_0, k_0 (n_1, p_1, k_1) for a given grade. There are several ways in which we can devise ensemble methods to combine these green and red flag rules in G and R to predict the class label (HIPO or non-HIPO) for any given employee $\mathbf{e}$ in D2. Let $n_g(\mathbf{e})$ and $n_r(\mathbf{e})$ respectively denote the number of green and red flag rules satisfied by the given employee $\mathbf{e} \in D2$.

1. **Voting-threshold**: This method uses only the green flag rules. The intuition is that more the number of green flag rules an employee satisfies, more likely she is to be a HIPO employee (class +1). Given a *minimum voting threshold* value m_0, classify $\mathbf{e}$ as HIPO if $n_g(\mathbf{e}) \geq m_0$ and as non-HIPO otherwise. Given ruleset G, a specific value for m_0 yields a simple binary classifier, for which we measure the precision and recall on dataset D2.
2. **Employee-ranking**: This method uses only the green flag rules. We rank the employees in D2 in descending values of $n_g(\mathbf{e})$, label top K employees as HIPO and label the remaining as non-HIPO.
3. **Combined-voting-threshold**: This method uses both green and red flag rules. The intuition is that larger the difference between the number of green and red flag rules an employee satisfies, more likely she is to be a HIPO employee (class +1). Given a *minimum combined voting threshold* value Δm, classify $\mathbf{e}$ as +1 (HIPO) if $n_g(\mathbf{e}) - n_r(\mathbf{e}) \geq \Delta m$ and as −1 otherwise. Given rulesets G and R, a specific value for Δm yields a simple binary classifier, for which we measure the precision and recall on dataset D2.
4. **Combined-employee-ranking**: This method uses both green and red flag rules. We rank the employees in D2 in descending values of $n_g(\mathbf{e}) - n_r(\mathbf{e})$, label top K employees as HIPO and label the remaining as non-HIPO.

We can define more ensemble methods. For example, we could weigh the vote of each rule in **Voting-threshold** or **Combined-employee-ranking** by the precision of that rule, and then define a cut-off based on the aggregation of weighted votes. For brevity, we omit these details; their results are very similar to the results of the above methods. To find the best values for the parameters, we search over the space of all possible parameter (discretized) value combinations. For employee ranking methods, we use the same value of K as used in Table 3. Table 6 shows the best results obtained for each of the above methods, along with the corresponding values for the parameters. Strictly speaking, we should search over dataset D1 to find the best values of the parameters (since we cannot really assume that class labels are available for D2) and use them to predict the results on dataset D2. However, we perform the search for best parameter values on D2 itself, in order to show the best possible results that above methods can potentially achieve on D2. The parameter settings identified on D1 and applied to D2 yield very slightly lower results.

Figure 1 shows some charts (for **Combined-voting-threshold** applied to grade G5 in D2) that depict the effect of varying one parameter and keeping all others constant on the oeverall accuracy. Increasing Δm initially increases the accuracy but after some point leads to reduced accuracy, because requiring a larger difference between green and red flags before an employee is classified as $+1$ is a stricter decision rule. Increasing k_0 leads to using lesser precision rules, which leads to increased accuracy but eventually leads to a slow reduction in the accuracy. Increasing p_0 eventually leads to reduced accuracy, because higher values of p_0 lead to using lesser number of rules. The effect of p_1 on accuracy

Table 6. Ensemble methods for HIPO identification on dataset D2.

Method	Grade	Fixed	Varied	#Predicted HIPO	#Actual HIPO	Precision P	Recall R	F
Voting-threshold		n_0	p_0, k_0, m_0					
	G3	10	0.2, 30, 2	312	101	0.324	0.471	0.384
	G4	15	0.3, 83, 5	351	151	0.402	0.824	0.540
	G5	15	0.3, 234, 78	161	82	0.509	0.891	**0.648**
Employee-ranking		n_0, K	p_0, k_0					
	G3	10, 469	0.22, 29	469	131	0.280	0.612	0.383
	G4	15, 336	0.30, 65	336	135	0.401	0.789	0.532
	G5	15, 130	0.30, 44	130	71	0.546	0.771	**0.640**
Combined-voting-threshold		n_0, n_1	$p_0, p_1, k_0, k_1, \Delta m$					
	G3	10,30	0.2,0.97,26,26,15	430	130	0.302	0.607	0.403
	G4	15,30	0.3,0.55,154,154,16	366	146	0.398	0.853	0.543
	G5	15,30	0.3,0.53,44,44,3	125	73	0.584	0.793	**0.673**
Combined-employee-ranking		n_0, n_1, K	p_0, p_1, k_0, k_1					
	G3	10,30	0.2,0.7,31,31	469	135	0.288	0.631	0.395
	G4	15,30	0.39,0.97,74,74	336	139	0.413	0.813	0.548
	G5	15,30	0.3,53,43,43	130	75	0.577	0.815	0.676

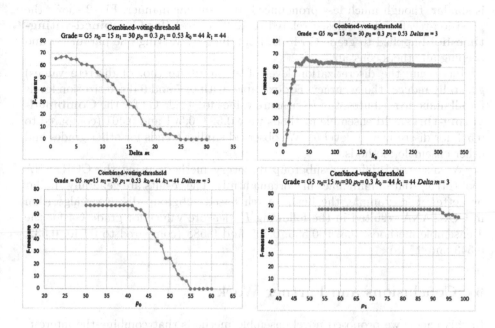

Fig. 1. Variations in F-measure with respect to some parameters.

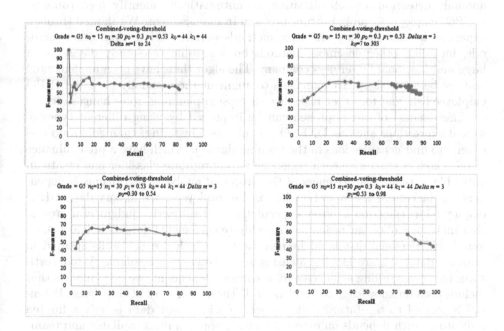

Fig. 2. Precision-recall curves using variations in some parameters.

is similar, though much less pronounced. In a similar manner, Fig. 2 shows the precision-recall curves (closed related to ROC curve) (for **Combined-voting-threshold** applied to grade G5 in D2) obtained by varying one parameter and keeping all others constant.

Increasing the discretization level of an attribute (no. of discrete values) generally reduces the accuracy. For example, with 6, 7 and 9 discretization levels for all numeric atributes, the F-measure values for grade G4 for the **Combined-voting-threshold** ensemble methods were 0.523, 0.501 and 0.461 (compared to 0.543 for discretization level 5). We observed similar trends for other grades and other ensemble methods.

We also tried different subgroup discovery methods using the CORTENA tool (http://datamining.liacs.nl/cortana.html). The results are generally similar or better. For example, the beam-search based subgroup discovery algorithm in CORTANA resulted in the following F-measure values for grade G4 for the 4 ensemble methods: 0.575, 0.556, 0.605 and 0.588 (compared to 0.540, 0.532, 0.543 and 0.548).

6   Conclusions and Further Work

In this paper, we proposed novel ensemble methods that combine the interesting subgroups (discovered using standard subgroup discovery techniques) into a single clsssifier. We applied these new ensemble methods, along with standard anomaly detection and classification, to automatically identify *high potential (HIPO)* employees - an important problem in management. We showed through experiments that the new ensemble methods perform better than other methods, including other ensemble methods on a real-life case-study dataset of a large multinational IT services company. The algorithms have been implemented and are being used by HR managers to augment the manually prepared HIPO employee lists and for other tasks like succession planning, role changes etc.

Effectiveness of our approach can be improved by using different types of logical expressions such as DNF. One could use clustering, than subgroup discovery, to find *dense* regions in the training datasets which have a predominance of one class, characterize each region by an appropriate classifier and create an ensemble of them. Effectiveness of the proposed analytics algorithms is dependent on the completeness and quality of the employee work history data. HIPO employee identification is often a speculative and subjective judgmental process that makes use of organizational citizenship (engagement) data as well as behavioural and personality related human factors, which are not well-captured in business data. We are looking at using such datasets, if available. What costitutes true outstanding achievements is context dependent; we need models that include domain knowledge to "understand" the actual work of employees. Potential is related to the future performance, of which past data is only a limited indicator. Much depends on external factors, opportunities available and teams involved. We are looking at building human performance prediction models.

Acknowlededgments. The authors thank Dr. Ritu Anand, Preeti Gulati for their support and our team members for much help.

References

1. Hewitt Associates: Getting to high potential: how organizations define and calibrate their critical talent. Hewitt Associates (2008). http://www.hewitt.com
2. Atzmüller, M., Puppe, F.: SD-Map – a fast algorithm for exhaustive subgroup discovery. In: Fürnkranz, J., Scheffer, T., Spiliopoulou, M. (eds.) PKDD 2006. LNCS (LNAI), vol. 4213, pp. 6–17. Springer, Heidelberg (2006)
3. Atzmuller, M.: Knowledge intensive subgroup mining: techniques for automatic and interactive discovery. Aka Akademische Verlagsgsellschaft (2007)
4. Azzara, J.: Identifying High Potential Employees. PeopleTalentSolutions (2007)
5. Barnett, R.: Identifying High Potential Talent. MDA Leadership Consulting Inc. (2008)
6. Breiman, L.: Bagging predictors. Mach. Learn. **24**(2), 123–140 (1996)
7. Breiman, L.: Random forests. Mach. Learn. **45**(1), 5–32 (2001)
8. Buckingham, M., Vosburgh, R.M.: The 21st century human resource function: it's the talent, stupid!. Hum. Resour. Plann. **24**(4), 17–23 (2001)
9. Bueno, C.M., Tubbs, S.L.: Identifying global leadership competencies: an exploratory study. J. Am. Acad. Bus. **5**(1–2), 80–87 (2004)
10. Corporate Leadership Council: Realizing the Full Potential of Rising Talent (Volume I): A Quantitative Analysis of the Identification and Development of High Potential Employees. Corporate Executive Board (2005)
11. Dries, N., Pepermans, R.: Using emotional intelligence to identify high potential: a metacompetency perspective. Leadersh. Organ. Dev. J. **28**(8), 749–770 (2007)
12. Freund, Y., Schapire, R.E.: Experiments with a new boosting algorithm. In: Proceedings of the Thirteenth International Conference on Machine Learning (ICML), pp. 148–156 (1996)
10. Friedman, J., Fisher, I.: Bump hunting in high-dimensional data. Stat. Comput. **9**, 123–143 (1999)
14. Gelens, J., Hofmans, J., Dries, N., Pepermans, R.: Talent management and organisational justice: employee reactions to high potential identification. Hum. Resour. Manag. J. **24**(2), 159–175 (2014)
15. Gemberger, D., Lavrac, N.: Expert guided subgroup discovery: methodology and application. J. Artif. Intell. Res. **17**, 501–527 (2002)
16. Jerusalim, R.S., Hausdorf, P.A.: Managers' justice perceptions of high potential identification practices. J. Manag. Dev. **26**(10), 933–950 (2007)
17. Kavsek, B., Lavrac, N., Jovanoski, V.: APRIORI-SD: adapting association rule learning to subgroup discovery. In: Berthold, M., Lenz, H.-J., Bradley, E., Kruse, R., Borgelt, C. (eds.) IDA 2003. LNCS, vol. 2810, pp. 230–241. Springer, Heidelberg (2003)
18. Klosgen, W.: Explora: a multipattern and multistrategy discovery assistant. In: Advances in Knowledge Discovery and Data Mining, pp. 249–271. MIT Press (1996)
19. Knorr, E.M., Ng, R.T., Tucakov, V.: Distance-based outliers: algorithms and applications. VLDB J. **8**(3–4), 237–253 (2000)
20. Lavrac, N., Cestnik, B., Gemberger, D., Flach, P.: Subgroup discovery with CN2-SD. Mach. Learn. **57**, 115–143 (2004)

21. Lavrac, N., Kavsek, B., Flach, P., Todorovski, L.: Subgroup discovery with CN2-SD. J. Mach. Learn. Res. **5**, 153–188 (2004)
22. Lombardo, M.M., Eichinger, R.W.: Do rising stars avoid risk?: status-based labels and decision making. High Potentials High Learn. **39**(4), 321–329 (2000)
23. Pepermans, R., Vloeberghs, D., Perkisas, B.: High potential identification policies: an empirical study among belgian companies. J. Manag. Dev. **22**(8), 660–678 (2003)
24. Ramaswamy, S., Rastogi, R., Shim, K.: Efficient algorithms for mining outliers from large datasets. In: Proceedings of SIGMOD 2000, pp. 162–172 (2000)
25. Rogers, R.W., Smith, A.B.: Finding future perfect senior leaders: spotting executive potential. In: Development Dimensions International (2007)
26. Scheffer, T., Wrobel, S.: Finding the most interesting patterns in a database quickly by using sequential sampling. J. Mach. Learn. Res. **3**, 833–862 (2002)
27. Scholtz, M.: Sampling based sequential subgroup mining. In: Proceedings of the 11th SIG KDD, pp. 265–274 (2005)
28. Spreitzer, G.M., McCall, M.W., Mahoney, J.D.: Early identification of international executive potential. J. Appl. Psychol. **82**(1), 6–29 (1997)
29. Wells, S.J.: Who's next: creating a formal program for developing new leaders can pay huge dividends, but many firms aren't reaping those rewards. HR Mag. **48**(11), 44–64 (2003)
30. Wrobel, S.: An algorithm for multi-relational discovery of subgroups. In: Komorowski, J., Żytkow, J.M. (eds.) PKDD 1997. LNCS, vol. 1263, pp. 78–87. Springer, Heidelberg (1997)
31. Zhi-Hua, Z.: Ensemble Methods: Foundations and Algorithms. Chapman and Hall/CRC, Boca Raton (2012)

Dynamic Grouped Mixture Models
for Intermittent Multivariate Sensor Data

Naoya Takeishi[1(✉)], Takehisa Yairi[1], Naoki Nishimura[2], Yuta Nakajima[2],
and Noboru Takata[2]

[1] The University of Tokyo, Tokyo, Japan
takeishi@ailab.t.u-tokyo.ac.jp
[2] Japan Aerospace Exploration Agency, Tsukuba, Japan

Abstract. For secure and efficient operation of engineering systems, it is of great importance to watch daily logs generated by them, which mainly consist of multivariate time-series obtained with many sensors. This work focuses on challenges in practical analyses of those sensor data: temporal unevenness and sparseness. To handle the unevenly and sparsely spaced multivariate time-series, this work presents a novel method, which roughly models temporal information that still remains in the data. The proposed model is a mixture model with dynamic hierarchical structure that considers dependency between temporally close batches of observations, instead of every single observation. We conducted experiments with synthetic and real dataset, and confirmed validity of the proposed model quantitatively and qualitatively.

Keywords: Multivariate time-series · Unevenly spaced time-series · Mixture models · Latent factor models · Sensor data

1 Introduction

For secure and efficient operation of engineering systems, such as industrial plants, vehicles and artificial satellites, it is of great importance to watch daily logs generated by them. Those logs are useful for various tasks including fault detection and isolation, maintenance prediction, and operation optimization. They mainly consist of multivariate time-series obtained with many sensors equipped to the system, and their amount keeps on increasing as the system goes on running. To deal with such sensor data with a machine learning approach, one can utilize a statistical model that represents their characteristic behavior. For example, state space models (SSM) have been widely used for multivariate time-series like the sensor data obtained from engineering systems (see e.g. [7,23]).

This work focuses on challenges in practical analyses of the sensor data: temporal unevenness and sparseness. Practically, observation grid of the sensor data is often unevenly and sparsely spaced as illustrated in Fig. 1(a), and it makes the traditional models such as the SSM inappropriate. One of the challenges,

© Springer International Publishing Switzerland 2016
J. Bailey et al. (Eds.): PAKDD 2016, Part II, LNAI 9652, pp. 221–232, 2016.
DOI: 10.1007/978-3-319-31750-2_18

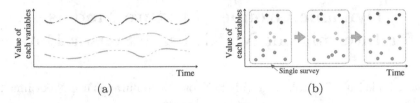

Fig. 1. Concepts of the intermittent multivariate time-series (IMT); (a) multivariate time-series with a large part of observations missing (represented with dotted lines), and (b) results of periodical surveys where observations in a single survey (enclosed by the dotted square) are not ordered. The different colors denote different variables (Color figure online).

the temporal unevenness, is often the case not only with the sensor data of the engineering systems but also with many kinds of scientific and industrial time-series. Many researchers have been studying on analyses of the unevenly spaced time-series (see e.g. [12,20,24]), though the range of application is limited. With regard to another challenge, the temporal sparsity, the situation of practical analyses can be tragic; the sampling rate of each sensor may vary from a few seconds to several days, the measurements may be unsynchronized, and the sensors may be stopped for a long time and restarted by event-driven measurements. Moreover, data that contain faulty behaviors may not be utilized, which leads to further loss of data and their temporal sequentiality. Consequently it is quite difficult to model the whole sensor data as successive time-series. In the following, we will refer to such unevenly and sparsely spaced sequential data as *intermittent multivariate time-series* (IMT).

One of our motivating examples of the IMT is telemetry data obtained from small artificial satellites. Generally, artificial satellites retrieve readings of hundreds of sensors such as voltmeters, thermometers, accelerometers, gyroscopes, and star sensors, as well as indicators of satellite's status including on/off of equipment, error flags, and operating modes. Hence the telemetry data from the artificial satellites would be generated as multivariate time-series. With regard to some satellites, however, the retrieved sensor readings are not always recorded nor transmitted to the ground due to limitation of memory size and transmission capability. Therefore the telemetry data that we finally obtain are often intermittent, that is, data series in interest are obtained at very low and uneven sampling rates. This is especially true for small satellites and microsatellites, and nanosatellites, which are expected to play a key role in space development. An instance of an observation grid of the telemetry data obtained from a small artificial satellite is shown in Fig. 2, whose vertical axis corresponds to different sensor types and horizontal axis is along time. One can see that the white cells, where measurement is recorded, are unevenly and sparsely placed.

An example of the IMT other than the sensor data is results of longitudinal studies or periodical surveys. These data generally have temporal sequentiality in some large scales such as years or decades, but observations in a single survey do not have any temporal ordering. This property suits the idea of the IMT,

variables

06:00 AM

Time

11:59 AM

Fig. 2. An observation grid of telemetry data obtained from a small satellite. The white cells denote timestamps where a measurement was recorded, and the black cells are ones where a measurement was *not* recorded. The rows represent different sensors, and the columns are along time. One can see the data are unevenly and sparsely spaced.

in the sense that there occurs certain amount of observations intermittently, as illustrated in Fig. 1(b).

This work presents a novel method to model the intermittent multivariate time-series, which roughly captures temporal information of the data. The proposed method is a mixture model with dynamic hierarchical structure that considers dependency between temporally close batches of observations, instead of every single observation. Our approach can be a useful option in practical sensor data analyses, because it models correlation along time as well as correlation among multiple variables (sensors), without evenly spaced measurements nor interpolation of the data.

In the rest of this paper, related work is briefly reviewed in Sect. 2, the proposed model is explained in Sect. 3, and experimental results are presented in Sect. 4. This paper is concluded in Sect. 5.

2   Related Work

2.1   Discrete-Time Latent Variable Models

With regard to *evenly* spaced multivariate time-series, one of the most widely accepted models would be discrete-time state space models (SSM), which are also referred to as linear dynamical system (LDS) or dynamic factor models (see e.g. [7, 23]). State space models assume (often low-dimensional) continuous latent factors with temporal correlation behind observations, and every observation is conditionally independent given the latent factors. Another model often used is hidden Markov models (HMM) and their variants such as factorial hidden Markov models [9] and factor analyzed hidden Markov models [18], which assume discrete latent variables with temporal dependency.

These models, in discrete time setting, are capable of dealing with the *unevenly* spaced time-series by skipping some measurement updates within filtering procedures [12]. However, skipping too many measurements can cause a significant bias on model estimation, especially when the sampling rate is very low or a large part of observations is missing.

2.2   Processing of Unevenly Spaced Time-Series

Many researches have explicitly focused on the unevenly spaced time-series. For example, Zumbach and Müller presented some basic operators such as the moving average [24], and Erdogan applied the autoregressive model [5]. Some other studies handled the unevenly spaced time-series with continuous-time models (see e.g. [3,12]). In astronomy, periodicity analysis of the unevenly spaced time-series has been intensely discussed (see e.g. [6,20]). Note that most of those studies are on basic operators and models for univariate and stationary time-series, and do not aim to handle the sensor data of engineering systems, which are multivariate and possibly nonstationary as well as intermittent.

Another major way to deal with the unevenly spaced time-series is interpolation.[1] However, it has been pointed out that interpolation of time-series generates a bias on statistic estimation (see e.g. [11,17]). Moreover, interpolation does not make much sense if an observation interval is too large.

2.3   Practical Alternative: Ignoring Temporal Dependencies

Practically there is a powerful alternative to model the IMT: just ignore the temporal dependency and regard them as a set of i.i.d. observations! If we decided to adopt this i.i.d. assumption, possible modeling approaches would be to use latent variable models such as mixtures of Gaussian, the principal component analysis (PCA), almost equivalently the factor analysis (FA), their mixture versions [8,22], the independent component analysis (ICA), the canonical correlation analysis (CCA), Gaussian process latent variable models (GPLVM) [14], restricted Boltzmann machines (RBM), and an autoencoder and its variants.

One justification of this compromise is that important information of the sensor data lies in inter-variable (inter-sensor) relationships than in temporal relationships. For example, if usually correlated sensors lose the correlation, a failure of the system can be easily suspected. Moreover, learning the models that assume i.i.d. observations is usually faster and less likely to be trapped in local minima. Of course, such models will completely miss temporal information of the data, but this drawback can be mitigated to some extend by methods like moving average. Actually, the models with the i.i.d. assumption have been widely used for sensor data analyses such as fault detection. See literatures [4,13] for example, though there would be numerous similar cases in practice.

3   Dynamic Grouped Mixtures of Factor Analyzers

In the previous section, we briefly introduced some methods that can be used for the unevenly spaced time-series. These would be useful in many applications, but they mainly treat univariate time-series with a moderate range of observation intervals, which is not the case with the IMT sensor data. We also mentioned the practical compromise, that is, i.i.d. assumption. Although this is acceptable in many cases, it wastes temporal information that still remains in the IMT.

In this section, we present a novel method that especially focuses on two important characteristics of the IMT sensor data of engineering systems.

---

[1] Detailed description on interpolation can be found in surveys such as [1,17].

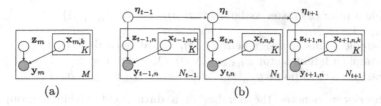

Fig. 3. Graphical models of (a) mixtures of factor analyzers (MFA), and (b) dynamic grouped mixtures of factor analyzers (DGMFA) for three time slices. Quantities other than probabilistic variables are omitted to simplify the figure.

First, the IMT data maintain temporal information at *some* scale; while the time-dependency at the finest level cannot be traced due to the sparseness, it can be observed in a larger scale. For example, physical condition of satellites in orbits around the earth gradually changes according to the revolution of the earth, which can be observed with sensor data even if only a single measurement occurs a day. Second, the sensor data of engineering systems often behave non-linearly because of physical phenomena, control laws and changes of operational modes. The nonlinearity can be handled with nonlinear models or mixtures of linear models, and we adopt the latter in this work.

The proposed model is a dynamical extension of mixture models. Our idea presented below works with any types of mixture models, though we focus on mixtures of factor analyzers (MFA) [8] for their simplicity and good performance in this work. The MFA is a latent variable model with two types of latent variables z and x as shown in Fig. 3(a), where z denotes a cluster to which each data point y belongs, and x is a low-dimensional representation behind each data point y. For the details, see the literature [8].

3.1 Data Grouping

An essential procedure of the proposed method is grouping of the time-series. We propose to partition the IMT data into multiple batches of data points, on a scale where we would like to model the temporal correlation of the IMT data. This time-series grouping can be formulated as follows: an observation y_m ($m = 1, \ldots, M$) is labeled with t ($t = 1, \ldots, T$) when the timestamp of y_m is within the range from τ_t^{begin} to τ_t^{end}, where $\tau_t^{\text{begin}} = \tau_{t-1}^{\text{end}}$. In this work, the grouping interval is set evenly $\tau_t^{\text{end}} = \tau_t^{\text{begin}} + \Delta$ for a fixed value of Δ. Thus, the amount of observations within the t-th group, N_t ($\sum_{t=1}^{T} N_t = M$), differs for each t generally. A suitable granularity of the partition depends on nature of the data and should be tuned empirically or with validation.

3.2 Generative Model

A data generation procedure of the proposed model for the t-th data-group can be described as follows:

1. Sample a prior of cluster assignment $\eta_t | \eta_{t-1} \sim \mathcal{N} (\eta_{t-1}, \Lambda)$.
2. For $n = 1, \ldots, N_t$:
   (a) Sample a cluster assignment $z_{t,n} | \eta_t \sim \text{Categorical} (\mathcal{S} (\eta_t))$.
   (b) Sample a latent factor $x_{t,n,k} \sim \mathcal{N} (0, I)$, where $k = z_{t,n}$.
   (c) Sample a data point $y_{t,n} | z_{t,n}, x_{t,n,k} \sim \mathcal{N} (L_k x_{t,n,k} + b_k, \Psi_k)$.

The subscript n denotes the number of a data point within a group ($n = 1, \ldots, N_t$), and k is the number of clusters or mixture components ($k = 1, \ldots, K$). The parameter Λ controls temporal transition of the cluster assignment priors η, while the parameters $\{L_k, b_k, \Psi_k\}$ are a loading matrix, bias and residual variance of the k-th factor analyzer, respectively. Note that $\mathcal{S}(\cdot)$ is a softmax function whose k-th element is denoted by $\mathcal{S}_k(\cdot)$. A graphical model for probabilistic variables of the proposed model is shown in Fig. 3(b). In the following, we term this model dynamic grouped mixtures of factor analyzers (DGMFA).

3.3 EM Algorithm with Variational Approximation

An objective function to learn the DGMFA is a incomplete-data likelihood:

$$p(y_{1:T, 1:N}) = \int d\eta_{1:T} \left[\prod_{t=1}^{T} \mathcal{N} (\eta_t; \eta_{t-1}, \Lambda) \right.$$
$$\left. \prod_{n=1}^{N_t} \sum_{k=1}^{K} \mathcal{S}_k (\eta_t) \int dx_{t,n,k} \mathcal{N} (0, I) \mathcal{N} (L_k x_{t,n,k} + b_k, \Psi_k) \right]. \quad (1)$$

As analytical maximization of this function is intractable, we adopt EM algorithm with variational approximation closely related to one presented by [2], which computes variational posteriors of η by Kalman filtering and smoothing with variational pseudo observations. The most part of the algorithm is the same with one presented in the literatures [2,8]. Readers should be careful only about an update procedure of variational parameters for z; they are updated with Gaussian likelihoods of observation given an expectation of other latent variables.

4   Experiments

We conducted three experiments with synthetic and real dataset: (1) Denoising of synthetic time-series and (2) visualization and (3) anomaly detection of the IMT sensor data of an artificial satellite.

4.1   Simulation: Denoising of Multivariate Time-Series

To confirm validity of the proposed model quantitatively, we conducted denoising experiments with synthetically-generated time-series. We prepared noise free time-series $\{\bar{y}_1, \ldots, \bar{y}_M\}$ and noised time-series $\{y_1, \ldots, y_M\}$, where

$$y_m = \bar{y}_m + e_m \quad \text{where} \quad e_m \sim \mathcal{N} (0, \nu^2 I), \quad (2)$$

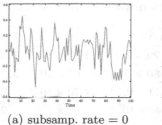 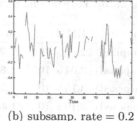

(a) subsamp. rate = 0   (b) subsamp. rate = 0.2

Fig. 4. One variable of the noised time-series generated by a linear dynamical system and a Gaussian noise. Note that data actually used in the experiment is 8-dimensional. (a) is the original time-series with no subsampling, and (b) is subsampled by rate of 0.2, i.e., 20 % of observations was disposed.

for $m = 1, \ldots, M$. Now the task is to recover the noise free time-series $\{\bar{\boldsymbol{y}}_1, \ldots, \bar{\boldsymbol{y}}_M\}$ from the noised time-series $\{\boldsymbol{y}_1, \ldots, \boldsymbol{y}_M\}$. We generated the noise-free time-series by a linear dynamical system as follows:

$$\boldsymbol{x}_m = \begin{bmatrix} -0.6 & 0.2 \\ -0.1 & 0.5 \end{bmatrix} \boldsymbol{x}_{i-1} + \boldsymbol{w}_m \quad \text{for} \quad m = 2, \ldots, M,$$

$$\bar{\boldsymbol{y}}_m = A\boldsymbol{x}_m,$$

where $\boldsymbol{w}_i$ is a noise that follows $\mathcal{N}(\boldsymbol{0}, 0.1^2 I)$, and A is an 8×2 emission matrix whose elements were sampled randomly from $\mathcal{N}(0, I)$ independently. Therefore the time-series were in 8 dimension, and the noised version was made following (2) with $\nu = 0.1$. We generated the data for 5000 timestamps and used 3000 timestamps for training, 1000 for validation and another 1000 for testing. The intermittent situation was simulated by randomly subsampling the original time-series, and its rate was varied from 0 (no subsampling) to 0.8 (80% of observations was disposed). We showed a part of the generated time-series at Figs. 4(a) and (b), with different subsampling rates.

The parameters of the proposed model and other baselines were learned with training set of data, model settings (the number of mixtures K and the number of observation groups T) were selected by a grid-search with the validation set. Note that we do not need to select the latent dimension d_x in this case, because we know $d_x = 2$ originally.

Denoising performances were evaluated using the test set of data. The performances in terms of root mean squared errors are shown in Table 1 for different models and different subsampling rates. As baseline, we tried some methods that are widely used in practice: the hidden Markov model (HMM), the linear dynamical system (LDS), the mixture of factor analyzers (MFA). The HMM and LDS are learned skipping missing observations. One can confirm that the proposed model, DGMFA, performed well even for high subsampling rates, while the LDS, with which the original data were generated, failed with the high subsampling rates. In Fig. 5, a part of denoising results and the ground truth are plotted for

Table 1. Denoising performances in RMS errors.

Subsampling rate	HMM	LDS	MFA	DGMFA
0 (no subsampling)	2.80×10^{-1}	$\mathbf{1.29 \times 10^{-1}}$	1.77×10^{-1}	$\mathbf{1.24 \times 10^{-1}}$
0.2 (20 % disposed)	3.04×10^{-1}	$\mathbf{1.30 \times 10^{-1}}$	1.69×10^{-1}	$\mathbf{1.26 \times 10^{-1}}$
0.4 (40 % disposed)	3.15×10^{-1}	1.39×10^{-1}	1.70×10^{-1}	$\mathbf{1.25 \times 10^{-1}}$
0.6 (60 % disposed)	3.19×10^{-1}	1.46×10^{-1}	1.98×10^{-1}	$\mathbf{1.24 \times 10^{-1}}$
0.8 (80 % disposed)	3.45×10^{-1}	2.13×10^{-1}	1.92×10^{-1}	$\mathbf{1.25 \times 10^{-1}}$

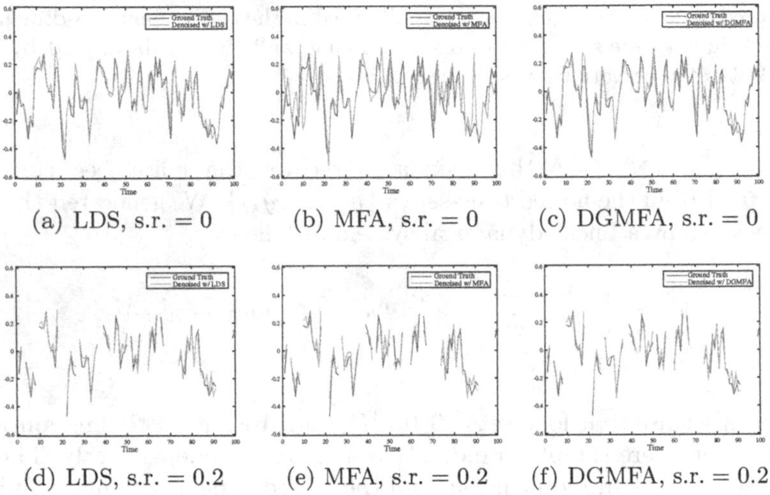

(a) LDS, s.r. = 0   (b) MFA, s.r. = 0   (c) DGMFA, s.r. = 0

(d) LDS, s.r. = 0.2   (e) MFA, s.r. = 0.2   (f) DGMFA, s.r. = 0.2

Fig. 5. A part of denoising results for test data with subsampling rate (s.r.) 0 and 0.2.

the LDS, MFA, and DGMFA. The simple MFA produced large errors at some timestamps, while the DGMFA achieved the compatible result with the LDS.

4.2 Application: Visualization of Sensor Data

The task addressed in this section is to visualize sensor data of an artificial satellite. It is indispensable to watch behaviors of an artificial satellite at all times for secure operation, though it is a tremendous task for human operators since the amount of the data is getting huge. Thus visualizing those data will be a great help to see the picture of satellite's behaviors. We adopted two unsupervised learning techniques, clustering and dimensionality reduction, and tried to use the MFA and the proposed model DGMFA.

The telemetry sensor data visualized here are obtained from SDS-4 [15], which is an working small satellite operated by Japan Aerospace Exploration Agency. The data contain sensor readings such as equipment's voltage, current, temperature. Their sampling rate varies approximately from one second to five minutes and differs by sensors. Moreover, the observations are often missing

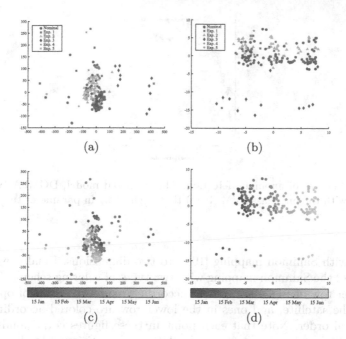

(a)   (b)

(c)   (d)

Fig. 6. Visualization of sensor data with cluster assignment rates of each day learned by MFA (left) and DGMFA (right). Each point corresponds to each day. The coloring scheme is by satellite's operational modes (upper) and chronological order (lower) (Color figure online).

for a long period, approximately from 3 to 12 h. The asynchrony was slightly compensated by a zero-order hold that lasted up to only 10 s. We used 92 types of sensor readings that took continuous values in this experiment, i.e., the data were 92-dimensional intermittent time-series. Remember the fact that satellite's operation consists of a nominal mode and five types of experimental modes, since it is emphasized in the following.

Model settings were empirically selected without any validation because the aim of this experiment is just to show visualization capability of the proposed model qualitatively. The data were partitioned into groups by days and the period of the data is from 1 January 2015 to 30 June 2015, hence $T = 183$. Consequently, the number of data points within a day was from 300 to 1300 approximately. The intrinsic dimensionality was set $d_x = 6$ with an intrinsic dimensionality estimator [10], and the number of the mixture components was chosen empirically $K = 10$ just for visualization clarity.

Results of the visualization are presented in Fig. 6, where the left column is by the simple MFA and the right is by its extension, DGMFA. Figure 6 was drawn as follows. First, we calculated a cluster assignment rate of each group (day) of the data: $\sum_{\tau_t^{\text{begin}} \leq \text{timestamp}(\boldsymbol{y}_m) < \tau_t^{\text{end}}} \mathbb{E}[\boldsymbol{z}_m]$ for MFA and $\mathbb{E}[\boldsymbol{\eta}_t]$ for DGMFA, from $t = 1$ to $t = 183$. It is a 10-dimensional quantity since the number of clusters K was set to 10. Second, we further reduced the dimensionality of these

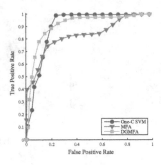

Fig. 7. ROC curves of anomaly detection. The proposed model, DGMFA (yellow), is compatible with the one-class SVM (blue) that is cheating in parameter tuning (Color figure online).

quantities with Sammon mapping [19] into two dimensions. Finally, we plotted the results of the Sammon mapping with two types of coloring scheme. The plots in the upper row of Fig. 6 are colored according to the six types of operational modes of the satellite, and ones in the lower row are colored according to the chronological order. Note that each point in these figures corresponds to each day of satellite's operation, because the data were partitioned by days.

In Fig. 6(d), the points along the horizontal axis are arranged by the color gradation, i.e. approximately ordered by date. On the other hand, the points in Fig. 6(c) are not ordered geometrically by the color (date) at all. This means that the proposed model was more successful at learning the dynamical behavior of the satellite than the classical MFA was. Since the horizontal axis in Figs. 6(b) and (d) would denote chronological order (left to right) and the vertical axis would correspond to operational modes of the satellite, it can be anticipated that one of the experimental operation (denoted by orange points, referred to as "Exp.1") was not conducted after the end of March.

4.3   Application: Anomaly Detection

Another experiment was conducted with the telemetry sensor data obtained from the satellite to quantitatively confirm availability of the proposed model. We focused on semi-supervised anomaly detection, where models were trained using data without anomalous behaviors, and another range of data that may contain anomalies were tested by the trained models. We used the telemetry sensor data during three months: a month for training, another month for validation, and still another month for testing. It has been known that two anomalous (novel) events occurred within the test range: one is an unexpected change of satellite's attitude, and another is a first-ever type of operation of the satellite. We expected to detect these events with as few false alarms as possible.

The proposed model were compared with two practical baselines: one-class SVM [21] and the MFA. The parameters of MFA and DGMFA were tuned using the validation data that contain no anomalies, while this is suboptimal for the detection task. On the other hand, the parameter of the one-class SVM was tuned with cheating, that is, one with the best detection performance for the test data was adopted. We investigated negative log likelihoods by the one-class SVM and an absolute value of reconstruction errors by the MFA and DGMFA, and a part of the data whose scores exceed a threshold was reported as an anomaly.

The detection performances in ROC curves are shown in Fig. 7, and AUC of each curve is 0.8998 for the one-class SVM, 0.8307 for the MFA, and 0.9105 for the DGMFA. Note that the proposed model was compatible with the one-class SVM that is cheating in parameter tuning.

5   Conclusion

In this paper, we introduced a novel way to handle the multivariate intermittent time-series (IMT), where observations were unevenly and sparsely spaced. The proposed method consists of partitioning the time-series into multiple batches of observations, and a model that roughly captures the temporal information of the batches of observations. The partition is done on a scale in which the temporal dependency can be observed even if finer-level information is lost. The proposed model performed better than simple time-series models such as HMM and LDS on a denoising task with the IMT data. Also, it successfully visualized the sensor data obtained from an small artificial satellite, which are difficult to model appropriately with standard time-series models or simple i.i.d. models.

One of the most demanding extensions to the method proposed in this work is automatic partition of the data. The constant interval was adopted in this work, though an optimal granularity of the partition is usually unknown and may change in different parts of time-series.

It is noteworthy that the idea to set the dynamic priors on the cluster assignment of batches of observations is widely known because it was introduced in the work on dynamic topic models (DTM) [2]. In the DTM, each word in a document follows a mixture of multinomials and each document has a prior over mixture component assignment, with the priors having temporal sequentiality.[2] It shall be interesting to combine the proposed model with the DTM, which would enable us to analyze the IMT sensor data and text simultaneously.

References

1. Adorf, H.M.: Interpolation of irregularly sampled data series - a survey. In: Astronomical Data Analysis Software and Systems IV (1995)
2. Blei, D.M., Lafferty, J.D.: Dynamic topic models. In: Proceedings of the 23rd International Conference on Machine Learning, pp. 113–120 (2006)
3. Brockwell, P.J.: Lévy-driven continuous-time ARMA processes. In: Mikosch, T., Kreiß, J.-P., Davis, R.A., Andersen, T.G. (eds.) Handbook of Financial Time Series, pp. 457–480. Springer, Heidelberg (2009)

[2] From the document-modeling point of view, it is also interesting to compare the MFA with the mixtures of unigrams [16].

4. Dunia, R., Qin, S.J., Edgar, T.F., McAvoy, T.J.: Identification of faulty sensors using principal component analysis. AIChE J. **42**(10), 2797–2812 (1996)
5. Erdogan, E.: Statistical models for unequally spaced time series. In: Proceedings of the 5th SIAM International Conference on Data Mining, pp. 626–630 (2005)
6. Foster, G.: Wavelets for period analysis of unevenly sampled time series. Astron. J. **112**, 1709–1729 (1996)
7. Geweke, J.: The dynamic factor analysis of economic time-series models. In: Aigner, D.J., Goldberger, A.S. (eds.) Latent Variables in Socio-Economic Models. North-Holland, New York (1977)
8. Ghahramani, Z., Hinton, G.E.: The EM algorithm for mixtures of factor analyzers. Technical report, University of Toronto (1996)
9. Ghahramani, Z., Jordan, M.I.: Factorial hidden markov models. Mach. Learn. **29**(2–3), 245–273 (1997)
10. Gupta, M.D., Huang, T.S.: Regularized maximum likelihood for intrinsic dimension estimation. In: Proceedings of the 26th Conference on Uncertainty in Artificial Intelligence (2010)
11. Hayashi, T., Yoshida, N.: On covariance estimation of non-synchronously observed diffusion processes. Bernoulli **11**, 359–379 (2005)
12. Jones, R.: Time series analysis with unequally spaced data. In: Hannan, E.J., Krishnaiah, P.R., Rao, M.M. (eds.) Handbook of Statistics, vol. 5. Elsevier Science, Amsterdam (1985)
13. Kermit, M., Tomic, O.: Independent component analysis applied on gas sensor array measurement data. IEEE Sens. J. **3**(2), 218–228 (2003)
14. Lawrence, N.: Probabilistic non-linear principal component analysis with gaussian process latent variable models. J. Mach. Learn. Res. **6**, 1783–1816 (2005)
15. Nakamura, Y., Nishijo, K., Murakami, N., Kawashima, K., Horikawa, Y., Yamamoto, K., Ohtani, T., Takhashi, Y., Inoue, K.: Small demonstration satellite-4 (SDS-4): development, flight results, and lessons learned in JAXAs microsatellite project. In: Proceedings of the 27th Annual AIAA/USU Conference on Small Satellites (2013)
16. Nigam, K., McCallum, A.K., Thrun, S., Mitchell, T.: Text classification from labeled and unlabeled documents using EM. Mach. Learn. **39**(2/3), 103–134 (2000)
17. Rehfeld, K., Marwan, N., Heitzig, J., Kurths, J.: Comparison of correlation analysis techniques for irregularly sampled time series. Nonlinear Proc. Geophys. **18**, 389–404 (2011)
18. Rosti, A.V.I., Gales, M.J.F.: Factor analysed hidden Markov models for speech recognition. Comput. Speech Lang. **18**(2), 181–200 (2004)
19. Sammon, J.W.: A nonlinear mapping for data structure analysis. IEEE Trans. Comput. **18**, 401–409 (1969)
20. Scargle, J.: Studies in astronomical time series analysis. II - Statistical aspects of spectral analysis of unevenly spaced data. Astrophys. J. **343**, 874–887 (1982)
21. Schölkopf, B., Platt, J.C., Shawe-Taylor, J.C., Smola, A.J., Williamson, R.C.: Estimating the support of a high-dimensional distribution. Neural Comput. **13**(7), 1443–1471 (2001)
22. Tipping, M., Bishop, C.: Mixtures of probabilistic principal component analysers. Neural Comput. **11**(2), 443–482 (1999)
23. Watson, M.W., Engle, R.F.: Alternative algorithms for the estimation of dynamic factor, mimic and varying coefficient regression models. J. Econometrics **23**, 385–400 (1983)
24. Zumbach, G., Müller, U.: Operators on inhomogeneous time series. Int. J. Theor. Appl. Fin. **4**, 147–177 (2001)

Parallel Discord Discovery

Tian Huang[1], Yongxin Zhu[1(✉)], Yishu Mao[1], Xinyang Li[1], Mengyun Liu[1],
Yafei Wu[1], Yajun Ha[2], and Gillian Dobbie[3]

[1] School of Microelectronics, Shanghai Jiao Tong University, Shanghai, China
{ian_malcolm,zhuyongxin,maoyishu,all-get,old-ant,wuyf0406}@sjtu.edu.cn
[2] Institute for Infocomm Research, A*STAR, Singapore, Singapore
ha-y@i2r.a-star.edu.sg
[3] Department of Computer Science, University of Auckland,
Auckland, New Zealand
g.dobbie@auckland.ac.nz

Abstract. Discords are the most unusual subsequences of a time series. Sequential discovery of discords is time consuming. As the scale of datasets increases unceasingly, datasets have to be kept on hard disk, which degrades the utilization of computing resources. Furthermore, the results discovered from segmentations of a time series are non-combinable, which makes discord discovery hard to parallelize. In this paper, we propose Parallel Discord Discovery (PDD), which divides the discord discovery problem in a combinable manner and solves its subproblems in parallel. PDD accelerates discord discovery with multiple computing nodes and guarantees the correctness of the results. PDD stores large time series in distributed memory and takes advantage of in-memory computing to improve the utilization of computing resources. Experiments show that given 10 computing nodes, PDD is seven times faster than the sequential method HOTSAX. PDD is able to handle larger datasets than HOTSAX does. PDD achieves over 90 % utilization of computing resources, nearly twice as much as the disk-aware method does.

Keywords: Time series discord · Parallel · Large scale · In-memory computing

1 Introduction

Time series discords are the subsequences of a time series that are maximally different to all the rest subsequences [8,11]. Discord can be found by computing the pair-wise distances among all subsequences of a time series. Recently, finding time series discord has attracted much attention. The definition, despite its simplicity, captures an important class of anomalies. Its relevance has been shown in several data mining applications [3,9,13,16,19].

Various memory based methods, e.g. the classical method HOTSAX [11], have been proposed to speed up discord discovery, but they are still time consuming when the datasets are large. For example, to discover the top discord

© Springer International Publishing Switzerland 2016
J. Bailey et al. (Eds.): PAKDD 2016, Part II, LNAI 9652, pp. 233–244, 2016.
DOI: 10.1007/978-3-319-31750-2_19

from an ECG time series that contains 648 K data instances, a Java implementation of HOTSAX on a modern PC costs about 20 min, which is close to the time period of the ECG time series. In fact, any dataset in existing literature of memory based methods contains no more than 64 K data instances. We believe the time consumption is one of the main limitations to discover the discords of larger time series.

Low utilization of computing resources is another issue when discovering discords from large datasets. A disk-aware method [20] discovers discords from 100-million scale time series, which can only be fitted onto hard disks. Disk I/Os take more than 50 % of the total time consumption, leaving the computing resources in idle state for most of the time during the discord discovery.

Parallel computing is a potential way to mitigate these issues. However discord discovery is hard to parallelize. The results discovered from segmentations of a time series are non-combinable [11], which means divide-and-conquer methodology may leads to incorrect results that do not conform to the original definition of discord.

To mitigate the above issues, we propose Parallel Discord Discovery (PDD), which divides discord discovery problem in a combinable manner and solves its sub-problems in parallel. PDD stores a long time series in distributed memory of multiple computing nodes. These nodes work together to reduce the time consumption of discord discovery. As far as we know, this is the first work that discovers time series discords in parallel. Our contributions are summarized as follows:

– We accelerate discord discovery by harnessing multiple computing nodes.
– We ensure the correctness of the results by dividing discord discovery problem in a combinable manner.
– We improve the utilization of computing resources in large scale discord discovery by using an in-memory computing framework.

We implement PDD using Apache Spark [18]. Experiments show that given 10 computing nodes, PDD achieves 7 times speedup against HOTSAX. PDD is able to handle larger datasets than traditional memory based methods. PDD achieves nearly twice the utilization of computing resources compared to the disk-aware method [20].

The rest of the paper is organized as follows. Section 2 presents related works and their issues. Section 3 analyzes the feasibility of parallelization and describes the detailed implementation of the Parallel Discord Discovery (PDD) method. Section 4 presents an empirical evaluation of PDD. Section 5 draws the conclusion.

2   Related Works and Issues

Discords can be discovered by comparing every pair of subsequences with two-layer nested for-loops. The outer loop considers each possible candidate subsequence, and the inner loop is a linear scan to identify the non-overlapping nearest neighbor of the candidates. The computational complexity of a naive discord discovery method is $O(m^2)$, where m is the size of the dataset. Various methods are proposed to accelerate discord discovery. Keogh et al. propose HOTSAX [11], which applies heuristic sorting techniques and early abandon technique to reduce the computational complexity. The heuristic order of the outer loop helps HOTSAX visit the most unusual subsequences in the first few iterations, while the heuristic order of the inner loop helps HOTSAX visit the subsequences that are similar to the current candidate subsequence. A conditional branch in the inner loop helps HOTSAX skip the calculations of the normal subsequences. HOTSAX speeds up by three orders of magnitude compared with the naive method.

Many efforts have been made to improve discord discovery from various aspects. [3,7,12] introduce different feature extraction methods to provide better heuristic inner and outer order of the nested loop. They reduce the dimensionality of time series data and the computational complexity of discord discovery. [1–3,7,10,14,15] mitigate the negative effects of improper determination of parameters to the computational efficiency of discord discovery. [4,5,12] use elaborate indexes to find nearest neighbor distance more efficiently and reduce the computational complexity of discord discovery.

Most existing discovery methods assume that datasets are small enough to be stored in the memory of a single computer. Yankov et al. propose a disk-aware method [20] to deal with the time series whose scale grows beyond the capacity of the memory of a single computing node. The method improves the efficiency of the disk I/O operation by linearly scanning the disk to obtain all subsequences. The disk-aware method eliminates the limitation of the memory based method in terms of the scale of datasets.

Previous methods are all sequential methods. They suffer from the limitation of computing power and storage of a single computing node. As the scale of time series increases unceasingly, the utility of discord discovery deteriorates. Besides, the disk-aware method suffers from the low utilization of computing resources.

Although parallel computing is a potential way to mitigate these issues, discord discovery is hard to parallelize. The results of discord discovery is non-combinable [11]. In other words, divide-and-conquer methodology may lead to incorrect results, which do not conform to the original definition of discord.

In this paper we mitigate the issues by proposing Parallel Discord Discovery (PDD). PDD enables accelerating discord discovery with multiple computing nodes. We implement PDD with Apache Spark so that PDD has better utilization of computing resources than the disk-aware method.

3   Parallel Discord Discovery

To parallelize discord discovery, we must divide the problem into independent sub-problems, and solve each sub-problem respectively in different computing nodes. [11] states that divide-and-conquer methodology may yield incorrect results. We need to find another feasible way to partition discord discovery problem. To better analyze the feasibility of parallel time series discord discovery, we review the concept of discord, analyze the communication computation ratio and the parallelism.

3.1 Dividing the Problem

Before analyzing the formal definition of discord, we describe a set of preliminary notations [11]. We use T to denote a *time series* $t_1, ..., t_m$, where m is the length of the time series, $t_i \in \mathbb{R}$. We denote a *subsequence* of a time series T as $C_{p,n} = t_p, ..., t_{p+n-1}$, where p is the starting position, n is the length of the subsequence. Because subsequences of the same length are compared in a discovery of discords, we use C_p as an abbreviation of $C_{p,n}$ in the rest of the paper. Since every subsequence could be a discord, we will use a *sliding window*, whose size is equal to the length of subsequences, to extract all possible subsequences from T. The process of time series discord discovery can be expressed by the following equation:

$$p^{(1)} = \underset{p}{\mathrm{argmax}}\{nnDist(C_p)|1 \leq p \leq m - n + 1\} \tag{1}$$

$p^{(1)}$ indicates the start position of the first discord. $nnDist(C_p)$ is the nearest neighbor distance of a subsequence C_p. n is the length of the sliding window. m stands for the length of the time series. argmax in Eq. (1) means the subsequence with the largest $nnDist$, which is the discord of the time series.

As shown in Eq. (1), the processes of solving each the nearest neighbor distance $nnDist(C_p)$ of each subsequence C_p is independent. Therefore, we divide discord discovery into independent sub-problems. Each sub-problem finds the nearest neighbor of the subsequence C_p and the distance between them. The results of all sub-problems can be combined. The subsequences with the largest $nnDist$ is the discords of the time series.

3.2 Improving Computation Communication Ratio

To find the nearest neighbor of a subsequence C_p, we transmit C_p and any other non-overlapping [6] subsequence to one or more computing nodes to calculate the distance between them. However, the time consumption of transmitting two subsequences and computing the distance between them is of the same order of magnitude. This will results in low Computation Communication Ratio (CCR) and therefore low utilization of computing resources. We have to improve CCR.

We utilize the overlapped region between two adjacent subsequences to improve CCR. We transmit continuous data instances so that the overlapped region can be reused by multiple subsequences. Taking the sliding window of length $n = 100$ as an example, the transmission of the first 100 data forms one subsequence. Each transmitted data instance forms a new subsequence with the previous transmitted 99 data instances. If we transmit 299 continuous data instances, we get $299 - 100 + 1 = 200$ subsequences of length $n = 100$. Therefore, CCR is improved by a factor of $\frac{200}{299} \div \frac{1}{100} \approx 67$.

3.3 Overview of PDD

We divide discord discovery in a combinable manner. Next we design Parallel Discord Discovery (PDD). We first present the overview of PDD.

Data: All subsequences $\mathbb{C}$ and #subsequences/bulk b

Result: Position $bsfPos$ and nearest neighbor distance $bsfDist$ of the discord

1 Initialize $bsfDist = 0$, $bsfPos = null$;

2 Find the estimation $\tilde{d}$ of $nnDist$ of discord;

3 // see Section 3.4 Update $bsfDist = \tilde{d}$;

4 **for** *each b subsequence $C_p, \ldots, C_{p+b-1}$ of* $\mathbb{C}$ **do**

5   // see Section 3.5 $i = \arg\max_i \{nnDist(C_{p+i})|0 \geq i \geq b-1\}$;

6   **if** $bsfDist \leq nnDist(C_{p+i})$ **then**

7   Update $bsfDist = nnDist(C_{p+i})$;

8   Update $bsfPos = p + i$;

9   **end**

10 **end**

Algorithm 1. The pseudocode of PDD

Algorithm 1 shows the overview of PDD. $\mathbb{C}$ is the set of all possible subsequences of a time series. b is the number of continuous subsequences in each transmission. This method has two steps. First, PDD estimates global $nnDist$ of the discord with Distributed Discord Estimation (DDE) method (line 2–3). Then PDD linearly scans $\mathbb{C}$ to find the $nnDist$ of the true discord of the time series. This linear scanning step consists of multiple rounds. Continuous subsequences are transmitted in batches among computing nodes for better CCR. In each round, computing nodes work separately and exchange intermediate results at the end of the round (line 5). After each round, the current best-so-far distance ($bsfDist$) is updated (line 6–9). DDE and linear scanning (line 4) are the most time consuming parts. We present the detailed implementation of these parts.

3.4 Distributed Discord Estimation

The first step of discord discovery is to estimate $nnDist$ of discord. The estimation together with an early abandon technique can efficiently reduce the number of calls to the distance function (Algorithm 1, line 2). The closer the estimation is to the ground truth, the less the distance function is invoked. Traditional methods usually set an index to achieve this. However, when a time series is divided into several segments and stored into non-unified memory spaces, creating a centralized index for these distributed data is inefficient because it degrades the CCR. We propose a method called Distributed Discord Estimation (DDE), which estimates the distance and minimizes the communication between computing nodes.

In Algorithm 2, $\mathbb{C}$ represents the set of all subsequences. $\mathbb{S}$ is the segment information of a time series. DDE outputs the estimated $nnDist$, which is $\tilde{d}$. If the real $nnDist$ of the discord is $nnDist(C_d)$, then $\tilde{d} \leq nnDist(C_d)$.

DDE approximates every subsequence with a symbolic representation (line 1–3). The approximated representation of a subsequence C_p is denoted as A_p. Similar subsequences have the same approximation symbol. DDE divides the subsequences into groups according to their symbolic representations. The number of subsequences in one group reflects the frequency of the group (line 4).

Data: All subsequences $\mathbb{C}$ and Segmenting Information $\mathbb{S}$
Result: Nearest neighbor distance estimation $\tilde{d}$ of discord

1 **for** *Each C_p in $\mathbb{C}$* **do**
2 | Calculate A_p of C_p;
3 **end**
4 Group C_p by A_p;
5 Find $A_{\tilde{d}}$ that is the A_p with smallest group of C_p;
6 **for** *each C_p in $A_{\tilde{d}}$* **do**
7   **for** *each S in $\mathbb{S}$* **do**
8 | Calculate local $nnDist$ of C_p in S;
9   **end**
10   Calculate global $nnDist$ of C_p by finding its min local $nnDist$;
11 **end**
12 $\tilde{d} = \max\{$global $nnDist(C_p)|C_p$ that can be approximated as $A_{\tilde{d}}\}$;

Algorithm 2. Distributed Discord Estimation

DDE selects the group with the least number of members, named $A_{\tilde{d}}$, as the candidates of discords. For each subsequence C_p in this $A_{\tilde{d}}$ group (line 6), DDE finds a local $nnDist(C_p)$ for each S. The minimum of all local $nnDist(C_p)$ is called global $nnDist(C_p)$. The maximum among all global $nnDist(C_p)$ is the estimation of the global discord $nnDist$ of the time series.

The number of subsequence in the $A_{\tilde{d}}$ group, denoted as $|A_{\tilde{d}}|$, affects the accuracy of the estimation. Sometimes $|A_{\tilde{d}}|$ is too small for DDE to get a reliable estimation. Empirically we select 2 to 10 groups of subsequences to get better precision for the estimation.

As we can see, the computing process of the approximation is independent, which means it can be concurrently executed on a computing cluster. Each computing node processes different subsequences at the same time. Then PDD aggregates all data to find the globally estimated $nnDist$.

3.5 Linearly Scanning the Entire Dataset

The second step of PDD is to calculate the $nnDist$ of every subsequence and update $bsfPos$ and $bsfDist$. Linear scanning is the most time consuming part of PDD. In order to improve the CCR, a bulk of continuous subsequences is transmitted and calculated in batch. During this process, early abandon technique is used to reduce the computational complexity. We describe the linear scanning step from two perspectives.

Life Cycle of a Bulk of Subsequences. Algorithm 3 explains the life cycle of a bulk of subsequences. The inputs of this pseudocode are formed by a bulk of subsequences $\mathbb{C}_b$, the set of segment $\mathbb{S}$, and $bsfDist$.

As the life cycle of a bulk of subsequences $\mathbb{C}_b$ starts at the beginning of Algorithm 3, the $nnDist$ of all subsequences in the bulk is initialized as positive infinity (line 1). $\mathbb{C}_b$ visits all computing nodes, each of which contains different segment of the time series (line 3). During this process, the local $nnDist$ of the

Data: A bulk of subsequence $\mathbb{C}_b$, Segmenting Information $\mathbb{S}$ and $bsfDist$
Result: Position $bsfPos$ and nearest neighbor distance $bsfDist$ of the discord
1 $nnDist[:] = $ positive infinity;
2 **for** *each S in* $\mathbb{S}$ **do**
3 $\quad$ **for** *each* C_p *in* $\mathbb{C}_b$ **do**
4 $\quad\quad$ **for** *each* C_q *in S* **do**
5 $\quad\quad\quad$ **if** $nnDist[p] > dist(C_p, C_q)$ **then**
6 $\quad\quad\quad\quad |$ $nnDist[p] = dist(C_p, C_q)$;
7 $\quad\quad\quad$ **end**
8 $\quad\quad\quad$ **if** $nnDist[p] < bsfDist$ **then**
9 $\quad\quad\quad\quad |$ $\mathbb{C}_b = \mathbb{C}_b/C_p$;
10 $\quad\quad\quad\quad$ continue to next C_p;
11 $\quad\quad\quad$ **end**
12 $\quad\quad$ **end**
13 $\quad$ **end**
14 $\quad$ **if** $nnDist[p] \geq bsfDist$ **then**
15 $\quad\quad |$ $bsfPos = p$;
16 $\quad$ **end**
17 **end**
18 $bsfDist = nnDist(bsfPos)$;

Algorithm 3. Linear scanning: life cycle of a bulk of subsequences

subsequence C_p with respect to the current segment is calculated (line 4–7). If the global best so far Distance ($bsfDist$) is larger than the $nnDist[p]$ (line 8), subsequence C_p could be abandoned from $\mathbb{C}_b$ (line 9), and the follow-up calculating about C_p is skipped (line 10). Heuristic access order improves the efficiency of the early abandon technique (line 5). After calculating the distance between C_p and all other subsequences in this segment without triggering the early abandon technique, the $nnDist[p]$ becomes the global $nnDist$ of C_p. Under this circumstance it must hold $nnDist[p] \geq bsfDist$. Hence $bsfPos$ should be pointed to current subsequences(line 14–15). $bsfDist$ must be synchronously shared by all computing nodes, and updated after a bulk visit to all computing nodes (line 18).

During the linear scanning, relevant intermediate results of bulk $\mathbb{C}_b$, including the $nnDist$ of every subsequence $nnDist[:]$, are transmitted between computing nodes (line 3). The DDE method mentioned above initiates the $bsfDist$ as much as possible in order to call the Early Abandon Technology more frequently (line 8–10). Therefore most subsequences of bulk $\mathbb{C}_b$ are dumped and the number of subsequences in bulk $\mathbb{C}_b$ continually decreases. So the number of data and intermediate results being transmitted is quite small.

Communication Among Computing Nodes. In each round, computing nodes receive bulk, work separately and exchange intermediate results with other computing nodes. We give an example to explain the details. The time series is divided into segments $S_0 \ldots S_{(a-1)}$ and restored in nodes $node_0 \ldots node_{(a-1)}$. The subsequences are grouped into bulks $b_0, b_1, b_2 \ldots$. Table 1 explains the timing of transmissions among computing nodes.

Table 1. The timing of transmission among computing nodes

	$node_0(S_0)$	$node_1(S_1)$	...	$node_{a-1}(S_{a-1})$
R_0	b_0	b_1	...	b_{a-1}
R_1	$b_a + b_{a-1}$	$b_{a+1} + b_1$	...	$b_{2a-1} + b_{a-2}$
R_2	$b_{2a} + b_{2a-1} + b_{a-2}$	$b_{2a+1} + b_a + b_{a-1}$	...	$b_{3a-1} + b_{2a-2} + b_{a-3}$
...	...	...	...	...
R_{a-1}	$b_{a(a-1)} + \sum\limits_{i=0}^{a-2} b_{ai+i+1}$	$b_{a(a-1)+1} + \sum\limits_{i=0}^{a-2} b_{ai+i+2}$	...	$b_{a(a-1)+a-1} + \sum\limits_{i=0}^{a-2} b_{ai+i+aa}$
...	...	...	...	...
R_t	$b_{at} + \sum\limits_{i=t-a}^{t-1} b_{ai+i+1}$	$b_{at} + \sum\limits_{i=t-a}^{t-1} b_{ai+i+2}$	...	$b_{at} + \sum\limits_{i=t-a}^{t-1} b_{ai+i+a}$

Allocating New Bulk. At the beginning of each round, each node receives a new bulk, which has never visited other nodes. The $nnDist$ of every subsequence in this new bulk is set as $+\inf$ (line 1 in Algorithm 3). For example, the node $node_{a-1}$ receives new bulk $b_{a-1}, b_{2a-1}, b_{3a-1}, \ldots$ at round $R_0, R_1, R_2, \ldots$. This process continues until all new bulks are allocated to nodes.

Transmission Between Computing Nodes. Every bulk is sent to all computing nodes to find the global $nnDist$ of every subsequence. After one node finishes the $nnDist$ of subsequences of a bulk against one segment (e.g. S_0), the bulk is sent to the next node (e.g. $node_1$). There are many paths to transmit between nodes, and we choose the rotation shift method. For example, at R_1, bulks $b_{a+1} + b_0$ are in the $node_1$. At R_2, the bulks are sent to $node_2$. At R_{a-1}, the bulks are sent to $node_{a-1}$. At this time bulk b_0 has traversed to all nodes (which means all data segments) and will not be sent to $node_0$ at R_a. The computing of b_0 finishes here.

3.6   Improving Utilization of Computing Resources

Load imbalance may occurs among computing nodes. Some computing nodes finish computations and exchanges earlier and enter an idle state before the round ends. The idle state degrades the utilization of computing resource.

The idle state can be alleviated by allocating extra bulks and using a shared input queue. In each round, PDD creates a shared queue which contains bulks several times the number of the computing nodes. Once a computing node finishes the processing of a bulk, the node is assigned to the next bulk from the shared queue. The round is finished when all bulks in the shared queue are processed.

3.7   Guarantee on Correctness of PDD

PDD discovers exact results, which completely conform to the definition of discord. PDD provides the guarantee on the correctness of results by discovering discords according to the definition of discord.

More specifically, PDD finds the $nnDist$ of a candidate subsequence C_p (line 3 in Algorithm 3) by calculating the distances (line 5–7 in Algorithm 3) of C_p and every other subsequence C_q (line 2 and line 4 in Algorithm 3) of a time series. PDD finds the $nnDist$ of all candidate subsequences (line 4 in Algorithm 1) and returns the subsequence with the maximum $nnDist$ (line 6–9 in Algorithm 1) as the top discords.

DDE improves the performance of PDD but has no impact on the correctness of PDD. PDD still produces correct results when it skips the step of DDE and sets the initial value of $bsfDist$ in Algorithm 3 to positive infinity. Other design details of PDD, such as transmission among nodes and allocation of bulks, do not affect the correctness of PDD.

4   Empirical Evaluation

In this section we empirically evaluate the practical performance of PDD. We use randomly generated time series datasets for the following reasons: (1) Random time series is generally more challenging than real-world time series in terms of the time consumption of discord discovery methods [3,7,11,12,14,15,20]. The experimental results of time consumption on random time series are more convincing. (2) Large scale random time series is easier to acquire. For better reproducibility we choose random time series for our datasets.

We implement PDD using Apache Spark [18]. PDD makes use of the in-memory-computing feature of Spark to accelerate the detection of discord. We establish a Spark cluster consisting of 10 computing nodes, each of which is equipped with 512 MB memory. The time series are initially stored in Hadoop distributed file system [17], which is accessible to all computing nodes.

4.1   Scalability

PDD method can be scaled to discover discords. In this paper, we evaluate the scalability of PDD in terms of parallelism and data sizes, and compare the time consumptions between PDD and the classic method HOTSAX. For better comparability, we implement HOTSAX with Java and run HOTSAX on one of the computing nodes of the Spark cluster.

Table 2 shows the scalability of PDD. First we use PDD with different parallelism to detect the top discord from a 1×10^5 dataset. PDD achieves a speed-up ranging from 1.54 to 6.75 compared to HOTSAX (column 5 and row 2–6). Then we detect the top discord from the datasets of sizes ranging from 1×10^5 to 1×10^6 with 10 computing nodes of the Spark cluster. PDD achieves a speed-up ranging from 6.75 to 8.04 compared to HOTSAX (column 5 and row 6–11).

The speed-up provided by each computing node indicates the scalability of PDD (column 6 of Table 2). On average, each additional computing node of PDD provides more than 0.7 times speed-up compared to HOTSAX. The speed-up per node is not sensitive to the size of the dataset or the parallelism of PDD.

Table 2. Scalability of PDD

Data size	HOTSAX time	PDD time	#node	Speedup	Speedup/node
1×10^5	3374 s	2184 s	2	1.54	0.77
		1190 s	4	2.84	0.71
		761 s	6	4.43	0.74
		570 s	8	5.92	0.74
		500 s	10	6.75	0.68
2×10^5	4.0 h	2058 s		7.00	0.7
4×10^5	11 h	1.6 h		6.88	0.69
6×10^5	28 h	3.9 h		7.18	0.71
8×10^5	60 h	8.0 h		7.5	0.75
1×10^6	82 h	10.2 h		8.04	0.80
1×10^7	NA	111 h		NA	NA

We also perform experiments of PDD and HOTSAX on a 1×10^7 datasets. HOTSAX collapses because of the limitation of memory capacity of a single computing node. PDD successfully discovers the top discord with non-united memory spaces, which is 10 times bigger than that of one computing node. This experiment indicates PDD relieves the limitation of single computer memory in large scale time series discord discovery.

4.2 Utilization of Computing Resources

In this part we evaluate the utilization of computing resources of PDD. We detect the top discord from a 1×10^5 dataset with 10 computing nodes of the Spark cluster. Table 3 shows the utilization of computing resources of PDD and the disk-aware method [20].

In the first row, PDD allocates 10 data blocks per round (BPR), which means each computing node gets one data block. The idling time caused by load imbalance accounts for 18.9 % of the total PDD time consumption. Along with the increment of BPR, idling time decreases. When BPR = 500, the idling time drops to 4.71 % of the total time consumption, and the total time consumption of PDD also drops from 580 s to 500 s.

When BPR equals to 1000, the total time rises to 547 s. This is because computing nodes receive all possible subsequences in a single round such that early abandon technique has less effect on reducing computation complexity (line 8–11 in Algorithm 3). Empirically, BPR × (#subsequences/bulk) should be set no larger than one tenth of the total number of subsequences of a time series so that early abandon technique functions adequately.

We also use Computing time to Running time Ratio (CRR) in the last two columns of Table 3 to evaluate the utilization. The CRR of PDD is about 95 % in different conditions of BPR. Besides computing time, the running time also

Table 3. Utilization of the computational resources

BPR	#subsequences/bulk	Total time	Idle	CRR of PDD	CRR of Yankov [20]
10	200	580 s	18.9 %	96 %	≤ 50 %
100		527 s	6.25 %	95 %	
500		500 s	4.71 %	95 %	
1000		547 s	3.04 %	94 %	

includes the time for backstage job of the Spark, such as scheduling, task deserialization, Java garbage collection, data transmission between computing nodes, etc. Compared with the disk-aware method [20], the utilization of computing resources of PDD is about twice more than that of [20].

5 Conclusion

Discords are subsequences that are maximally different to all the other subsequences of a time series. Existing methods of discord discovery are all sequential. They suffer from the limitation of computing power and storage of a single computing node. Also, because the results discovered from segmentations of a time series are non-combinable, discord discovery is hard to parallelize. In this paper, we propose Parallel Discord Discovery (PDD), which divides discord discovery in a combinable manner and solves its sub-problems in parallel. Experiments show that given 10 computing nodes, PDD is 7 times faster than the classical discord discovery method HOTSAX. PDD handles larger datasets, which cannot be handled by the memory based methods. Experiments indicate the computing time accounts for more than 90 % of the execution time, and reduces the negative effects on performance caused by disk I/O operations.

Acknowledgments. This paper is sponsored by National Natural Science Foundation of China (No. 61373032), the National Research Foundation Singapore under its Campus for Research Excellence and Technological Enterprise (CREATE) program and the National High Technology and Research Development Program of China (863 Program, 2015AA050204).

References

1. Ameen, J., Basha, R.: Higherrarchical data mining for unusual sub-sequence identifications in time series processes. In: Second International Conference on Innovative Computing, Information and Control, 2007. ICICIC 2007, p. 177. IEEE (2007)
2. Basha, R., Ameen, J.: Unusual sub-sequence identifications in time series with periodicity. Int. J. Innovative Comput. Inf. Control **3**(2), 471–480 (2007)
3. Bu, Y., Leung, O.T.W., Fu, A.W.C., Keogh, E.J., Pei, J., Meshkin, S.: Wat: finding top-k discords in time series database. In: SDM, pp. 449–454. SIAM (2007)

4. Buu, H.T.Q., Anh, D.T.: Time series discord discovery based on isax symbolic representation. In: 2011 Third International Conference on Knowledge and Systems Engineering (KSE), pp. 11–18. IEEE (2011)
5. Camerra, A., Palpanas, T., Shieh, J., Keogh, E.: isax 2.0: Indexing and mining one billion time series. In: 2010 IEEE 10th International Conference on Data Mining (ICDM), pp. 58–67, December 2010
6. Chiu, B., Keogh, E., Lonardi, S.: Probabilistic discovery of time series motifs. In: Proceedings of the Ninth ACM SIGKDD International Conference on Knowledge Discovery and Data Mining, pp. 493–498. ACM (2003)
7. Fu, A.W., Leung, O.T.-W., Keogh, E.J., Lin, J.: Finding time series discords based on haar transform. In: Li, X., Zaïane, O.R., Li, Z. (eds.) ADMA 2006. LNCS (LNAI), vol. 4093, pp. 31–41. Springer, Heidelberg (2006)
8. Fu, T.C.: A review on time series data mining. Eng. Appl. Artif. Intell. **24**(1), 164–181 (2011)
9. Huang, T., Zhu, Y., Wu, Y., Bressan, S., Dobbie, G.: Anomaly detection and identification scheme for VM live migration in cloud infrastructure. Future Gener. Comput. Syst. **56**, 736–745 (2016)
10. Jones, M., Nikovski, D., Imamura, M., Hirata, T.: Anomaly detection in real-valued multidimensional time series. In: International Conference on Bigdata/Socialcom/Cybersecurity. Stanford University, ASE (2014). ASE@360 Open Scientific Digital Library. http://www.ase360.org/bitstream/handle/123456789/56/submission34.pdf?sequence=1&isAllowed=y
11. Keogh, E., Lin, J., Fu, A.: Hot sax: efficiently finding the most unusual time series subsequence. In: Fifth IEEE International Conference on Data Mining, p. 8. IEEE (2005)
12. Li, G., Bräysy, O., Jiang, L., Wu, Z., Wang, Y.: Finding time series discord based on bit representation clustering. Knowl.-Based Syst. **54**, 243–254 (2013)
13. Lin, J., Keogh, E., Fu, A., Van Herle, H.: Approximations to magic: finding unusual medical time series. In: 18th IEEE Symposium on Computer-Based Medical Systems, 2005. Proceedings, pp. 329–334. IEEE (2005)
14. Luo, W., Gallagher, M.: Faster and parameter-free discord search in quasi-periodic time series. In: Huang, J.Z., Cao, L., Srivastava, J. (eds.) PAKDD 2011, Part II. LNCS, vol. 6635, pp. 135–148. Springer, Heidelberg (2011)
15. Luo, W., Gallagher, M., Wiles, J.: Parameter-free search of time-series discord. J. Comput. Sci. Technol. **28**(2), 300–310 (2013)
16. Miller, C., Nagy, Z., Schlueter, A.: Automated daily pattern filtering of measured building performance data. Autom. Constr. **49**, 1–17 (2015)
17. Shvachko, K., Kuang, H., Radia, S., Chansler, R.: The hadoop distributed file system. In: 2010 IEEE 26th Symposium on Mass Storage Systems and Technologies (MSST), pp. 1–10. IEEE (2010)
18. Spark, A.: Apache spark–lightning-fast cluster computing (2014)
19. Wei, L., Keogh, E.J., Xi, X.: Saxually explicit images: finding unusual shapes. In: ICDM, vol. 6, pp. 711–720 (2006)
20. Yankov, D., Keogh, E., Rebbapragada, U.: Disk aware discord discovery: finding unusual time series in terabyte sized datasets. Knowl. Inf. Syst. **17**(2), 241–262 (2008)

Dboost: A Fast Algorithm for DBSCAN-based Clustering on High Dimensional Data

Yuxiao Zhang[1,4], Xiaorong Wang[2], Bingyang Li[3(✉)], Wei Chen[4,5],
Tengjiao Wang[1,4,5], and Kai Lei[1]

[1] School of Electronics and Computer Engineering (ECE), Peking University,
Shenzhen 518055, China
{zhangyuxiao,tjwang,leik}@pku.edu.cn
[2] Technology and Strategy Research Center,
China Electric Power Research Institute, Beijing 100192, China
xrwang@epri.sgcc.com.cn
[3] School of Information Science and Technology,
University of International Relations, Beijing 100091, China
beyondlee1982@163.com
[4] Key Laboratory of High Confidence Software Technologies,
Peking University, Ministry of Education, Beijing 100871, China
pekingchenwei@pku.edu.cn
[5] School of Electronics Engineering and Computer Science, Peking University,
Beijing 100871, China

Abstract. DBSCAN is a classic density-based clustering technique,
which is well known in discovering clusters of arbitrary shapes and
handling noise. However, it is very time-consuming in density calculation when facing high dimensional data, which makes it inefficient in
many areas, such as multi-document summarization, product recommendation, etc. Therefore, how to efficiently calculate the density on
high dimensional data becomes one key issue for DBSCAN-based clustering technique. In this paper, we propose a fast algorithm for DBSCAN-
based clustering on high dimensional data, named Dboost. In our algorithm, a ranked retrieval technique adaption named $WAND^{\#}$ is novelly
applied to improving the density calculations without accuracy loss, and
we further improve this acceleration by reducing the invoking times of
$WAND^{\#}$. Experiments were conducted on wire voltage data, Netflix
dataset and microblog corpora. The results showed that an acceleration
of over 50 times were achieved on wire voltage data and Netflix dataset,
and 100 more times can be expected on microblog data.

Keywords: DBSCAN · High dimensionality · WAND

This work was supported by Natural Science Foundation of China (Grant No.
61572043, 61300003, 61502115), State Grid Basic Research Program (DZ71-15-
004), the Fundamental Research Funds for the Central Universities (Grant No.
3262014T75).

© Springer International Publishing Switzerland 2016
J. Bailey et al. (Eds.): PAKDD 2016, Part II, LNAI 9652, pp. 245–256, 2016.
DOI: 10.1007/978-3-319-31750-2_20

1 Introduction

DBSCAN is a classic density-based clustering technique, which uses the density property of data points to identify groups or connected structures. The density in DBSCAN can be defined as the number of the neighbours (ϵ-neighborhood) within a predefined distance ϵ to a given point. By utilizing the density information, DBSCAN is not limited to finding only compact and spherical shaped clusters [1], nor does it need a predefined k as total cluster numbers or be susceptible to noisy outliers. However, DBSCAN requires much time in its running due to the density calculation, which becomes much worse when facing high dimensional data.

High dimensional data is usually referred to the data with hundreds of dimensions. They are widely appearing in many areas of data analysis such as textual data, DNA microarray data and daily voltage data on wires. Due to the high dimensionality, the ϵ-neighborhood querying of DBSCAN becomes very time consuming.

For example, in our work with National Grid, we analysis electric wire voltage data to find the similar power consuming wire regions (by DBSCAN based clustering). We find a bare 1-year record can get over 1000 dimensions, let alone the huge number of records in a big area, which makes a DBSCAN process consume nearly a whole day to assure only one possible result.

However, most current accelerating approaches for DBSCAN have not yet fitted in circumstances on such high dimensionality. Usually, they either work when dimensionality is lower than 50 [2] or fail in guaranteeing the generality of application, which usually leads to unacceptable clustering results [3,4].

In this paper, we propose an algorithm named Dboost, to accelerate DBSCAN-based clustering without accuracy loss on high dimensional data. Dboost aims at reducing and improving the density calculations of DBSCAN. To achieve this, it first uses an adaption ($WAND^{\#}$) of a ranked retrieval technique WAND [5] to speedily fetch a larger neighborhood for a checking point p, in its density calculation, then it checks all p's ϵ-neighbors' density based on only p's larger neighborhood, instead of giving each member in p's ϵ-neighborhood a global searching as the original DBSCAN does.

Major contributions of Dboost are from two aspects: Firstly, we novelly apply the idea of the ranked retrieval technique WAND [5] in information retrieval (IR) field to handle the density-calculation task of DBSCAN, which dramatically increases DBSCAN's speed performance on high dimensional data. Second, we reduce the invoking times of our WAND adaption by taking advantages of DBSCAN's characteristics, to further improve the acceleration substantially. Dboost can be used in a single-machine setting, or at individual node in partitioned implementations for parallel computing.

Experiments were conducted on Netflix data [6], wire voltage data and microblog corpus, which are typical high dimensional data. We compared Dboost with DBSCAN, LSH-based DBSCAN and k-means on the speed efficiency. The results showed that over 50 times speeding up without accuracy loss were achieved on wire voltage data and Netflix dataset. We also found that about

99.9 % of data would be filtered out during data's density calculation in the microblog corpus, which implies that much more than 100 times acceleration should be achieved with the data size growing.

2   Related Work

DBSCAN DBSCAN was first proposed by Ester et al. [7] and has been used as a base for many other techniques [2]. There are much work on enhancing the speed of DBSCAN, which can be divided into two categories. Accelerating without loss in clustering accuracy and accelerating with accuracy loss.

For the first category, a popular method is by partition [8]. R-trees and R*-trees can reduce the complexity to O (NlogN), but they work well only when the dimensionality is lower than 50 [8]. [9] was proposed to accelerated by clustering in parallel, but it was still inefficient with each executing node on high dimensional data.

The second category attempts to speed up by sacrificing effectiveness, e.g. sampling [4]. Recently, Zhang et al. proposed Linear DBSCAN algorithm based on locality sensitive hashing [10]. The main problems in this technique locates in adding more input parameters [3] and lacking of generality.

Ranked Retrieval. Ranked retrieval strategies are used as query evaluation strategies to reduce data in the calculations for a query search. The strategies can be divided into two main classes: Term at a time (TAAT) and document at a time (DAAT). A recent survey and a comparative study of in-memory TAAT and DAAT algorithms was reported in [11]. A large study of known TAAT and DAAT algorithms was conducted by [12]. They found that the Moffat TAAT algorithm [13] had the best performance, but it came at a tradeoff of loss of precision compared to naive TAAT approaches and other approaches.

Compared with TAAT strategies, DAAT strategies require smaller run-time memory and they exploit I/O parallelism more effectively. So, we choose DAAT. Our method finally chooses WAND [5], because the characteristics of WAND is well fit for the large scale data and numbers of features in the query [11].

3   Algorithms

In this section we describes the very details of Dboost.

The global routine and cluster identifying standards of Dboost is the same as DBSCAN, for it only focuses on reducing and improving the unnecessary density calculations. Consequently, it preserves DBSCAN's clustering accuracy. The parts Dboost differs locate in the density-fetch related subroutines:

Traditional DBSCAN fetches each point's density by giving each point the distance comparisons with all of the rest points, to get their corresponding ϵ-neighborhood. However, Dboost treats each ϵ-neighborhood search as a querying issue in IR, and novelly applied the ranked retrieval technique WAND for such search implementation (our $WAND^{\#}$). Later, we figure out there is no need to treat each ($WAND^\#$) search as an independent new search, based on DBSCAN's characteristics. Thus we further improve ($WAND^\#$) to reduce its invocation times.

To better explain our work, we first introduce the distance metric we choose, then, the background and soundness of our WAND adaption ($WAND^\#$), which is the base idea Dboost takes advantage of. Finally, in Sect. 3.3, we describe the improvements of ($WAND^\#$) as well as the whole Dboost algorithm. Algorithms 1–3 gives a basic description of Dboost and they will be explained in Sect. 3.3.

3.1   The Distance Metric

We choose our distance metric based on the Cosine similarity in order to make usage of WAND as well as get fast comparison speed. For two points $p, p' \in$ D, their standard cosine similarity is calculated as:

$$cossim(p, p') = abs\left(\frac{p \bullet p'}{\|p\| \, \|p'\|}\right),$$

where $\bullet$ denotes the vector dot product, and $\|\|$ denotes the vector's length. Recall that, since the *cossim* is always between 0 and 1, so **we define $f(p, p')$ = 1-$cossim(p, p')$ as the distance metric**, which is non-negative, symmetric and satisfies the triangle inequality.

The usage of Cosine similarity based metric has the same effect as using Euclidean distance metric. For the fact that, when $cossim(a, c) > cossim(a, b)$, the Euclidean distance between a and c is exactly closer than a and b, which is easy to prove. As for the situations when $cossim(a, c) = 0$, then a and c are too dissimilar (sharing no common features) to be neighbors and they shouldn't be considered as neighbors under Euclidean distance metric, too.

3.2   The Background and Soundness of $WAND^\#$

$WAND^\#$ is adapted by the WAND algorithm [5], which is a method allowing for fast retrieval of the top-k ranked documents for a query in the IR field. Although the task of $WAND^\#$ is to fetch a neighborhood within a given distance d, which is different from the top-k retrieval issue, the conversion of the task is not difficult for WAND:

WAND makes comparisons by Cosine similarity. A global view of WAND mechanism can be seen in Fig. 1(a). When searching, WAND keeps a heap H of nearest top-k points among the points it checked so far. The similarity score of the k-th best result acts as a threshold θ. WAND works by repeatedly calling its subroutine *next* to fetch a next point (in order of point ID) whose similarity score surpasses θ and using it to update H. The final results in H is then the top-k neighbors.

To convert WAND to achieve the task of $WAND^\#$, we can change the heap H in Fig. 1(a) into a set with unlimit size, and change the θ to be fixed to the distance d. This is just the previous version of our $WAND^\#$, it is still the base of our final $WAND^\#$ version.

From above, the soundness of $WAND^{\#}$ locates in the subroutine *next* of WAND, which is also the core idea of WAND. WAND aims at using an upper bound of each term's similarity contribution to remove points that are too dissimilar from the query to become a member of the top-k list. To explain this better, we first introduce some preliminaries, then the details of WAND logic.

In IR, the concept of **document** and **query** correspond to the concept of **point** in our situation, and **term** corresponds to **feature** or **dimension**. If a document doesn't contain a term, its corresponding weight under that term (dimension) is zero (otherwise, above zero). WAND runs on top of *inverted index*, which is an efficient indexing way for high dimensional sparse data. In *inverted index*, each term (dimension) has an associated *points list* which contains all of the documents (points) that contain this term. The document (point) are expressed as its vector value the same as the point in our clustering issue and they are sorted by their ID in ascending order in *points list*.

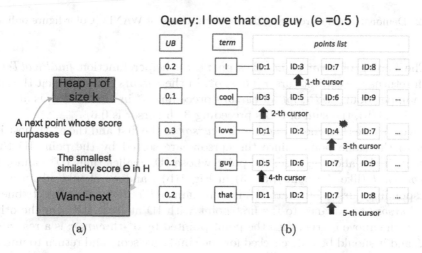

Fig. 1. (a) A global view of WAND mechanism, (b) Demonstration of pivot point fetching

To make *next* work, WAND keeps one pointer called a *cursor* for each of the query terms (dimensions) that points at an entry in the corresponding *points list* (see Fig. 1(b)). During the searching of the next point in H, the *cursors* are kept sorted by the document (point) ID they point to, in a list called *cursors list*.

During the initialization, for each corresponding *points list* of the query term (non-zero weight dimension), WAND fetches the upper bound of its entry points' weights under this dimension (term), and uses it to calculate the similarity contribution upper bound UB_t for each *cursor* with respect to their corresponding terms. Next, all *cursors* will be initialized to point at the first entry in their corresponding *points lists* (the one having the minimum point ID in the list).

An intuition of *next* is, if we can tell the similarity score of a point cannot be over θ, then the point can never be a true member in the top-k list, and we should skip to a next possible point. The first point that has the upper bound of its similarity score higher than θ is denoted as the *pivot point*.

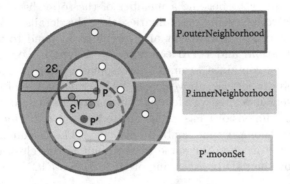

Fig. 2. Demonstration of reducing the invoking times of WAND (Color figure online)

The finding of *pivot point* relies on *next's* helper function *findPivotTerm*, which returns the earliest cursor index, o*, in the *cursors list*, such that the sum of the upper bound UB_t, for all *cursors* preceding o* in *cursors list*, is at least θ (See Fig. 1(b), the sum of UB_ts preceding 3-th cursor is 0.6)

Given o*, we then check whether the *cursor* in the first and the o*-th position point to the same point. Since the cursors are sorted by the point ID they point at, if the above is false, the point whose ID smaller than ID* cannot be a *pivot point* (like the point (id = 3) in Fig. 1(b)), and we should advance the corresponding cursors. In *next*, a function named *PickTerm* then selects one of these *cursors* to advance to the first point with ID at least ID*. On the other hand, if the above is true, then the point pointed by o*-th *cursor* is a real *pivot point*, and it should be fully checked for the similarity score and return to update heap H if the score really surpasses.

The WAND algorithm is fully described in [5].

3.3 Improvements for $WAND^{\#}$

The improvements for $WAND^{\#}$ is in two aspects: to reduce its invocation times and to diminish its searching scope. All of which based on the characteristics of DBSCAN.

Invocation Reduce. A short description of DBSCAN can be seen in Sect. 2. During the clustering of DBSCAN, each point in the data set shall be checked for density, which means a $WAND^{\#}$ search of overall data shall be repeated as many times as the number of data points. On the other hand, many of these searching results are overlapped: In Fig. 2, P and P' are two points in the data set, with their ϵ-neighborhood within circle yellow and circle grey. The blue points are the common neighbors for both P and P', yet they have been searched from overall data for at least two times during the density check for P and P' and other points in Fig. 2.

An intuitive solution for above is to fetch a larger scale of neighborhood of P first, then check the density of each P's ϵ-neighbor within just the larger neighborhood. This relies on Theorem 1 which is easy to prove, since the distance metric we use here follows the triangular inequality (see Sect. 3.1).

Theorem 1. *For any point $p^{\#}$ in point p's ϵ-neighborhood, the ϵ-neighborhood of $p^{\#}$ is within p's 2ϵ-neighborhood.*

Algorithm 1. Dboost

Input: the data points' fetch scope S; distance ϵ; density threshold ν;
Output: clustering result sets of data;
1: **for** each point p in S **do**
2:   $WAND^{\#}(p,\epsilon,S)$;
3:   **if** $p.innerNeighborNum <= \nu$ **then**
4:     clear the states of p as well as the points in p's extended neighborhood
5:   **else**
6:     create next cluster set C and add p to C;
7:     ExpandCluster(p,C);
8:     **for** each point p' in C **do**
9:       remove p' from S; # for the sake of reducing searching scope
10:    **end for**
11: **end if**
12: **end for**

Algorithm 2. ExpandCluster

Input: initial point p; Cluster C;
Output: final cluster set C;
1: init the expand set $CES = \{\}$; ## CES is the cluster expanding set
2: **repeat**
3:   **for** each unlabeled point p' in $p.innerNeighborhood$ **do**
4:     add p' to cluster C and mark p' as labeled;
5:     **if** $p'.innerNeighborNum > \nu$ **then**
6:       $CES = CES \cup p'.moonSet$;
7:     **end if**
8:   **end for**
9:   pick a point p from CES; $CES = CES - p$;
10: $WAND^{\#}(p,\epsilon,S)$;
11: **until** CES is empty

When checking a point p's density, we first fetch an expanded set ES which is two times the range of ϵ. If p is a core point, then we should check all its ϵ-neighbor's density as described in DBSCAN. Based on Theorem 1, all ϵ-neighbors of p's ϵ-neighbors are now within ES. So all p's ϵ-neighbors' checking can be operated in the scope of ES, instead of the entire data space. To make this better, we can combine density checking with the ES querying process:

Algorithm 3. $WAND^{\#}$

Input: point p; distance ϵ; the data points' fetch scope S;
Output: find p's extended Neighborhood and update them;
 1: threshold = $1 - 2\epsilon$; ##the threshold here is cosine similarity
 2: **loop**
 3:     $p' = next(p, threshold, S)$;
 4:     **if** p'==null **then**
 5:         return ;
 6:     **end if**
 7:     $p.ES = p.ES \cup p'$;
 8:     **if** $f(p',p) < \epsilon$ **then**
 9:         $p.innerNeighborhood = p.innerNeighborhood \cup p'$;
10:         **for** each point c in $p.ES$ **do**
11:             **if** $f(p',c) < \epsilon$ **then**
12:                 $p'.innerNeighborNum = p'.innerNeighborNum + 1$
13:                 **if** $c \in p'.outerNeighborhood$ **AND** c is unlabeled **then**
14:                     $p'.moonSet = p'.moonSet \cup c$;
15:                 **else**
16:                     $c.innerNeighborNum = c.innerNeighborNum + 1$
17:                 **end if**
18:             **end if**
19:         **end for**
20:     **else**
21:         $p.outerNeighborhood = p.outerNeighborhood \cup p'$
22:         **for** each point c in $p.innerNeighborhood$ **do**
23:             **if** $f(p',c) < \epsilon$ **AND** p' is unlabeled **then**
24:                 $c.innerNeighborNum = c.innerNeighborNum + 1$
25:                 $c.moonSet = c.moonSet \cup p'$;
26:             **end if**
27:         **end for**
28:     **end if**
29: **end loop**

When we fetch a ES member m, we compare it with ϵ, and identity it as the *inner-neighbor* (neighbor whose distance$< \epsilon$) or the *outer-neighbor* (distance$\geq \epsilon$). We then update the ϵ-neighbor number (density info) of p's *inner-neighbors* by calculating their distance scores with m and plus 1 if the score exceeds ϵ. As for the core points among p's *inner-neighbors*, we use a special data structure called *moonSet* to store their ϵ-neighbors for later cluster expansion (see Sect. 2). The *moonSet* is described in Fig. 2. Notice the *moonSet* of P' in Fig. 2 only stores the *outer-neighbors* of P, for the *inner-neighbors* are all checked by the end of ES fetching and got added into the corresponding cluster.

Scope Diminishing. This improvement is simple: To make WAND more effective, we reduce its searching scope by repeatedly eliminating the points in one cluster from the global search scope, once the cluster is fully established. Because now the clustering problem subjects to the sub problem of the rest of the points. The diminishing step is in Algorithm 1 (Lines 8, 9).

Our final $WAND^{\#}$ is described in Algorithm 3. We passed the diminishing searching scope to it every time it is invoked, and so as to its subroutine *next* (Lines 3). Each time it fetches a 2ϵ-neighbor of the query point, it identifies the neighbor as inner-neighbor or outer-neighbor and updates the density info of query point's inner-neighbors (Lines 8–28). Notice the function $f()$ in Line 8 is our distance metric defined in Sect. 3.1.

Rest Detail of Dboost. The top level of Dboost is described in Algorithm 1, once it identifies a cluster and expands it (Lines 6, 7), it removes the points in the cluster from the global searching scope (Lines 8, 9). Algorithm 2 is a DBSCAN expanding subroutine, but it only adds the points in the *moonSet* to the expanding set (Lines 6, 9–11) as described above.

4 Experiments

In this section, we will first give a brief introduction to the experiment setup, including the experiment environment, datasets, comparison approaches. We then evaluate the performance of Dboost over other approaches in different aspects. Finally we will display the experimental results and make some discussion.

4.1 Experiment Setup

The experiment is built on a common PC with an Intel(R) Core(TM) i3–2120 CPU, and 8 G RAM. The operating system is a 64bit-windows 7.

Since our experiment is to test the acceleration on high dimensional data, we use three datasets in our experiment, wire voltage data, Netflix dataset and microblog corpus.

The daily voltage data contains 49 k (thousand) records with 23 k dimensions on time series. The Netflix dataset consists of movie rating score data from Netflix Prize competition [6]. We use its rating info of 420 thousand customers over 13 thousand movies (dimensions), and we cluster customers by their movie ratings. The microblog corpus consists of 120 k articles from Sina weibo and it contains 305 thousand terms (dimensions).

To compare the performance, we choose several classic density-based clustering methods.

- DBSCAN: is the classic density-based clustering approach, we make it run under *inveted index*.
- k-means: is the classic centroid-based clustering approach, we make it run under *inveted index*.

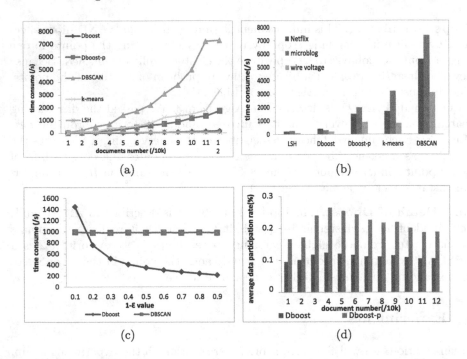

Fig. 3. (a) Comparison of time consuming on different data scales on microblog data, (b) Comparison of time consuming on three full datasets, (c) Comparison of time consuming with the increase of $1 - \epsilon$, (d) Comparison of data participation in density calculation on 120 k (thousand) size microblogdata

- LSH: is the current fastest DBSCAN approach but with accuracy loss.
- Dboost-p: is a poor implemention of Dboost, which directly uses WAND for each point's ϵ-neighborhood searching.
- Dboost: the standard Dboost in this paper.

The DBSCAN and k-means methods are implemented with *inverted index* as a result of the high dimensionality of our experiment data sets, the original version of the two algorithms cannot run the data with their original data structures. Due to the same reason, the indexing or partition based methods, e.g. R-tree, are not suitable as well, but they cannot be applied on *inverted index* and only work when the dimension is below 50. We implement LSH with the hash technique mentioned in [14].

To establish a fair comparisons, we set the $\epsilon = 0.4$ (which means the similarity threshold for WAND is $1 - \epsilon = 0.6$. See first head in Sect. 3), $\nu = 10$ for all the DBSCAN-based algorithms, and set $k = 10$ for k-means with randomly initial centers. The k-means here is used for speed comparison only, and its predefined cluster numbers k is equal to the smallest number of DBSCAN's results. The extra parameters of LSH were set as the value to achieve the best performance.

We doesn't show the accuracy comparison in this paper, as the outputs on experiment data of Dboost are exactly the same to DBSCAN and accuracy is not our aim. As described in Sect. 3.2, Dboost doesn't change the identifying logic for clusters of DBSCAN.

4.2 Comparison for Speeding up

We first compared the efficiency on the three datasets, and the experimental results were shown in Fig. 3(a) and (b).

In Fig. 3(a), we could find that our approach scaled well with the growing of data size. On the contrary, DBSCAN was with the temporal complexity O (N^2), so the time it cost was increasing dramatically. The performance of k-means only defeated DBSCAN, when the cluster numbers limited to only 10, however, it performed poor when the k became larger. Meanwhile the cluster numbers for DBSCAN varied from 10~306 with different data scale.

Dboost performed better than DBSCAN and k-means, but still a little less than the LSH. Note that, Dboost achieved acceleration without loss in accuracy while LSH cannot guarantee the generality of quality.

In Fig. 3(b), we see the speedup not so obvious on wire voltage data, that is because the wire voltage data is much denser than the other two data sets, which lows down the effect of *inverted index*.

Influence of ϵ. Since WAND uses Cosine similarity threshold $(1 - \epsilon)$ to avoid points' similarity calculations, the smaller the threshold, the less the dissimilar points get skipped. Thus the setting of $1 - \epsilon$ do influent the accelerating effect of Dboost. Figure 3(c) shows a tint that Dboost's performance may not be restricted too much by $1 - \epsilon$ when $1 - \epsilon$ value is in a reasonable range. The results in above section shows a significant speeding up even with $1 - \epsilon$ set to be 0.6.

4.3 Discussion

To provide an insight of Dboost, Fig. 3(d) demonstrates the average ratio of the promising candidate points that participate a fully similarity calculation, which explains Dboost's high acceleration effect. The results show that our improved Dboost can achieve around 0.1 %, and the ratio is empirically independent to the growth of the data size. That means 99.9 % data can be removed from density calculation. Note that compared with the real world data, the size of the corpus with 120 k microblogs is quite low. Since the condition judgment weights less and less significant with the growing of data size. It implied that much more than 100 times speeding up against the traditional DBSCAN would be achieved.

5   Conclusion and Future Work

In this paper, we target to improve the efficiency of DBSCAN-based approach on high dimensional data. An efficient approach, named Dboost, is proposed to accelerate clustering speed through the ranked retrieval strategy WAND.

We further improve the acceleration by reducing the invoking times of our WAND adaption. The experimental results showed that 70 times speeding up without accuracy loss were achieved on the Netflix dataset. We also found that about 99.9 % of data would be filtered out during data's density calculation in the microblog corpus, which implies much more than 100 times acceleration will achieved with the data size growing.

Though Dboost is adaptable for many kinds of high dimensional data, it was originally proposed to fast detect similar wire regions in our work with National Grid. In the future, we will focus more on its performance on denser data.

References

1. Viswanath, P., Pinkesh, R.: l-dbscan: a fast hybrid density based clustering method. In: 18th International Conference on Pattern Recognition, ICPR 2006, vol 1, pp. 912–915. IEEE (2006)
2. Dharni, C., Bansal, M.: Survey on improved dbscan algorithm. Int. J. Comput. Sci. Technol. **4** (2013)
3. Ali, T., Asghar, S., Sajid, N.A.: Critical analysis of dbscan variations. In: 2010 International Conference on Information and Emerging Technologies (ICIET), pp. 1–6. IEEE (2010)
4. Borah, B., Bhattacharyya, D.: An improved sampling-based dbscan for large spatial databases. In: Proceedings of International Conference on Intelligent Sensing and Information Processing, 2004, pp. 92–96. IEEE (2004)
5. Broder, A.Z., Carmel, D., Herscovici, M., Soffer, A., Zien, J.: Efficient query evaluation using a two-level retrieval process. In: Proceedings of the Twelfth International Conference on Information and Knowledge Management, pp. 426–434. ACM (2003)
6. Corporation of netflix: the netflix prize (1997-2009). http://www.netflixprize.com/
7. Ester, M., Kriegel, H.P., Sander, J., Xu, X.: A density-based algorithm for discovering clusters in large spatial databases with noise. Kdd **96**, 226–231 (1996)
8. El-Sonbaty, Y., Ismail, M., Farouk, M.: An efficient density based clustering algorithm for large databases. In: 16th IEEE International Conference on Tools with Artificial Intelligence, ICTAI 2004, pp. 673–677. IEEE (2004)
9. Patwary, M.M.A., Palsetia, D., Agrawal, A., Liao, W.k., Manne, F., Choudhary, A.: A new scalable parallel dbscan algorithm using the disjoint-set data structure. In: 2012 International Conference for High Performance Computing, Networking, Storage and Analysis (SC), pp. 1–11. IEEE (2012)
10. Cheu, E.Y., Keongg, C., Zhou, Z.: On the two-level hybrid clustering algorithm. In: International Conference on Artificial Intelligence in Science and Technology, pp. 138–142 (2004)
11. Fontoura, M., Josifovski, V., Liu, J., Venkatesan, S., Zhu, X., Zien, J.: Evaluation strategies for top-k queries over memory-resident inverted indexes. Proc. VLDB Endowment **4**(12), 1213–1224 (2011)
12. Lacour, P., Macdonald, C., Ounis, I.: Efficiency comparison of document matching techniques. In: European Conference for Information Retrieval Efficiency Issues in Information Retrieval Workshop, pp. 37–46 (2008)
13. Moffat, A., Zobel, J.: Self-indexing inverted files for fast text retrieval. ACM Trans. Inf. Syst. (TOIS) **14**(4), 349–379 (1996)
14. Wu, Y.P., Guo, J.J., Zhang, X.J.: A linear dbscan algorithm based on lsh. In: 2007 International Conference on Machine Learning and Cybernetics, vol. 5, pp. 2608–2614. IEEE (2007)

A Precise and Robust Clustering Approach Using Homophilic Degrees of Graph Kernel

Haolin Yang[1]($\boxtimes$), Deli Zhao[2], Lele Cao[1], and Fuchun Sun[1]

[1] State Key Laboratory on Intelligent Technology and Systems,
Tsinghua National Laboratory for Information Science and Technology (TNList),
Department of Computer Science and Technology,
Tsinghua University, Beijing, China
{yang-hl13,caoll12}@mails.tsinghua.edu.cn
[2] HTC Beijing Advanced Technology and Research Center, Beijing, China
zhaodeli@gmail.com

Abstract. To address the difficulties of "data noise sensitivity" and "cluster center variance" in mainstream clustering algorithms, we propose a novel robust approach for identifying cluster centers unambiguously from data contaminated with noise; it incorporates the strength of *homophilic degrees* and *graph kernel*. Exploiting that in-degrees can breed the homophilic distribution if ordered by their associated sorted out-degrees, it is easy to separate clusters from noise. Then we apply the diffusion kernel to the graph formed by clusters so as to obtain graph kernel matrix, which is treated as the measurement of global similarities. Based on local data densities and global similarities, the proposed approach manages to identify cluster centers precisely. Experiments on various synthetic and real-world databases verify the superiority of our algorithm in comparison with state-of-the-art algorithms.

1 Introduction

Clustering plays a fundamental role in computer vision, machine learning, pattern recognition, and data mining. There are a variety of clustering algorithms available, one classic algorithm of which is k-means that works well for discovering spherical or near-spherical clusters in relatively small-scaled databases. Single, average, and complete linkages are the hierarchical agglomerative algorithms, which take the minimal, average, and maximum distances between points as the inter-cluster similarity measurement, respectively. Shi and Malik [20] proposed the Normalized Cuts (N-Cuts) algorithm that casts the clustering process as a graph partition problem; they partitioned the graph into N parts by eigenvectors of a graph-Laplacian matrix, providing N clusters to be identified. The nonparametric clustering method Affinity Propagation (AP) [6] updates the affinity between points via a messaging process. Algorithms based on graph theories have the appealing property that a graph can portray the inherent structures of given data. Inspired by PageRank [15], Minsu and Kyoung defined graph authority scores [3], and used authority node traversals and authority propagation to identify clusters. Then authors of [4] re-defined the graph representation to strengthen

© Springer International Publishing Switzerland 2016
J. Bailey et al. (Eds.): PAKDD 2016, Part II, LNAI 9652, pp. 257–270, 2016.
DOI: 10.1007/978-3-319-31750-2_21

relevancy between neighbors and weaken that among non-neighbors; and its performance is eventually improved by applying the authority-ascent kernel to this kind of graph representation. In spite of the aforementioned algorithms, there still remain two major difficulties in handling data with complex structures: "data noise sensitivity" and "cluster center variance". Data noise sensitivity affects algorithms such as k-means, Linkage, N-Cuts, AP, and Authority-Ascent Shift (AAS) [3], making them fail to correctly separate clusters from noise (Figs. 4(c), (d), (e), (f) and (g)). Cluster center variance requires multiple execution of k-means to obtain precise and stable results due to its dependence on initial partitions.

Rodriguez and Laio [18] proposed a simple-and-effective clustering algorithm: Density Peaks (DP), which used simple ideas instead of complicated mathematics to tackle the aforementioned problems. They defined the concepts of local density and the minimum distance to the higher density neighbor, and illustrated the relationship between these two concepts in the form of *decision graph*. The decision graph provides a visualization of candidates of cluster centers, shedding light on the research of clustering from noise. However, the weak algorithmic assumptions make their approach constantly result in incorrect clusters. They built their algorithm on the belief that cluster structures and centers are characterized by the distances between points. In fact, measuring similarity according to distance can be easily affected by data structure and distribution, which fails to identify centers from the decision graph. Moreover, the criteria for deciding the cluster border can also be easily affected by the pre-defined cutoff distance (or the distances between clusters), jeopardizing the overall robustness.

In this paper, we propose a novel robust clustering approach to effectively address the challenges of data noise sensitivity and cluster center variance. It has been found that points are aggregated to form homophilic layers if they are re-organized by their high-order in-degrees and corresponding sorted out-degrees, and noisy points generate the weakest layer due to low densities, making it easy to separate clusters from noise [24]. Comparing to the state-of-the-art methods, we highlight the main advantages of our approach as follows:

- established upon *degree homophily* [24], we use a robust cluster-to-noise estimator to separate clusters from noise, leading to a potentially superior clustering quality;
- with extracted clusters, our method takes advantage of *graph kernel* to provide an accurate global similarity measurement between points as opposed to [18];
- the minimum distance to higher density neighbor is re-defined with global similarity, thus providing a much better visualization of center candidates in decision graph.

With all cluster centers identified from the decision graph, the remaining points are categorized to the same clusters as their most similar neighbors with higher densities. The rest of this paper is organized as follows. In Sect. 2, we review and evaluate the DP algorithm in detail. Sections 3 and 4 present our proposed way of integrating *degree homophily* and *graph kernel*, respectively. A comprehensive evaluation on both synthetic and real-world databases is conducted in Sect. 5; conclusions are drawn in Sect. 6.

2   The Algorithm of Density Peaks (DP)

The assumption of the DP algorithm is that local densities of cluster centers are higher than those of their neighbors, and cluster centers are far apart from each other [18]. The point i embodies two variables, ρ_i and δ_i, to characterize its structural attributes. ρ_i is the local density of that point, which is defined as follows:

$$\rho_i = \sum_j \chi(D_{ij} - d_c),\qquad(1)$$

where $\chi(x) = 1$ if $x < 0$; otherwise $\chi(x) = 0$ [18]. D is the distance matrix, and d_c is a cutoff distance [18]. δ_i denotes the distance between point i and its nearest neighbor with higher local density (Eq. (2) Left) [18]; for the point which has the highest local density, δ_i is assigned with the maximum in distance matrix D (Eq. (2) Right) [18]:

$$\delta_i = \min_{j:\rho_j > \rho_i} (D_{ij}) \qquad \text{or} \qquad \delta_i = \max(D).\qquad(2)$$

The variables ρ_i and δ_i serve the purpose of finding cluster centers and border regions.

2.1   Clustering Process

According to Rodriguez and Laio's assumptions, cluster centers are more distinctive with larger ρ and δ. As described in Eq. (2), δ_i is a function of ρ_i; and this relationship is shown in a *decision graph* (Fig. 1(e)), which visualizes the candidates of cluster centers. Colored points with relatively larger ρ and δ are recognized as cluster centers, while points near axes are either with lower densities or close with neighbors with higher densities. The remaining points are categorized into the clusters as their nearest neighbors with higher densities [18]. We formally define r as the vector containing the nearest neighbors with higher density, r_i as nearest neighbor with higher densities than i, b as the labels of points, b_i as the label of point i, C_{center} as indexes of centers, and n_c as the number of clusters. We summarize the specific steps for obtaining the decision graph in Algorithm 1, and the procedure of cluster-label assignment in Algorithm 2.

Algorithm 1. Decision Graph (r: nearest neighbors with higher densities, init.: initialize, min.: minimum)

Input: distance matrix D, cutoff distance d_c.
1: Init. local densities of points, $\rho \leftarrow 0$.
2: **for** each point i **do**
3:    Calculate ρ_i by Eq. (1).
4: **end for**
5: Init. min. distance: $\delta \leftarrow 0$.
6: Init. $r \leftarrow 0$.

7: **for** each point i **do**
8:    Calculate δ_i by Eq. (2).
9:    $r_i = \arg \min_{j:\rho_j > \rho_i} D_{ij}$.
10: **end for**
   {For point i with highest density:}
11: $\delta_i = \max(D)$, $r_i = 0$.
Output: Figure: plot(ρ, δ); r.

Algorithm 2. Cluster-label Assignment (ρ: local densities, init.: initialize)

Input: r, cluster centers $\mathcal{C}_{center}$, ρ.
1: Sort ρ in descending order.
{Assign for cluster centers:}
2: Init. $j \leftarrow 0$.
3: **for** each $i \in \mathcal{C}_{center}, j \leqslant n_c$ **do**
4:     $b_i = j, j = j + 1$.

5: **end for**
{Assign for remaining points:}
6: **for** each point i of ρ_i **do**
7:     $b_i = b_{r_i}$.
8: **end for**
Output: Labels of points, b.

The *border region* [18], or *cluster halo* (a.k.a. *outliers*), is the set of points which are categorized to one cluster but close to other clusters within a cutoff distance d_c [18]. Points whose densities are higher than the maximum density (denoted by ρ_b) of border region are considered as members of clusters, while others are treated as noise. We summarize the procedure of labeling the halo in Algorithm 3, where we further denote $\mathcal{B}$ as the cluster border region and h as the labels of halo which is labeled as zero. The clustering result is shown in Fig. 1(a).

Algorithm 3. Halo-label Assignment (agg.: aggregate, init.: initialize, $\mathcal{C}$: the set of clusters)

Input: D, ρ, d_c, cluster labels b.
1: Init. the cluster border region, $\mathcal{B} = \emptyset$.
2: Init. the labels of halo, $h \leftarrow b$.
3: Agg. points to $\mathcal{C}$ according to b.
4: **for** each cluster $\mathcal{C}_t, t \in [1, n_c]$ **do**
5:     **for** each point $i \in \mathcal{C}_t, \mathcal{C}_t \subseteq \mathcal{C}$ **do**
6:         **if** $(D_{ij} < d_c), j \in \mathcal{C} \setminus \mathcal{C}_t$ **then**
7:             $\mathcal{B}_t \leftarrow \mathcal{B}_t \cup i$.
8:         **end if**
9:     **end for**

10: **end for**
11: **for** each cluster $\mathcal{C}_t, t \in [1, n_c]$ **do**
12:     $\rho_b = \max_{j \in \mathcal{B}_t}(\rho_j)$.
13:     **for** each point $i \in \mathcal{C}_t$ **do**
14:         **if** $(\rho_i < \rho_b)$ **then**
15:             $h_i = 0$.
16:         **end if**
17:     **end for**
18: **end for**
Output: Halo labels, h.

2.2   Analysis of DP Algorithm

The appealing novelties of DP are the definition of δ and the decision graph which provides us a useful tool to visualize candidates of cluster centers. The computation of δ depends on D and local density. Therefore, δ will be affected by distances between points, shapes or distributions of clusters, and the local densities. Figure 1(b) shows that DP fails to recognize five cluster centers if the distance between centers of clusters 1 and 2 is shortened; from its corresponding decision graph (Fig. 1(f)), we observe that there are only four points with relatively higher ρ and δ; and the center of cluster 1 becomes indistinguishable in decision graph as its δ_i decreases. Besides, in the case which is demonstrated in Fig. 1(d), another center candidate shows up in decision graph if there exist two local density maxima in cluster 3. These examples indicate that it is difficult to identify cluster centers without sufficiently large δ_i. From Algorithm 1, we perceive that for point i, the assignment for its nearest neighbor with higher density depends on its δ_i. As shown in Fig. 4(b), even if centers are identified correctly from decision graphs, clusters in strip-shaped and circle-shaped are divided into several parts, meaning that points are assigned to wrong neighbors. Therefore, δ_i is important not only in locating cluster centers but also in clustering process. Furthermore, the estimator ρ_b for cluster border region is easily affected by d_c and distances between clusters. If we manually locate five cluster centers in Fig. 1(g), clusters 1 and 2 shrink severely (Figs. 1(b)–(c) because many points are categorized to border region and recognized as noise.

Attempting to address these weaknesses, we propose a robust clustering algorithm based on graph techniques to eliminate data noise sensitivity and cluster center variance.

3 Apply Degree Homophily

Although *homophily* is a social concept which depicts the individual behavior preference, it has been adopted by computer science to solve many problems, such as modeling evolving networks [16]. In [24], *homophilic degree* is used as a type of centralized measure of points. Meanwhile, a robust estimator is proposed to eliminate noise based on *degree homophily* [24]. Therefore, we make use of

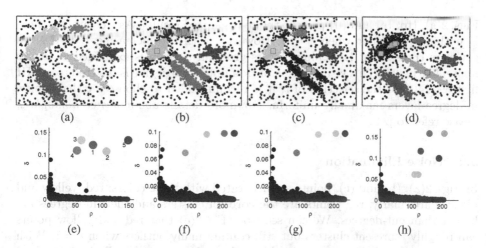

Fig. 1. (a) Clustering results of DP. The synthetic data is generated according to [18], but not exactly identical to that data. (b) The distance between centers of clusters 1 and 2 is shortened. (c) The clustering result after five cluster centers are manually separated. (d) There are two local maxima in cluster 3. (e)–(h) The decision graphs of (a)–(d), respectively.

degree homophily to separate clusters from noise so as to guarantee a superior clustering quality.

3.1   Homophilic In-Degree Figure

The basis of *degree homophily* is a directed asymmetric graph. If we construct a directed graph $\mathcal{G}$ by creating links from points to their k nearest neighbors (denoted by $\mathcal{N}$), then $\mathcal{G}$ is also asymmetric owing to the asymmetric neighborhood of points: point i is possibly not contained in $\mathcal{N}_j$ even if point j is one of k nearest neighbors of point i. The weights of links can be measured by the distance exponential $e^{-D_{ij}^2/\sigma^2}$ (parameterized by σ), or other similarity measurements. Thus, each element W_{ij} in similarity matrix $\boldsymbol{W}$ of $\mathcal{G}$ is set to zero unless:

$$W_{ij} = \begin{cases} e^{-D_{ij}^2/\sigma^2} & j \in \mathcal{N}_i \\ 1 & i = j \end{cases}. \tag{3}$$

Each point in a directed graph $\mathcal{G}$ has two attributes describing its connections with others: in-degree and out-degree; we denote the t-order in-degrees and out-degrees with $\boldsymbol{d}^{in} = \boldsymbol{W}^t\boldsymbol{1}$ and $\boldsymbol{d}^{out} = (\boldsymbol{W}^t)^\top\boldsymbol{1}$ (the scripts "t" in $\boldsymbol{d}^{in}$ and $\boldsymbol{d}^{out}$ have been left out), respectively. Here comes the most interesting part of *degree homophily*. Points can be re-organized according to the following steps:

- define a coordinate whose horizontal and vertical axes are indexes and degrees of points, respectively;
- sort $\boldsymbol{d}^{out}$ in descending order and plot the sorted $\boldsymbol{d}^{out}$ as an out-degree curve;
- for each $\boldsymbol{d}_i^{out}$, draw its corresponding $\boldsymbol{d}_i^{in}$ as a point in the same coordinate.

Figs. 2(a), (b), (c) and (d) present the re-organized results. As we can see, there is nothing particular when $t = 1$. However, with the increasing t, in-degrees generate transparent homophilic layers gradually, which is explained as *degree homophily* in [24]. This visualization of Homophilic In-degree is referred to as "HI figure" [24]. For the selection of t, it is recommended to choose a t that makes in-degree layers maximally uniform. For more details about *degree homophily*, please refer to [24].

3.2   Noise Elimination

In Figs. 2(a), (b) and (c), points are decorated with three colors: red, yellow and black. Particularly, red points are the cores of clusters and their in-degrees are higher than out-degrees. We can see from Fig. 2(e) that red and yellow points can roughly represent clusters but still contain many outliers when $t = 1$. With the growth of t, clusters represented by colorized points gradually become clean and clear (Figs. 2(f)-(g)). There are two types of noise. The first type of noise generates connections between clusters and noise. This noise always lie at the bottom of the HI figure due to low in-degree links to these noisy points. Another one is uniformly distributed among points. Connections formed by these kind

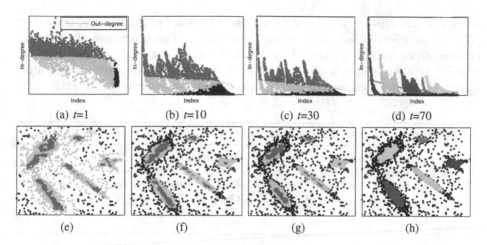

Fig. 2. (a)–(d) HI figures with increasing t. (e)–(h) Clusters extracted from HI figures. Red points are cluster cores; both red and orange points are cluster members. (d) Colored HI figure (Color figure online).

of noisy points are so weak that the homophilic layer formed by them will be the first one to fall below the out-degree curve with the growth of t. They will finally lie on the bottom together with the first type of noise for a moderate t. Although noise will not disappear by increasing t, we can separate clusters from them easily with local densities of points.

A popular method of estimating local density is to use the distance/similarity of point i to its k-th nearest neighbor [1,23], or the average distance of k nearest neighbors [14,22]. Rodriguez and Laio defined the local density using the number of neighbors (Eq. (1)) [18], which is similar to the ideas of using k nearest neighbors in density estimation. We estimate the local density $\hat{\rho}_i$ of point i by the average of similarities [24], i.e.

$$\hat{\rho}_i = \frac{1}{k} \sum_{j \in \mathcal{N}_i} e^{-\frac{D_{ij}^2}{\sigma^2}}, \tag{4}$$

where $\hat{\rho}_i$ is decoupled from d_c, and hence becomes more amendable to control. Next, we look for the minimum local density in the set of largest cores (red points) of clusters (denoted by $\mathcal{C}_{max}$) whose in-degrees are higher than out-degrees:

$$\hat{\rho}_{\mathcal{C}_{max}}^{min} = \min_{i \in \mathcal{C}_{max}} \hat{\rho}_i. \tag{5}$$

Then the set of clusters can be obtained by extracting points whose densities are larger than $\hat{\rho}_{\mathcal{C}_{max}}^{min}$:

$$\mathcal{C}_{cluster} = \{i | \hat{\rho}_i \geq \hat{\rho}_{\mathcal{C}_{max}}^{min}, i \in \mathcal{X}\}, \tag{6}$$

where $\mathcal{X}$ is the set of points. Here, yellow points are identified and merged into $\mathcal{C}_{cluster}$. Thus, the remaining points are categorized to the set of noise $\mathcal{C}_{noise}$

naturally. The clusters separated from noise shown in Fig. 2(h) demonstrate that $\hat{\rho}_{C_{max}}^{min}$ is an effective estimator to eliminate noise. The procedure of detecting clusters from noise is presented in Algorithm 4. Though DP uses ρ_b (similar to $\hat{\rho}_{C_{max}}^{min}$) as an estimator of cluster border region, ρ_b is sensitive to the cutoff distance d_c and sometimes affected by distances between clusters (Figs. 1(b)-(c)).

Algorithm 4. Detect Clusters from Noise with Homophily (init.: initialize)

Input: similarity matrix $\boldsymbol{W}$.
1: Init. $\mathcal{C}_{max} = \emptyset$, $\mathcal{C}_{cluster} = \emptyset$, $\mathcal{C}_{noise} = \emptyset$.
2: Calculate high-order $\boldsymbol{d}^{in}$ and $\boldsymbol{d}^{out}$.
3: $\mathcal{C}_{max} \leftarrow \mathcal{C}_{max} \cup$ points with $\frac{d_i^{in}}{d_i^{out}} > 1$.

4: Calculate $\hat{\rho}_{C_{max}}^{min}$ in $\mathcal{C}_{max}$ by Eq. (5).
5: Separate $\mathcal{C}_{cluster}$ from noise by Eq. (6).
6: $\mathcal{C}_{noise} \leftarrow \mathcal{C}_{noise} \cup$ the remaining points.
Output: cluster set $\mathcal{C}_{cluster}$, noise set $\mathcal{C}_{noise}$.

---

4   Apply Graph Kernel

After separating clusters from noise by *degree homophily*, we can divide $\mathcal{G}$ into sub-graphs $\mathcal{G}_{cluster}$ and $\mathcal{G}_{noise}$, whose corresponding similarity matrices are $\boldsymbol{W}_{cluster}$ and $\boldsymbol{W}_{noise}$, respectively. Thereafter we apply the *graph kernel* to $\boldsymbol{W}_{cluster}$ and obtain the global similarity matrix to re-compute δ for each point.

4.1   Similarity Propagation by Diffusion Kernel

In a graph, the similarity between points can be estimated by the linkage along the paths connecting them, which is regarded as long-range relationship. It has been verified that the long-range relationships can be used to enhance the affinities in network communities [11,13]. The graph kernel technique is useful in capturing the long-range relationships between points induced by the local graph structure [12]. Exploiting the fact that W_{ij}^t (we have left out the script of "cluster" in W_{ij}^t) of $\boldsymbol{W}_{cluster}$ in t-order represents the semantic similarity along the paths of length t, we can take advantage of such properties of diffusion kernel. Here, we use the *von Neumann diffusion kernel*:

$$\boldsymbol{K} = (\boldsymbol{I} - \alpha \boldsymbol{W}_{cluster})^{-1} = \sum_{t=0}^{\infty} \alpha^t \boldsymbol{W}_{cluster}^t, \tag{7}$$

where $\alpha < 1/\lambda$, and λ is the spectral radius of $\boldsymbol{W}_{cluster}$ [9]; α is a damping parameter that controls the long-range relationships. The *von Neumann diffusion kernel* can also be written in the form of infinite series of matrix powers. Therefore, K_{ij} presents the semantic similarity, which is calculated by summating the similarities of all possible paths in different lengths from point i to j. In this way, points in the cluster become more similar and pairs in the same cluster with long-range relationships can be drawn together from the viewpoint of structural similarity.

4.2   Center Location and Cluster Aggregation

One of the appealing novelties in [18] is the definition of δ, which essentially measures the similarity between point i and its nearest neighbor with higher density directly by distance. However, as mentioned in Sect. 2.2, δ_i is unable to enlarge the gap between centers and the remaining points in the decision graph. By the virtue of *graph kernel*, we re-define δ_i to make it more plausible and robust. For point i, we compute a quantity $\hat{s}_i$ that is the maximum similarity between itself and other points with higher densities:

$$\hat{s}_i = \max_{j \in \mathcal{S}_i, \hat{\rho}_i < \hat{\rho}_j} \{K_{ij}, K_{ji}\}, \tag{8}$$

where $\mathcal{S}_i$ is the set of points that have relation with point i. After noise elimination by *degree homophily*, there are no connections between clusters, and K is a block matrix, meaning that K_{ij} or K_{ji} is non-zero only when points i and j are connected directly or indirectly. Note that $\mathcal{N}_i \subseteq \mathcal{S}_i$ since each point i is also associated with indirectly connected points (via long-range relationships) according to Eq. (7). Although K is asymmetric as well, our concern is the similarities between pairs of points instead of the direction. Hence, the asymmetry of K_{ij} and K_{ji} is simultaneously taken care of in computing $\hat{s}_i$; and we set $\hat{s}_i = 0$ for the point with the highest density. Here, we compute $\hat{\delta}_i$ by the reciprocal of $\hat{s}_i$ added with a small number ϵ to avoid diving by zeroes:

$$\hat{\delta}_i = (\hat{s}_i + \epsilon)^{-1}, \tag{9}$$

where $\hat{\delta}_i$ is regarded as a measure of structural length of the path from point i to its most similar neighbor with higher density in graph $\mathcal{G}$.

As shown in Figs. 3(d), (e), (f), (g), (h) and (i), cluster centers densely concentrate at the upper right corner, which shows that centers are well characterized by relatively higher $\hat{\rho}_i$ and $\hat{\delta}_i$, making it easy to identify all centers unambiguously. There are larger margins between center candidates and the remaining points in decision graphs, clearly validating the superiorities of $\hat{\delta}$ and *degree homophily*. Clustering is straightforward based on the recognized centers: the remaining points are categorized into the same clusters as their most similar neighbors with higher densities, which resembles the Algorithm 2.

The space complexity of *graph kernel* based clustering algorithm is $\mathcal{O}(\frac{(n-n_s)^2}{n_c})$, where n is the number of points; and n_s is the number of noisy points. After noise elimination, the number of points to handle is reduced to $(n-n_s)$, and each point i in $\mathcal{G}_{cluster}$ is generally related to points of the same cluster in $\mathcal{S}_i$, which may contain the entire points in that cluster in the extreme case. For t-order degrees, the time complexity is $\mathcal{O}(tn)$, where t is much lower than n in practice. The label assignment stage can be completed in linear time $\mathcal{O}(n)$.

Algorithm 5. Clustering with Graph Kernel (r: nearest neighbors with higher densities)

Input: Similarity matrix W.
1: Apply Algorithm 4 to get $\mathcal{C}_{cluster}$ and $\mathcal{C}_{noise}$.
2: Extract the similarity sub-matrix $W_{cluster}$ by $\mathcal{C}_{cluster}$.
3: Calculate the local density $\hat{\rho}$ by Eq. (4).
4: Apply the diffusion kernel to $W_{cluster}$ to get $K_{cluster}$.
5: Calculate $\hat{s}$ by Eq. (8) and obtain r.
6: Locate cluster centers in decision graph manually, or specify zero $\hat{s}$ as centers automatically.
7: Assign each point to the same cluster as its r_i and get labels b of clusters.
Output: labels of points b, noise $\mathcal{C}_{noise}$.

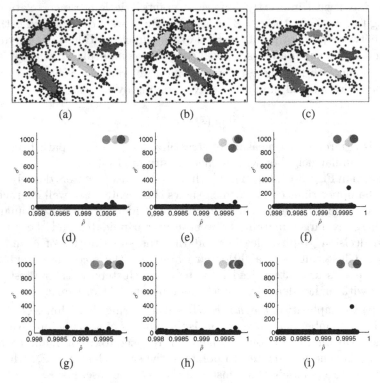

Fig. 3. (a)-(c) Clustering results of our algorithm. (d)–(f) and (g)–(i) are decision graphs of (a)-(c) before and after noise elimination. Cluster centers lie in the position where $\delta = 1/\epsilon$ in (g)–(i). Margins between center candidates and the remaining points are large enough to single out centers in decision graphs, meanwhile the superiorities of $\hat{\delta}$ and *degree homophily* are evident.

5   Experimental Evaluations on Synthetic and Real Problems

We evaluate our approach in relation to several state-of-the-art methods on two synthetic (generated following [3,18]) and six real-world databases [2,5,7,8,10, 17,19], which represent different clustering problems with various complexities.

5.1   Synthetic Data

We first apply our approach to the same synthetic database (Figs. 1(a), (b) and (d)) to compare decision graphs with DP algorithm. In Fig. 3(b), although two clusters are close to each other, their centers can be conveniently identified from the decision graph in Fig. 3(h). With *degree homophily*, clusters and noise can be separated completely, thereby producing better decision graphs. Meanwhile, Fig. 3(b) demonstrates again that $\hat{\rho}_{C_{max}}^{min}$ is more effective to detect clusters from noise in comparison with ρ_b and d_c for DP algorithm (Fig. 1(c)). For the cluster that contains two local density maxima, our algorithm can handle it as well, as demonstrated in Figs. 3(c) and (i).

Concerning the second synthetic database [3], we take three challenging patterns for our experiments. Different from [3], 20 % noise is added to test the robustness of algorithms. Besides DP, we also compare our algorithm with other mainstream algorithms such as k-means, Linkage, N-Cuts, AP, and AAS; and our approach is hereafter referred to as "HK" (Homophily-degree + graph-Kernel). Observed from the first row in Fig. 4, we find that methods working on the measure of distance directly (i.e. k-means, Linkage, AP, and DP) fail to capture the line-shaped structure, thus resulting in incorrect results. However, the graph-based algorithms (i.e. AAS, N-Cuts, and HK) produce much better results. Particularly, the result of HK turns out to be the best of all in clustering and noise elimination. A similar conclusion can also be drawn from the results on the second and third rows. We plot the decision graphs of HK and DP algorithms in Fig. 5, illustrating the evident superiority of HK.

5.2   Real-World Problems

We also carry out performance evaluation of clustering on six real-world databases. The FRGC (Face Recognition Grand Challenge) database [17] contains 466 persons/clusters embodying 16,028 facial images; the number of members ranges from 2 to 80 in each cluster. The Yale face database [7] contains 165 gray-scale images belonging to 15 individuals. The ORL database [19] contains 400 face images of 40 subjects. The MNIST [8] and USPS [2,10] databases are collected for handwritten-digit recognition; MNIST includes 10,000 handwritten digits of 10 classes; USPS involves 8-bit gray-scale images from "0" to "9" (each class has 1,100 samples). Besides, we also test on the Isolet [2,5], which is a spoken letter recognition database.

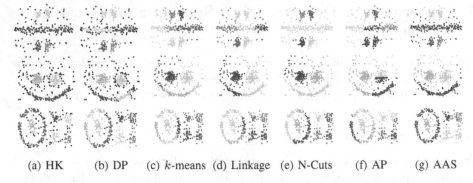

(a) HK   (b) DP   (c) k-means (d) Linkage (e) N-Cuts   (f) AP   (g) AAS

Fig. 4. Clustering results of different algorithms on databases [3]; we add 20 % noise to all databases.

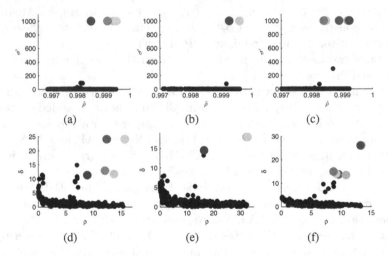

Fig. 5. (a)–(c) are the corresponding decision graphs of data shown in Fig. 4(a), and (d)–(f) show the corresponding decision graphs of Fig. 4(b).

We empirically find that the parameters of AP are difficult to control, which forces us leaving it out and merely performing the comparison with other algorithms. The performance of algorithms is evaluated by Normalized Mutual Information (NMI) [21]. As we can see from Table 1, on relatively simple databases (e.g. FRGC), all algorithms except Single Linkage manage to perform well; and HK obtains the best result of all. However, when performed on Yale, MNIST, and USPS databases that are more challenging, HK algorithm is significantly superior than DP. Generally, points in high-dimensional spaces or on curved manifolds are far from each other, hence δ of centers are generally not big enough to generate large margins on decision graphs for DP. On the opposite, our algorithm based on graph kernel shows satisfying robustness quality on these databases, and yields the best results across all six databases.

Table 1. Performance on real-world problems. We show the best results of algorithms with selected parameters to maximize the resulting performance, and build the same graph for N-Cuts and HK. Symbol '-' indicates computationally in-feasible for relevant cases.

Databases [Reference(s)]	k-means	Linkage			N-Cuts	AAS	DP	HK
		Single	Average	Complete				
FRGC [17]	0.8995	0.7719	0.9544	0.9189	0.9575	-	**0.9554**	**0.9703**
Yale [7]	0.6183	0.3162	0.3698	0.5417	0.6145	0.551	**0.3691**	**0.6843**
ORL [19]	0.7518	0.623	0.7561	0.754	0.825	0.731	**0.8174**	**0.8446**
MNIST [8]	0.4461	0.0144	0.2988	0.3608	0.6234	-	**0.246**	**0.6785**
USPS [2,10]	0.5971	0.0153	0.1875	0.404	0.7911	0.309	**0.6596**	**0.8114**
Isolet [2,5]	0.7098	0.1254	0.6724	0.7095	0.7804	0.73	**0.5740**	**0.803**

6 Conclusion

To address the difficulties of "data noise sensitivity" and "cluster center variance" in mainstream clustering algorithms, we propose a novel approach that can identify cluster centers unambiguously from data contaminated with noise. The *degree homophily* is applied to eliminate outliers, which effectively simplifies the clustering process greatly. Built upon the concepts of *graph kernel*, our approach can capture long-range relationships in data and discover clusters of manifold-structured shapes. The comprehensive experimental evaluations on synthetic and real-world databases verify the effectiveness and robustness of our algorithm comparing to several mainstream state-of-the-art methods.

Acknowledgments. This work was supported by the grants from China National Natural Science Foundation under Grant No. 613278050 & 61210013.

References

1. Byers, S., Raftery, A.E.: Nearest-neighbor clutter removal for estimating features in spatial point processes. J. Am. Stat. Assoc. **93**(442), 577–584 (1998)
2. Cai, D., He, X., Han, J., Huang, T.S.: Graph regularized nonnegative matrix factorization for data representation. IEEE Trans. Pattern Anal. Mach. Intell. **33**(8), 1548–1560 (2011)
3. Cho, M., MuLee, K.: Authority-shift clustering: hierarchical clustering by authority seeking on graphs. In: pp. 3193–3200. IEEE (2010)
4. Cho, M., Lee, K.M.: Mode-seeking on graphs via random walks. In: pp. 606–613. IEEE (2012)

5. Dietterich, T.G., Bakiri, G.: Error-correcting output codes: a general method for improving multiclass inductive learning programs. In: AAAI, pp. 572–577. Citeseer (1991)

6. Frey, B.J., Dueck, D.: Clustering by passing messages between data points. Science **315**(5814), 972–976 (2007)

7. Georghiades, A.S., Belhumeur, P.N., Kriegman, D.J.: From few to many: illumination cone models for face recognition under variable lighting and pose. IEEE Trans. Pattern Anal. Mach. Intell. **23**(6), 643–660 (2001)

8. Guyon, I., Gunn, S., Ben-Hur, A., Dror, G.: Result analysis of the nipps 2003 feature selection challenge. In: Advances in Neural Information Processing Systems, pp. 545–552 (2004)

9. Horn, R.A., Johnson, C.R.: Matrix Analysis. Cambridge University Press, Cambridge (1985)

10. Hull, J.J.: A database for handwritten text recognition research. IEEE Trans. Pattern Anal. Mach. Intell. **16**(5), 550–554 (1994)

11. Katz, L.: A new status index derived from sociometric analysis. Psychometrika **18**(1), 39–43 (1953)

12. Kondor, R.I., Lafferty, J.: Diffusion kernels on graphs and other discrete input spaces. In: pp. 315–322. Morgan Kaufmann (2002)

13. Newman, M.E.J., Girvan, M.: Finding and evaluating community structure in networks. Phys. Rev. E Stat. Nonlinear Soft Matter Phys. **69**(2 Pt 2), 026113 (2004)

14. O'Sullivan, D., Unwin, D.: Geographic Information Analysis. Wiley, Hoboken (2010)

15. Page, L., Brin, S., Motwani, R., Winograd, T.: The pagerank citation ranking: Bringing order to the web (1999)

16. Papadopoulos, F., Kitsak, M., Serrano, M., Bogun, M., Krioukov, D.: Popularity versus similarity in growing networks. Nature **489**(7417), 537 (2012)

17. Phillips, P.J., Flynn, P.J., Scruggs, T., Bowyer, K.W., Chang, J., Hoffman, K., Marques, J., Min, J., Worek, W.: Overview of the face recognition grand challenge. In: IEEE Computer Society Conference on Computer Vision and Pattern Recognition, CVPR 2005, vol. 1, pp. 947–954. IEEE (2005)

18. Rodriguez, A., Laio, A.: Clustering by fast search and find of density peaks. Science **344**(6191), 1492 (2014)

19. Samaria, F.S., Harter, A.C.: Parameterisation of a stochastic model for human face identification. In: Proceedings of the Second IEEE Workshop on Applications of Computer Vision, pp. 138–142. IEEE (1994)

20. Shi, J., Malik, J.: Normalized cuts and image segmentation. IEEE Trans. Pattern Anal. Mach. Intell. **22**(8), 888–905 (2000)

21. Strehl, A., Ghosh, J.: Cluster ensembles - a knowledge reuse framework for combining multiple partitions. J. Mach. Learn. Res. **3**, 583–617 (2002)

22. Wang, F.: Quantitative Methods and Applications in GIS. CRC Press, Boca Raton (2006)

23. Wong, M.A., Lane, T.: A kth nearest neighbour clustering procedure. J. R. Stat. Soc. Ser. B (Methodol.) **45**(3), 362–368 (1983)

24. Zhao, D., Tang, X.: Homophilic clustering by locally asymmetric geometry. Eprint Arxiv (2014)

Constraint Based Subspace Clustering for High Dimensional Uncertain Data

Xianchao Zhang[✉], Lu Gao, and Hong Yu

Dalian University of Technology, Dalian, China
{xczhang,hongyu}@dlut.edu.cn, gaolu@mail.dlut.edu.cn

Abstract. Both uncertain data and high-dimensional data pose huge challenges to traditional clustering algorithms. It is even more challenging for clustering high dimensional uncertain data and there are few such algorithms. In this paper, based on the classical FINDIT subspace clustering algorithm for high dimensional data, we propose a constraint based semi-supervised subspace clustering algorithm for high dimensional uncertain data, UFINDIT. We extend both the distance functions and dimension voting rules of FINDIT to deal with high dimensional uncertain data. Since the soundness criteria of FINDIT fails for uncertain data, we introduce constraints to solve the problem. We also use the constraints to improve FINDIT in eliminating parameters' effect on the process of merging medoids. Furthermore, we propose some methods such as sampling to get an more efficient algorithm. Experimental results on synthetic and real data sets show that our proposed UFINDIT algorithm outperforms the existing subspace clustering algorithm for uncertain data.

1 Introduction

Clustering provides a better understanding of the data by dividing data sets into clusters so that objects in the same cluster are more similar than those in different clusters. It is one of the most important techniques of data mining and has been studied for decades and many algorithms have been proposed [14]. Nevertheless, uncertain data and high-dimensional data pose huge challenges to traditional clustering algorithms.

High dimensional data clusters are formed in subspaces, and the search for subspace and the detection of clusters are circular dependent. Subspace clustering algorithms for high dimensional data can be divided into two groups according to the search technique: top-down subspace clustering methods and bottom-up subspace clustering methods. CLIQUE [4], ENCLUS [9], SUBACLU [15], MAFIA [18] and so on are all typical methods of bottom-up subspace clustering methods. And representative algorithms of top-down subspace clustering methods are like PROCLUS [1], ORCLUS [2], FINDIT [19], etc.

In many clustering practices, constraints, such as must-link and cannot-link, are accessible [6,7]. Recent works have integrated constraints into subspace clustering, such as SC-MINER [11], CDCDD [21], CLWC [10] and so on.

© Springer International Publishing Switzerland 2016
J. Bailey et al. (Eds.): PAKDD 2016, Part II, LNAI 9652, pp. 271–282, 2016.
DOI: 10.1007/978-3-319-31750-2_22

With the emergence of new application domains such as location-based services and sensor monitoring, uncertainty is ubiquitous due to reasons such as outdated sources or imprecise measurement [3]. As uncertainty is ubiquitous in many cases, uncertain data clustering causes more and more attention. In uncertain data, the attribute value may vary in a certain range, so that traditional clustering algorithms do not work. Uncertain clustering algorithms also can be divided into two groups: partition-based clustering methods such as UK-means [8], UK-medoids [12], etc. and density-based clustering methods such as FDBSCAN [17], FOPTICS [16], PDBSCANi [20], etc.

It is even more challenging for clustering high dimensional uncertain data. To our knowledge, there exists only one subspace clustering algorithm for high dimensional uncertain data [13], which is an extension of a bottom-up subspace clustering algorithm for handling data uncertainty.

In this paper, based on the classical FINDIT [19] subspace clustering algorithm for high dimensional data, we propose a constraint based subspace clustering algorithm for high dimensional uncertain data, UFINDIT. We extend both the distance functions and dimension voting rules of FINDIT to deal with high dimensional uncertain data. We also use the constraints to improve FINDIT both in eliminating parameters' effect on the process of merging medoids and in improving the effectiveness of soundness. Furthermore, we propose some methods such as sampling to get an more efficient algorithm. Experimental results on synthetic and real data sets show that our proposed UFINDIT algorithm outperforms the existing subspace clustering algorithm for uncertain data.

2   Related Work

FINDIT adopts dimension-oriented distance (dod) measure as its dissimilarity measure and determines the correlated dimensions for each cluster by dimension voting. FINDIT is composed of three phases: sampling phase, cluster forming phase and data assigning phase. In the first phase, sample set S and medoid set M (representatives of original clusters) are generated by a random sampling method. And in the second phase, correlated dimensions of all medoids are determined by the V nearest neighbors' voting and the neighbors of each medoid are calculated in dod measure. The medoids which are near from each other in dod measure are grouped together. As different medoid cluster sets are generated for each iteration on ϵ, an evaluation criteria is used to choose the best ϵ and the corresponding medoid cluster set MC_ϵ. Then the selected medoid cluster set is given to the following phase as the best summary of the original clusters. In data assignment phase, all points are assigned to their nearest medoid clusters, and the points not assigned to any medoid cluster are regarded as outliers.

The inputs for FINDIT are the dataset and two user parameters $C_{minsize}$ and $D_{mindist}$. The first parameter reflects the users wish about the minimum size of clusters and the second one is the merge threshold between two resultant clusters. That is, if the dod distance between two clusters is smaller than $D_{mindist}$ for the selected ϵ, they are regarded as one cluster by FINDIT and are subsequently merged.

FINDIT has a good performance on high dimensional clustering problem, but it is sensitive to parameter $D_{mindist}$ and ϵ. Though it selects the best ϵ from 25 different ϵ values by its soundness criteria, the soundness criteria itself is flawed. In view of its merit and demerit, this paper extends it to uncertain data clustering and give solutions to overcome its demerit.

3   Constraint Based Subspace Clustering for Uncertain Data

3.1   Problem Formulation

The attribute of each object in the database of uncertain data is described by a probability density function (pdf) which limits the range the attribute value varies in. We are given a data set DB in a D-dimensional space. And the pdf $p_i(x)$ with $x \in R^D$ denotes the i-th data point which satisfies that its integral is equal to 1. To address the components of a vector x, we use the notation $p_i(x_1, \cdots, x_D)$ instead of $p_i(x)$.

As our aim is to find clusters and the subspace each cluster is relevant to, we define the projection of $p_i(x)$ to a subspace Sub as $p_i^{Sub}(x)$ with $x \in R^{|Sub|}$ which is defined by:

$$p_i^{Sub}(x) = \int \cdots \int_{x_{s+1}, \dots, x_D} p_i(x_1, \cdots, x_D) \tag{1}$$

where $Sub = \{1, \cdots, s\}$. We define a subspace cluster as a set of $p_i^{Sub}(x)s$.

Given two types of constraints: must-link and cannot-link. $C = \{c_1, c_2, ..., c_c\}$ denotes the set of constraints, and c is the number of constraints, i.e., $c = |C|$. $ML = (x_i, x_j)$ denotes a must-link between data points x_i and x_j, $CL =< x_i, x_j >$ denotes a cannot-link between data points x_i and x_j. The points which are linked by must-links and cannot-links are called medoids. And M denotes a points set which contains all medoids in C.

3.2   Algorithm Framework

The inputs for our method are the data set X with N points in D-dimensional space, a user parameter $C_{minsize}$ and constraints set C based on the actual ground-truth clusters. According to C, we can easily get M. The parameter $C_{minsize}$ reflects the user's wish about the minimum size of clusters, in another word, our method does not report the clusters smaller than $C_{minsize}$.

Similar to FINDIT, UFINDIT is composed of three phases: sampling phase, cluster forming phase and data assigning phase. In the first phase, we randomly generate the set of sampled points S. The second phase includes five parts: determining key dimensions for each medoid, assigning every sampled point to its nearest medoid, merging similar medoids to get medoid clusters, refining medoid clusters and evaluation measuring medoid clusters. And the last phase is a process that assign every point in dataset to each medoid cluster in the best medoid clusters to obtain the finally clustering result. The algorithm framework is described in Algorithm 1.

Algorithm 1. UFINDIT Algorithm

Input: Data set X, $C_{minsize}$, C, M
Output: Result of clustering
$//E$:set of the 25 epsilons
1.**begin**
2.   Random generate the set of sampled points S by using $C_{minsize}$ to calculate $|S|$
3.   Use constraints to calculate the best subset E_{best} of E
4. $bestepsilon := 0; MC_{bestepsilon} := 0$;
5.   **for** each ϵ in E_{best} do begin
6.   Determine key dimensions for each medoid m in M;
7.   Assign every point p in S to the nearest medoid in M;
         $//$Merge similar medoids in M to make medoid clusters set MC_{ϵ}
8. $MC_{\epsilon} = MergeMedoids(M)$;
9.   Refine medoid clusters in MC_{ϵ}
10.   **if** $Soundness(MC_{\epsilon}) \geq Soundness(MC_{bestepsilon})$
11.   **then** $bestepsilon := \epsilon$;
12. **end**
13. Assign every point in the dataset to the nearest medoid in medoid
    cluster which belongs to $MC_{bestepsilon}$
14.**end**

3.3   Dimension-Oriented Distance for Uncertain Data

Dimension-oriented distance (dod) is an unique distance measure proposed in FINDIT which utilizes dimensional difference information and value difference information together. We extend this distance measure to solve uncertain data clustering problems, and we call it $udod$. We proposed two different methods to extend this distance measure. In case 1, we mainly compute the probability that the distance between point p and q is less than distance threshold ϵ. If the probability is greater than the given δ, then we believe that the two points are 'equal' on that dimension. In this case, we define directed probability-based $udod$ from a point p to a point q as:

$$udod_{\epsilon}(p \rightarrow q) = |D_p| - \{d|P(|p(d) - q(d)| \leq \epsilon) \geq \delta, d \in D_p \cap D_q\} \quad (2)$$

$$P(|p(d) - q(d)| \leq \epsilon) = \int_{|x-y| \leq \epsilon} p_p^d(x) \cdot p_q^d(y) \, dx \, dy \quad (3)$$

where $|p(d) - q(d)|$ denotes the Manhattan distance between point p and q in dimension d, D_p is the subspace of point p, $P(|p(d) - q(d)| \leq \epsilon)$ denotes the probability that the distance between point p and q is less than distance threshold ϵ, $p_p^d(x)$ is the pdf of point p in dimension d and δ denotes probability threshold.

In case 2, we define directed distance-based *udod* from a point p to a point q as:

$$udod_\epsilon(p \rightarrow q) = |D_p| - \{d|D_d(p,q) \leq \epsilon, d \in D_p \cap D_q\} \qquad (4)$$

$$D_d(p,q) = \int |x - y| \cdot p_p^d(x) \cdot p_q^d(y) \, dx \, dy \qquad (5)$$

where $D_d(p,q)$ denotes the distance integral between point p and q.

And we define $udod_\epsilon(p,q)$ as the larger value out of two directed *udod* values between p and q. That is:

$$udod_\epsilon(p,q) = \max(udod_\epsilon(p \rightarrow q), udod_\epsilon(q \rightarrow p)) \qquad (6)$$

3.4   Sampling, Cluster Forming and Data Assigning

Sampling Phase. In this phase, we randomly generate samples set S. According to FINDIT, any original cluster larger than $C_{minsize}$ should have more than a certain number of points in S and have at least one point in M. According to FINDIT, we also use *Chernoff bounds* to solve the minimum size problem of sampled set S. The minimum size assures that every cluster in S have more than ξ sampled points by the probability of $1 - \delta$, and is computed by the following equation:

$$Chernoff - bounds(S) = \xi k \rho + k \rho \log(\frac{1}{\delta}) + k \rho \sqrt{(\log(\frac{1}{\delta}))^2 + 2\xi \log(\frac{1}{\delta})} \qquad (7)$$

where constant k is the number of clusters, and ρ is a value satisfying $C_{minsize} = N/k\rho$ ($\rho \geq 1$), N is the size of the dataset and $C_{minsize}$ is the size of the smallest cluster in the dataset.

To obtain *Chernoff bounds* for a sample set S, typically, we use the setting $\xi = 30$, $\delta = 0.01$, $k = N/C_{minsize}$ and $\rho = 1$ according to FINDIT. We use the notation $S_{minsize}$ to indicate the value used for ξ, so $S_{minsize} = 30$. The expected number of points in constraints (M) is also computed by using *Chernoff bounds* with $\xi = 1$.

Cluster Forming Phase. Cluster forming phase is iterated several times with an increasing ϵ. Throughout the whole process, ϵ is a very important parameter which determines the computation of *udod*. To obtain an optimal ϵ, according to FINDIT, we try 25 different values from $(1/100)valuerange$ to $(25/100)valuerange$ where *valuerange* is a normalized value for all dimensions.

Determining Key Dimensions: The purpose of this step is to determine the related dimensions for all medoids. For each medoid in set M, we compute its V nearest neighbors in set S by calculating $udod_\epsilon(p,m)$ with $|D_p| = |D_m| = |D|$. As mentioned in FINDIT, we regard this V nearest neighbors as voters and dimension voting is a process for D independent questions. For each dimension of each medoid, in distance-based case, we count the number of voters which vote for 'YES'. While in the probability-based case, we compute the sum of probability that voters vote for 'YES'. If a neighbor's d-th dimensional distance from m is less than or equal to the given ϵ, it is considered that it votes for 'YES' in distance-based case. And in probability-based case, we regard the probability that a neighbor's d-th dimensional distance is less than or equal to the given ϵ as the probability that voter votes for 'YES'.

As the maximum number of reliable voters is $S_{minsize}$ (mentioned in [19]) which is equal to 30, we set $V = 20$. For each dimension of each medoid, as mentioned in FINDIT, if the computation value is greater than 12, then we believe this dimension is a relevant dimension for this medoid.

Assigning Sampled Points: Since the set of key dimensions have been generated for each medoid in last step, each medoid is meaningful in its own subspace composed of its related dimensions. So to verify whether there is really a cluster in the subspace, we assign points in S to medoids in M. We refer to the point assigned to medoid m as member of m. For each point p in S and each medoid m in M, p is assigned to the nearest medoid which satisfies the member assignment condition as follows:

$$udod_\epsilon(m \to p) = 0 \tag{8}$$

Merging Medoids: In this phase, we improve the medoid clustering step of FINDIT by using constraints. After the last step, we get medoids with their own key dimensions and members. We group similar medoids together gradually. And the grouped medoid clusters are original clusters used for finally clustering.

To cluster medoids, we also use the groupwise average clustering method (gac) [19]. It starts by regarding all medoids as independent medoid clusters. And the distance between any two medoid clusters is computed as the weighted average distance between the medoids belonging to them as follows:

$$udod_\epsilon(mc_A, mc_B) = \frac{\sum_{m_i \in mc_A, m_j \in mc_B}(|m_i||m_j|udod_\epsilon(m_i, m_j))}{(\sum_{m_i \in mc_A}|m_i|)(\sum_{m_j \in mc_B}|m_j|)} \tag{9}$$

where $|m_i|$ means the number of m_is members which is used as weight and mc_A means a medoid cluster A. To improve the effectiveness of gac, for any two medoids, if the number of dimensions on which they are considered equal is no more than 2, the distance is set to D as a penalty(note that D is the maximum distance). The process stops until there is a cannot-link between the closest pair. We believe that each cluster should have at least one must-link, so we remove the medoid clusters which do not contain a must-link before later step. The detailed process is described in Algorithm 2.

Refining Medoid Clusters: Since each medoid cluster obtained in last step has information inherited from its medoids and each medoid m in mc can have slightly different related dimensions, we obtain the finally subspace for each medoid cluster mc by averaging information of each medoid in mc. Similar to FINDIT, the average selection ratio for a dimension d is obtained as follows:

$$avg_d = \frac{\sum_{m_i} \delta_i|m_i|}{\sum_{m_i}|m_i|} \tag{10}$$

Algorithm 2. MergeMedoids Algorithm

Input: Medoid set M
Output: Medoid cluster set MC_ϵ
1.begin
2.   Merge each two medoids linked by a must-link and add the result to MC_ϵ;
3.   Merge medoid clusters in MC_ϵ which have common medoids;
4.   Get mc_A and mc_B between which the distance is shortest from MC_ϵ;
5.   **while** There is not a cannot-link between medoid in mc_A and medoid mc_B
     do **begin**
6.   Merge mc_A and mc_B and push the merged result into MC_ϵ;
7.   Get mc_A and mc_B between which the distance is shortest from MC_ϵ;
8.   **end**
9.   **for** each medoid cluster mc in MC_ϵ do **begin**
10.   **if** There is not a must-link between medoids in mc
11.   **then** Remove mc from MC_ϵ;
12. **end**
13. **return** MC_ϵ;
14.**end**

---

where δ_i is a binary function which is set to 1 when the dimension d is a key dimension of m_i and set to 0 otherwise. In this paper, we regard d as a key dimension of the medoid cluster, if avg_d is lager than 0.95. When the finally subspace is obtained, all medoids in mc are only meaningful in this common subspace rather than their own subspaces.

In addition, before going to the evaluation phase, some of the medoid clusters whose number of key dimensions is smaller than 2 are removed because their corresponding subspace is unreasonable.

Evaluation: Because the cluster forming phase is iterated several times, so we should select the best medoid cluster set MC_ϵ as the final result. According to FINDIT, the soundness criteria of MC_ϵ is measured by the following equation:

$$Soundness(MC_\epsilon) = \sum_{mc \in MC_\epsilon} (|mc| \cdot |KD_{mc}|) \tag{11}$$

where $|mc|$ is the size of a medoid cluster mc and $|KD_{mc}|$ is the number of common key dimensions of mc. And the MC_ϵ with the greatest soundness is chosen as the best one ($MC_{bestepsilon}$) which has the most dimension and member information.

Data Assigning Phase. In this phase, the points in the dataset are assigned to either a medoid cluster in $MC_{bestepsilon}$ or a noisy cluster. The principle of point assignation in this phase is the same as that used in member assignation step. If a point is assigned to a medoid m in medoid cluster mc, it is equal to that it is assigned to mc. The only difference is that all medoids are meaningful in the subspace of the medoid cluster mc rather than their own subspaces.

3.5 Pruning Based on Constraints and Sampling

Pruning. To solve the soundness's problem of FINDIT and reduce the run-time, we propose a pruning algorithm by using constraints. The principle is that the average distance between points linked by cannot-link should be greater than distance between points linked by must-link. So we compute the difference between the mean of cannot-link distances and the mean of must-link distances using *udod* as *diff*. Then select n epsilons whose values of *diff* are greater than unselected ones as the epsilons set rather than 25 epsilons. In this paper, we set n as 5. What's more, as *udod* costs is time-consuming, we propose a pruning method to reduce its runtime. Theoretically, if the max distance between point p and q in dimension d is less than ϵ, the real distance in dimension d must be less than ϵ and if the min distance between point p and q is more than ϵ, the real distance must be more than ϵ. So if anyone of the two conditions is fulfilled, there is no need for us to compute $P(|p(d) - q(d)| \leq \epsilon)$ or $D_d(p, q)$. Then we can reduce some computation of *udod*.

Sampling. In some scenarios, it might be possible to compute the exact probability by solving the integral of probability-based *udod*. However, the probability may be incomputable. To solve this problem, we use Monte-Carlo sampling to approximately calculate the integrals. Each pdf p_{ij} of point i in dimension j is represented by a set of objects, drawn from the distribution obtained by p_{ij}. Given two pdfs $p_{ij}1$ and $p_{ij}2$ and their sampled objects set $I1, I2$, we compute the distance between each object in $I1$ and each object in $I2$. Then compute the percentage of the objects pairs that have a smaller distance than ϵ as the probability that the distance between the two points on dimension j is less than ϵ.

For the integral of distance, we use an approximate method called Riemann integral to compute it. As we need to compute the integral with two variables, we divide both variables intervals into several equal sub-intervals, and the integral can be approximately regarded as the sum of several cuboids' volumes. The integral is computed as follows:

$$\int_c^d dy \int_a^b f(x, y) \, dx = \sum_{i=1}^n \sum_{j=1}^m f(t_{x_i}, t_{y_j}) \cdot (\triangle i) \cdot (\triangle j)$$

$$t_{x_i} = \frac{x_{i-1} + x_i}{2} \ and \ \triangle i = \frac{b-a}{n} \tag{12}$$

$$t_{y_j} = \frac{y_{j-1} + y_j}{2} \ and \ \triangle j = \frac{d-c}{m}$$

4 Experiments

4.1 Datasets and Baselines

We conduct a series of experiments to evaluate the performance of UFINDIT in terms of accuracy, efficiency and sacability. These experiments were performed

Table 1. Parameters of synthetic datasets

Parameter	Value	Interpretation
N	1000, 10000, 100000	Data size
D	20, 30, 50	Number of dimensions
AD	D/3, D/2, 2D/3	Average number of correlated dimensions
K	5,5,5	Clusters numbers
s	[2,4], [2,4], [2,4]	Cluster point distribution range
PO	0.1,0.1,0.1	Outliers percentage

Table 2. UCI datasets

Datasets	N	D	K
Glass	214	9	6
Wine	178	13	3
Pendigits	10992	16	10

on several synthetic dataset and real world datasets. The synthetic datasets are generated based on the method described in [1,13], but we make some modifications to generate high-dimensional uncertain data. Firstly, we generate high-dimensional data as described in [1]. Then we generate uncertain range for the dataset with a random length between 0 and 0.1 data range in the relevant dimensions and a random length between 0 and 0.5 data range in the non-relevant ones.

The coefficient factors are listed in Table 1. We use three UCI datasets [5] and summarize the details in Table 2. We generate uncertain range for the UCI dataset with a random length between 0 and 0.1 data range in all dimensions. Two kinds of uncertainty, uniform distribution and Guassian distribution are conducted on synthetic datasets and UCI datasets. Our two algorithms are respectively called UFINDIT_PRO which denotes the probability-based case, and UFINDIT_DIS which denotes the distance-based case. We call the algorithm proposed in [13] USC, and FINDIT algorithm using data removing uncertainty by computing expectation of uncertain data as a real value FINDIT+EXP. We compare the accuracy, efficiency and scalability of UFINDIT_PRO and UFINDIT_DIS with USC and FINDIT+EXP. As there are 6 algorithms in USC, we experiment on all the 6 algorithms and report the best result. We use the parameters suggested in the original paper except α in USC. We set $\alpha = 0.05$. And for UFINDIT, we use $|C| = 100$ according to Chernoff bounds, $C_{minsize} = 0.05 \cdot |DB|$ according to FINDIT.

4.2 Accuracy

We measure the accuracy by using F1-value criterion which is one of the most often used criterion to evaluate the accuracy of high dimensional data clustering

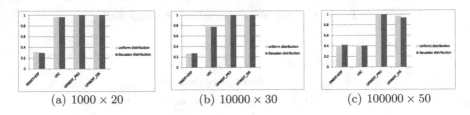

| (a) 1000 × 20 | (b) 10000 × 30 | (c) 100000 × 50 |

Fig. 1. Accuracy comparison on synthetic datasets

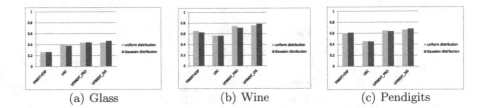

| (a) Glass | (b) Wine | (c) Pendigits |

Fig. 2. Accuracy comparison on UCI datasets

algorithms. Figure 1 shows the accuracy comparison on synthetic datasets. Results show that FINDIT+EXP performs worst while both two UFINDIT performs best. The soundness of FINDIT becomes invalid which makes FINDIT worst and as we improve the soundness of FINDIT by using constraints, the result of UFINDIT performs well. USC performs well when data size and dimensionality are small, but it has a bad performance on high-dimensional vast dataset. Figure 2 shows the accuracy comparison on UCI datasets. Results show that USC has the worst performance, while UFINDIT still performs best. In this case, the soundness of FINDIT performs well which makes results different. Comparing the result on synthetic datasets with that on real datasets, we find that the soundness of FINDIT is not always invalid. Overall we conclude that UFINDIT (UFINDIT_PRO and UFINDIT_DIS) performs better than the other two algorithms on both synthetic datasets and real datasets.

4.3 Efficiency and Scalability

In this subsection, we compare the efficiency of our two algorithms UFINDIT_PRO and UFINDIT_DIS with USC algorithm on a series of synthetic datasets. As FINDIT+EXP algorithm does not contain uncertain information, it will be much faster than uncertain algorithms, we do not compare with it here. Figure 3 shows the time comparison on different datasets. Figure 3(a) shows the time scalability with the increasing of data set size. The runtime of USC increases sharply with the increasing of dataset size which shows that it is not scalable. Whereas our algorithms scale well with dataset size. Our algorithms have advantages over USC on clustering large quantity of data. Figure 3(b) shows time scalability with respect to increasing of dimensionality. It can be seen that the

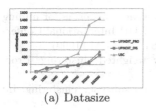
(a) Datasize

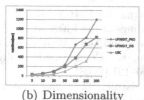
(b) Dimensionality

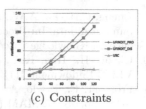
(c) Constraints

Fig. 3. Time scalability

slopes of all algorithms are similar, but USC increases faster than our algorithms. Figure 3(c) shows time scalability with respect to the increasing of constraints. As USC using no constraints, its runtime does not change with the increasing of constraints. Our algorithms almost increase linearly as the number of constraints increases.

5   Conclusion

In this paper, based on the classical subspace clustering algorithm FINDIT, we propose an effective subspace clustering algorithm UFINDIT for uncertain high dimensional data. We introduce a pruning algorithm to improve FINDIT by using constraints and also modify the medoids clustering process of FINDIT. To the best of our knowledge, this is the second (and the first top-down) subspace clustering algorithm for uncertain high dimensional data. Experimental results have shown the superiority of our method in terms of accuracy, efficiency and scalability.

References

1. Aggarwal, C.C., Wolf, J.L., Yu, P.S., Procopiuc, C., Park, J.S.: Fast algorithms for projected clustering. In: ACM SIGMoD Record, vol. 28, pp. 61–72. ACM (1999)
2. Aggarwal, C.C., Yu, P.S.: Finding generalized projected clusters in high dimensional spaces. In: Proceedings of the 2000 ACM SIGMOD Conference, pp. 70–81. ACM (2009)
3. Aggarwal, C.C., Yu, P.S.: A survey of uncertain data algorithms and applications. IEEE Trans. Knowl. Data Eng. **21**(5), 609–623 (2009)
4. Agrawal, R., Gehrke, J., Gunopulos, D., Raghavan, P.: Automatic subspace clustering of high dimensional data for data mining applications. In: Proceedings of the 1998 ACM SIGMOD Conference, pp. 94–105. ACM (1998)
5. Asuncion, A., Newman, D.: Uci machine learning repository (2007)
6. Bar-Hillel, A., Hertz, T., Shental, N., Weinshall, D.: Learning a mahalanobis metric from equivalence constraints. J. Mach. Learn. Res. **6**(6), 937–965 (2005)
7. Basu, S., Davidson, I., Wagstaff, K.: Constrained Clustering: Advances in Algorithms, Theory, and Applications. CRC Press, New York (2008)
8. Chau, M., Cheng, R., Kao, B., Ng, J.: Uncertain data mining: an example in clustering location data. In: Ng, W.-K., Kitsuregawa, M., Li, J., Chang, K. (eds.) PAKDD 2006. LNCS (LNAI), vol. 3918, pp. 199–204. Springer, Heidelberg (2006)

9. Cheng, C.H., Fu, A.W., Zhang, Y.: Entropy-based subspace clustering for mining numerical data. In: Proceedings of the Fifth ACM SIGKDD International Conference on Knowledge Discovery and Data Mining, pp. 84–93. ACM (1999)
10. Cheng, H., Hua, K.A., Vu, K.: Constrained locally weighted clustering. Proc. VLDB Endowment **1**(1), 90–101 (2008)
11. Fromont, E., Prado, A., Robardet, C.: Constraint-based subspace clustering. In: SDM, pp. 26–37. SIAM (2009)
12. Gullo, F., Ponti, G., Tagarelli, A.: Clustering uncertain data via K-medoids. In: Greco, S., Lukasiewicz, T. (eds.) SUM 2008. LNCS (LNAI), vol. 5291, pp. 229–242. Springer, Heidelberg (2008)
13. Günnemann, S., Kremer, H., Seidl, T.: Subspace clustering for uncertain data. In: SDM, pp. 385–396. SIAM (2010)
14. Jain, A.K.: Data clustering: 50 years beyond k-means. Pattern Recogn. Lett. **31**(8), 651–666 (2010)
15. Kailing, K., Kriegel, H.P., Kröger, P.: Density-connected subspace clustering for high-dimensional data. In: Proceedings of the SDM, vol. 4, pp. 246–257. SIAM (2004)
16. Kriegel, H.P., Pfeifle, M.: Hierarchical density-based clustering of uncertain data. In: Fifth IEEE International Conference on Data Mining, p. 4. IEEE (2005)
17. Kriegel, H.P., Pfeifle, M.: Density-based clustering of uncertain data. In: Proceedings of the Eleventh ACM SIGKDD International Conference on Knowledge Discovery in Data Mining, pp. 672–677. ACM (2005)
18. Nagesh, H.S., Goil, S., Choudhary, A.N.: Adaptive grids for clustering massive data sets. In: SDM, pp. 1–17. SIAM (2001)
19. Woo, K.G., Lee, J.H., Kim, M.H., Lee, Y.J.: Findit: a fast and intelligent subspace clustering algorithm using dimension voting. Inf. Softw. Technol. **46**(4), 255–271 (2004)
20. Zhang, X., Liu, H., Zhang, X., Liu, X.: Novel density-based clustering algorithms for uncertain data. In: Proceedings of the Twenty-Eighth AAAI Conference on Artificial Intelligence, 27–31 July 2014, Québec City, Québec, Canada, pp. 2191–2197 (2014). http://www.aaai.org/ocs/index.php/AAAI/AAAI14/paper/view/8185
21. Zhang, X., Wu, Y., Qiu, Y.: Constraint based dimension correlation and distance divergence for clustering high-dimensional data. In: 2010 IEEE 10th International Conference on Data Mining (ICDM), pp. 629–638. IEEE (2010)

A Clustering-Based Framework
for Incrementally Repairing Entity Resolution

Qing Wang$^{(\boxtimes)}$, Jingyi Gao, and Peter Christen

Research School of Computer Science, The Australian National University,
Canberra, ACT 0200, Australia
{qing.wang,jingyi.gao,peter.christen}@anu.edu.au

Abstract. Although entity resolution (ER) is known to be an important problem that has wide-spread applications in many areas, including
e-commerce, health-care, social science, and crime and fraud detection,
one aspect that has largely been neglected is to monitor the quality of
entity resolution and repair erroneous matching decisions over time. In
this paper we develop an efficient method for incrementally repairing
ER, i.e., fix detected erroneous matches and non-matches. Our method
is based on an efficient clustering algorithm that eliminates inconsistencies among matching decisions, and an efficient provenance indexing data
structure that allows us to trace the evidence of clustering for supporting
ER repairing. We have evaluated our method over real-world databases,
and our experimental results show that the quality of entity resolution
can be significantly improved through repairing over time.

Keywords: Data matching · Record linkage · Deduplication · Data
provenance · Data repairing · Consistent clustering

1 Introduction

Entity resolution (ER) is the process of deciding which records from one or more
databases correspond to the same entities. Typically, two types of matching
decisions are involved in the ER process: *match* (two records refer to the same
entity) and *non-match* (two records refer to two different entities). Based on
matching decisions, a clustering algorithm is often used to group records into
different *entity clusters*, each of which consists of a set of records and corresponds
to one entity [5]. Due to its importance in practical applications, ER has been
extensively studied in the past [5,8]. Nonetheless, the ER process is still far from
satisfactory in practice. For example, when data is dirty (i.e., incorrect, missing,
or badly formatted), an ER result may contain errors that cannot be detected at
the time of performing the ER task. Instead, such errors are often detected later
on by users, particularly when users their entities in various ER applications.
If we repair these errors incrementally whenever being detected, the quality of
ER can be improved over time, which would accordingly improve the quality of
applications that use ER.

© Springer International Publishing Switzerland 2016
J. Bailey et al. (Eds.): PAKDD 2016, Part II, LNAI 9652, pp. 283–295, 2016.
DOI: 10.1007/978-3-319-31750-2_23

However, in many real-life situations, repairing ER is complicated and labour-intensive. For example, we may find that two records r_3 and r_5 in an entity cluster $\langle r_1, r_2, r_3, r_4, r_5 \rangle$ are an erroneous match, and therefore should be split into two different entities. Since there are many possible ways of splitting r_3 and r_5 into two entity clusters, such as $\{\langle r_3 \rangle, \langle r_1, r_2, r_4, r_5 \rangle\}$, $\{\langle r_5 \rangle, \langle r_1, r_2, r_3, r_4 \rangle\}$, and $\{\langle r_3 \rangle, \langle r_5 \rangle, \langle r_1, r_2, r_4 \rangle\}$, repairing this error would require a manual review on all the records in the entity cluster, their match or non-match decisions, and possible inconsistencies. Such repairing tasks can become even more difficult if some constraints, such as *hard matches* and *hard non-matches* (often dictated by domain experts or users), are required to be preserved in the repairing process. For example, suppose that we have another entity cluster $\langle r_6, r_7, r_8, r_9 \rangle$ and know that (r_3, r_8) is an erroneous non-match, but (r_3, r_1) is a hard match and (r_1, r_9) is a hard non-match. In this case, it is not easy to repair the erroneous match (r_3, r_5) and non-match (r_3, r_8), while still preserving hard matches and non-matches.

This paper aims to develop an efficient method for incrementally repairing ER. Our observation is that repairing errors in an ER result should look at how an entity was produced, e.g., matches used to determine its entity cluster over time and the confidences of these matches. Thus, in order to establish an efficient procedure and also a reasonable level of credibility for performing ER repairing, we propose to build a provenance index that allows us to trace the evidences of matches used in the ER process, and interact with the evidence of non-matches. As illustrated in Fig. 1, such a provenance index can be built during the ER clustering process, then iteratively maintained in the repairing process. Based on this provenance index, our repairing algorithms can leverage the evidence of matches and non-matches to repair errors accurately and efficiently.

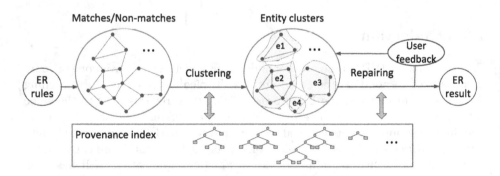

Fig. 1. A general framework for repairing ER based on provenance indexing

Contributions. In summary, our key contributions are as follows:

(1) We develop an algorithm that can solve the inconsistencies of matching decisions to generate entity clusters.

(2) We design a provenance index structure which can keep track of matches that are relevant to each entity and capture the relationships among these matches as they occur in the clustering process.
(3) Based on the provenance index structure, we propose a novel approach to automatically and efficiently repair errors in ER whenever they are detected.
(4) We experimentally validate the feasibility of the proposed provenance index and the benefits of our repairing technique using real-world databases.

2 Related Work

Our work lies in the intersection of two active lines of research: entity resolution and database repair. Below we discuss the related works from these two areas.

Entity resolution (ER) has long been central to the studies of data quality and integration, and has many important applications [5,8,15]. Previous research has studied various aspects of the ER process, such as blocking, similarity comparison, classification, evaluation and training data [5,9,14]. Recently, several works incorporated constraints into similarity-based ER techniques to improve the quality of ER [2,10–12]. Nonetheless, these works focused on preventing errors in ER results, rather than repairing ER results that contain errors.

Database repair has been extensively studied in the recent years [1,6,16], which was concerned with data consistency and accuracy. Existing approaches mostly deal with data errors either by obtaining consistent answers to queries over an inconsistent database [1] or by developing automated methods to find a repair that satisfies required constraints and also "minimally" differs from the original database [6]. These approaches are often computationally expensive and not applicable to repairing ER, particularly when ER is used in applications where the processing time is critical.

To date, few work has been reported on the entity repairing problem [13]. In this paper we show how to use a provenance index structure to incrementally and automatically repair errors in ER results. We consider not only matches but also non-matches in the ER process. Our work not only generalizes the clustering algorithm in [2] for supporting efficient clustering in the presence of both matches and non-matches, but also improve the work in [13] by automating the repairing process. In [13], the authors only considered ER rules that generate matches, and their work on ER repairing requires the manual review of suspicious matches by domain experts.

3 Problem Statement

Let $\mathbf{R}$ be a database that contains a set of relations, each having a finite set of records. An *entity relation* $R^* \in \mathbf{R}$ has three attributes, two of which refer to records in $\mathbf{R} - \{R^*\}$ and the third refers to values in $[-1, 1]$. That is, $(r_1, r_2, a) \in R^*$ indicates a *match* (r_1, r_2) with the confidence value $|a|$ if $a > 0$, or a *non-match* (r_1, r_2) with the confidence value $|a|$ if $a < 0$. A match or non-match is *hard* if it has $|a| = 1$, and *soft* otherwise.

Accordingly, there exist two kinds of ER rules: match rules and non-match rules. A *match rule* (resp. *non-match rule*) is a function that, given $\mathbf{R}$ as input, generates a set of matches (resp. non-matches). Any ER algorithm that generates matches (resp. non-matches) may be considered as a match (resp. non-match) rule in our work, including pairwise similarity and machine-learning algorithms [5,7]. Each ER rule associates with a weight, i.e., for a match rule u, $w(u) \in (0,1]$; for a non-match rule u, $w(u) \in [-1,0)$. In practice, not every ER rule is equally important. The more important a rule is, the higher its weight value is. Similarly, we call a rule u with $|w(u)| = 1$ as a *hard rule*. We use $u(r_i, r_j)$ to indicate that an ER rule u generates a match or non-match (r_i, r_j).

Given a set U of ER rules, applying U over $\mathbf{R}$ generates matches and non-matches in R^*. For each pair (r_i, r_j) of records, we use $U_{ij}^+ = \{u|u(r_i, r_j) \wedge w(u) > 0\}$ and $U_{ij}^- = \{u|u(r_i, r_j) \wedge w(u) < 0\}$ to denote the set of match or non-match rules that generate (r_i, r_j), respectively. With a slight abuse of notations, we will use $|\cdot|$ to denote the absolute value with lower-case letters and the cardinality of a set with upper-case letters. If there exists a hard rule $u \in U$ that generates (r_i, r_j), then we have $(r_i, r_j, w(u)) \in R^*$ with $|w(u)| = 1$. Otherwise, we have $(r_i, r_j, a) \in R^*$ where the confidence value a is determined by:

$$|a| = max \left(\sum_{u \in U_{ij}^+} \frac{|w(u)|}{|U_{ij}^+|}, \sum_{u \in U_{ij}^-} \frac{|w(u)|}{|U_{ij}^-|} \right). \tag{1}$$

We have $a > 0$ if $\sum_{u \in U_{ij}^+} \frac{|w(u)|}{|U_{ij}^+|} \geq \sum_{u \in U_{ij}^-} \frac{|w(u)|}{|U_{ij}^-|}$, and $a < 0$ otherwise.

An *entity cluster* c is a set of records in $\mathbf{R}$. An *ER clustering* C is a set $\{c_1, \ldots, c_k\}$ of entity clusters which are pairwise disjoint. We use $C(r)$ to denote the cluster of C to which a record r belongs. A clustering C is *valid* if, (i) for each hard match $(r_i, r_j, 1) \in R^*$, $C(r_i) = C(r_j)$, and (ii) for each hard non-match $(r_i, r_j, -1) \in R^*$, $C(r_i) \neq C(r_j)$.

Users can provide feedback on erroneous matches or non-matches, and accordingly user feedback may be viewed as hard matches or hard non-matches. In this paper, we address two related problems: (1) *entity clustering* and (2) *entity repairing*. The *entity clustering* problem is, given a set R^* of matches and non-matches, to find a valid clustering w.r.t. R^*. The *entity repairing* problem is, given a valid clustering C w.r.t. R^* and a collection R_f of user feedback, to find a valid clustering C' w.r.t. $R^* \cup R_f$. We say C' *repairs* C w.r.t. R_f.

We assume no conflicts among hard matching decisions because such conflicts have to be manually resolved by a domain expert. Our focus here is to automatically resolve conflicts among soft matches and soft non-matches.

4   Clustering Algorithms

We first present clustering graphs to describe how matches and non-matches are related in a graphical structure. Then we develop an efficient clustering algorithm, which generalizes the work in [2], to cluster records under consistency.

4.1   Clustering Graphs

Conceptually, each clustering graph is a graph with labelled edges. Let $L = L_P \cup L_N$ be a set of labels consisting of a subset L_P of *positive labels* and a subset L_N of *negative labels*, and $L_P \cap L_N = \emptyset$. A *clustering graph* $G = (V, E, \lambda)$ over L has a set V of vertices where each vertex represents a record, a set $E \subseteq V \times V$ of edges where each edge represents a match or non-match, and a labelling function λ that assigns each edge $e \in E$ with a label $\lambda(e) \in L$. For the labels in L, we have $L_P \supseteq (0, 1]$ and $L_N \supseteq [-1, 0)$ which correspond to the confidence values of matches and non-matches represented by edges, respectively. For convenience, we use $E^{(1)} = \{e | e \in E \text{ and } \lambda(e) = 1\}$ and $E^{(-1)} = \{e | e \in E \text{ and } \lambda(e) = -1\}$ to refer to the subsets of edges in a clustering graph G which represent hard matches and non-matches.

A clustering graph $G = (V, E, \lambda)$ is *consistent* if it corresponds to an ER clustering C_G over V such that for each edge $(r_i, r_j) \in E$, if $\lambda(r_i, r_j) \in L_P$, then r_i and r_j are in the same entity cluster; otherwise (i.e., $\lambda(r_i, r_j) \in L_N$) r_i and r_j are in two different entity clusters. Such an ER clustering C_G is *valid* w.r.t. R^* if every hard match (resp. non-match) in R^* is represented by an edge with a positive (resp. negative) label in G.

Given a clustering graph G that represents matches and non-matches in R^*, two questions naturally arise: (1) Can we efficiently decide whether G is consistent? (2) If G is not consistent, can we efficiently generate a consistent clustering graph from G to provide a valid ER clustering w.r.t. R^*? To answer these questions, we will discuss two clustering algorithms. The first algorithm was introduced in [2] which checks the consistency of clustering through triangles and, as proven by the authors, is a 3-factor approximation algorithm for the correlation clustering problem [3]. The second algorithm is proposed by us, which generalizes the first one to improve the efficiency by checking through cycles.

4.2   Triangle-Based Clustering (TriC)

The central idea of TriC is to eliminate occurrences of inconsistent triangles (as depicted in Fig. 2(a)) in a clustering graph. Thus, two transitive closure (TC) rules are used in the clustering process:

– **TC1**: If (r_1, r_2) and (r_1, r_3) are two hard matches, then (r_2, r_3) must also be a hard match, as depicted in Fig. 2(b).

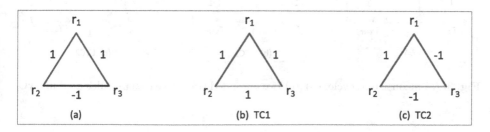

Fig. 2. (a) inconsistent triangle, (b)–(c) two transitive closure rules used in TriC

– **TC2**: If (r_1, r_2) is a hard match and (r_1, r_3) is a hard non-match, then (r_2, r_3) must be a hard non-match, as depicted in Fig. 2(c).

Algorithm 1. *Triangle-based clustering (TriC)*

Input: A clustering graph $G_1 = (V, E_1, \lambda_1)$
Output: A clustering graph $G_2 = (V, E_2, \lambda_2)$ that is consistent and only labelled by $\{1, -1\}$
1:   $G_2 := G_1$ // Initialize the clustering graph G_2
2:   Apply TC1 and TC2 on all possible edges in G_2 (i.e., edges in $V \times V$)
3:   Determine a queue **Q** of soft edges $E_1 - (E_2^{(1)} \cup E_2^{(-1)})$
4:   Do the following until **Q** $= \emptyset$:
5:       $e := $ **Q**.pop()   // Get the first soft edge from the queue
6:       **if** $\lambda_1(e) \in L_P$, **then:**   // Harden the edge according to its label
7:           $\lambda_2(e) := 1$
8:       **else:**
9:           $\lambda_2(e) := -1$
10:      Apply TC1 and TC2 all possible edges in G_2 (i.e., edges in $V \times V$)
11:      Remove the hard edges in $E_2^{(1)} \cup E_2^{(-1)}$ from **Q**
12: Return G_2

Algorithm 1 describes the main steps of TriC. Firstly, the TC1 and TC2 rules are applied to harden all possible edges over V, including the edges that do not exist in the initial clustering graph G_2 (line 2). Then, a queue **Q** of soft edges is determined (line 3), which reflects the relative importance of hardening soft edges according to their labels (i.e., hardened as a match if the label is positive, or as a non-match if the label is negative). Different strategies may be used to determine such a queue. We will discuss and evaluate two of such strategies in Sect. 6. Each time when an edge is hardened, the TC1 and TC2 rules are applied again to further harden other edges if any (lines 5–10). In each iteration, newly hardened edges are removed from the queue (line 11). The iterations continue until the queue **Q** is empty. Then the algorithm returns the clustering graph G_2 as output (line 12).

As TriC ensures the consistency of clustering through triangles, it unavoidably leads to a clique over V after the clustering (i.e., for any $v_1, v_2 \in V$ and $v_1 \neq v_2$, we have $(v_1, v_2) \in E_2$ in G_2). The complexity of the algorithm is $O(|V|^3)$ in the worst case (i.e., it needs $\frac{|V|!}{2 \times (|V|-3)!}$ steps of checking edges), which is computationally expensive when clustering over large databases.

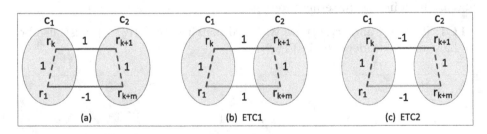

Fig. 3. (a) inconsistent cycle, (b)–(c) two extended transitive closure rules used in CyC

Algorithm 2. *Cycle-based clustering (CyC)*

Input: A clustering graph $G_1 = (V, E_1, \lambda_1)$
Output: A clustering graph $G_2 = (V, E_2, \lambda_2)$ that is consistent and only labelled by $\{1, -1\}$
1: $G_2 := G_1$ // Initialize the clustering graph G_2
2: Apply ETC1 and ETC2 on edges in E_2
3: Determine a queue $\mathbf{Q}$ of soft edges $E_1 - (E_2^{(1)} \cup E_2^{(-1)})$
4: Do the follow until $\mathbf{Q} = \emptyset$:
5:      $e := \mathbf{Q}.\text{pop}()$ // Get the first soft edge from the queue
6:      **if** $\lambda_1(e) \in L_P$, **then:**   // Harden the edge according to its label
7:          $\lambda_2(e) := 1$
8:      **else:**
9:          $\lambda_2(e) := -1$
10:     Apply ETC1 and ETC2 on edges in E_2
11:     Remove the hard edges $E_2^{(1)} \cup E_2^{(-1)}$ from $\mathbf{Q}$
12: Return G_2

4.3   Cycle-Based Clustering (CyC)

To improve the efficiency of clustering, we now propose an optimized algorithm (i.e. CyC) which ensures the consistency based on cycles, instead of triangles. In a nutshell, CyC generalizes TriC by eliminating inconsistent cycles, as depicted in Fig. 3(a). This is motivated by the observation that, for entity clustering, it is impossible to have a cycle $(r_1, \ldots, r_k, r_{k+1}, \ldots, r_{k+m}, r_1)$ in which all edges are hard matches except for one edge (r_1, r_{k+m}) that is a hard non-match. Thus, to eliminate such inconsistent cycles, two extended transitive closure rules are used in the clustering process:

- **ETC1**: If each (r_i, r_{i+1}) for $i \in [1, \ldots, k + m - 1]$ is a hard match, then (r_1, r_{k+m}) must also be a hard match, as depicted in Fig. 3(b).
- **ETC2**: If (r_k, r_{k+1}) is a hard non-match, and other edges from r_1 to r_{k+m} are hard matches, then (r_1, r_{k+m}) must be a hard non-match, as depicted in Fig. 3(c).

Algorithm 2 describes the main steps of CyC. Similar to TriC, the rules ETC1 and ETC2 are first applied to harden soft edges in G_2 (line 2). After that, a queue $\mathbf{Q}$ of soft edges is determined, and each of these soft edges is hardened according to their orders (lines 3–9). Again the queue reflects the relative importance of hardening different soft edges. Next, CyC applies the ETC1 and ETC2 rules to ensure the consistency of clustering (line 11). Newly hardened edges in each iteration are removed from the queue (line 12). The iterations continue until the queue is empty. Finally, the resulting graph G_2 is returned (line 13).

CyC applies the rules only on the existing edges of a clustering graph G, which has the complexity $O(|E| \times |V|)$ in the worst case. It is more efficient than TriC because $|E|$ is often much smaller than $|V|^2$ in a clustering graph. In practice, we may simply merge records that are connected by hard matches and treat each of such merged records as a single record.

5   Repairing Algorithms

In this section, we discuss an efficient provenance index structure and repairing algorithms for entities.

5.1   Provenance Index

In our work, a provenance index is constituted by a number of entity trees. Each entity tree describes how the records of an entity are identified from a number of matches. We use $leaf(t)$ to refer to the set of all leaves of an entity tree t, $root(t)$ to the root vertex of t, $parent(v)$ to the parent vertex of a vertex v, $subtree(v)$ to the subtree rooted at v, and $subtree(v/v_1, t_1)$ to a subtree that replaces $subtree(v_1)$ by t_1 in $subtree(v)$, where v_1 is a child vertex of v. An *entity tree* t is a binary tree with a labelling function ℓ such that (1) each leaf represents a record, (2) each internal vertex represents a match, and (3) each edge $(parent(v), v)$ is labelled by a record that participates in the match represented by $parent(v)$ such that $\ell(parent(v), v) \in leaf(t')$ for $t' = subtree(v)$.

Entity trees are constructed from bottom up during the clustering process. In terms of Algorithm 2, hard matches are first added into an entity tree and represented by vertices that are close to leaves, and these vertices must be under vertices for soft matches (line 2 of Algorithm 2). Then each time when a soft edge (r_1, r_2) is hardened as a match (i.e., line 5 of Algorithm 2), a vertex v representing that match is added as the root of the corresponding entity tree t if $\{r_1, r_2\} \nsubseteq leaf(t)$. Otherwise, we discard the match because this match has already been implicitly captured in the provenance index. Thus, entity trees enjoy two nice properties to support efficient repairing. (1) *Distinct leaves*: Records represented by leaves of an entity tree are distinct. This provides an efficient index for finding matches on records. (2) *Ordered vertices*: Internal vertices are ordered to reflect the importance of matches, which is determined by the queue **Q** in Algorithm 2, e.g., the closer a vertex is to a leaf (in terms of the number of edges between them), the more important the match represented by it is.

5.2   Entity Tree Splitting Algorithm

Algorithm 3 describes the proposed splitting algorithm for repairing erroneous matches. Given an erroneous match (r_1, r_2) detected by a user, according to the property of ordered vertices, the algorithm first finds the lowest common parent vertex v of r_1 and r_2 (line 1). Then the child vertices of v are split into two subtrees (lines 2–3). After that, to propagate the effect of splitting two child vertices of v, each vertex in the path from v to the root is checked. Depending on the edge label, the vertex is added to the top of one of the two subtrees (lines 6–9). The complexity of Algorithm 3 is $O(|V|)$ which is linear in the number $|V|$ of vertices in an entity tree t.

Algorithm 3. *Entity Tree Splitting Algorithm*

Input: An entity tree t, an erroneous match (r_1, r_2) with $r_1, r_2 \in leaf(t)$
Output: Two entity trees t_1 and t_2 with $r_1 \in leaf(t_1)$ and $r_2 \in leaf(t_2)$
1: Find the lowest common parent vertex v of r_1 and r_2 in t
2: $t_1 := subtree(v_1)$ where v_1 is the left child vertex of v
3: $t_2 := subtree(v_2)$ where v_2 is the right child vertex of v
4: $v' := v$
5: Do the following for each vertex v' in the path from v to $root(t)$:
6:    **if** $\ell(parent(v'), v') \in leaf(t_1)$, **then:**
7:        $t_1 := subtree(parent(v')/v', t_1)$
8:    **else:**
9:        $t_2 := subtree(parent(v')/v', t_2)$
10:   $v' := parent(v')$
11: Return t_1 and t_2

Algorithm 4. *Entity Tree Merging Algorithm*

Input: Two entity trees t_1 and t_2, an erroneous non-match (r_1, r_2) with $r_i \in leaf(t_i)$ for i=1,2, and
    a set of hard non-matches between t_1 and t_2 in R^*
Output: A set T of entity trees which includes an entity tree $t \in T$ and $r_1, r_2 \in leaf(t)$
1: $T := \{t\}$ where t is created with one vertex that has two child vertices r_1 and r_2
2: $v_i^1 := r_i$, $v_i := parent(r_i)$, $v_i^2 := otherchild(v_i, v_i^1)$ and $len_i := 1$ for i=1,2
3: Do the following until $len_1 = 0$ and $len_2 = 0$:
4:    **if** $len_1 \leq len_2$
5:        $k := 1$
6:    **else:**
7:        $k := 2$
8:    Find the tree $t' \in T$ satisfying $\ell(v_k, v_k^1) \in leaf(t')$
9:    **if** there exists $(r_i, r_j, -1) \in R^*$ with $r_i, r_j \in leaf(t') \cup leaf(subtree(v_k^2))$
10:       $T := T \cup \{subtree(v_k^2)\}$
11:   **else:**
12:       $t' := subtree(v_k/v_k^1, t')$
13:   **if** $parent(v_k)$ exists
14:       $v_k^1 := v_k$, $v_k := parent(v_k)$, $v_k^2 := otherchild(v_k, v_k^1)$ and $len_k := len_k + 1$
15:   **else:**
16:       $len_k := 0$
17: Return T

5.3   Entity Tree Merging Algorithm

Algorithm 4 describes the proposed merging algorithm for repairing erroneous non-matches, in which we use $otherchild(v, v_i)$ to denote the other child vertex of v in addition to v_i. Given an erroneous non-match (r_1, r_2) detected by a user, the algorithm first creates an entity tree t with one internal vertex representing (r_1, r_2) (line 1). Then for each vertex on the paths from the leaf r_1 (resp. r_2) to $root(t_1)$ (resp. $root(t_2)$) (lines 3–16), if there is no violation of hard non-matches, the vertex is added into an entity tree in T based on its closeness to r_1 or r_2 in t_1 and t_2 (line 12); otherwise, a new entity tree is created for the subtree at the other child vertex of the vertex (line 11). The complexity of this algorithm is $O(|V_1| + |V_2|)$ which is linear in the number of vertices in t_1 and t_2.

6   Experiments

We have implemented our framework to evaluate how efficiently our repairing algorithms can improve the quality of ER based on the provenance indexing.

6.1   Experimental Setup

Our implementation is written in Python. The experiments were performed on a Linux machine with Intel Core i7-3770 CPU at 3.40 GHz, and 8 GBytes RAM.

Data Sets. We used two real-world data sets in our experiments: CORA and NCVR. The CORA data set contains 1,878 machine learning publications and is publicly available together with its ground truth[1]. The NCVR data set is a public voter registration data set from North Carolina[2], which contains the names and addresses of over 8 million voters. Each record includes a voter registration number, providing us with the true match status of record pairs. The number of true duplicate records in the full NCVR data set is below 2 % of all records. We therefore extracted a smaller subset containing 448,134 records that includes all duplicates as well as randomly selected individuals (i.e. singleton clusters). Table 1 presents some characteristics of these two data sets.

ER Rules. For these data sets, we used the ER rules as described in Table 2 to obtain an initial ER clustering result, which nonetheless only serves as a baseline for evaluating our ER repair algorithms. The better quality an initial clustering has, the less repairs we would need for achieving a high-quality clustering. Thus, we used simple ER rules that can be easily setup by domain experts.

To our knowledge, this work is the first report on developing automatic techniques for repairing errors in ER. There are no baseline techniques which we can compare our repairing approach with.

Table 1. Characteristics of data sets used in our experiments

Data set names	Number of records	Number of unique weight vectors	Number of clusters	Min. size of clusters	Max. size of clusters	Avg. size of clusters
CORA	1,878	77,004	185	1	239	10.15
NCVR	448,134	1,556,184	296,431	1	6	1.51

6.2   Quality of Repairing

In our experiments, we used two different strategies to determine the queue in Algorithm 2: (a) *Random*: i.e., randomly selecting soft edges from a clustering graph and adding them to the queue; (b) *Weight*: i.e., adding soft edges to the queue in terms of their weights in descending order. We simulated user feedback by randomly selecting erroneous matches or non-matches in accordance with the clustering result after each repair, and used six metrics to evaluate the quality of ER: (1) *Precision*, *Recall* and *Fmeasure*, which are based on pairs of records

---

[1] Available from: http://www.cs.umass.edu/~mccallum/.
[2] Available from: ftp://alt.ncsbe.gov/data/.

Table 2. ER rules used in our experiments, where $sim(A)$ indicates the similarity of values in the attribute A of two records.

Data set	Rule	Rule description	Weight
CORA	u_1	$sim(title) > 0.6 \wedge sim(author) > 0.3 \wedge sim(date) > 0.3 \wedge$ $sim(name) > 0.1$	1
	u_2	$sim(author) > 0.5 \wedge sim(name) > 0.5$	0.36
	u_3	$sim(title) > 0.7 \wedge sim(author) > 0.7$	0.74
	u_4	$sim(title) > 0.1 \wedge sim(name) > 0.9 \wedge sim(vol) > 0.7$	0.9
	u_5	$sim(title) < 0.4 \wedge sim(author) < 0.3$	-1
NCVR	u_6	$sim(gender) = 1 \wedge sim(first\_name) \geq 0.8 \wedge$ $sim(last\_name) \geq 0.8 \wedge sim(age) \geq 0.7 \wedge$ $sim(phone\_number) = 1$	1
	u_7	$sim(first\_name) \geq 0.6 \wedge sim(last\_name) \geq 0.6 \wedge$ $sim(age) \geq 0.5$	0.8
	u_8	$sim(first\_name) \geq 0.7 \wedge sim(zip\_code) \geq 0.7 \wedge$ $sim(phone\_number) \geq 0.7$	0.7
	u_9	$sim(first\_name) \geq 0.5 \wedge sim(last\_name) \geq 0.5 \wedge$ $sim(zip\_code) \geq 0.8$	0.8
	u_{10}	$sim(gender) < 1 \wedge sim(phone\_number) < 1 \wedge$ $sim(first\_name) < 0.2 \wedge sim(last\_name) \leq 0.2$	-1

as in the traditional process [5], and (2) *Closest Cluster Precision, Recall* and *Fmeasure* (written as *CC-Precision*, *CC-Recall* and *CC-Fmeasure*, respectively), which compare clusterings by counting completely correct clusters [4].

In Figs. 4 and 5, the *Precision* and *Fmeasure* values are improved over both data sets, although the *Recall* values for NCVR are dropping slightly before going up. The *CC-Precision*, *CC-Recall* and *CC-Fmeasure* values also increase with increasing repairs for both data sets, and particularly repairing over time can lead to ER clustering results that are close to the ground truth. Compared with NCVR, CORA has clusters of larger sizes and its initial clustering result contains more conflicts. Thus, the difference between the random-based and weight-based strategies in CORA is also larger than in NCVR.

6.3 Efficiency of Repairing

We have also evaluated the efficiency of repairing over NCVR since it contains nearly half a million records. We measured the memory usage and time resources required during the repairing process. As depicted in Fig. 6, the weighted-based strategy is less time efficient than the random-based strategy because it requires time to sort edges based on their weights. Nonetheless, compared with the random-based strategy, the weighted-based strategy has more uneven but generally less memory usage requirement.

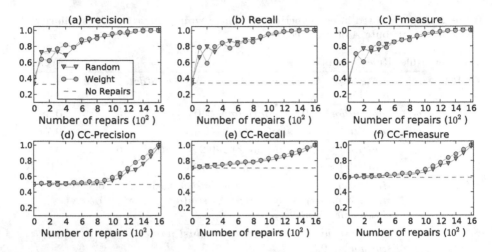

Fig. 4. Quality of the clustering over CORA after repairing errors

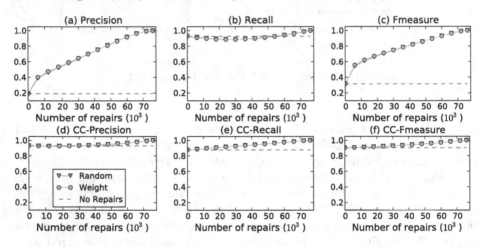

Fig. 5. Quality of the clustering over NCVR after repairing errors

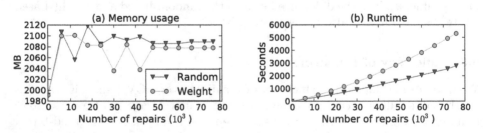

Fig. 6. Memory usage and time of repairing over NCVR

7 Conclusion

We have developed a framework for incrementally repairing errors existing in an ER clustering. During the clustering process, we establish a provenance index for capturing how records are clustered into each entity. This index provides useful information which enables us to efficiently repair errors when they are detected later on. In the future, we plan to continue this line of research by looking into how the provenance index structure can be generalized to support repairing for entity resolution that collectively resolves entities of different types.

References

1. Afrati, F.N., Kolaitis, P.G.: Repair checking in inconsistent databases: algorithms and complexity. In: ICDT, pp. 31–41 (2009)
2. Arasu, A., Ré, C., Suciu, D.: Large-scale deduplication with constraints using dedupalog. In: ICDE, pp. 952–963 (2009)
3. Bansal, N., Blum, A., Chawla, S.: Correlation clustering. Mach. Learn. 56(1–3), 89–113 (2004)
4. Barnes, M.: A practioner's guide to evaluating entity resolution results (2014)
5. Christen, P.: Data Matching: Concepts and Techniques for Record Linkage, Entity Resolution, and Duplicate Detection. Springer, Heidelberg (2012)
6. Cong, G., Fan, W., Geerts, F., Jia, X., Ma, S.: Improving data quality: consistency and accuracy. In: PVLDB, pp. 315–326 (2007)
7. Elmagarmid, A.K., Ipeirotis, P.G., Verykios, V.S.: Duplicate record detection: a survey. IEEE TKDE 19, 1–16 (2007)
8. Fellegi, I., Sunter, A.: A theory for record linkage. J. Am. Stat. Assoc. 64(328), 1183–1210 (1969)
9. Fisher, J., Christen, P., Wang, Q., Rahm, E.: A clustering-based framework to control block sizes for entity resolution. In: KDD, pp. 279–288 (2015)
10. Schewe, K.-D., Wang, Q.: A theoretical framework for knowledge-based entity resolution. TCS 549, 101–126 (2014)
11. Shen, W., Li, X., Doan, A.: Constraint-based entity matching. In: AAAI, pp. 862–867 (2005)
12. Shen, Z., Wang, Q.: Entity resolution with weighted constraints. In: Manolopoulos, Y., Trajcevski, G., Kon-Popovska, M. (eds.) ADBIS 2014. LNCS, vol. 8716, pp. 308–322. Springer, Heidelberg (2014)
13. Wang, Q., Schewe, K.-D., Wang, W.: Provenance-aware entity resolution: leveraging provenance to improve quality. In: Renz, M., Shahabi, C., Zhou, X., Cheema, M.A. (eds.) DASFAA 2015. LNCS, vol. 9049, pp. 474–490. Springer, Heidelberg (2015)
14. Wang, Q., Vatsalan, D., Christen, P.: Efficient interactive training selection for large-scale entity resolution. In: Cao, T., Lim, E.-P., Zhou, Z.-H., Ho, T.-B., Cheung, D., Motoda, H. (eds.) PAKDD 2015. LNCS, vol. 9078, pp. 562–573. Springer, Heidelberg (2015)
15. Whang, S.E., Garcia-Molina, H.: Entity resolution with evolving rules. VLDB 3(1–2), 1326–1337 (2010)
16. Wijsen, J.: Database repairing using updates. TODS 30(3), 722–768 (2005)

Adaptive Seeding for Gaussian Mixture Models

Johannes Blömer and Kathrin Bujna[(⊠)]

Paderborn University, 33098 Paderborn, Germany
{bloemer,kathrin.bujna}@uni-paderborn.de

Abstract. We present new initialization methods for the expectation-maximization algorithm for multivariate Gaussian mixture models. Our methods are adaptions of the well-known K-means++ initialization and the Gonzalez algorithm. Thereby we aim to close the gap between simple random, e.g. uniform, and complex methods, that crucially depend on the right choice of hyperparameters. Our extensive experiments indicate the usefulness of our methods compared to common techniques and methods, which e.g. apply the original K-means++ and Gonzalez directly, with respect to artificial as well as real-world data sets.

1 Introduction

Gaussian mixture modelling is an important task, e.g., in the field of cluster analysis. A common approach is the method of maximum likelihood for which the Expectation-Maximization (EM) algorithm [13] can be applied. The EM algorithm iteratively tries to improve a given initial mixture model and converges to a stationary point of the likelihood function. Unfortunately, the likelihood function is generally non-convex, possessing many stationary points [23]. The initial model determines to which of these points the EM algorithm converges [5].

1.1 Maximum Likelihood Estimation for Gaussian Mixtures

A Gaussian mixture model (K-GMM) over $\mathbb{R}^D$ can be described by a parameter $\theta = \{(w_k, \mu_k, \Sigma_k)\}_{k=1,\ldots,K}$, where $w_k \in \mathbb{R}$ is the mixing weight ($\sum_{k=1}^{K} w_k = 1$), $\mu_k \in \mathbb{R}^D$ is the mean, and $\Sigma_k \in \mathbb{R}^{D \times D}$ is the covariance matrix of the k-th mixture component. Its probability density function is given by $\mathcal{N}(x|\theta) = \sum_{k=1}^{K} w_k \mathcal{N}(x|\mu_k, \Sigma_k)$, where we denote the D-variate Gaussian distribution by $\mathcal{N}(\cdot|\mu, \Sigma)$. Given a data set $X \subset \mathbb{R}^D$, the Maximum Likelihood Estimation (MLE) problem is to find a K-GMM θ that maximizes the likelihood $\mathcal{L}(\theta|X) = \prod_{x \in X} \mathcal{N}(x|\theta)$. For $K = 1$, there is a closed-form solution [8]. For $K > 1$, the EM algorithm, whose outcome heavily depends on the initial model, can be applied.

1.2 Related Work

A common way to initialize the EM algorithm is to first draw means uniformly at random from the input set and then to approximate covariances and weights

© Springer International Publishing Switzerland 2016
J. Bailey et al. (Eds.): PAKDD 2016, Part II, LNAI 9652, pp. 296–308, 2016.
DOI: 10.1007/978-3-319-31750-2_24

[7,21,24,25]. To compensate for the random choice of initial means, several candidate solutions are created and the one with the largest likelihood is chosen. Often, few steps of the EM, Classification EM, or Stochastic EM algorithm are applied to the candidates. Similarly, the K-means algorithm may be used [8, p. 427]. Due to the random choice of the initial means, all these methods are better suited for spherical and well-separated clusters. Furthermore, testing several candidates is computationally expensive.

Other popular initializations are based on hierarchical agglomerative clustering (HAC). For instance, in [21,24,25], HAC (with different distance measures) is used to obtain mean vectors. Since HAC is generally very slow, it is usually only executed on a random sample [24]. However, the size of any reasonable sample depends on the size of the smallest optimal component. Moreover, it is often outperformed by other methods (e.g. [24,25]). Another approach using HAC is presented in [21]. It aims at finding the best local modes of the data set in a reduced m^*-dimensional space and applies HAC only on these modes. However, this method is time-consuming and the choice of m^* is crucial [21, p. 5, 13]. Moreover, in [22] it is outperformed by simple random methods.

[25] presents a density based approach which not only determines an initial solution but also the number of components. It initializes the means by points which have a "high concentration" of neighbors. To this end, the size m of the neighborhood of a point (i.e., the minimum number of points in a cluster) has to be fixed in advance. In our experiments, we found that the performance crucially depends on the choice of m. Hence, we ignore this method in this paper.

In [27], a greedy algorithm is presented which constructs a sequence of mixture models with 1 through K components. Given a model θ_k with k components, it constructs several new candidates with $k + 1$ components. Each candidate is constructed by adding a new component to θ_k and executing the EM algorithm. Hence, this method is only useful if several values of K need to be considered.

In [20] a modification of the Gonzalez algorithm for GMMs is presented. Furthermore, there are some practical applications using the K-means++ algorithm for the initialization of GMMs (e.g., in [19] GMMs are used for speech recognition). Additional initialization methods can be found e.g. in [6,15,21,26].

1.3   Our Contribution

Clearly, there is no way to determine the *best* initialization algorithm that outperforms all other algorithms on all instances. The performance of an initialization depends on the given data and the allowed computational cost. Nonetheless, the initializations presented so far (except the simple random initializations) face mainly two problems: Firstly, they are rather complex and time consuming. Secondly, the choice of hyperparameters is crucial for the outcome.

In this paper, we present new methods that are fast and do not require choosing sensitive hyperparameters. These methods can be seen as adaptions of the K-means++ algorithm [3] and the Gonzalez algorithm [17] and as an extension of the initial work in [19,20]. We present experiments indicating the superiority of our methods compared to a large number of alternative methods.

2   Baseline Algorithms

The most widely used initializations start by choosing K data points:

Unif draws K points independently and uniformly at random from X.

HAC computes a uniform sample S of size $s \cdot |X|$ of the input set X and executes hierarchical clustering with average linkage cost on S.

G executes the algorithm given in [17], which yields a 2-approximation for the discrete radius K-clustering problem. Iteratively, it chooses the point with the largest Euclidean distance from the already chosen points.

KM++ executes the K-Means++ algorithm [3], which has been designed for the K-means problem. In each round, KM++ samples a data point (i.e. the next mean) from the given data set X with probability proportional to its K-means cost (with respect to the points chosen so far). In expectation, the resulting K-means costs are in $\mathcal{O}(\log(K) \cdot \mathrm{opt})$. KM++ is particularly interesting since the K-means algorithm is a special case of the EM algorithm [8].

Then, given K data points, Algorithm 1 is used to create a GMM, which is then the initial solution that is fed to the EM algorithm.

Algorithm 1. Means2GMM($X \subset \mathbb{R}^D$, $\mathcal{C} \subset \mathbb{R}^D$, $|\mathcal{C}| = k$)

1: Derive a partition $\{C_1, \ldots, C_k\}$ by assigning each $x \in X$ to a closest point in $\mathcal{C}$.
2: **for** $l = 1, \ldots, k$ **do**
3:     Set $\mu_l := 1/|C_l| \sum_{x \in C_l} x$, $w_l := |C_l|/|X|$, $\Sigma_l := 1/|C_l| \sum_{x \in C_l} (x - \mu_l)(x - \mu_l)^T$.
4:     If Σ_l is not positive definite, set $\Sigma_l := 1/(D \cdot |C_l|) \sum_{x \in C_l} \|x - \mu_l\|^2 \cdot I_D$
5:     If Σ_l is still not positive definite, set $\Sigma_l := I_D$
6: **return** $\theta = \{(w_l, \mu_l, \Sigma_l)\}_{l=1,\ldots,k}$.

A popular alternative is to apply the K-means algorithm with the chosen data points before executing Algorithm 1. The main idea behind this is that starting the EM algorithm with a coarse initial solution (where e.g. not all clusters are covered well) might impose a high risk of getting stuck at a poor local minimum. To avoid this problem, one first runs a different algorithm that optimizes a function similar to the likelihood, i.e. the K-means costs (cf. [8, p. 427, 443]). We refer to the K-means algorithm as an *intermediate* algorithm and indicate its use by the postfix "$_{km}$".

3   Adaptive Seeding for GMMs

Our new adaptive methods construct a sequence of models with $k = 1$ through $k = K$ components adaptively. Given a $(k-1)$-GMM θ_{k-1}, our methods try to choose a point from the data set that is not described well by the given θ_{k-1} and which is hopefully a good representative of a component of an optimal k-GMM. The idea behind is that this point can lead us to a significant refinement of θ_{k-1}.

Choosing a Point. The negative log-likelihood of a point $x \in \mathbb{R}^D$, given the GMM θ_{k-1}, measures how well x is described by θ_{k-1}.[1] Unfortunately, it may take negative values and does not scale with the data set and the GMM[2]. This also applies to the minimum component-wise negative log-likelihood

$$\min\left\{-\log\left((2\pi)^{D/2}|\Sigma_l|^{1/2}\right) + \tfrac{1}{2}(x-\mu_l)^T\Sigma_l^{-1}(x-\mu_l) \mid (w_l,\mu_l,\Sigma_l) \in \theta_{k-1}\right\},$$

due to the first summand. Hence, we use the minimum Mahalanobis distance

$$m(x|\theta_{k-1}) := \min\left\{(x-\mu_l)^T\Sigma_l^{-1}(x-\mu_l) \mid (w_l,\mu_l,\Sigma_l) \in \theta_{k-1}\right\}.$$

Our first method chooses the point $x \in X$ maximizing $m(x|\theta_{k-1})$. Since these points are more likely to be outliers, we also consider choosing a point only from a uniform sample of X, which is chosen in advance (cf. Algorithm 2).

Our second method chooses point $x \in X$ with probability $\propto m(x|\theta_{k-1})$ (cf. Algorithm 3). In order to reduce the probability to choose an outlier, we also consider adding an α portion of uniform distribution, i.e. drawing x with probability

$$m_\alpha(x|\theta_{k-1}) := \alpha \cdot m(x|\theta_{k-1})/\sum_{y \in X} m(y|\theta_{k-1}) + (1-\alpha) \cdot 1/|X|.$$

Constructing a GMM. Then, given a point $x \in X$ and the means of θ_{k-1}, we construct a k-GMM. In our first experiments, we used Algorithm 1 to construct a k-GMM. However, it turned out that estimating only spherical covariance matrices (with variable variances) yields a better performance than estimating full covariance matrices. We assume that this is due to the fact that θ_{k-1} is only a very coarse estimate of $(k-1)$-components of an optimal k-GMM. Formally, we replace the covariance update in Line 3 of Algorithm 1 by $\Sigma_l = 1/(D \cdot |C_l|)\sum_{x \in C_l}\|x-\mu_l\|^2 \cdot I_D$. We denote this version of Algorithm 1 as **Means2SphGMM**. Given the resulting k-GMM, our methods then again choose a new point from X as already described above.

Intermediate Algorithm. Recall that some baselines use the K-means algorithm as an intermediate algorithm (cf. Sect. 2). Since we do not only construct means but GMMs, we apply a hard-clustering variant of the EM algorithm, i.e. the Classification EM algorithm (CEM) [10], and let it only estimate spherical covariances. We indicate its use by the postfix "$_{cem}$".

Algorithms 2 and 3 summarize our methods. Note that we do not optimize the hyperparameters α and s in our experiments.

Comparison to Baselines. Our adaptive initializations can be seen as adaptions of the **Gonzalez** and **Kmeans++** algorithm. Simply speaking, these methods assume that each component is represented by a Gaussian with *the same*

---

[1] The (inverse) pdf is unsuited due to the exponential behavior (over-/underflows).
[2] Even wrt. a single Gaussian, i.e. $\log\mathcal{N}(c \cdot x|c \cdot \mu, c^2 \cdot \Sigma) = \log\mathcal{N}(x|\mu,\Sigma) - D\ln(c)$.

fixed spherical covariance matrix and fixed uniform weights. In contrast, our goal is to estimate also the covariance matrices *adaptively*. Furthermore, in [20] another adaption of the Gonzalez algorithm is presented, which we denote by KwedlosGonzalez (KG). Unlike our method, it chooses weights and covariance matrices randomly and independently of the means (and of each other).

Algorithm 2. SphericalGonzalez (SG)

Require: $X \subset \mathbb{R}^D, K \in \mathbb{N}, s \in (0, 1]$
1: $\theta_1 :=$ optimal 1-MLE wrt. X
2: If $s < 1$, let S be a uniform sample of X of size $\lceil s \cdot |X| \rceil$. Otherwise, set $S = X$.
3: **for** $k = 2, \ldots, K$ **do**
4: $\quad p := \arg\max_{x \in S} m(x|\theta_{k-1})$
5: $\quad M_k := \{\mu|(\cdot, \mu, \cdot) \in \theta_{k-1}\} \cup \{p\}$
6: $\quad \theta_k :=$ Means2SphGMM(X, M_k)
7: (optional) Run CEM algorithm
8: **return** θ_K

Algorithm 3. Adaptive (Ad)

Require: $X \subset \mathbb{R}^D, K \in \mathbb{N}, \alpha \in [0, 1]$
1: $\theta_1 :=$ optimal 1-MLE wrt. X
2: **for** $k = 2, \ldots, K$ **do**
3: $\quad$ Draw p from X with probability $m_\alpha(p|\theta_{k-1})$.
4: $\quad M_k := \{\mu|(\cdot, \mu, \cdot) \in \theta_{k-1}\} \cup \{p\}$
5: $\quad \theta_k :=$ Means2SphGMM(X, M_k)
6: (optional) Run CEM algorithm
7: **return** θ_K

---

4 Experiments

We evaluated all presented methods with respect to artificial as well as real world data sets. Our implementation as well as the complete results are available at [9]. Due to space limitations, we omit the results of those algorithms that are consistently outperformed by others. These results are available at [9] as well.

Quality Measure. Recall that the goal of our paper (and the EM algorithm) is to find a maximum likelihood estimate (MLE). Thus, the *likelihood* is not only the common but also the appropriate way of evaluating our methods.

Other measures need to be treated with caution: Some authors consider their methods only with respect to *some specific tasks* where fitting a GMM to some data is part of some framework. Hence, any observed effects might be due to several reasons (i.e. correlations). In particular, GMMs are often compared with respect to certain *classifications*. As already pointed out by [14], the class labels of real world data sets do not necessarily correspond to the structure of an MLE. The same holds for data sets and classifications generated according to some GMM. Moreover, a *cross-validation*, that examines whether methods overfit models to training data, is not reasonable, since our methods do *not* aim at finding a model that does not fit too well to the given data set. Finally, one should not generate data sets according to some *"ground truth"* GMM θ_{gt} and compare GMMs with θ_{gt} because in many cases (e.g. small $|X|$) one cannot expect θ_{gt} to be a good surrogate of the MLE.

Setup. Recall that in Algorithms 2 and 3 hyperparameters α and s are used. We do not optimize them, but test reasonable values, i.e. $\alpha \in \{0.5, 1\}$ and

$s \in \{0.1, 1\}$. We execute each method with 30 different seeds. On the basis of some initial experiments, we decided to execute the intermediate algorithms for 25 rounds and the EM algorithm for 50 rounds. If only the EM algorithm is applied, then we execute it for 75 rounds.

4.1 Artificial Data Sets

Data Generation. We create 192 test sets, each containing 30 data sets that share certain characteristics [9]. For each data set, we first create a GMM[3] at random but control the following properties: First, the components of a GMM can either be spherical or elliptical. We describe the eccentricity of Σ_k by $e_k = \frac{\max_d \lambda_{kd}}{\min_d \lambda_{kd}}$, where λ_{kd}^2 denotes the d-th eigenvalue of Σ_k. Second, components can have different sizes, in terms of the smallest eigenvalue of the corresponding covariance matrices. Third, components have different weights.

Fourth, components can overlap more or less. Following [12], we define the separation parameter $c_\theta = \min_{l \neq k} \|\mu_l - \mu_k\| / \sqrt{\max\{\mathrm{trace}(\Sigma_l), \mathrm{trace}(\Sigma_k)\}}$. In high dimension $D \gg 1$, $c_\theta = 2$ corresponds to almost completely separated clusters (i.e. points generated by the same component), while $c_\theta \in \{0.5, 1\}$ indicates a slight but still negligible overlap [11]. However, in small dimension, $c_\theta \in \{0.5, 1\}$ corresponds to significant overlaps between clusters, while $c_\theta = 2$ implies rather separated clusters (cf. Fig. 1).

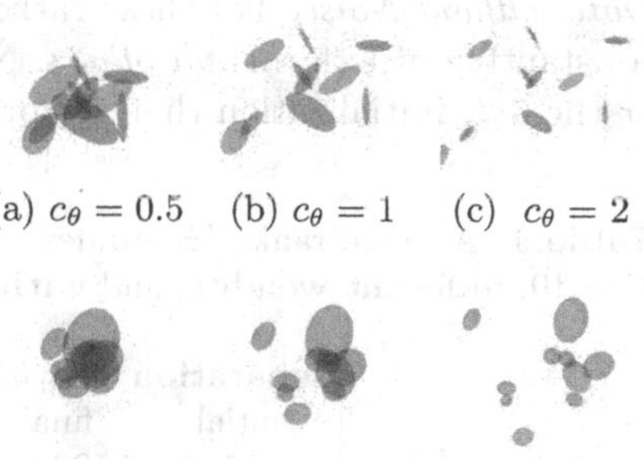

(a) $c_\theta = 0.5$ (b) $c_\theta = 1$ (c) $c_\theta = 2$

(d) $c_\theta = 0.5$ (e) $c_\theta = 1$ (f) $c_\theta = 2$.

Fig. 1. Examples for different separation parameters. Figures show orthogonal projections to random plane. Data sets in (a)–(c) have $D = 3$. (d)–(f) have $D = 10$.

We generate random GMMs as follows. Initially, we draw means uniformly at random from a cube with a fixed side length. For the weights, we fix some $c_w \geq 0$, construct a set of weights $\{2^{c_w \cdot i} / \sum_{j=1}^{K} 2^{c_w \cdot j}\}_{i=1,\ldots,K}$ and assign these weights randomly. To control the sizes and the eccentricity, we fix the minimum and maximum eigenvalue and draw the remaining values uniformly at random from the interval. Then, we set $\Sigma_k = Q^T \mathrm{diag}(\lambda_{k1}^2, \ldots, \lambda_{kD}^2) Q$ for a random $Q \in \mathrm{SO}(D)$. Finally, the means are scaled as to fit the predefined separation parameter. Given the resulting GMM θ, we first draw some points according to θ. Then, we construct a bounding box, elongate its side lengths by a factor 1.2, and draw noise points uniformly at random from the resized box.

We created a test set (i.e. 30 data sets) for each combination of the following parameters: $K = 20$, $|X| \in \{1000, 5000\}$, $D \in \{3, 10\}$, $c_\theta \in \{0.5, 1, 2\}$, $c_w \in \{0.1, 1\}$, different combinations of size and eccentricity (i.e., equal size and $e_k = 10$,

---

[3] As explained before, our goal is *not* to identify these GMMs.

equal size and $e_k \in [1, 10]$, different size and $e_k = 1$, different size and $e_k \in [1, 10]$), and without or with 10 % noise points.

Evaluation Method. We consider the *initial solutions* produced by the initialization (possibly followed by an intermediate algorithm) and the *final solutions* obtained by running the EM algorithm afterwards. For each data set, we compute the average log-likelihood of the initial and final solution, respectively. Based on these averages, we create rankings of the algorithms[4]. Then, we compute the average rank (and standard deviation of the rank) of each algorithm over all datasets matching certain properties.

Results. In general, we observe that one should use an intermediate algorithm before applying the EM algorithm. Thus, we omit the results of some methods [9].

Data without Noise. For these rather simple data sets, there is no method that constantly outperforms *all others*. Nonetheless, it is always one of our adaptive or the G_{km} initialization that performs best.

Table 1. Average ranks ($\pm$ std.dev.) for generated data with $K = 20$, $|X| = 1000$, $D = 10$, **different weights**, and **without noise**.

	separation $c_\theta = 0.5$		separation $c_\theta = 1$		separation $c_\theta = 2$	
	initial	final	initial	final	initial	final
$SG\left(s = \frac{1}{10}\right)$	7.53±1.08	3.58±1.93	7.29±0.86	5.08±2.95	7.14±0.57	7.28±2.31
$SG(s = 1)$	8.00±1.52	3.26±2.27	8.77±0.68	5.53±3.42	8.72±0.53	7.44±2.60
$KG(s = 1)$	10.00±0.00	9.75±0.72	10.00±0.00	7.68±2.52	10.00±0.00	2.38±1.34
$Unif_{km}$	1.56±0.74	8.39±1.15	2.19±0.61	7.46±1.88	2.98±0.13	7.08±2.14
G_{km}	3.34±1.33	6.03±2.32	2.38±0.86	5.16±2.79	1.23±0.46	1.99±1.31
$KM++_{km}$	1.85±0.64	7.87±1.14	1.43±0.64	6.22±2.60	1.78±0.41	3.75±2.25
$SG\left(s = \frac{1}{10}\right)_{cem}$	6.15±0.60	3.95±1.24	6.30±0.68	4.71±2.54	6.10±0.40	6.65±2.13
$SG(s = 1)_{cem}$	6.47±1.51	3.20±2.48	7.32±1.26	5.12±3.31	8.03±0.61	7.80±2.61
$Ad(\alpha = 1)_{cem}$	5.16±1.84	4.31±1.77	4.59±0.64	3.88±1.75	4.47±0.50	5.06±1.40
$Ad\left(\alpha = \frac{1}{2}\right)_{cem}$	4.93±2.11	4.67±1.93	4.72±0.80	4.15±1.77	4.53±0.50	5.58±1.71

The results depicted in Tables 1 and 2 suggest that, regardless of the weights, the performance is determined by the separation. Furthermore, a good initial solution does not imply a good final solution. Given overlap ($c_\theta = 0.5$) or moderate separation ($c_\theta = 1$), $SG_{cem}(s = 1)$ and $Ad_{cem}(\alpha = 1)$ work best, even though their initial solutions have low average ranks compared to $KM++_{km}$. Given higher separation ($c_\theta = 2$), we expect it to be easier to identify clusters and that skewed covariance matrices do not matter much if means are assigned properly in the first place. Indeed, the simple G_{km} and KG do the trick.

---

[4] Averaging the (average) log-likelihood values over different data sets is not meaningful since the optimal log-likelihoods may deviate significantly.

Table 2. Average ranks (± std.dev.) for generated data with $K = 20$, $|X| = 1000$, dimension $D = 10$, **equal weights**, and **without noise**.

	separation $c_\theta = 0.5$		separation $c_\theta = 1$		separation $c_\theta = 2$	
	initial	final	initial	final	initial	final
$\text{SG}\left(s = \frac{1}{10}\right)$	7.58±0.98	3.98+1.87	7.36±0.73	5.02±2.97	7.08±0.41	7.35±2.24
$\text{SG}(s = 1)$	8.11±1.53	3.62±2.58	8.67±0.85	5.67±3.20	8.79±0.43	7.77±2.59
$\text{KG}(s = 1)$	10.00±0.00	9.54±0.96	10.00±0.00	7.97±2.40	10.00±0.00	2.38±1.23
Unif_{km}	1.44±0.70	8.36±1.25	2.14±0.61	7.28±1.83	2.98±0.16	7.00±1.98
G_{km}	3.39±1.34	6.12±2.17	2.53±0.83	6.04±2.76	1.27±0.50	1.82±1.08
KM++_{km}	1.91±0.55	7.82±1.30	1.35±0.56	5.89±2.77	1.75±0.43	3.77±2.50
$\text{SG}\left(s = \frac{1}{10}\right)_{cem}$	6.18±0.62	3.58±1.31	6.44±0.87	4.26±2.43	6.02±0.13	6.55±1.95
$\text{SG}(s = 1)_{cem}$	6.49±1.44	3.17±2.73	7.35±1.13	5.30±3.23	8.12±0.45	8.04±2.37
$\text{Ad}(\alpha = 1)_{cem}$	5.03±1.75	4.12±1.79	4.49±0.64	3.48±1.61	4.42±0.50	4.90±1.35
$\text{Ad}\left(\alpha = \frac{1}{2}\right)_{cem}$	4.87±1.99	4.69±1.91	4.67±0.65	4.09±1.73	4.58±0.50	5.42±1.38

Table 3. Average ranks (± std.dev.) for generated data with $K = 20$, $|X| = 1000$, dimension $D = 10$, and **without noise**. Only final solutions.

	equal weights			different weights		
	spherical	elliptical	both	spherical	elliptical	both
$\text{SG}\left(s = \frac{1}{10}\right)$	6.13±2.63	5.22±2.80	5.45±2.78	6.03±2.65	5.07±2.90	5.31±2.87
$\text{SG}(s = 1)$	6.64±2.99	5.37±3.30	5.69±3.27	6.04±3.22	5.20±3.28	5.41±3.28
$\text{KG}(s = 1)$	6.78±3.19	6.58±3.59	6.63±3.49	6.81±3.20	6.53±3.65	6.60±3.54
Unif_{km}	7.67±1.48	7.50±1.91	7.54±1.81	7.64±1.64	7.64±1.92	7.64±1.85
G_{km}	3.03±2.26	5.20±2.92	4.66±2.92	2.83±2.33	4.91±2.78	4.39±2.82
KM++_{km}	5.44±2.60	5.96±2.87	5.83±2.81	5.57±2.66	6.07±2.69	5.95±2.69
$\text{SG}\left(s = \frac{1}{10}\right)_{cem}$	4.77±2.61	4.81±2.23	4.80±2.33	5.68±2.19	4.91±2.35	5.10±2.33
$\text{SG}(s = 1)_{cem}$	6.53±3.16	5.16±3.46	5.51±3.43	6.07±3.30	5.14±3.40	5.37±3.39
$\text{Ad}(\alpha = 1)_{cem}$	3.62±1.61	4.34±1.69	4.16±1.69	4.02±1.82	4.55±1.66	4.42±1.71
$\text{Ad}\left(\alpha = \frac{1}{2}\right)_{cem}$	4.38±1.78	4.86±1.75	4.74±1.77	4.30±1.86	4.97±1.88	4.80±1.89

Table 3 shows that $\text{Ad}_{cem}(\alpha = 1)$ works well for elliptical data, while G_{km} should be chosen for spherical data. Recall that there are no noise points yet. We expect that the performance of G_{km} degenerates in the presence of noise since it is prone to choose outliers. Overall, given data sets without noise, $\text{Ad}_{cem}(\alpha = 1)$ performs best.

Noisy Data. When introducing noise, our adaptive methods are still among the best methods, while the performance of some others degenerates significantly. Tables 4 and 5 show that SG_{cem} and Ad_{cem} still work well for $c_w \leq 1$ and, in contrast to data without noise, also for separated instances ($c_w = 2$). KG and G_{km} are now among the methods with the lowest average rank. This is not a surprise since our noise contains outliers. From the results depicted in Table 6 one can draw the same conclusion, i.e. KG and G_{km} can not handle noisy data. For noisy data, our Ad_{cem} methods outperform the others.

Table 4. Average ranks (± std.dev.) for generated data with $K = 20$, $|X| = 1000$, dimension $D = 10$, **different weights**, and **10 % noise**.

	separation $c_\theta = 0.5$		separation $c_\theta = 1$		separation $c_\theta = 2$	
	initial	final	initial	final	initial	final
$SG(s = \frac{1}{10})$	8.41±0.68	3.40±1.75	8.22±0.64	4.44±2.45	8.06±0.68	5.46±2.42
$SG(s = 1)$	8.25±0.98	3.46±2.60	8.67±0.47	4.13±3.02	8.74±0.44	5.93±2.87
$KG(s = 1)$	10.00±0.00	9.95±0.22	10.00±0.00	9.72±0.76	10.00±0.00	9.02±1.49
$Unif_{km}$	1.98±0.89	8.65±1.03	1.05±0.25	7.89±1.71	1.19±0.49	7.34±1.87
G_{km}	4.29±1.29	5.31±1.76	4.17±0.98	6.59±1.43	3.97±0.96	7.16±1.38
$KM++_{km}$	3.23±0.98	6.45±1.40	2.27±0.60	6.83±1.62	2.17±0.60	6.69±1.69
$SG(s = \frac{1}{10})_{cem}$	6.06±0.55	4.26±1.36	6.04±0.20	3.90±0.90	6.01±0.091	3.65±1.27
$SG(s = 1)_{cem}$	6.31±1.43	3.49±2.64	7.08±0.39	3.80±2.89	7.19±0.42	5.00±3.01
$Ad(\alpha = 1)_{cem}$	3.64±1.88	4.61±2.26	4.05±0.90	3.57±2.03	3.76±1.26	2.06±1.39
$Ad(\alpha = \frac{1}{2})_{cem}$	2.83±2.26	5.42±2.70	3.46±0.89	4.12±2.38	3.92±0.78	2.69±1.45

Table 5. Average ranks (± std.dev.) for generated data sets with $K = 20$, $|X| = 1000$, dimension $D = 10$, **equal weights**, and **10 % noise**.

	separation $c_\theta = 0.5$		separation $c_\theta = 1$		separation $c_\theta = 2$	
	initial	final	initial	final	initial	final
$SG(s = \frac{1}{10})$	8.57±0.62	3.38±1.92	8.18±0.65	4.17±2.28	7.94±0.77	5.02±2.44
$SG(s = 1)$	8.05±1.08	3.19±2.23	8.68±0.47	3.77±2.68	8.73±0.44	5.47±2.96
$KG(s = 1)$	10.00±0.00	9.93±0.35	10.00±0.00	9.62±0.87	10.00±0.00	7.85±2.22
$Unif_{km}$	1.92±0.87	8.83±0.77	1.02±0.13	8.39±1.22	1.11±0.31	8.18±1.63
G_{km}	4.47±0.99	5.53±1.75	4.01±1.01	6.74±1.51	3.53±0.83	7.18±1.28
$KM++_{km}$	3.20±1.07	6.66±1.22	2.08±0.31	7.04±1.46	1.93±0.37	7.66±1.56
$SG(s = \frac{1}{10})_{cem}$	6.08±0.53	4.47±1.31	6.03±0.18	3.92±0.97	6.00±0.00	3.64±1.11
$SG(s = 1)_{cem}$	6.20±1.43	3.14±2.43	7.10±0.40	3.88±2.75	7.33±0.47	5.13±2.76
$Ad(\alpha = 1)_{cem}$	3.62±1.85	4.42±2.35	4.20±0.79	3.54±2.11	4.15±0.82	2.27±1.59
$Ad(\alpha = \frac{1}{2})_{cem}$	2.89±2.45	5.45±2.47	3.69±0.74	3.92±2.29	4.28±0.66	2.60±1.51

Table 6. Average ranks (± std.dev.) for generated data sets with $K = 20$, $|X| = 1000$, dimension $D = 10$, and **10 % noise**. Only final solutions.

	equal weights			different weights		
	spherical	elliptical	both	spherical	elliptical	both
$SG(s = \frac{1}{10})$	5.38±2.26	3.79±2.20	4.19±2.32	5.53±2.30	4.07±2.29	4.43±2.38
$SG(s = 1)$	4.98±2.79	3.87±2.76	4.14±2.81	5.53±3.14	4.17±2.90	4.51±3.02
$KG(s = 1)$	8.61±2.03	9.31±1.49	9.13±1.66	9.32±1.28	9.64±0.95	9.56±1.05
$Unif_{km}$	8.36±1.36	8.50±1.26	8.47±1.28	7.63±1.72	8.07±1.63	7.96±1.66
G_{km}	6.70±1.69	6.41±1.67	6.49±1.68	6.42±1.87	6.33±1.66	6.35±1.71
$KM++_{km}$	7.09±1.71	7.13±1.39	7.12±1.48	6.46±1.71	6.72±1.53	6.66±1.58
$SG(s = \frac{1}{10})_{cem}$	4.00±1.45	4.01±1.09	4.01±1.18	4.03±1.34	3.90±1.17	3.94±1.22
$SG(s = 1)_{cem}$	5.04±2.79	3.72±2.69	4.05±2.77	4.88±2.98	3.84±2.86	4.10±2.92
$Ad(\alpha = 1)_{cem}$	2.01±1.29	3.88±2.27	3.41±2.22	2.31±1.57	3.78±2.25	3.41±2.19
$Ad(\alpha = \frac{1}{2})_{cem}$	2.83±1.72	4.37±2.50	3.99±2.42	2.88±1.78	4.48±2.57	4.08±2.50

Table 7. Average ranks ($\pm$ std.dev.) for generated data ($K = 20$, $|X| = 1000$, $D = 3$).

	without noise		noisy	
	initial	final	initial	final
$\text{SG}\left(s = \frac{1}{10}\right)$	7.31±0.63	7.94±1.39	8.05±0.70	7.93±1.31
$\text{SG}(s = 1)$	8.90±0.39	8.56±1.98	8.72±0.45	7.99±2.19
$\text{KG}(s = 1)$	9.94±0.42	3.28±1.98	10.00±0.00	8.09±1.46
Unif_{km}	2.82±0.58	4.63±1.40	3.38±0.94	2.76±1.38
G_{km}	1.93±1.04	2.80±1.97	4.43±1.59	6.01±1.59
KM++_{km}	1.51±0.60	1.99±1.21	2.96±1.12	2.35±1.57
$\text{SG}\left(s = \frac{1}{10}\right)_{cem}$	6.10±0.42	7.46±1.14	5.75±0.82	6.07±1.16
$\text{SG}(s = 1)_{cem}$	7.65±0.94	8.83±1.74	7.04±0.86	8.16±2.10
$\text{Ad}(\alpha = 1)_{cem}$	4.35±0.83	4.66±1.48	2.54±1.43	3.02±1.13
$\text{Ad}\left(\alpha = \frac{1}{2}\right)_{cem}$	4.49±0.74	4.85±1.51	2.12±1.36	2.62±1.28

Low Dimensional or High Sample Size Data. We expect that, if the dimension is low or the sample size is large enough, it is generally easier to identify clusters. Indeed the results differ significantly from our previous results. In general, the KM++_{km} and Unif_{km} perform best. For data sets with $D = 3$ and $|X| = 1000$, Table 7 shows that the KM++_{km} method works well even in the presence of noise. However, if we are given noise *and* small separation, the simple Unif_{km} does well. We also increased the sample size to $|X| = 5000$ *and* the dimension to $D = 10$, expecting that the higher sample size can make up for the higher dimension (results available in [9]). Indeed, for data sets without noise, where clusters can presumably be identified easier, KM++_{km} still suffices. However, given noise or too small a separation, our Ad_{cem} methods and the simple Unif_{km} work better.

4.2 Real World Data Sets

We use four publicly available data sets: *Covertype* ($|X| = 581\,012$, $D = 10$ real-valued features) [4]; two *Aloi* data sets ($|X| = 110\,250$, $D \in \{27, 64\}$) based on color histograms in HSV color space [18] from data provided by the ELKI project [2] and the Amsterdam Library of Object Images [16]; *Cities* ($|X| = 135\,082$, $D = 2$) is a projection of the coordinates of cities with a population of at least 1000 [1]; *Spambase* ($|X| = 4601$, $D = 10$ real-valued features) [4].

The results are depicted in Fig. 2: For *Aloi* ($D = 27$) and *Spambase* ($K = 3$), $\text{SG}_{cem}(s = 1)$ is considerably better than the other methods. For *Cities* and *Spambase* ($K = 10$), $\text{SG}(s = 1)$ does better (without running the CEM). For *Aloi* ($D = 64$) and the *Covertype*, $\text{Ad}_{cem}(\alpha = 1)$ works better than the others.

4.3 Time Measurement

The run times of the compared methods match our expectation (cf. Table 8). First, (intermediate) steps of the CEM algorithm are faster than (more) steps of the EM algorithm. Second, sampling and running methods on a random subset of the data should in general reduce the run time.

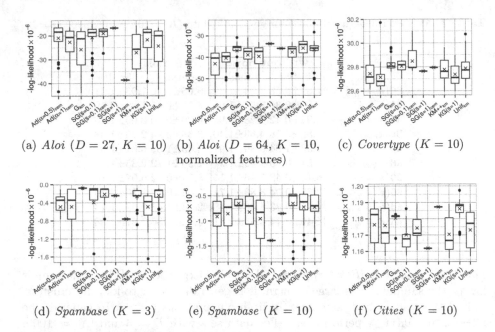

(a) *Aloi* ($D = 27$, $K = 10$)　(b) *Aloi* ($D = 64$, $K = 10$, 　(c) *Covertype* ($K = 10$)
normalized features)

(d) *Spambase* ($K = 3$)　　(e) *Spambase* ($K = 10$)　　(f) *Cities* ($K = 10$)

Fig. 2. Results for the real world data sets depicted as boxplots (final solutions only).

Table 8. Average run times (in seconds) over 12 data sets with $|X| = 10^3$ and $D = 10$ and different runs per data set using an Intel Core i7-3770 CPU (3.40 GHz, 8 GB RAM).

SG($s = 1$)	0.314	SG($s = 0.1$)	0.307	Ad$_{cem}$($\alpha = 1$)	0.226	Unif$_{km}$	0.206
SG$_{cem}$($s = 1$)	0.225	SG$_{cem}$($s = 0.1$)	0.197	Ad$_{cem}$($\alpha = 0.5$)	0.253	HAC($s = 1$)	1.588
G$_{km}$	0.205	KG($s = 1$)	0.315				

5　Conclusion and Future Work

If you need a fast and simple method, then we suggest to use one of the following methods: Given a data set with a large number of points per cluster or low dimension, the K-means++ initialization followed by the K-means algorithm and **Means2GMM** should do well. Otherwise, we recommend our new methods **Ad** and **SG** followed by the spherical CEM algorithm, especially if your data is presumably noisy. Last but not least, whatever you prefer, we suggest trying intermediate steps of the spherical CEM or K-means algorithm.

For the K-means++ algorithm and the Gonzalez algorithm there are provable guarantees. We hope that our results are a good starting point for a theoretical analysis that will transfer these results to the MLE problem for GMMs.

References

1. GeoNames geographical database. http://www.geonames.org/
2. Achtert, E., Goldhofer, S., Kriegel, H.-P., Schubert, E., Zimek, A.: Evaluation of Clusterings - Metrics and Visual Support. http://elki.dbs.ifi.lmu.de/wiki/DataSets/MultiView
3. Arthur, V.: k-means++: The advantages of careful seeding. In: SODA 2007 (2007)
4. Asuncion: UCI machine learning repository (2007). http://www.ics.uci.edu/mlearn/MLRepository.html
5. Baudry, J.-P., Celeux, G.: EM for mixtures. Stat. Comput. **25**(4), 713–726 (2015)
6. Biernacki, C.: Initializing EM using the properties of its trajectories in Gaussian mixtures. Stat. Comput. **14**(3), 267–279 (2004)
7. Biernacki, C., Celeux, G., Govaert, G.: Choosing starting values for the EM algorithm for getting the highest likelihood in multivariate Gaussian mixture models. Comput. Stat. Data Anal. **41**(3–4), 561–575 (2003)
8. Bishop, C.: Pattern Recognition and Machine Learning. Information Science and Statistics. Springer, Secaucus (2006)
9. Bujna, K., Kuntze, D.: Supplemental Material. http://www-old.cs.upb.de/fachgebiete/ag-bloemer/forschung/clusteranalyse/adaptive_seeding_for_gmms.html
10. Celeux, G., Govaert, G.: A classification EM algorithm for clustering and two stochastic versions. Comput. Stat. Data Anal. **14**(3), 315–332 (1992)
11. Dasgupta, S.: Experiments with random projection. In: UAI 2000 (2000)
12. Dasgupta, S.: Learning mixtures of gaussians. In: FOCS 1999 (1999)
13. Dempster, A.P., Laird, N.M., Rubin, D.B.: Maximum likelihood from incomplete data via the EM algorithm. J. R. Stat. Soc. Ser. B Stat. Methodol. **39**(1), 1–38 (1977)
14. Färber, I., Günnemann, S., Kriegel, H., Kröger, P., Müller, E., Schubert, E., Seidl, T., Zimek, A.: On using class-labels in evaluation of clusterings. In: MultiClust 2010 (2010)
15. Fayyad, U., Reina, C., Bradley, P.S.: Initialization of iterative refinement clustering algorithms. In: KDD 1998 (1998)
16. Geusebroek, J.M., Burghouts, G.J., Smeulders, A.W.M.: The Amsterdam library of object images. Int. J. Comput. Vis. **6**(1), 103–112 (2005)
17. Gonzalez, T.F.: Clustering to minimize the maximum intercluster distance. Theor. Comput. Sci. **38**, 293–306 (1985)
18. Kriegel, H.-P., Schubert, E., Zimek, A.: Evaluation of multiple clustering solutions. In: MultiClust 2010 (2010)
19. Krüger, A., Leutnant, V., Haeb-Umbach, R., Ackermann, M., Blömer, J.: On the initialization of dynamic models for speech features. In: Sprachkommunikation 2010 (2010)
20. Kwedlo, W.: A new random approach for initialization of the multiple restart EM algorithm for Gaussian model-based clustering. Pattern Anal. Appl. **18**(4), 757–770 (2015)
21. Maitra, R.: Initializing partition-optimization algorithms. IEEE/ACM Trans. Comput. Biol. Bioinform. **6**(1), 144–157 (2009)
22. Maitra, R., Melnykov, V.: Simulating data to study performance of finite mixture modeling and clustering algorithms. J. Comput. Graph. Stat. **19**(2), 354–376 (2010)

23. McLachlan, G.J., Krishnan, T.: The EM Algorithm and Extensions. Wiley Series in Probability and Statistics, 2nd edn. Wiley-Interscience, New York (2008)
24. Meilă, M., Heckerman, D.: An experimental comparison of several clustering and initialization methods. In: UAI 1998. Morgan Kaufmann Inc., San Francisco (1998)
25. Melnykov, V., Melnykov, I.: Initializing the EM algorithm in Gaussian mixture models with an unknown number of components. Comput. Stat. Data Anal. **56**, 1381–1395 (2011)
26. Thiesson, B.: Accelerated quantification of Bayesian networks with incomplete data. University of Aalborg (1995)
27. Verbeek, J.J., Vlassis, N., Kröse, B.: Efficient greedy learning of Gaussian mixture models. Neural Comput. **15**(2), 469–485 (2003)

A Greedy Algorithm to Construct L1 Graph with Ranked Dictionary

Shuchu Han$^{(\boxtimes)}$ and Hong Qin

Stony Brook University (SUNY), Stony Brook, USA
{shhan,qin}@cs.stonybrook.edu

Abstract. $\mathcal{L}_1$ graph is an effective way to represent data samples in many graph-oriented machine learning applications. Its original construction algorithm is nonparametric, and the graphs it generates may have high sparsity. Meanwhile, the construction algorithm also requires many iterative convex optimization calculations and is very time-consuming. Such characteristics would severely limit the application scope of $\mathcal{L}_1$ graph in many real-world tasks. In this paper, we design a greedy algorithm to speed up the construction of $\mathcal{L}_1$ graph. Moreover, we introduce the concept of "Ranked Dictionary" for $\mathcal{L}_1$ minimization. This ranked dictionary not only preserves the locality but also removes the randomness of neighborhood selection during the process of graph construction. To demonstrate the effectiveness of our proposed algorithm, we present our experimental results on several commonly-used datasets using two different ranking strategies: one is based on Euclidean metric, and another is based on diffusion metric.

Keywords: Sparse graph · Clustering

1 Introduction

For graph-oriented learning tasks, a quality graph representation [4] of input data samples is the key to success. In the past few decades, researchers in machine learning area propose many different methods to solve such tasks, for example, k-nearest neighbor (kNN) graph and ϵ-ball graphs. These methods are very straightforward and proved to be efficient for general data. The reason of these methods' success is that their construction algorithm acts as a local smooth "filter" which sets the weight between faraway data points and source point to zero. The built graph is constructed by many such local star-shape patches (or subgraphs). However, both of them need a user-specified parameter such as k or ϵ which is chosen empirically. Considering the versatility and uncertainty of the real world data, a bad selection of parameter k and ϵ will lead to an inaccurate conclusion for subsequent machine learning tasks. Recently, a nonparametric graph called $\mathcal{L}_1$ graph is proposed by Cheng et al. [2]. Based on existing sparse representation frameworks [10,12], the construction algorithm of $\mathcal{L}_1$ graph can be described as follows: Given an input data samples $\mathbf{X} = [\mathbf{x_1}, \mathbf{x_2}, \cdots, \mathbf{x_n}]$,

© Springer International Publishing Switzerland 2016
J. Bailey et al. (Eds.): PAKDD 2016, Part II, LNAI 9652, pp. 309–321, 2016.
DOI: 10.1007/978-3-319-31750-2_25

where each $x_i, i \in [1, \cdots, n]$ is a vector that represents one single data sample. The $\mathcal{L}_1$ graph of X is built by finding a sparse coding [11] of each x_i with a dictionary constructed from all data samples except x_i itself. The coefficient of sparse coding is used as the edge weight of resulted $\mathcal{L}_1$-graph. The mathematical definition of sparse coding is:

$$(\textbf{P1}) \quad \min_{\alpha_i} \|\alpha_i\|_1 \text{ subject to } x_i = \Phi^i \alpha_i, \tag{1}$$

where dictionary $\Phi^i = [x_1, \cdots, x_{i-1}, x_{i+1}, \cdots, x_n]$, and $\alpha_i \in \mathbb{R}^{n-1}$ is the sparse code of x_i. The coefficients of α_i could be negative, depending on the choices of $\mathcal{L}_1$ minimization solvers. To make them have the physical meaning of "Similarity", the absolute value or nonnegative constraints are employed.

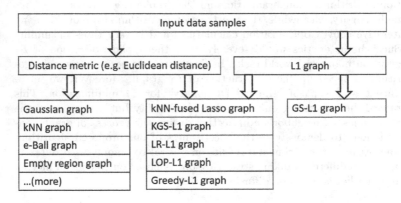

Fig. 1. Connection of Greedy $\mathcal{L}_1$ graph to other graphs. Several of them are: kNN-fused Lasso graph [16], Group Sparse (GS) $\mathcal{L}_1$ graph, Kernelized Group Sparse (KGS) $\mathcal{L}_1$ graph [6], Laplacian Regularized (LR) $\mathcal{L}_1$ graph [14] and Locality Preserving (LOP) $\mathcal{L}_1$ graph [7].

As we could see from the above description, the $\mathcal{L}_1$ graph construction algorithm is nonparametric and the user is not required to input any parameters except for the solver. The construction algorithm is a pure numerical process based on convex optimization. Cheng et al. [2] show that $\mathcal{L}_1$ graph has three advantages comparing to traditional graph construction methods. They are: (1) robustness to data noise; (2) sparsity; (3) datum-adaptive neighborhood. Their experimental results also prove that $\mathcal{L}_1$ graph has significant performance improvement in many machine learning applications such as spectral clustering, subspace learning, semi-supervised learning, etc. [2]. Nevertheless, just like each sword has double edges, $\mathcal{L}_1$ graph also bears some disadvantages such as: (1) sensitive to duplications. For example, if every data sample has a duplication, the resulted $\mathcal{L}_1$ graph will only have edge connections between the data sample and its duplication; (2) randomness, the edge and edge weight are highly dependent on the solver; (3) high computational cost [2]; (4) lost of the locality [6,7,15]. To overcome these disadvantages, many improved algorithms have

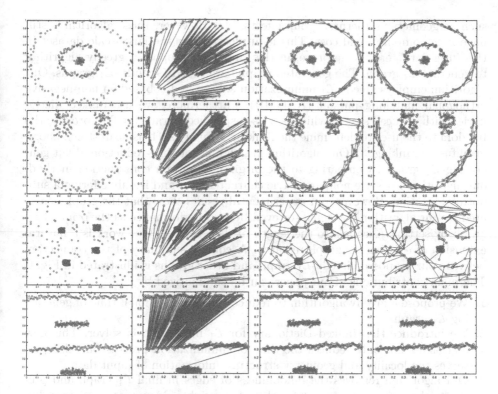

Fig. 2. $\mathcal{L}_1$ graphs generated by different construction algorithms. From left to right: 2D toy dataset; $\mathcal{L}_1$ graph; Greedy-$\mathcal{L}_1$ graph with Euclidean metric (K = 15); Greedy-$\mathcal{L}_1$ graph with Diffusion metric (K = 15).

been proposed in recent years. Now, we would like to classify them into two categories: *soft-modification* and *hard-modification* (Fig. 2).

1. *Soft-modification* algorithms. Algorithms in this category usually add one or more regularization terms to the original $\mathcal{L}_1$ minimization objective function in Eq. (1). For example, the structure sparsity [16] preserves the local structure information of input data, the auto-grouped sparse regularization [6] adds the group effect to the final graph, and the Graph Laplacian regularization [13,14] lets the closed data samples have similar sparse coding coefficients (or α_i).
2. *Hard-modification* algorithms. These algorithms define a new dictionary for each data sample during $\mathcal{L}_1$ minimization. By reducing the solvers' solution space for each data sample into a local space, the locality of input data is preserved and the computational time of $\mathcal{L}_1$ minimization (Eq. (1)) is reduced. For example, the locality preserved (LOP) $\mathcal{L}_1$ graph is utilizing k-nearest neighbors as dictionaries [7].

The *soft-modification* algorithms preserve the nonparametric feature and improve the quality of $\mathcal{L}_1$ graph by exploiting the intrinsic data information such as geometry structure, group effects, etc. However, those algorithms still have high computational cost. This is unpleasant for the large-scale dataset in this "Big-data" era. To improve, in this paper we propose a greedy algorithm to generate $\mathcal{L}_1$ graph. The generated graphs are called **Greedy-$\mathcal{L}_1$ graphs**. Our algorithm employs greedy $\mathcal{L}_1$ minimization solvers and is based on non-negative orthogonal matching pursuit (NNOMP). Furthermore, we use ranked dictionaries with reduced size K which is a user-specified parameter. We provide the freedom to the user to determine the ranking strategy such as nearest neighbors, or diffusion ranking [3]. Our algorithm has significant time-reduction about generating $\mathcal{L}_1$ graphs. Comparing to the original $\mathcal{L}_1$ graph construction method, our algorithm loses the nonparametric characteristics and is only offering a sub-optimal solution. However, our experimental results show that the graph generated by our algorithm has equal (or even better) performance as the original $\mathcal{L}_1$ graph by setting K equals to the length of data sample. Our work is a natural extension of existing $\mathcal{L}_1$ graph research. A concise summary of the connection between our proposed Greedy-$\mathcal{L}_1$ graph and other graphs is illustrated in Fig. 1. The main contributions of our paper can be summarized by

1. We propose a greedy algorithm to reduce the computational time of generating $\mathcal{L}_1$ graph.
2. We introduce the Ranked Dictionary for $\mathcal{L}_1$ minimization solver. This new dictionary not only reduces the time of minimization process but also preserves the locality and geometry structure information of input data.
3. Our algorithm removes the randomness of edges in final $\mathcal{L}_1$ graph and preserves the uniqueness except for the edge weights. Moreover, our algorithm can generate $\mathcal{L}_1$ graphs with lower sparsity.
4. We present experiment and analysis results by applying our algorithm to spectral clustering application with different datasets. Our experimental results show that the graphs generated by our proposed greedy algorithm have equal clustering performance even though it is only providing a sub-optimal solution.

The organization of our paper is as follows. First, an overview of the disadvantages of original $\mathcal{L}_1$ graph construction algorithm will be presented in Sect. 2. Second, we will introduce our proposed greedy algorithm in Sect. 3. After that, we will give a review of existing works on how to improve the quality of $\mathcal{L}_1$ graph. Finally, we will present our experimental results in Sect. 5 and draw conclusion in Sect. 6.

2   Overview

In this section, we make our attempts to address two problems of original $\mathcal{L}_1$ graph construction algorithm. They are: (1) curse of dictionary normalization, and (2) non-local edges.

2.1   Curse of Dictionary Normalization

While solving $\mathcal{L}_1$ minimization, the atoms of dictionary are normalized to have unit length. The goal of this step is to satisfy the theoretic requirement of Compressive Sensing. The less-ideal part about this normalization is that it is not

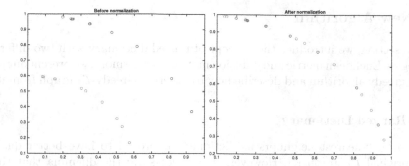

Fig. 3. Demonstration of dictionary normalization for a toy dataset. The red and blue points represent two different clusters. Left: before normalization; right: after normalization. We can see that the neighborhood relationship is changed after normalization (Color figure online).

preserving neighborhood information of input data. This can be illustrated in Fig. 3. To illustrate this phenomenon, we manually create a toy dataset in 2D and it has two clusters visually. After normalization, we can see that the neighbors of a node are changed. This normalization step projects all data samples onto a unit hypersphere and the original geometry structure information is lost.

2.2 Non-local Edges

During the construction of $\mathcal{L}_1$ graph, an over-complete dictionary is required for each data sample. The original method simply selects all other data samples as the dictionary. This strategy affords the *nonparametric* property of $\mathcal{L}_1$ graph. However, it also introduces non-local edges. In other words, it doesn't preserve the **locality** of input data [7]. This phenomenon can be illustrated in Fig. 4,

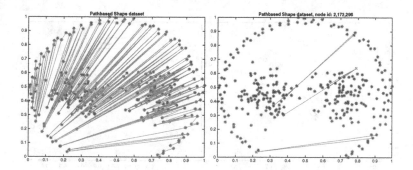

Fig. 4. $\mathcal{L}_1$-graph of path-based dataset. Left: the entire graph; right: edges of three selected points. We can see the existence of non-local edges.

3 New Algorithm

In this section, we introduce the concept of ranked dictionary with two different strategies: Euclidean metric and diffusion metric. Furthermore, we present our proposed greedy algorithm and describe how to generate Greedy-$\mathcal{L}_1$ graph from it.

3.1 Ranked Dictionary

The use of k-nearest neighbors as dictionary is proved to have better quality than original $\mathcal{L}_1$ graph [7]. However, it can not solve the dilemma that there might exist data samples with the same direction but different length in input data. The dictionary normalization process will project them onto to the same location at hypersphere. Since they have the same values, the $\mathcal{L}_1$ minimization solver will choose one of them randomly. To avoid this randomness, we need to rank those atoms (or data samples) of dictionary (Fig. 5).

Euclidean Metric. Using Euclidean metric to rank atoms of dictionary is quite straightforward. We rank them by distance. The shorter distance will have a higher rank score. The Euclidean distance is defined as:

$$dist(\boldsymbol{x}_i, \boldsymbol{x}_j) = \|\boldsymbol{x}_i - \boldsymbol{x}_j\|_2 \tag{2}$$

Diffusion Metric. As pointed out by Yang et al. [14], many real-world datasets are similar to an intrinsic low dimensional manifold embedded in high dimensional ambient space, and the geometry structure of manifold can be used to improve the performance of learning algorithms. We now present a strategy to search dictionaries following the geometry structure of input data. Based on the diffusion theory [3,5], we rank the atoms of dictionary through diffusion matrix. A diffusion process has three stages [5]: (1) initialization; (2) definition of transition matrix; (3) definition of the diffusion process. In our setting, the first stage

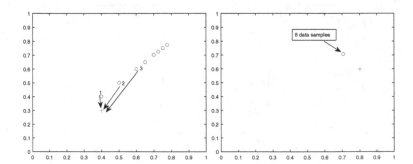

Fig. 5. Ranked dictionary. Left: eight data samples have the same direction but with different length. Red cross is the target data sample for calculating sparse coefficients. Right: after normalization, those eight data samples have the same location (Color figure online).

is to build an affinity matrix A from the input dataset X. We use Gaussian kernel to define the pairwise distance:

$$A(i,j) = \exp\left(-\frac{\|x_i - x_j\|}{-2\sigma^2}\right), \tag{3}$$

where $A(i,j)$ is the distance between data sample x_i and data sample x_j, and σ is a normalization parameter. In our configuration, we use the median of K nearest neighbors to tune σ. The second stage is to define the transition matrix P:

$$P = D^{-1}A, \tag{4}$$

where D is a $n \times n$ degree matrix defined as

$$D(i,j) = \begin{cases} \sum_{j=1}^{n} A(i,j) & \text{if } i = j, \\ 0 & \text{otherwise.} \end{cases} \tag{5}$$

Now the diffusion process can be defined as:

$$W_{t+1} = PW_t P^{'}, \tag{6}$$

where $W_0 = A$ and t is the number of steps for diffusion steps. Each row of W_t is the diffusion ranking scores. In this paper, we let t equal to K for the sake of simplicity. Once W_t is calculated, the first K data samples with top scores of each row is selected as dictionary. The algorithmic details can be documented as follows:

Algorithm 1. Diffusion Dictionary

    Input   : Data samples $X = [x_1, x_2, \cdots, x_n]$, where $x_i \in X$;
               Size of dictionary: K;
    Output: Diffusion dictionary index matrix Φ_K.

1 Calculate Gaussian similarity graph A;
2 $P = D^{-1}A$;
   /* calcualte diffusion process iteratively.   */
3 for $t = 1 : K$ do
4 | $W_t = PW_{t-1}P^{'}$
5 end
   /* sort each row in descend order.   */
6 for $i = 1 : n$ do
7 | sort($W_t(i,:)$)
8 end
   /* fetch the index of highest K values in each row of W_t */
9 for $i = 1 : n$ do
10 | $\Phi(i,:) =$index($W_t(i, 1 : k)$)
11 end

3.2   Greedy-$\mathcal{L}_1$ Graph

We now propose a greedy algorithm to build $\mathcal{L}_1$ graph. Our proposed algorithm is based on non-negative orthogonal matching pursuit (NNOMP) [1,9]. By using this solver, we switch the $\mathcal{L}_1$ minimization problem (**P1**) back to the original $\mathcal{L}_0$ optimization with non-negative constraints (**P2**) as:

$$(\textbf{P2}) \quad \min_{\alpha_i} \|\alpha_i\|_0 \text{ subject to } x_i = \Phi^i \alpha_i, \alpha_i \geq 0. \tag{7}$$

The main difference between our algorithm and the original NNOMP [1] is that the atoms of dictionary are ranked. We let the solver choose and assign higher coefficient values to atoms that are closer to source data sample. The detailed processes are described in Algorithm 2.

Algorithm 2. Greedy Solver

Input   : Data sample x;
              Ranked dictionary Φ_K;
              Residual threshold $\theta_{threshold}$

Output: Sparse coding α of x.

1 **for** $i = 1 : \|x\|_1$ **do**
2   **if** $i == 0$ **then**
3   Temporary solution: $\alpha^i = 0$;
4   Temporary residual: $r^i = x - \Phi_K \alpha^i$;
5   Temporary solution support: $S^i = Support\{\alpha^i\} = \emptyset$;
6   **else**
7   **for** $j = 1 : k$ **do**
         /* ϕ_j is the j-th atom of Φ_K */
8           $\epsilon(j) = \min_{\alpha_j \geq 0} \|\phi_j \alpha_j - r^{i-1}\|_2^2 = \|r^{i-1}\|_2^2 - \max\{\phi_j^T r^{i-1}, 0\}^2$.
9   **end**
10   Find j_0 such that $\forall j \in S^c, \epsilon(j_0) \leq \epsilon(j)$, if there are multiple j_0 atoms, choose the one with smallest index value.;
11   Update support: $S^i = S^{i-1} \cup \{j_0\}$;
12   Update solution: $\alpha^i = \min_z \|\Phi_K \alpha - x\|_2^2$ subject to $Support\{\alpha^i\} = S^i$ and $\alpha^i \geq 0$;
13   Update residual: $r^i = x - \Phi_K \alpha^i$;
14   **if** $\|r^i\|_2^2 < \theta_{threshold}$ **then**
15   Break;
16   **end**
17   **end**
18 **end**
19 Return α^i;

---

4   Related Works

Original $\mathcal{L}_1$ graph [2] is a pure numerical result and doesn't exploit the physical and geometric information of input data. To improve the quality of $\mathcal{L}_1$ graph, several research works are proposed to use the intrinsic structure information of data by adding one or several regularization terms to the $\mathcal{L}_1$ minimization **P1**. For example, consider the elastic net regularization [6], OSCAR regularization [6], and Graph Laplacian regularization [14].

Another research direction of $\mathcal{L}_1$ graph is to reduce its high computational cost. Zhou et al. [16] propose a kNN-fused Lasso graph by using the idea of k-nearest neighbors in kernel feature space. With a similar goal, Fang et al. [6] propose an algorithm which transfers the data into reproducing kernel Hilbert space and then projects them into a lower dimensional subspace. By these operations, the dimension of the dataset is reduced and the computational time is reduced.

5   Experiments

We now present our experimental results. We first document our configuration of parameters and datasets. Second, we evaluate the effectiveness of our proposed graph construction methods through spectral clustering application. To satisfy the input of spectral clustering algorithm, we transform the adjacency matrix of $\mathcal{L}_1$ graph $\boldsymbol{W}$ into a symmetry matrix $\boldsymbol{W}'$ by $\boldsymbol{W}' = (\boldsymbol{W} + \boldsymbol{W}^T)/2$. All analyses and experiments are carried out by using Matlab on a PC with Intel 4-core 3.4 GHz CPU and 16 GB RAM.

5.1   Experimental Setup

Datasets. To demonstrate the performance of our proposed algorithm, we evaluate it on seven UCI benchmark datasets including three biological data sets (BreastTissue, Iris, Soybean), two vision image data sets (Vehicle, Image), one chemistry data set (Wine), and one physical data set (Glass), whose statistics

Table 1. Data set statistics.

Name	#samples	#attributes	#clusters
BreastTissue (BT)	106	9	6
Iris	150	4	3
Wine	178	13	3
Glass	214	9	6
Soybean	307	35	19
Vehicle	846	18	4
Image	2100	19	7

Table 2. NMI comparison of graph construction algorithms. M is the number of attributes.

Name	$\mathcal{L}_1$	Gaussian			Greedy-$\mathcal{L}_1$ graph (Euclidean)			Greedy-$\mathcal{L}_1$ graph (Diffusion)		
		K=1*M	K=2*M	K=3*M	K=1*M	K=2*M	K=3*M	K=1*M	K=2*M	K=3*M
BT	0.4582	0.3556	0.4909	0.4722	0.5473	0.4517	0.5024	0.4197	0.4073	0.3839
Iris	0.5943	0.4557	0.5923	0.7696	0.3950	0.4623	0.4070	0.5106	0.4626	0.4640
Wine	0.7717	0.8897	0.8897	0.8897	0.8943	0.9072	0.8566	0.6925	0.4291	0.6093
Glass	0.3581	0.1598	0.2941	0.2614	0.2569	0.3688	0.3039	0.2991	0.3056	0.2918
Soybean	0.7373	0.6839	0.6911	0.6541	0.6919	0.6833	0.6775	0.5788	0.5493	0.5432
Vehicle	0.1044	0.1528	0.1519	0.1341	0.1512	0.2121	0.2067	0.1438	0.1035	0.1244
Image	0.4969	0.2461	0.3382	0.0486	0.5821	0.6673	0.6649	0.4866	0.4483	0.3155
Average	0.5030	0.4205	0.4926	0.4614	0.5170	**0.5361**	0.5170	0.4473	0.3865	0.3903

Table 3. AC comparison of different graph construction algorithms. M is the number of attributes.

Name	$\mathcal{L}_1$	Gaussian			Greedy-$\mathcal{L}_1$ graph (Euclidean)			Greedy-$\mathcal{L}_1$ graph (Diffusion)		
		K=1*M	K=2*M	K=3*M	K=1*M	K=2*M	K=3*M	K=1*M	K=2*M	K=3*M
BT	0.5472	0.3208	0.5189	0.5472	0.6698	0.4811	0.5943	0.4528	0.4906	0.4717
Iris	0.7400	0.6667	0.6867	0.9090	0.6933	0.7200	0.6800	0.7200	0.6533	0.64
Wine	0.9326	0.9719	0.9719	0.9719	0.9719	0.9719	0.9551	0.8989	0.7865	0.8596
Glass	0.4206	0.4206	0.4486	0.4206	0.4579	0.4533	0.4346	0.4626	0.4813	0.5187
Soybean	0.6156	0.5440	0.5570	0.5505	0.5244	0.4853	0.5016	0.4430	0.3746	0.4876
Vehicle	0.3713	0.3983	0.3983	0.4066	0.4539	0.4243	0.4090	0.3664	0.3522	0.3605
Image	0.5629	0.3262	0.3919	0.1895	0.6348	0.7181	0.7043	0.5190	0.5524	0.3505
Average	0.6105	0.5546	0.5757	0.5746	0.6227	**0.6288**	0.6141	0.5683	0.5334	0.5362

are summarized in Table 1. All of these data sets have been popularly used in spectral clustering analysis research. The diverse combinations of data sets are necessary for our comprehensive studies.

Parameters Setting. In our experiments, we use the *l1_ls* solver [8] for original $\mathcal{L}_1$ graph construction algorithms. We set the solver's parameter λ to 0.1. The *threshold* $\theta_{threshold}$ of Greedy solver Algorithm 2 is set to $1e - 5$. For Gaussian graph and Greedy-$\mathcal{L}_1$ graph, we select three different K values and document their clustering performance results respectively. The K is set to be the multiple of data attribute size.

5.2 Spectral Clustering Performance

Baseline. To evaluate the quality of our algorithms, we compare the spectral clustering performance with Gaussian similarity graph, and original $\mathcal{L}_1$ graph. The results are documented in Tables 2 and 3.

Evaluation Metrics. We evaluate the spectral clustering performance with Normalized Mutual Information (NMI) and Accuracy (AC). NMI value ranges from 0 to 1, with higher values meaning better clustering performance. AC is another metric to evaluate the clustering performance by measuring the fraction of its clustering result that are correct. It's value also ranges from 0 to 1, and the higher the better.

Greedy-$\mathcal{L}_1$ Graph vs. Gaussian Graph. Overall, the Greedy-$\mathcal{L}_1$ graph using Euclidean metric has better average spectral clustering performance than Gaussian graphs. However, since the Gaussian graph we used are not tuned, the best clustering performance of Gaussian graphs may not occur in our experiments.

Greedy-$\mathcal{L}_1$ Graph vs. $\mathcal{L}_1$ Graph. Greedy-$\mathcal{L}_1$ graph has better clustering performance than $\mathcal{L}_1$ graph on average. However, for iris and soybean datasets, the $\mathcal{L}_1$ graph shows the best clustering result: Iris (NMI = 0.5943, AC = 0.74); Soybean (NMI = 0.7373, AC = 0.6156). The best result of Greedy-$\mathcal{L}_1$ graph are: Iris (NMI = 0.5106, AC = 0.72); Soybean (NMI = 0.6919, AC = 0.5244).

Euclidean Metric vs. Diffusion Metric. The Euclidean metric appears to have better clustering performance than that of diffusion metric in general. This is rather a surprising result to us. Only for Iris dataset, the result of diffusion metric is better than that of Euclidean metric.

5.3   Discussions

Running Time. We report the running time of generating $\mathcal{L}_1$ graphs using different construction algorithms. As we can see from Fig. 6, the Greedy-$\mathcal{L}_1$ graphs have consumed significantly less construction time than that in original $\mathcal{L}_1$ graphs.

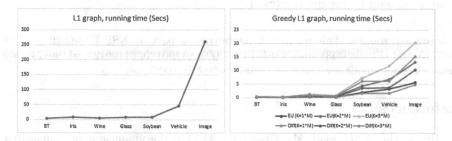

Fig. 6. Running time of different $\mathcal{L}_1$ graph construction algorithms. Left: original $\mathcal{L}_1$ graph construction algorithm. Right: the construction of $\mathcal{L}_1$ graph using greedy solver.

Graph Sparsity. We check the sparsity of graphs by calculating the edge density:

$$sparsity(G) = \frac{|E|}{|V| * (|V| - 1)}. \tag{8}$$

The results are reported in Table 4. We can see that Greedy-$\mathcal{L}_1$ graphs with diffusion metric are more sparse than that with Euclidean metric.

Table 4. Graph sparsity comparison of different graph construction algorithms. M is the number of attributes.

Name	$\mathcal{L}_1$	Gaussian			Greedy-$\mathcal{L}_1$ graph (Euclidean)			Greedy-$\mathcal{L}_1$ graph (Diffusion)		
		K=1*M	K=2*M	K=3*M	K=1*M	K=2*M	K=3*M	K=1*M	K=2*M	K=3*M
BT	0.0604	1	1	1	0.0457	0.0615	0.0705	0.0341	0.0442	0.0548
Iris	0.0403	1	1	1	0.0217	0.0288	0.0311	0.0203	0.0237	0.0265
Wine	0.0600	1	1	1	0.0413	0.0496	0.0552	0.0347	0.0409	0.0437
Glass	0.0369	1	1	1	0.0242	0.0308	0.0349	0.0188	0.0204	0.0239
Soybean	0.030	1	1	1	0.0286	0.0317	0.0346	0.0258	0.0299	0.034
Vehicle	0.0135	1	1	1	0.0104	0.0124	0.0135	0.0062	0.0074	0.0084
Image	0.0039	1	1	1	0.0034	0.004	0.0044	0.0026	0.0029	0.0027

6   Conclusion

In this paper, we have devised a greedy algorithm to construct $\mathcal{L}_1$ graph. Moreover, we introduced the concept of ranked dictionary for our greedy solver. Except for the Euclidean metric and diffusion metric that have been discussed in this paper, the user can choose other ranking methods such as manifold ranking that could be more appropriate for specific dataset in real applications. Our greedy algorithm can generate sparse $\mathcal{L}_1$ graph faster than the original $\mathcal{L}_1$ graph construction algorithm, and the resulting graphs have better clustering performance on average than original $\mathcal{L}_1$ graph. Nevertheless, our algorithm could be generalized in a straightforward way by introducing regularization terms such as elastic net into the current solver, which would indicate the quality of generated $\mathcal{L}_1$ graphs could be further improved.

Acknowledgments. This research is supported in part by NSF (IIS-0949467, IIS-1047715, and IIS-1049448), and NSFC (61532002, 61190120, 61190125, 61190124). We thank the anonymous reviewers for their constructive critiques.

References

1. Bruckstein, A.M., Elad, M., Zibulevsky, M.: On the uniqueness of nonnegative sparse solutions to underdetermined systems of equations. IEEE Trans. Inf. Theor. **54**(11), 4813–4820 (2008)
2. Cheng, B., Yang, J., Yan, S., Fu, Y., Huang, T.S.: Learning with-graph for image analysis. IEEE Trans. Image Process. **19**(4), 858–866 (2010)
3. Coifman, R.R., Lafon, S.: Diffusion maps. Appl. Comput. Harmonic Anal. **21**(1), 5–30 (2006)
4. Correa, C.D., Lindstrom, P.: Locally-scaled spectral clustering using empty region graphs. In: Proceedings of the 18th ACM SIGKDD International Conference on Knowledge Discovery and Data mining, pp. 1330–1338. ACM (2012)
5. Donoser, M., Bischof, H.: Diffusion processes for retrieval revisited. In: IEEE Conference on Computer Vision and Pattern Recognition, pp. 1320–1327. IEEE (2013)
6. Fang, Y., Wang, R., Dai, B., Wu, X.: Graph-based learning via auto-grouped sparse regularization and kernelized extension. IEEE Trans. Knowl. Data Eng. **27**(1), 142–154 (2015)

7. Han, S., Huang, H., Qin, H., Yu, D.: Locality-preserving l1-graph and its application in clustering. In: Proceedings of the 30th Annual ACM Symposium on Applied Computing, pp. 813–818. ACM (2015)
8. Koh, K., Kim, S.J., Boyd, S.P.: An interior-point method for large-scale l1-regularized logistic regression. J. Mach. Learn. Res. **8**(8), 1519–1555 (2007)
9. Lin, T.H., Kung, H.: Stable and efficient representation learning with nonnegativity constraints. In: Proceedings of the 31st International Conference on Machine Learning (ICML 2014), pp. 1323–1331 (2014)
10. Tibshirani, R.: Regression shrinkage and selection via the lasso. J. R. Stat. Soc. Ser. B (Methodological) **58**, 267–288 (1996)
11. Tropp, J.A., Wright, S.J.: Computational methods for sparse solution of linear inverse problems. Proc. IEEE **98**(6), 948–958 (2010)
12. Wright, J., Yang, A.Y., Ganesh, A., Sastry, S.S., Ma, Y.: Robust face recognition via sparse representation. IEEE Trans. Pattern Anal. Mach. Intell. **31**(2), 210–227 (2009)
13. Yang, Y., Wang, Z., Yang, J., Han, J., Huang, T.: Regularized l1-graph for data clustering. In: Proceedings of the British Machine Vision Conference. BMVA Press (2014)
14. Yang, Y., Wang, Z., Yang, J., Wang, J., Chang, S., Huang, T.S.: Data clustering by laplacian regularized l1-graph. In: Proceedings of the Twenty-Eighth AAAI Conference on Artificial Intelligence, pp. 3148–3149 (2014)
15. Zhang, Y.M., Huang, K., Hou, X., Liu, C.L.: Learning locality preserving graph from data. IEEE Trans. Cybern. **44**(11), 2088–2098 (2014)
16. Zhou, G., Lu, Z., Peng, Y.: L1-graph construction using structured sparsity. Neurocomputing **120**, 441–452 (2013)

Novel Models and Algorithms

A Rule Based Open Information Extraction Method Using Cascaded Finite-State Transducer

Hailun Lin[1], Yuanzhuo Wang[2], Peng Zhang[1(✉)], Weiping Wang[1],
Yinliang Yue[1], and Zheng Lin[1]

[1] Institute of Information Engineering, Chinese Academy of Sciences, Beijing, China
{linhailun,pengzhang}@iie.ac.cn
[2] Institute of Computing Technology, Chinese Academy of Sciences, Beijing, China

Abstract. In this paper, we present R-OpenIE, a rule based open information extraction method using cascaded finite-state transducer. R-OpenIE defines contextual constraint declarative rules to generate relation extraction templates, which frees from the influence of syntactic parser errors, and it uses cascaded finite-state transducer model to match the satisfied relational tuples. It is noted that R-OpenIE creates inverted index for each matched state during the matching process of cascaded finite-state transducer, which improves the efficiency of pattern matching. The experimental results have shown that our R-OpenIE can achieve good adaptability and efficiency for open information extraction.

Keywords: Relation extraction · Contextual constraint rule · Declarative definition · Knowledge base population

1 Introduction

In the past years, a large number of knowledge bases such as Freebase [4], YAGO [12], Probase [16], Knowledge Valut [7] have been constructed for purposes such as semantic search and question answering. Although a typical knowledge base may contain millions of entities and billions of relational facts, it is usually far from complete. Due to the explosive generation speed of big Web data, new knowledge is emerging rapidly everyday in the latest news and online social media. Therefore, to access that knowledge, information extraction and knowledge fusion methods are necessary. In this paper, we focus on the task of open information extraction, that is to automatically extract relational tuples from natural language text, which can be exploited for knowledge base population.

In recent years, open information extraction has received a lot of research interests, and many techniques have been proposed (e.g., [2,6,8,9,13,14]). Most of them do not require any background knowledge or manually labeled training data. For example, TextRunner [2], Reverb [9] and R2A2 [8] make use of natural language processing techniques, such as part-of-speech tagging and chunking. They focus on efficiency and high precision for high-confidence relations, but at low points of recall. Other techniques such as OLLIE [13], OpenIE [6] and

© Springer International Publishing Switzerland 2016
J. Bailey et al. (Eds.): PAKDD 2016, Part II, LNAI 9652, pp. 325–337, 2016.
DOI: 10.1007/978-3-319-31750-2_26

ClausIE [14] additionally use dependency parsing to improve precision and recall but at the cost of efficiency. These methods depend on the performance of syntactic parsing technology, which unavoidably involves in errors caused by parsers and costs a lot of time to obtain parsing results.

In order to overcome the limitations, in this paper, we present R-OpenIE, a rule based open information extraction method using cascaded finite-state transducer [11]. The method defines contextual constraint rules to generate relation extraction templates, and uses cascaded finite-state transducer model to match the satisfied relational tuples. It is noted that R-OpenIE creates inverted index for each matched state during the matching process of cascaded finite-state transducer, which can improve the efficiency of template matching to relation extraction. In general, the main contributions of this paper are three-fold:

- We propose R-OpenIE, a rule base open information extraction method using cascaded finite-state transducer, which can implement relations extraction without domain tuning and perform efficiently on large number of text.
- We present a context-free grammar based declarative rule definition language, which makes the relation extraction out of context and is easy to expand new rules, therefore it can improve the adaptability and scalability of R-OpenIE.
- We introduce cascaded finite-state transducer, which can realize complicated rules based relations extraction.

The rest of this paper is organized as follows. In Sect. 2, we give a review of related work. Section 3 formulates the problem and presents the R-OpenIE framework. Section 4 introduces the contextual constraint rules and relation templates for information extraction. Section 5 describes our method for information extraction based on cascaded finite-state transducer. Section 6 is devoted to the experimental results. Finally, the paper is concluded in Sect. 7.

2   Related Work

Open information extraction is aimed at automatically extracting relational tuples from natural language text. It was first introduced by the seminal work of Banko et al. [2], they proposed TextRunner system for open information extraction. Following that, a large variety of extraction techniques have been developed, such as Reverb [9], R2A2 [8], OLLIE [13], OpenIE [6] and ClausIE [14]. They can be divided into two categories:

The first category of open information extraction techniques only exploits shallow syntactic parsing, such as part-of-speech (POS) and chunking. TextRunner, Reverb and R2A2 belong to this category. TextRunner [2] first labeled a small corpus sample data as positive or negative, and used the labeled data to train a Naive Bayes classifier in an offline phase, and then applied the classifier to efficiently extract relations in an online phase, which used part-of-speech and lightweight noun phrase chunker to heuristically extracting relations. Reverb [9] was a simple method, it introduced two simple syntactic and lexical constraints on binary relations to reduce incoherent, uninformative and over-specific relation extractions. Based on Reverb, R2A2 [8] added an argument identifier, it used a number of classifiers to identify the arguments of a verbal phrase, and was able to extract relations that contain arguments that are not noun phrases.

The second category of open information extraction techniques exploits heavier natural language process technology, generally they not only used POS and chunking but also used dependency parsing. OLLIE, OpenIE, ClausIE belong to this category. OLLIE [13] trained data to learn extraction patterns on the dependency tree. OpenIE [6] was the upgraded version of TextRunner, it handled both verb-mediated relations and noun-mediated relations. ClausIE [14] exploited a set of domain independent lexica, it firstly decomposed sentence into clauses, and then extracted relations from these clauses based on dependency parsing.

In addition to the above works, Bast and Haussmann [3] proposed a contextual sentence decomposition based open information extraction method CSD-IE, it first decomposed a sentence into the parts that semantically belong together, and then identified the implicit or explicit verb in each such part to extract relations based on dependency parsing technique. Augenstein et al. [1] presented a distantly supervised class-based relation extraction method, it exploited large knowledge bases to automatically label the corpus data and use the annotated text to extract features and train a classifier, and then applied the classifier to extract relation mentions across sentence boundaries.

Based on the above analysis, we can conclude that although some open information extraction methods do not need any hand-labeled data for training, and can process very large amounts of data, these methods depends on the performance of syntactic parsing technology, which unavoidably involves in errors caused by parsers and costs a lot of time to obtain parsing results. In addition, it has been documented that these methods often produce uninformative as well as incoherent extractions.

Different from these methods, R-OpenIE is a rule based method, which is easy to expand new rules based on declarative rule definition language. R-OpenIE generates a unified cascaded finite-state transducer for all rules, and traverses the transducer from the bottom to up which can avoid redundant sub-rules matching, and during the matching process, the inverted index created from the bottom to up can accelerate the transducer matching efficiency.

3   R-OpenIE

In this section, we will firstly give some notations and formulate the problem of open information extraction, and then introduce the framework of R-OpenIE.

In the knowledge bases, a relational tuple typically takes the form of *subject-predicate-object* (SPO) according to the RDF model [15]. For example, considering the sentence: "Michael I. Jordan was born in Ponchatoula, Louisiana", it contains one relational tuple: (*Michael I. Jordan*)-(*was born in*)-(*Ponchatoula, Louisiana*). Now we can formulate the problem of open information extraction.

Open Information Extraction. Given a collection of documents D, the task of open information extraction is to automatically extract relational tuples from D in the form of $SPOs$ without manually labeled training data.

Based on the definition, we propose R-OpenIE, a new rule based open information extraction method using cascaded finite-state transducer. The framework of R-OpenIE contains three modules as follows:

- **Entity Recognition.** To extract all the relational tuples contained in the given document $d \in D$, R-OpenIE firstly exploits this module to extract all entities in the document. In this paper, this module directly adopts the most well-known tool, i.e., Stanford NER [10], to recognize entities.
- **Relation Templates Generation.** In this module, we define a context-free grammar based declarative rule definition language, and present a set of contextual constraint rules to generate relation templates.
- **Relational Tuples Extraction.** This module introduces cascaded finite-state transducer to model relation templates, where each constraint rule in the relation template is represented as a state in the transducer. The transducer takes the extracted entities as input, and starts to check whether the state is matched from initial states via bottom-up method. If all the states of a relation template are matched, a relational tuple is extracted.

In the following sections, we will introduce relation templates generation module and relational tuples extraction module in details.

4   Relation Templates Generation

The aim of this section is to generate proper relation templates for information extraction. Essentially, a relation template is composed of three basic types of constraint rules, i.e., entity type constraint, sign word constraint and contextual constraint. In what follows, we will discuss how the rules are defined.

(1) *Entity Type Constraint Rule*: This rule is used to confine the range of subject (S) or object (O) in relational tuples that can be matched by a relation template. In this paper, we mainly focus on four entity type, i.e., person, organization, location and date, which is denoted as PER, ORG, LOC and DATE individually. The entity type constraint rule is defined as follows:

$$RULE : [ENTITY\_TYPE] : [INSTANCE]$$

where $RULE$ is the constraint rule indicator; $ENTITY\_TYPE=\{PER, ORG, LOC, DATE\}$ is the constraint type; $INSTANCE$ is an entity mention of $ENTITY\_TYPE$. For example, considering the following sentence, which will be our running example throughout the paper:

Michael I. Jordan received his MS in Mathematics in 1980 from the Arizona State University and his PhD in Cognitive Science in 1985 from the University of California, San Diego.

It can be seen that the sentence can be matched by the following entity type constraint rules:

- RULE:PER:[INSTANCE], the matched result is *Michael I. Jordan.*
- RULE:ORG:[INSTANCE], the matched results are *Arizona State University* and *University of California, San Diego.*
- RULE:DATE:[INSTANCE], the matched results are *1980* and *1985.*

(2) *Sign Word Constraint Rule*: This rule is used to define the range of predicate (P). In this paper, we exploit a set of domain independent lexica derived from Wikipedia[1] for sign word rule definition. The sign word constraint rule is defined as follows:

$$RULE : [SIGN\_WORD] : [INSTANCE]$$

where *SIGN_WORD* is the indicator of relation type; *INSTANCE* is a mention of *SIGN_WORD*. For example, our example sentence above can be matched by the following sign word constraint rule:

- RULE:GRADUATE:[INSTANCE], the matched result is *received from.*
- RULE:DEGREE_GET_OF:[INSTANCE], the matched result is *from.*
- RULE:DEGREE:[INSTANCE], the matched results are *MS* and *PhD.*

(3) *Contextual Constraint Rule*: This rule defines the logical restriction between entity type or sign word constraint rules, or defines the instances restriction in context. In this paper, we define four type contextual constraint rules as illustrated in Table 1:

Table 1. The contextual constraint rules

Type	Definition
AND	All parts in a rule clause must be matched
OR	At least one part in a rule clause be matched
SENT	All parts in a rule clause occur at the same sentence
ORD	All parts in a rule clause occur at the pre-defined ordering
DIST_n	Distance between some parts is no more than n words

Taking rule *ORD* as an example, we can define the contextual constrain rule between two sign word constraint rules as follows: (*ORD, DEGREE, DEGREE_GET_OF*), which defines that the sign word of *DEGREE* must occur before *DEGREE_GET_OF*.

Combined these rules, in the following we will elaborate on the process of relation templates generation.

---

[1] https://www.wikipedia.org/.

Relation Templates Generation. In this paper, we adopt distant supervised method to automatically generate relation templates, which is an offline phrase. A relation template is a composition of constraint rules. It is defined as follows:

$$Template : [name] : ([variables]) : ([contextual\_rule],$$
$$[variable\_v\{entity\_type\_rule\}], [sign\_word\_rule])$$

where *name* is the relation type that the relation template represents; *variables* is the output variables list that the template can match; *contextual_rule*, *entity_type_rule* and *sign_word_rule* are the constraints rules defined as above; *variable_v* is one of the output variable assignment. Next, we elaborate on the generation of relation templates:

- **Label the Corpus using the Existing Knowledge Base.** R-OpenIE makes use of YAGO [12] relation facts to label sentences from Wikipedia documents.
- **Determine the Sets of Clauses using Dependency Parsing.** R-OpenIE firstly exploits the Stanford dependency parser [5] to analyze structure of an input sentence, and then identifies the clauses in the input sentences, based on mapping the dependency relations to relation facts constituents. Specifically, we first identify the dependency clause for subject (S) in relation facts, meanwhile the dependency clause should contain a governor verb (P), and then find other constituents of the clause based on the governor verb.
- **Generate Relation Templates using Clauses Generalization.** R-OpenIE firstly tries to identify the type of each clause, and finds proper entity type or sign word constraint rules to represent the clauses. After that, R-OpenIE counts the occurrence of these clauses in sentences, and then generalizes the occurrence of these clauses in terms of contextual constraint rules. R-OpenIE finally combines the generalized contextual constraint rules with the above entity type and sign word constraint rules to generate the relation templates.

For example, from our sentence above, we can generate the following "graduated and graduation time" relation templates:

- Template:PERSON_COLLEGE1(person, graduate_time, college):(SENT, person_v{PER}, (DIST_10, graduate_time_v{DATE}, GRADUATE, college_v {ORG}))
- Template:PERSON_COLLEGE2(person, graduate_time, college):(SENT, person_v{PER}, graduate_time_v{DATE}, college_v{ORG}, (ORD, DEGREE, DEGREE_GET_OF))

Based on the generated relation templates, in the following, we will elaborate on the process of relational tuples extraction.

5   Relational Tuples Extraction

In this section, we will present how R-OpenIE extracts relational tuples based on relation templates. Given a relation template and a document text, R-OpenIE takes the following steps to extract relational tuples:

(1) Based on the dependencies between rules in a relation template, a rule based cascaded finite-state transducer is constructed, where:

- State: each rule is represented as a state, where entity type and sign word constraint rules are represented as lower layer states, and contextual constraint rules are represented as higher layer states.
- Transition: if and only if the extracted instances matched with lower layer states satisfy the constraints of higher layer state, the transition that these lower layer states are transformed into higher layer state is fired.

For example, from our generated "graduated and graduation time" relation template, the constructed cascaded finite-state transducer is illustrated in Fig. 1.

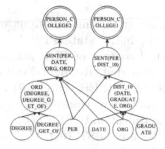

Fig. 1. The example of relation template cascaded finite-state transducer.

From Fig. 1, we can see that there are four layers in the transducer. Layer 1 models entity type and sign word constraint rules; Layer 2 models basic contextual constraint rules; Layer 3 models combination constraint rules; Layer 4 models constraint rules for relation templates of interest.

(2) Computing the matching results of the given text using the constructed transducer: it takes the extracted entities from the given text as input, and takes the lower layer states in the transducer as starting points to check whether the states are matched via bottom-up methods. According to the dependency relationship between states, checking whether their parent states are reachable or not, until there is no state can be reached, the matching is terminated. If all the states corresponding a relation template are matched, then a relational tuple is extracted. During the matching process, R-OpenIE creates inverted index for each matched state, which can improve the efficiency of template matching.

For example, from our sentence above, in the following, we will match the sentence with transducer illustrated in Fig. 1. Based on the cascaded finite-state transducer, starting from initial states, the inverted index on matched content of PER, DEGREE, DEGREE_GET_OF, DATE, GRADUATE and ORG states are created. The matched initial state transducer is shown in Fig. 2(a), and its corresponding inverted index is shown in Fig. 2(b). It is noted that R-OpenIE only creates inverted index for the matched content of states.

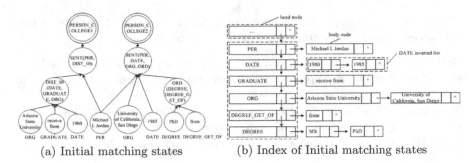

(a) Initial matching states   (b) Index of Initial matching states

Fig. 2. (a) The initial matching states of transducer. (b) The inverted index of initial matching states.

Specifically, in the transducer, taking an input entity instance "Michael I. Jordan" for example, if the initial state PER is matched, R-OpenIE creates an index for the matched content of PER state. The inverted index for PER is shown in Fig. 2(b). After that, R-OpenIE will check the next potential matched higher layer states: SENT(PER, DIST_10) and SENT(PER, DATE, ORG, ORD), it checks whether these transitions to these states are fired or not. For example, if the state transition to SENT(PER, DIST_10) state is fired, it requires that besides PER state, DIST_10(DATE, GRADUATE, ORG) has also been matched. R-OpenIE will search the inverted index, and it will find that DIST_10(DATE, GRADUATE, ORG) does not exist which means that this state has not been matched, so the transition to SENT(PER, DIST_10) is terminated. The checking process of SENT(PER, DATE, ORG, ORD) is similar to SENT(PER, DIST_10).

After matching PER state, R-OpenIE starts to match DATE state from bottom-up. Similar to PER state matching, if DATE state is matched, R-OpenIE will create an inverted index for the matched content of DATE state, and it will check the next potential matched higher layer states: DIST_10(DATE, GRADUATE, ORG) and SENT(PER, DATE, ORG, ORD), and then sequentially checks whether the transition conditions to these states are satisfied: for state DIST_10(DATE, GRADUATE, ORG), R-OpenIE searches DATE, GRADUATE, ORG in the inverted index, if the matched instances of DATE, GRADUATE and ORG are existed, the transition conditions to state DIST_10(DATE, GRADUATE, ORG) are satisfied, and R-OpenIE will check whether the distance between GRADUATE and ORG is no more than 10 words, if it holds, the state DIST_10(DATE, GRADUATE, ORG) is matched, and the created inverted index for the state is shown in Fig. 3. R-OpenIE takes the similar way to check the matching status of state SENT(PER, DATE, ORG, ORD) based on the inverted index.

From above, it can be seen that according to the transducer layers, the process of R-OpenIE extracting relational tuples is an iterative process. Until the states checking process of R-OpenIE can reach to a termination state, a relational tuple is extracted. In Fig. 2(a), given a text, if R-OpenIE can reach to

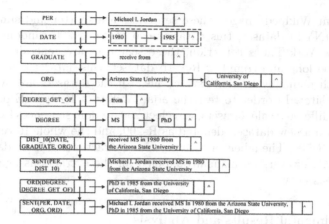

Fig. 3. The inverted index for example sentence.

state PER_COLLEGE1 or PERSON_COLLEGE2, it identifies that a relational tuple satisfied "graduated and graduation time" relation template is extracted. After that the matching process finishes, and the output results of the matching are: person="Michael I. Jordan" (PER), graduate_time="1980" (DATE), college="Arizona State University" (ORG). In addition, we can continue to look up the other triggered transitions. The final created inverted index structure for our sentence matching is shown in Fig. 3.

6   Experiments

In this section, we will evaluate the performance of our method R-OpenIE for open information extraction. We will: (1) evaluate the adaptability of our method compared with four state-of-the-art baseline methods in terms of accuracy on different benchmark datasets; (2) evaluate the efficiency of our method compared with the baseline methods in terms of time on the benchmark datasets.

6.1   Experimental Settings

All experiments were conducted on a server running 64-bit Linux OS, with 16 core 2GHz AMD Opteron(tm) 6128 Processors and 32GB RAM. And we compared our method R-OpenIE with three state-of-the-art methods described in Sect. 2, namely, Reverb, OLLIE and ClausIE. Specifically, we evaluated the effectiveness using the metric of accuracy based on the number of relations correctly extracted, and we evaluated the efficiency using run time. Accuracy is computed as the rate of the number of correctly extracted relations to the total number of extracted relations.

In the experiments, we used the same three datasets with different size which are used in ClausIE [6]: Wikipedia dataset, this dataset contains 200 random sentences from Wikipeda pages, these sentences are shorter and simpler; New York Times (NYT) dataset, this dataset also contains 200 random sentences from the New York Times collection, these sentences are generally very clean but tend to be long and complex; Reverb dataset, this dataset consists of 500 sentences with manually labeled extractions. All the datasets are available on ClausIE web site[2]. In order to test the adaptability of R-OpenIE on different datasets and different scale datasets, we select 200 random sentences from Reverb dataset to form a new dataset, denoted as R-200, and the whole Reverb dataset is denoted as R-500. The labels of their relation extractions of these datasets are also available. However, we manually labeled all the extractions again for the reason of consistency.

6.2   Experimental Results and Analysis

Effectiveness Evaluation. In this experiment, we conducted two experiments to test the adaptability of R-OpenIE and the baseline methods. Firstly, we conducted experiments on Wikipedia, NYT and R-200 datasets to test these methods on different datasets with same size. Secondly, we conducted experiments on R-200 and R-500 datasets to test these methods on same dataset with different size. Tables 2 and 3 show the results individually.

Table 2. Results on different datasets with same size

Approach	Wikipedia			NYT			R-200		
	#correct	#total	accuracy	#correct	#total	accuracy	#correct	#total	accuracy
Reverb	188	249	0.755	177	271	0.653	168	292	0.575
OLLIE	230	408	0.564	200	358	0.559	260	493	0.527
ClausIE	421	610	0.690	412	644	0.640	603	948	0.636
R-OpenIE	**474**	**583**	**0.813**	**529**	**719**	**0.736**	**683**	**952**	**0.717**

Table 3. Results on datasets with different size

Approach	R-200			R-500		
	#correct	#total	accuracy	#correct	#total	accuracy
Reverb	168	292	0.575	411	727	0.565
OLLIE	260	493	0.527	542	1240	0.437
ClausIE	603	948	0.636	1502	2375	0.632
R-OpenIE	**683**	**952**	**0.717**	**1704**	**2379**	**0.716**

From Table 2, it can be seen that on the Wikipeida, NYT and R-200 datasets, our method R-OpenIE extracts the largest number of correct relations (see column of #correct) and obtains the highest accuracy on these datasets. In addition,

---

[2] http://www.mpi-inf.mpg.de/departments/databases-and-information-systems/software/clausie/.

comparing the results of Wikipedia dataset with the results of NYT dataset, we can see that the results of the later dataset are slightly worse than the former ones for all the methods. The most likely reason for this is that sentences from news are deeply nested and typically longer the sentences from Wikipedia, and this involves in more broken parses.

From Table 3, we can see that on the R-200 and R-500 datasets, our method R-OpenIE extracts the largest number of correct relations and obtains the highest accuracy on these datasets. In addition, as the results shown in Table 3, R-OpenIE exhibits more steady accuracy on the datasets with different scales.

In general, we can conclude that our method R-OpenIE can obtain better adaptability compared with the baseline methods. This is because that our method does not exploit dependency parsing, it uses patterns to match relations. It demonstrates that dependency parser introduces errors which affects the accuracy of relation extraction systems.

Efficiency Evaluation. In this section, we evaluate the efficiency of our method R-OpenIE and the baseline methods on the datasets. We took the average time per relation extraction on the Wikipedia, NYT and R-200 datasets, and took the total run time of n facts on the R-500 datasets as the metric to measure the efficiency. The results are graphically depicted in Fig. 4.

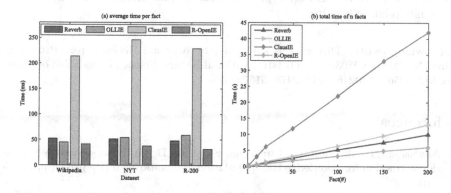

Fig. 4. The run time on the datasets

From Fig. 4(a), it can be seen that on these datasets, the average time per relation extraction of R-OpenIE is the least among these methods. And from Fig. 4(b), we can see that R-OpenIE spends the least run time to extract the same size of relation facts. In addition, with the increasing of relation facts extraction, R-OpenIE takes minimal run time growth compared with the baseline methods. This is benefit from that R-OpenIE does not using syntactic parsing, which may cost a lot of time to obtain parsing results, and during the extraction process, R-OpenIE exploits inverted index for cascaded finite-state transducer matching to improve the extraction efficiency.

As was mentioned above, it can be seen that R-OpenIE can not only obtain better accuracy, but also obtain better efficiency in relation extraction, which demonstrates the good adaptability and efficiency of R-OpenIE for open information extraction. Moreover, from the accuracy of knowledge engineering perspective, a slightly improvement is of great importance to applications. Therefore, our method R-OpenIE is a useful technique for open information extraction.

7   Conclusions

In this paper, we presented R-OpenIE, a new rule based open information extraction method using cascaded finite-state transducer. The experimental results have shown that our R-OpenIE can achieve good adaptability and efficiency for open information extraction. We will explore the following further work:

- Relation normalization method. As relational tuples extracted by R-OpenIE may be uninformative and incoherent, they are hard to map to knowledge bases for population. In future, we may explore a semantic based method using Markov clustering for relation normalization.
- Relation quality evaluation method. As relational tuples extracted from different data sources may exist errors and conflicts, it will bring errors to knowledge bases if these tuples are directly populated into knowledge bases. In future, we may explore a multi-layer probabilistic graphic model based method to evaluate relation quality.

Acknowledgments. This work is supported by National HeGaoJi Key Project of China (No. 2013ZX01039-002-001-001), National Natural Science Foundation of China (No. 61303056, 61402464, 61502478, 61572469).

References

1. Augenstein, I., Maynard, D., Ciravegna, F., et al.: Distantly supervised web relation extraction for knowledge base population. Semantic Web (Preprint), pp. 1–15 (2015)
2. Banko, M., Cafarella, M.J., Soderland, S., Broadhead, M., Etzioni, O.: Open information extraction for the web. IJCAI **7**, 2670–2676 (2007)
3. Bast, H., Haussmann, E.: Open information extraction via contextual sentence decomposition. In: 2013 IEEE Seventh International Conference on Semantic Computing (ICSC), pp. 154–159. IEEE (2013)
4. Bollacker, K., Evans, C., Paritosh, P., Sturge, T., Taylor, J.: Freebase:a collaboratively created graph database for structuring human knowledge. In: Proceedings of the 2008 ACM SIGMOD International Conference on Management of Data, pp. 1247–1250 (2008)
5. Chen, D., Manning, C.D.: A fast and accurate dependency parser using neural networks. In: Proceedings of the 2014 Conference on Empirical Methods in Natural Language Processing, vol. 1, pp. 740–750 (2014)
6. Del Corro, L., Gemulla, R.: Clausie: clause-based open information extraction. In: Proceedings of the 22nd International Conference on World Wide Web, pp. 355–366 (2013)

7. Dong, X., Gabrilovich, E., Heitz, G., Horn, W., Lao, N., Murphy, K., Strohmann, T., Sun, S., Zhang, W.: Knowledge vault: a web-scale approach to probabilistic knowledge fusion. In: The 20th ACM SIGKDD International Conference on Knowledge Discovery and Data Mining, pp. 601–610. ACM (2014)
8. Etzioni, O., Fader, A., Christensen, J., Soderland, S., Mausam, M.: Open information extraction: the second generation. IJCAI **11**, 3–10 (2011)
9. Fader, A., Soderland, S., Etzioni, O.: Identifying relations for open information extraction. In: Proceedings of the Conference on Empirical Methods in Natural Language Processing, pp. 1535–1545 (2011)
10. Finkel, J.R., Manning, C., Manning, C.: Incorporating non-local information into information extraction systems by gibbs sampling. In: Proceedings of the 43rd Annual Meeting of the Association for Computational Linguistics. pp. 363–370 (2005)
11. Hobbs, J.R., Appelt, D., Bear, J., Israel, D., Kameyama, M., Stickel, M., Tyson, M.: 13 fastus: a cascaded finite-state transducer for extracting information from natural-language text. Finite-state language processing, p. 383 (1997)
12. Hoffart, J., Suchanek, F.M., Berberich, K., Weikum, G.: Yago2: A spatially and temporally enhanced knowledge base from wikipedia. In: Proceedings of the Twenty-Third International Joint Conference on Artificial Intelligence, pp. 3161–3165 (2013)
13. Schmitz, M., Bart, R., Soderland, S., Etzioni, O., et al.: Open language learning for information extraction. In: Proceedings of the 2012 Joint Conference on Empirical Methods in Natural Language Processing and Computational Natural Language Learning, pp. 523–534 (2012)
14. Soderland, S., Gilmer, J., Bart, R., Etzioni, O., Weld, D.S.: Open information extraction to kbp relations in 3 hours (2013)
15. Wood, P.T.: Query languages for graph databases. ACM SIGMOD Rec. **41**(1), 50–60 (2012)
16. Wu, W., Li, H., Wang, H., Zhu, K.Q.: Probase: a probabilistic taxonomy for text understanding. In: Proceedings of the ACM SIGMOD International Conference on Management of Data, pp. 481–492 (2012)

Active Learning Based Entity Resolution Using Markov Logic

Jeffrey Fisher[✉], Peter Christen, and Qing Wang

Research School of Computer Science, Australian National University,
Canberra, ACT 0200, Australia
{jeffrey.fisher,peter.christen,qing.wang}@anu.edu.au

Abstract. Entity resolution is a common data cleaning and data integration problem that involves determining which records in one or more data sets refer to the same real-world entities. It has numerous applications for commercial, academic and government organisations. For most practical entity resolution applications, training data does not exist which limits the type of classification models that can be applied. This also prevents complex techniques such as Markov logic networks from being used on real-world problems. In this paper we apply an active learning based technique to generate training data for a Markov logic network based entity resolution model and learn the weights for the formulae in a Markov logic network. We evaluate our technique on real-world data sets and show that we can generate balanced training data and learn and also learn approximate weights for the formulae in the Markov logic network.

1 Introduction

Entity resolution (ER) is a common data cleaning and data integration task that involves determining which records in one or more data sets refer to the same real-world entities. It has numerous applications for commercial, academic and government organisations including matching customer databases following a corporate merger, combining different data sets for research purposes, and detecting persons of interest for national security [4].

In many applications, a domain expert can perform the ER task manually, albeit in a very time-consuming fashion [4]. If a domain expert is presented with two records and their context, they can usually determine whether or not the two records refer to the same real-world entity. As a result, one approach that has been used successfully for ER is active learning [1, 2, 23, 29], where an active learning algorithm selects pairs of records to present to an expert who then manually classifies them as either matches or non-matches. The results are then used to build an automated classification model.

However, in domains such as group linkage [11, 20] or population reconstruction [6], a simple pair-wise classification model may be insufficient to accurately perform ER and the classification model may need to be more complex to capture the characteristics of the data sets involved. Collective classification techniques such as

© Springer International Publishing Switzerland 2016
J. Bailey et al. (Eds.): PAKDD 2016, Part II, LNAI 9652, pp. 338–349, 2016.
DOI: 10.1007/978-3-319-31750-2_27

those proposed by Bhattacharya and Getoor [3], Kalashnikov and Mehrotra [15], and Markov logic networks (MLNs) [22,27], are all effective at capturing the intricacies of complex ER problems such as matching bibliographic data, group linkage and population reconstruction. In this paper we demonstrate how we can use active learning to incorporate the domain knowledge of experts into an active learning based framework to generate and refine the rules for an MLN based ER model. While the expert still classifies pairs of records as either matches or non-matches, the rule-based approach of MLNs allows a much more sophisticated ER model to be developed.

The rest of this paper is structured as follows: In Sect. 2 we describe recent literature relating to active learning, MLNs and ER, and provide a short background on MLNs. In Sect. 3 we describe our notation and formally define the problem we are attempting to solve. In Sect. 4 we then present our approach for using active learning to create the rules and weights for an MLN based ER model. In Sect. 5 we evaluate our approach and in Sect. 6 we present our conclusions and some directions for future work.

2   Related Work and Background

Entity resolution has been the subject of a large amount of literature, and several recent surveys have been conducted [9,17]. In this section we briefly outline recent techniques relating to MLN based ER and active learning based ER.

Markov logic networks have attracted considerable research and been applied to a range of problems. Proposed by Richardson and Domingos [22] they combine first-order logic and probabilistic graphical models into a single approach. First-order logic allows a compact representation and expression of knowledge, and probabilistic graphical models allow treatment of uncertainty. Combining these two techniques allows MLNs to perform inference on a wide variety of problems and to handle contradictory knowledge and uncertainty.

A number of techniques have been proposed to determine the parameters of an MLN using a variety of different methods [13,16,26]. However, these techniques all rely on training data being available which is often not the case for ER [4]. Moreover, due to the fact that the classes of matches and non-matches are typically unbalanced, it can become very expensive to generate training data [29]. For example, a random sampling might require reviewing hundreds or thousands of non-matches for every match found.

Markov logic networks have successfully been used to perform ER [27]. The scalability of MLNs is a problem which has been partially overcome by Rastogi et al. [21]. Their experiments showed that MLN based techniques can achieve very high quality ER in situations where training data exists and the scalability problems can be overcome. As a result, an approach that uses active learning to generate training data can be used to learn the rules and weights for an MLN. In doing so it would allow the full power of MLNs to be applied to many more entity resolution applications where training data does not exist.

Active learning has been used in many fields as a way to generate training data so that supervised classification techniques can be applied. It has been extensively studied and a survey of relevant literature has been conducted by Settles [24]. In its traditional form, it involves presenting examples to an expert for manual classification and the features of these examples are used to build a classification model.

Active learning has been successfully applied to the ER problem [1,2,23,29]. As described above, one particular problem when using active learning for entity resolution is that the classes of matches and non-matches are typically very unbalanced. As a result, a challenge with each of these techniques is to select representative examples of each class to present to the expert for manual classification. This has been addressed in several ways including exploiting cluster structure in the data [29] and using a multi-stage algorithm that prunes redundant pairs [7]. However, active learning has not been used to generate a training set for use in creating an MLN based ER model.

Active learning techniques vary based on the way examples are selected for manual classification. Of particular relevance to this paper are techniques dealing with variance reduction and density weighted methods. Variance reduction based techniques aim to reduce generalisation errors by minimising output variance and have been applied to classification models such as neural networks [18] and conditional random fields [25]. Density based techniques such as that proposed by Settles and Craven [25] aim to ensure that the examples chosen for manual classification are not only uncertain, but also somehow representative of the underlying distribution. This means that the expert does not waste time classifying outliers, which may be uncertain, but have minimal impact on the overall classification quality.

Finally, while traditional active learning techniques make use of an all-knowing domain expert to perform the manual classification, from a practical perspective this may not always be possible. Variations have been proposed which deal with noisy oracles, i.e. experts who return the wrong classification result [8]. Instead of a domain expert, other techniques make use of crowd-sourcing to perform the labelling [28]. While our technique does not explicitly deal with these variations, there is no reason why different forms of oracles could not be used instead of the traditional domain expert.

2.1 Markov Logic Networks

An MLN consists of a set of weighted first-order logical formulae (rules). In a traditional first-order knowledge base, a single violation of a formula invalidates the entire knowledge base. However, for an MLN, the weight of a formula indicates the relative strength of the formula, i.e. its likelihood of being true. As a result, an MLN can handle noisy or inconsistent data without requiring a huge number of very specific formulae. Formally, given a set of first-order formulae and their weights, $\{(u_i, w_i)\}$, and a finite set of constants, we can instantiate an MLN as a Markov random field where each node is a grounding of a predicate and each feature is a grounding of one of the formulae. The joint probability distribution is given by:

$$P\left(\mathbf{X}=x\right)=\frac{1}{\mathbf{Z}}exp\left(\sum_{i}w_{i}n_{i}\left(x\right)\right) \tag{1}$$

where n_i is the number of times the ith formula is satisified in world x, and $\mathbf{Z}$ is a normalisation constant [22].

The formulae in an MLN can be defined by an expert or determined using inductive logic programming [16], and a number of techniques have been developed to determine their weights using training data, both in batch mode [13, 26] and online [14, 19].

3 Notation

We briefly describe the notation we will use in this paper. We begin with a data set $\mathbf{R}$. We assume a finite list of *rules* $\mathbf{U} = \langle u_1, u_2, \ldots u_k \rangle$, where each rule $u \in \mathbf{U}$ is a function that takes as input two records $r_i, r_j \in \mathbf{R}$ and returns either *True* or *False*. We denote the number of rules in $\mathbf{U}$ as $|\mathbf{U}|$. Each rule $u \in \mathbf{U}$ has a real valued weight denoted $w(u)$. An example of a rule is:

$u(r_1, r_2) = HasName(r_1, n_1) \wedge HasName(r_2, n_2) \wedge SameName(n_1, n_2)$
which returns *True* if records r_1 and r_2 have values n_1 and n_2 respectively in the *Name* attribute, and n_1 and n_2 are the same, and *False* otherwise. Rules with positive weights indicate positive evidence that records r_i and r_j refer to the same entity (match), while rules with negative weights indicate negative evidence that r_i and r_j refer to the same entity (non-match). By creating rules in this format and relating them to whether the records are a match or non-match, these rules can later be easily converted to formulae for the MLN model [27]. We assume that the initial list of rules is provided by a domain expert who will be performing the manual classification in the active learning process.

We define a *rule vector* as a binary vector v, and the set of such rule vectors as $\mathbf{V}$. We generate a rule vector by applying each $u \in \mathbf{U}$ to a pair of records $r_i, r_j \in \mathbf{R}$ using a function $\Psi(\mathbf{U}, r_i, r_j)$ which returns a rule vector v of length $|\mathbf{U}|$ where $v[k] = 1$ if $u_k(r_i, r_j) = True$ and $v[k] = 0$ if $u_k(r_i, r_j) = False$ for $1 \leq k \leq |\mathbf{U}|$. Since different pairs of records may produce the same rule vector, for each rule vector v_x we define a set of record pairs $P(v_x) = \{\langle r_y, r_z \rangle : r_y, r_z \in \mathbf{R}, \Psi(\mathbf{U}, r_y, r_z) = v_x\}$. The cardinality of $P(v_x)$ is denoted $|P(v_x)|$. Since $\mathbf{V}$ only includes rule vectors that are produced by a pair of records in $\mathbf{R}$, its size is limited to the smaller of $O(|\mathbf{R}|^2)$ and $O(2^{|\mathbf{U}|})$ and in practice it is significantly smaller than both these limits. The weight of each v is defined as $w(v) = (\sum w(u_k) : 1 \leq k \leq |\mathbf{U}|$ and $v[k] = 1)$. In essence, $w(v)$ is the sum of the weights of those rules that are set to *True* in v.

Our technique uses a *domain expert* $\mathbf{E}$ (or an oracle, crowd sourcing, etc. as per other active learning techniques [24]) to perform manual classification as part of our approach. However, typically the expert can only classify a certain number of record pairs so we assume a total *budget* of manual classifications b with $b \geq 1$. The budget is divided into q manual classifications per round, over n rounds, where $q * n = b$. In future work we intend to investigate ways to adaptively distribute the budget [29]. We denote the output of the manual classification of records r_j and r_k as o_{jk} where $o_{jk} = True$ if r_j and r_k are classified as a match by the expert and $o_{jk} = False$ if they are classified as a non-match.

During the active learning process we need a way of determining which record pairs $\langle r_i, r_j \rangle$ to present to the expert for manual classification. We define a *strategy* S as an ordering of the rule vectors in $\mathbf{V}$. After ordering $\mathbf{V}$ by S, we select one candidate pair from each of $P(v_1) \ldots P(v_q)$ to be presented to the expert, where q is the number of manual classifications each round. Some possible strategies for S include selecting: (1) the most definite positive examples ($S^+ :=$ descending sort of $w(v)$), (2) the most definite negative examples ($S^- :=$ ascending sort of $w(v)$), (3) the most definite examples of either type ($S^d :=$ descending sort of $|w(v)|$), (4) the most ambiguous cases ($S^a :=$ ascending sort of $|w(v)|$, since the most ambiguous cases have a total weight close to 0), (5) the most commonly occuring vectors ($S^c :=$ descending sort of $|P(v)|$) and (6) $S^r :=$ random ordering.

Problem Statement: Given a data set $\mathbf{R}$, an initial set of rules $\mathbf{U}_s$, a budget b, a strategy S, and an expert $\mathbf{E}$ who can make manual classifications, the problem is to generate a balanced set of training examples for an MLN based ER model, a final set of rules $\mathbf{U}_n$, and values of $w(u_i)$ for each $u_i \in \mathbf{U}_n$.

4   Methodology

The basic idea of our approach is that we apply traditional active learning techniques [23] by presenting pairs of records to an expert for manual classification. However, in addition to classifying each pair as either a match or a non-match we ask the expert to specify the reason for their decisions - i.e. the rule(s) that most influenced their decisions. In the case where the expert's classification of a pair contradicts the weight of the corresponding rule vector, i.e. a record pair which produces a rule vector with $w(v) > 0$ but the expert classifies the pair as a non-match, or $w(v) < 0$ but the expert classifies the pair as a match, we allow the expert to specify a new rule that would cover the particular pair. We could also use inductive logic programming based techniques [16] to specify the new rule instead. This new rule is then added to the current list of rules and is used to evaluate record pairs in the next iteration of the active learning process.

Our approach is described in Algorithm 1. We begin with an initial list of rules $\mathbf{U}_s$ created by the domain expert, along with a selected strategy S. Throughout the process, we refine the list of rules along with the weights in order to end up with a final list of rules that will be converted to formulae in an MLN based ER model. Our algorithm assumes the budget is split into n rounds, with q manual classifications per round, such that $q * n = b$. Each iteration of the main loop (lines 2 – 22) is one round of the algorithm.

Within each round, we start by creating an empty set of rule vectors (line 3). Because calculating a rule vector for each pair $r_i, r_j \in \mathbf{R}$ has time complexity

---

Algorithm 1. Active Weight Learning

Input:
- Set of records: $\mathbf{R}$
- Initial list of rules: $\mathbf{U}_s$
- Strategy: S
- Budget: b split into q, n such that $q * n = b$ // Questions per round, number of rounds
- Expert: $\mathbf{E}$

Output:
- Final list of rules: $\mathbf{U}_n$

```
1:  i := 1, U₀ := ⟨ ⟩                          // Set U₀ to an empty list
2:  while i ≤ n and Uᵢ ≠ Uᵢ₋₁ do:
3:      V := ∅, Uᵢ := Uᵢ₋₁                     // Create Uᵢ, for i = 1, U₁ := Uₛ
4:      B := GenerateBlocks(R)                 // Create a set of blocks
5:      foreach bₐ ∈ B do:          // In practice we only do this for a proportion of blocks in B
6:          foreach ⟨rⱼ, rₖ⟩ ∈ bₐ do:
7:              vⱼₖ := Ψ(Uᵢ, rⱼ, rₖ)    // vⱼₖ is generated by applying the rules in Uᵢ to rⱼ, rₖ
8:              if vⱼₖ ∈ V do:
9:                  UpdateRuleVector(V, vⱼₖ)         // Update P(vⱼₖ) if we have seen vⱼₖ before
10:             else do:
11:                 V.add(vⱼₖ)           // Otherwise add vⱼₖ to V with P(vⱼₖ) = {⟨rⱼ, rₖ⟩}
12:     V* := Order(V, S)                       // Order V based on strategy S
13:     for vⱼ ∈ V* and 1 ≤ j ≤ q do:          // For the first q rule vectors in V*
14:         ⟨rₐ, r_b⟩ := P(vⱼ).random           // Randomly select a candidate pair from P(vⱼ)
15:         oₐb := ManualClassify(rₐ, r_b, E)   // Get the true match status from expert
16:         UpdateRuleWeights(Uᵢ, oₐb)          // Update rule weights based on expert
17:         if (oₐb = True and w(vⱼ) < 0) or (oₐb = False and w(vⱼ) > 0) do:
18:             uₙₑw := GetNewRule                   // Get a new rule to cover the pair
19:             Uᵢ.append(uₙₑw)
20:     while |Uᵢ| > |Uₛ| do:                   // Check if we added some rules
21:         RemoveLowestPredictiveRule(Uᵢ)      // And if we did, remove the least useful rules
22:     i + +
23: return Uₙ
```

$O(|\mathbf{R}|^2)$, we make use of blocking to reduce the number of rule vectors calculated. Blocking is the process whereby a data set is split into subsets called blocks and only records within the same block are compared [5]. We assume the blocking process is a black box where we pass a data set and get back a set of blocks (line 4). Within each block, we evaluate the rules on each pair of records to generate the rule vectors (lines 5–11).

We apply the strategy S to the rule vectors in $\mathbf{V}$ to produce an ordering $\mathbf{V}^*$ (line 12). Then, we select a candidate pair from each of the first q rule vectors in $\mathbf{V}^*$ to be presented to the expert $\mathbf{E}$ for manual classification (lines 14–15). Based on the output of this manual classification we update the weights of the rules in $\mathbf{U}_i$, and if the manual classification result is contrary to what the weights of the rule vectors indicate, i.e. $w(v_{ij}) > 0$ but $o_{ij} = False$ or $w(v_{ij}) < 0$ but $o_{ij} = True$, we ask the expert to create a new rule that would cover the incorrectly classified pair (line 18). Alternatively, techniques such as inductive logic programming based learning [16] could also be used to refine the rules in $\mathbf{U}_i$ so that the incorrectly classified pair is fixed.

Finally, if our number of rules has increased in the current round (line 20), we look at removing the least informative rules from $\mathbf{U}_i$ based on a combination of the absolute values of their weights and their coverage. For each rule $u_j \in \mathbf{U}_i$, we calculate a score $s = |w(u_j)| * log(|T(u_j)|)$ where $T(u_j) = \{r_a, r_b : r_a, r_b \in \mathbf{R}, u_j(r_a, r_b) = True\}$. We then remove the rules with the lowest score (line 21) which allows us to avoid overfitting and keeps rules with a balance of high predictive power (both positive and negative) and high coverage.

4.1   Building Blocks of the Algorithm

We discuss some aspects of the approach in more detail, namely blocking, the choice of strategy S, and some considerations to prevent overfitting.

Blocking. In our approach we treat the blocking step in line 4 as a black box. However, there are some blocking techniques that are more appropriate to our approach than others. One practical consideration is that ideally we will use the same blocks from our algorithm in the MLN based ER model. Since the scalability of MLNs is typically very poor, this means we need to limit the maximum block size by using sorted neighbourhood based blocking [12] or a size-constrained clustering approach similar to Fisher et al. [10].

In addition, the rules in $\mathbf{U}_i$ can be used to inform the blocking. If we sort the rules in $\mathbf{U}_i$ based on $w(u)$ and use the rules with the highest weights to generate blocks, we typically get a blocking approach that is well aligned with an MLN based ER model. For example, if having the same surname is strong evidence that the records are a match (as indicated by a large positive weight to the associated rule), then by placing records with the same surname into blocks, we know that during the ER process, the records being compared have a higher likelihood of being matches since they satisfy at least one rule with a large positive weight. After we have generated the set of blocks $\mathbf{B}$, we need to use the candidate pairs in each $b \in \mathbf{B}$ to generate our rule vectors.

Even when blocking is used, computing $\Psi(\mathbf{U}, r_i, r_j)$ for all candidate pairs within all blocks is equivalent to the complete matching step in a traditional ER approach [4]. Since it is impossible to present all the pairs to the expert for manual classification anyway, we instead sample a large number of pairs. While this does not mathematically guarantee that every rule vector is generated, in practice it means that the frequently ocurring ones are present. In addition, after the first round (i.e. once $i > 1$) if a rule vector v_k is missing but was present in an earlier round, we can apply $\mathbf{U}$ to the pairs in $P(v_k)$ from the earlier round to ensure that v_k is also represented in the current round.

Selection Strategy. In Sect. 3 we described several possible strategies for S based on $w(v)$ and $|P(v)|$. In practice we also wish to record the number of manually classified examples for each $v \in \mathbf{V}$, along with their counts of *True* and *False* from the manual classifications. This is because if we keep the same S for each round, it is likely the ordering of $\mathbf{V}$ does not change significantly between rounds. As a result, we end up sampling candidate pairs for the same rule vectors in each round. We can avoid this problem by incorporating previous results into the ordering. If the manual classification for a rule vector always returns the same value of *True* or *False* we can lower its priority in the ordering. However, if a rule vector produces a mix of *True* and *False* values when being manually classified, we may wish to increase its priority in the ordering. It is also possible to change the strategy S between rounds. For example, we can start by looking at the most frequently ocurring rule vectors (highest values of $|P(v)|$) and once we have a good list of rules for the frequent pairs we can change strategies to deal with the difficult cases (lowest values of $|w(v)|$). In the future we intend to investigate ways to determine the optimal choice of strategy.

Overfitting. In practice care needs to be taken to prevent overfitting. In the implementation of our approach described in Algorithm 1, we assume the starting number of rules is the desired number for the final model. However, there is no reason this needs to be the case and we can use a different mechanism to limit the number of rules such as specifying a different maximum number of rules or a minimum score s_{min}, using the scoring method described in Sect. 4.0. In practice it can be the case that the newly added rule has the lowest score and we end up removing it straight away. This essentially means that the expert **E** has ended up manually classifying an outlier or exception, so using it to inform other matching decisions will lead to overfitting.

5   Experimental Evaluation

We have evaluated our approach on two data sets. (1) Cora: This is a public bibliographic data set of scientific papers that has previously been used to evaluate ER techniques [27]. This data set contains 1,295 records and truth data is available. (2) UKCD: This data set consists of census data for the years 1851 to 1901 in 10 year intervals for the town of Rawtenstall and surrounds in the United Kingdom. It contains approximately 150,000 individual records of 32,000 households. A portion of this data (nearly 5,000 records) has been manually linked by domain experts. Fu et al. [11] have used this data set for household based group linkage where the task was to link households across time.

Instead of presenting candidate pairs to the expert **E** for manual classification, we limited our evaluation to the portions of the two data sets that truth data was available and o_{ij} was known for all pairs of $r_i, r_j \in \mathbf{R}$, i.e. the entirety of Cora and about 5,000 records in UKCD. Instead of allowing the expert to select significant rules and altering their weights appropriately we adjusted the weights automatically. For each rule vector v_{ij}, where a candidate pair $\langle r_i, r_j \rangle \in P(v_{ij})$ was selected for manual classification, we adjusted the weights for each $u_k \in \mathbf{U}$ where $u_k(r_i, r_j) = True$ as follows: if $o_{ij} = True$ and $w(u_k) > 0$ we increased $w(u_k)$ by 10 % and if $o_{ij} = True$ and $w(u_k) < 0$ we decreased $w(u_k)$ by 10 %. If $o_{ij} = False$ we performed the opposite adjustments. Essentially this meant that rules that were $True$ and correctly predicted the classification result had their weights increased while those that were $True$ but incorrectly predicted the classification result had their weights decreased.

All our experiments were performed on a Macbook Air with an Intel I5 1.3 Ghz CPU, 4 GBytes of memory and running OS X. All programs were written in Python 3. None of our experiments took longer than five minutes to run 20 rounds of 10 manual classifications each (200 manual classifications in total). Since in a practical ER application each manual classification is likely to take a

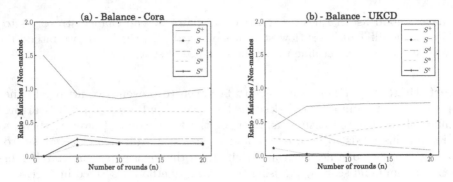

Fig. 1. The ratio of matches / non-matches for different strategies and different values of n.

few minutes, our approach runs fast enough that the slowest part of the process will be most likely be the manual classification step.

For our evaluation we consider two measures, the ratio of candidate pairs classified as matches to the candidate pairs classified as non-matches by the expert $\mathbf{E}$, and the rule weights generated by our techniques. To generate a balanced training set the ratio between matches and non-matches should be as close to 1 as possible. We also examine the weights that are generated for the rules in $\mathbf{U}_n$ at the end of our algorithm as ideally these will be used as the weights for the corresponding formulae in the MLN. We consider both the total sum of the weights calculated as $|\sum(w(u_i) : u_i \in \mathbf{U}_n)|$ as well as the maximum value of $|w(u_i)| : u_i \in \mathbf{U}_n$.

For both data sets, we created a list of rules based on simple attribute equality and assigned weights of 1 to those rules that were positive evidence that the two records might be the same (i.e. the values in the attribute were the same) and weights of -1 to those rules that were negative evidence that the two records might be the same (the values in the attribute were different). For Cora we used a total of eight rules, four with positive weights and four with negative weights, and for UKCD we used 10 rules, five with positive weights and five with negative weights. For both data sets we set $q = 10$ and varied the value of n to simulate different budgets b.

Figure 1 shows the ratio of candidate pairs classified as matches vs non-matches in the manual classification step for the strategies described in Sect. 3. Figure 2 shows the sum of the rule weights for the rules in $\mathbf{U}_n$ and Fig. 3 shows the largest value of $|w(u_i)|$ for $u_i \in \mathbf{U}_n$. As can be seen in the results, the strategies that produce the most balanced splits of the training data between matches and non-matches are S^+ and S^a with the other strategies generally performing poorly. The strategy S^+ orders the rule vectors by descending walue of $w(v)$ which means that a candidate pair from rule vectors with the highest positive weights are always presented to the expert for manual classification in each round. Since apriori these are the rule vectors most likely to be matches, this strategy is effective at overcoming the class imbalance between matches and

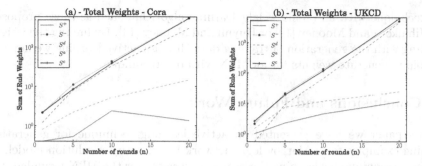

Fig. 2. The total weight of the rules in $\mathbf{U}_n$ for different strategies and different values of n.

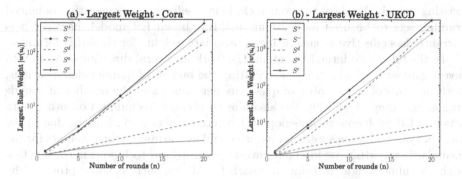

Fig. 3. The maximum value of $|w(u_i)|$ for $u_i \in \mathbf{U}_n$ for different strategies and different values of n.

non-matches. While S^a selects candidate pairs from rule vectors that are hard to classify, i.e. $w(v)$ is close to 0, it means that although it starts by selecting a higher proportion of non-matches, as the algorithm proceeds the negative rule weights get further away from 0. This means that the candidate pairs that are presented to the expert are more likely to have satisfied positive rules and thus be matches which leads to a degree of balance in the classifications. This is also true of the weights generated by these two strategies which do not get very large, either in terms of the sum of the weights or the largest absolute weight.

However, strategies S^-, S^d and S^c all produced very poor results overall. This is because after the initial rounds, these strategies mean that the expert is almost exclusively being presented with non-matches which creates a compounding feedback loop where the weights on the negative rules get more and more negative. As a result, none of these strategies produce balanced training sets and the weights they produce for the rules are not useful at all for an MLN based entity resolution model.

While the weights generated by the strategies S^+ and S^a are plausible for an MLN based ER model, they do not resolve to an equilibrium around the correct weights like we had hoped for. As a result, in the future we intend to combine our technique with an online weight learning algorithm such as those proposed by Mihalkova and Mooney [19] or Huynh and Mooney [14]. By linking the weight learning with the generation of training data through active learning we hope to be able to generate weights for an MLN with our technique.

6   Conclusions and Future Work

In this paper we have presented an active learning technique for generating training examples for a Markov logic network based entity resolution model, as well as a technique for learning the necessary weights for the MLN formulae. We have also presented a method which allows a domain expert to add new rules to the MLN to capture pairs of records that are not being correctly classified by the existing model. We show that our technique is effective at generating balanced training sets to be used for learning an MLN based ER model, however it is currently less effective at generating appropriate weights for the MLN formulae.

In the future we intend to extend the work in several directions. We aim to investigate ways of adaptively distributing the budget of manual classifications, both in terms of the number of questions per round and the number of rounds similar to Wang et al. [29]. We also aim to test our techniques on other data sets and ER problems such as population reconstruction [6]. Finally, due to the fact that our techniques were not as successful at learning correct weights for the formulae in the MLN, we intend to investigate using our technique in conjunction with an online weight learning approach for MLNs, such as those proposed by Mihalkova and Mooney [19] or Huynh and Mooney [14].

References

1. Arasu, A., Götz, M., Kaushik, R.: On active learning of record matching packages. In: ACM SIGMOD, pp. 783–794, Indianapolis (2010)
2. Bellare, K., Iyengar, S., Parameswaran, A.G., Rastogi, V.: Active sampling for entity matching. In: ACM SIGKDD. ACM (2012)
3. Bhattacharya, I., Getoor, L.: Collective entity resolution in relational data. ACM TKDD 1(1), 5 (2007)
4. Christen, V.: Data Matching: Concepts and Techniques for Record Linkage, Entity Resolution, and Duplicate Detection. Data-Centric Systems and Applications. Springer, Heidelberg (2012)
5. Christen, P.: A survey of indexing techniques for scalable record linkage and deduplication. IEEE TKDE 24(9), 1537–1555 (2012)
6. Christen, P., Vatsalan, D., Fu, Z.: Advanced record linkage methods and privacy aspects for population reconstruction - a survey and case studies. In: Bloothooft, G., Christen, P., Mandemakers, K., Schraagen, M. (eds.) Population Reconstruction, pp. 87–110. Springer, Switzerland (2015)
7. Dal Bianco, G., Galante, R., Gonalves, M., Canuto, S., Heuser, C.: A practical and effective sampling selection strategy for large scale deduplication. IEEE KDE 27(9), 2305–2319 (2015)
8. Du, J., Ling, C.: Active learning with human-like noisy oracle. In: IEEE ICDM, pp. 797–802 (2010)

9. Elmagarmid, A.K., Ipeirotis, P.G., Verykios, V.S.: Duplicate record detection: a survey. IEEE TKDE **19**(1), 1–16 (2007)
10. Fisher, J., Christen, P., Wang, Q., Rahm, V.: A clustering-based framework to control block sizes for entity resolution. In: ACM SIGKDD (2015)
11. Fu, Z., Christen, P., Zhou, J.: A graph matching method for historical census household linkage. In: Tseng, V.S., Ho, T.B., Zhou, Z.-H., Chen, A.L.P., Kao, H.-Y. (eds.) PAKDD 2014, Part I. LNCS, vol. 8443, pp. 485–496. Springer, Heidelberg (2014)
12. Hernandez, M.A., Stolfo, S.J.: Real-world data is dirty: Data cleansing and the merge/purge problem. DMKD **2**(1), 9–37 (1998)
13. Huynh, T.N., Mooney, R.J.: Discriminative structure and parameter learning for Markov logic networks. In: ACM ICML (2008)
14. Huynh, T.N., Mooney, R.J.: Online max-margin weight learning for Markov logic networks. In: SDM, pp. 642–651 (2011)
15. Kalashnikov, D., Mehrotra, S.: Domain-independent data cleaning via analysis of entity-relationship graph. ACM TODS **31**(2), 716–767 (2006)
16. Kok, S., Domingos, P.: Learning the structure of Markov logic networks. In: ACM ICML (2005)
17. Köpcke, H., Rahm, E.: Frameworks for entity matching: a comparison. Data Knowl. Eng. **69**(2), 197–210 (2010)
18. MacKay, D.J.: Information-based objective functions for active data selection. Neural Comput. **4**(4), 590–604 (1992)
19. Mihalkova, L., Mooney, R.: Learning to disambiguate search queries from short sessions. In: Buntine, W., Grobelnik, M., Mladenić, D., Shawe-Taylor, J. (eds.) ECML PKDD 2009, Part II. LNCS, vol. 5782, pp. 111–127. Springer, Heidelberg (2009)
20. On, B.W., Elmacioglu, E., Lee, D., Kang, J., Pei, J.: Improving grouped-entity resolution using quasi-cliques. In: IEEE ICDM, pp. 1008–1015 (2006)
21. Rastogi, V., Dalvi, N., Garofalakis, M.: Large-scale collective entity matching. VLDB Endowment **4**, 208–218 (2011)
22. Richardson, M., Domingos, P.: Markov logic networks. Mach. Learn. **62**(1–2), 107–136 (2006)
23. Sarawagi, S., Bhamidipaty, A.: Interactive deduplication using active learning. In: ACM SIGKDD (2002)
24. Settles, B.: Active learning literature survey. Computer Sciences Technical Report 1648, University of Wisconsin, Madison (2010)
25. Settles, B., Craven, M.: An analysis of active learning strategies for sequence labeling tasks. In: ACL Empirical methods in NLP (2008)
26. Singla, P., Domingos, P.: Discriminative training of Markov logic networks. AAAI **5**, 868–873 (2005)
27. Singla, P., Domingos, P.: Entity resolution with Markov logic. In: IEEE ICDM, pp. 572–582 (2006)
28. Wang, J., Kraska, T., Franklin, M.J., Feng, J.: CrowdER: crowdsourcing entity resolution. Proc. VLDB Endow. **5**(11), 1483–1494 (2012)
29. Wang, Q., Vatsalan, D., Christen, P.: Efficient interactive training selection for large-scale entity resolution. In: Cao, T., Lim, E.-P., Zhou, Z.-H., Ho, T.-B., Cheung, D., Motoda, H. (eds.) PAKDD 2015. LNCS, vol. 9078, pp. 562–573. Springer, Heidelberg (2015)

Modeling Adversarial Learning as Nested Stackelberg Games

Yan Zhou$^{(\boxtimes)}$ and Murat Kantarcioglu

Computer Science Department, The University of Texas at Dallas, Richardson, USA
{yan.zhou2,muratk}@utdallas.edu

Abstract. Many data mining applications potentially operate in an adversarial environment where adversaries adapt their behavior to evade detection. Typically adversaries alter data under their control to cause a large divergence of distribution between training and test data. Existing state-of-the-art adversarial learning techniques try to address this problem in which there is only a single type of adversary. In practice, a learner often has to face multiple types of adversaries that may employ different attack tactics. In this paper, we tackle the challenges of multiple types of adversaries with a nested Stackelberg game framework. We demonstrate the effectiveness of our framework with extensive empirical results on both synthetic and real data sets. Our results demonstrate that the nested game framework offers more reliable defense against multiple types of attackers.

1 Introduction

In a class of learning problems known as adversarial learning, learning models are designed to not only solve standard learning tasks, but improve their resilience to adversarial attacks. For instance, in e-mail spam filtering spammers often conceal spam messages in a seemingly legitimate e-mail to evade detection. A well-known technique used by spammers is referred to as "good word" attack [1]. "Good word" attacks change word distributions in spam messages by inserting "good" words that frequently appear in legitimate e-mail. Such attacks change data distributions by transforming instances at test time, and therefore invalidate the standard assumption that training data and test data are identically distributed. What makes the problem even more complex is that a single learner frequently has to face many adversaries and adversary types in the real world. For example, spammers have unlimited options to generate spam messages; they may employ techniques as simple as random word permutation or as advanced as social engineering. In addition, spammers may need to customize attacks to their specific targets.

The research reported herein was supported in part by AFOSR awards FA9550-12-1-0082, NIH awards 1R0-1LM009989 &1R01HG006844, NSF awards #1054629, Career-CNS-0845803, CNS-0964350, CNS-1016343, CNS-1111529, &CNS-1228198, ARO award W911NF-12-1-0558.

© Springer International Publishing Switzerland 2016
J. Bailey et al. (Eds.): PAKDD 2016, Part II, LNAI 9652, pp. 350–362, 2016.
DOI: 10.1007/978-3-319-31750-2_28

The adversarial learning problem resembles a two-player game in many respects. First, one side takes initiatives to maximize the threat to its opponent, then the opponent answers by deploying the most effective countermeasures. If this competition continues, it becomes an arms race between the two players. This would lead to constant demands for model enhancement given that the adversary's attack strategies are inexhaustible. When one player's gain is the other player's loss, the players are said to play a zero-sum game. A zero-sum game is typically solved by playing a minimax strategy [2–8]. In practice, the zero-sum assumption is often too pessimistic. When the two players are not strictly competitive, a Nash equilibrium solution can be found so that neither of the players would benefit by unilaterally changing its strategy [9,10]. For each player, it is optimal to play the Nash equilibrium only when the other player is rational and has no knowledge about its opponent's strategy before taking its own actions. In many applications, the assumption that the learner's strategy is unknown to the adversary is invalid since the adversary can often obtain this information by probing the learner's strategy models. A Stackelberg game is widely used to model problem domains in which the learner plays a strategy that is known to the adversary before the adversary commits to its own strategy [9,11,12]. The game is played between a single leader and a single type of follower. The leader commits to its strategy ahead of the follower's response. The follower plays its optimal strategy to maximize its utility after fully observing the leader's strategy. The leader cannot change its strategy after knowing the follower's response strategy. In real world applications, a single leader may have to face followers of multiple types, and the leader is uncertain which follower type is participating in the game. The problem can be modeled as a Bayesian Stackelberg game. The solution to a Bayesian Stackelberg game is an optimal mixed strategy for the leader to play in the game.

In this paper, we consider a malicious data modification problem where adversaries may use different strategies to corrupt the data. We present a nested Stackelberg game to handle both data corruption and unknown types of adversaries. The game framework consists of a set of *single leader single follower* (SLSF) Stackelberg games and a *single leader multiple followers* (SLMF) Bayesian Stackelberg game. We first solve a SLSF Stackelberg game for each adversary type. This level of Stackelberg game takes into consideration that training and test data are not necessarily identically distributed in practice. Given the learner's learning model, the adversary responds to the learner's strategy by optimally transforming data to maximize the learner's predictive error. The Stackelberg equilibrium solution consists of optimal learning parameters for the learner and data transformations for the adversary. The optimal solutions will be used as pure strategies in the Bayesian Stackelberg game. The Bayesian Stackelberg game consists of one learner and multiple adversaries of various types. When facing adversaries of multiple types, instead of settling on one learning model by playing a pure strategy, it is more practical for the learner to play a mixed strategy consisting of a set of learning models with assigned probabilities. The optimal solution to the Bayesian Stackelberg game introduces randomness to the solution, and hence increases the difficulty of attacking the underlying learning models via reverse engineering.

The rest of the paper is organized as follows. Section 2 presents the related work on game-theoretic solutions to adversarial learning problems. Section 3 formally defines the adversarial learning problem and presents our nested Stackelberg game framework. Section 4 presents experimental results on both artificial and real datasets. Section 5 concludes our work.

2   Related Work

The connection between adversarial learning and a two-player game has been extensively researched recently. The problem is often modeled as a game between a classifier and an adversary that modifies the distributions of data. In earlier research in this direction, the problem is modeled as a single step minimax game in which the worst case loss is minimized over all possible data points that satisfy pre-defined constraints. For example, Globerson and Roweis [2] solve an optimal SVM learning problem to deal with test-time malicious feature deletion. They search for the zero-sum minimax strategy that minimizes the hinge loss of the support vector machine. El Ghaoui et al. [3] present a minimax strategy for a similar problem in which training data is bounded by hyper-rectangles. The work by Zhou et al. [4] also fits in this line of research. Several other techniques have also been studied for handling classification-time feature modification [5–8].

In many real applications, the adversarial learning problem is more appropriately modeled as a sequential game between two players. One player must commit to its strategy before the other player responds. The advantage the responding player has is partial or complete information of the first player. The responding player can therefore play its optimal strategy against its opponent. This type of game is known as a Stackelberg game in which the first player is the leader and its opponent is the follower. Adversarial learning researched in this area falls into two categories, depending on who plays the role of the leader: the classifier or the adversary. Kantarcioglu et al. [9] solve for a Nash equilibrium using simulated annealing and the genetic algorithm to discover an optimal set of attributes. Similar work has also been done by Liu and Chawla [10]. The difference between their research lies in that the former assumes both players know each other's payoff function, while the latter relaxed the assumption and only the adversary's payoff function is required. In both cases, the adversary is the leader whose strategies are stochastically sampled while the classifier is the follower that searches for an equilibrium given its knowledge about the adversary. In a more realistic scenario, the classifier is more likely to commit to a strategy before the adversary takes its actions. The adversary's response is optimal given that it has some knowledge about the classifier's strategy. Example solutions to this type of Stackelberg game are presented by Brückner and Scheffer [12].

There are many other research directions in adversarial learning, for example, considering training-time attacks such as poisoning attacks [13], applying an ensemble of a set of adversarial learning models [14], and using a divide-and-conquer approach [15]. Unlike existing work, in this paper we consider adversarial learning problems where there are multiple types of attackers.

3   Nested Bayesian Stackelberg Games

We first define the adversarial learning problem. Next, we present strategies to construct component SLSF learning models given adversary types. Finally, we solve the SLMF Stackelberg game with the component SLSF models to counter adversaries of various types.

3.1   Adversarial Learning

Assume we are given input samples $x_{i|i=1,...,n} \in \mathcal{X}$ from some input space $\mathcal{X}$ and we want to estimate $y_i \in \mathcal{Y}$ where $\mathcal{Y} = \{+1, -1\}$ in a binary classification problem. In the standard settings of a learning problem, given a learning function $g : \mathcal{X} \to \mathbb{R}$ and a feature vector $\phi(x \in \mathcal{X}) \in \mathbb{R}^N$, we estimate

$$\hat{y}_i = g(w, x_i) = w^T \cdot \phi(x_i).$$

Our estimate $\hat{y}_i = g(w, x_i)|_{w \in \mathbb{R}^N}$ is obtained by optimizing an objective function L. Let $\ell : \mathcal{Y} \times \mathcal{Y} \to \mathbb{R}_+$ be the learner's loss function. Let the objective function L be the learner's loss with L_2 regularization:

$$L = \sum_{i=1}^{n} \ell(\hat{y}_i, y_i) + \lambda ||w||^2 \tag{1}$$

where λ is a regularization parameter that weighs the penalty of reduced norms. In cost-sensitive learning where the loss of data in different categories is considered unequal, a cost vector c is used to weigh the loss in Eq. (1), and the learner solves the following optimization problem:

$$\underset{w}{\text{argmin}} \ L = \underset{w}{\text{argmin}} \ \sum_{i=1}^{n} c_i \cdot \ell(\hat{y}_i, y_i) + \lambda ||w||^2 \tag{2}$$

The definition of the adversarial learning problem follows naturally from (2). Assume there exists an adversary that influences the learning outcome by modifying the data. The classification task becomes an estimate of $\hat{y}_i$ on the transformed data:

$$\hat{y}_i = w^T \cdot \phi(f_t(x_i, w)),$$

where $f_t(x_i, w)$ is a data transformation function used by the adversary to modify the data:

$$f_t(x_i, w) = x_i + \delta_x(x_i, w)$$

where δ_x returns the displacement vector for x_i. Therefore the adversarial learning problem can be defined as the following optimization problem:

$$\operatorname*{argmin}_{w} \operatorname*{argmax}_{\delta_x} L(w, x, \delta_x). \tag{3}$$

Note that the above learning problem becomes a disjoint bilinear problem with respect to w and δ_x assuming that the adversary plays its optimal attack $\delta_x^*(w)$.

3.2   A Single Leader Single Follower Stackelberg Game

Each component learning model in our framework is obtained by solving a Stackelberg game between the learner and the adversary. The learner first commits to its strategy that is observable to the adversary and the adversary plays its optimal strategy to maximize the learner's loss while minimizing its own loss. Therefore, the adversarial learning problem of this *single leader single follower* (SLSF) game is:

$$\operatorname*{argmin}_{w^*} \operatorname*{argmax}_{\delta_x^*} \quad L_\ell(w, x, \delta_x)$$

$$s.t. \qquad\qquad \delta_x^* \in \operatorname*{argmin}_{\delta_x} L_f(w, x, \delta_x)$$

where L_ℓ is the leader's loss as defined in Eq. (1):

$$L_\ell = \sum_{i=1}^{n} c_{\ell,i} \cdot \ell_\ell(\hat{y}_i, y_i) + \lambda_\ell ||w||^2 \tag{4}$$

and L_f is the follower's loss where the second term penalizes for the L_2 norm of data transformation:

$$L_f = \sum_{i=1}^{n} c_{f,i} \cdot \ell_f(\hat{y}_i, y_i) + \lambda_f \sum_{i=1}^{n} ||\phi(x_i) - \phi(f_t(x_i, w))||^2. \tag{5}$$

λ_ℓ, λ_f, c_ℓ, and c_f are the weights of the penalty terms and the costs of data transformation. ℓ_ℓ and ℓ_f are the classification loss functions of the leader and the follower. A Stackelberg equilibrium solution exists if the adversary's loss is convex and continuously differentiable [12].

3.3   Component Learning Models and Adversary Types

There are two issues we need to resolve in our adversarial learning framework: defining a set of component learning models for the learner to randomly choose from and the types of adversaries the learner has to face at application time.

Component Learning Models. The Stackelberg equilibrium solution discussed in Sect. 3.2 can be used to generate a set $\mathcal{G} = \{g_s, g_{f_1}, \ldots, g_{f_i}, \ldots\}$ of learning functions. Learning function g_s is the Stackelberg equilibrium solution

$$g_s(w_s, x) = w_s^T \cdot \phi(x)$$

where w_s is the Stackelberg solution for the leader. The rest of the learning functions are obtained using the follower's solution in the Stackelberg equilibrium. Recall that the follower's solution is the optimal data transformation $\delta_x(w_s)$ given w_s. If we switch roles and let the adversary be the leader and disclose the data transformation δ_x to the learner, we can train a learning function g_f by solving Eq. (3) as a simple optimization problem. Of course, g_f by itself does not constitute a robust solution to the adversarial learning problem since the adversary could easily defeat g_f with a different data transformation. g_f is better understood as a special case solution under the umbrella of the general Stackelberg solution. However, when the adversary does use δ_x as the data transformation model, g_f typically performs significantly better than the Stackelberg solution in terms of classification error. This is the motivation that we include such learning models in the learner's strategy set in our framework. To determine when and how often to use such learning models during test time, we resort to the solution to the *single leader multiple followers* game discussed later.

We can train a set of g_{f_i} by varying how much impact a data transformation has on the follower's loss function defined in Eq. (5). When λ_f is very large, the penalty of data transformation becomes dominant and the adversary's best strategy is not to transform data at all. When λ_f is relatively small, the adversary has more incentives to perform data transformation. However, the diminishing cost of data transformation is counterbalanced by the increase in the cost of misclassification. This prevents the adversary modifying training data arbitrarily. Therefore, we can define a spectrum of learning models specifically characterizing various types of adversaries, from the least aggressive to the most aggressive ones. The equilibrium data transformations involve data in both classes. However, how much data can be transformed in either class can be controlled by the cost factors c_f^+ and c_f^- of the adversary. Pessimistic component learning models are built to countermeasure aggressive adversaries that make significant data modifications. Optimistic component models, on the other hand, are more suitable for more moderate attacks.

Adversary Types. Adversaries may come in many different types. For example, in e-mail spam filtering some spammers' objectives are to successfully send spam contents to the end user, others aim to clog the network by performing denial-of-service attacks; some spammers can modify both spam and legitimate e-mail, others have little privilege and are not entitled to access to legitimate e-mail. It is impossible to implement a spam filter with a single learning model to effectively counter every possible type of adversary. This motivates us to take into account different types of adversaries in our adversarial learning framework.

We model different adversary types by applying different constraints on data transformation. We consider three cases in which the adversary can:

- attack both positive and negative data (e.g. insider attack)
- attack positive data only (e.g. good word attack in spam)
- attack randomly by transforming data freely in the given domain (e.g. malware induced data corruption).

In the first two scenarios, the adversary can transform positive, or negative, or both classes of data with specific goals—transforming data in one class to another; while in the last case, the adversary can transform any data freely in the given domain. For each type of adversary, we define three pure attack strategies for the adversary: mild, moderate, and aggressive attacks. Each pure strategy has a different degree of impact on how much the transformed data diverges from the original data distribution.

Setting Payoff Matrices for the Single Leader Multiple-Followers Game. The solutions to the *single leader and single follower* Stackelberg games produce pure strategies (learning parameters) for the learner: the equilibrium predictor, a pessimistic predictor built on the equilibrium data transformation of an aggressive adversary, and an optimistic predictor built on the equilibrium data transformation of a mild adversary. We use this information to compute the payoff matrices of the learner and the adversaries, and pass them to the *single leader multiple followers* game discussed in the next section.

The learner's payoffs are the predictive accuracies of its component learning models on the data transformation corresponding to each pure strategy of an adversary. We refer to the three component learning models as *Equi**—the Stackelberg equilibrium predictor, SVM*[*1]—the pessimistic SVM model trained on aggressive attacks, and SVM*[*2]—the less pessimistic SVM model trained on moderate attacks. We refer to the three pure strategies of the adversary as mild, moderate, and aggressive, respectively. The mild adversary transforms data by adding a small displacement vector. The moderate adversary transforms data more aggressively than the mild adversary, but less than the aggressive adversary. The payoff matrix for the learner is shown in Table 1. R_{ij} denotes the predictive accuracy of predictor i tested on the optimal data transformation of an adversary of type j.

Table 1. The payoff matrix of the learner.

	Mild	Moderate	Aggressive
Equi*	R_{11}	R_{12}	R_{13}
SVM*1	R_{21}	R_{22}	R_{23}
SVM*2	R_{31}	R_{32}	R_{33}

The adversary's payoff matrix is similar to Table 1 except that the rewards of the adversary is the false negative rate discounted by the amount of data movement that is responsible for the false negatives. The amount of movement is measured as the L_2 distance summed over all data points before and after the data transformation. The payoff matrices of the learner and the followers are taken as input to a *single leader multiple followers* game which we discuss next.

3.4 A Single Leader Multi-followers Stackelberg Game

In a *single leader multiple followers* (SLMF) game [16], the leader makes its optimal decision prior to the decisions of multiple followers. The Stackelberg game played by the leader is:

$$\min_{x,y^*} F(x,y^*)$$
$$s.t. \quad G(x,y^*) \leq 0$$
$$H(x,y^*) = 0$$

where F is the leader's objective function, constrained by G and H; x is the leader's decision and y^* is in the set of the optimal solutions of the lower level problem:

$$y^* \in \left\{ \begin{array}{ll} \operatorname*{argmin}_{y_i} f_i(x, y_i) \\ s.t. \quad g_i(x, y_i) \leq 0 \\ \quad\quad h_i(x, y_i) = 0 \end{array} \right\} \forall i = 1, \ldots, m$$

where m is the number of followers, f_i is the ith follower's objective function constrained by g_i and h_i. For the sake of simplicity, we assume the followers are not competing among themselves. This is usually a valid assumption in practice since adversaries rarely affect each other through their actions. In a Bayesian Stackelberg game, the followers may have many different types and the leader does not know exactly the types of adversaries it may face when solving its optimization problem. However, the distribution of the types of adversaries is known or can be inferred from past experience. The followers' strategies and payoffs are determined by the followers' types. The followers play their optimal responses to maximize the payoffs given the leader's strategy. The Stackelberg equilibrium includes an optimal mixed strategy of the learner and corresponding optimal strategies of the followers.

Problem Definition. *Given the payoff matrices R^ℓ and R^f of the leader and the m followers of n different types, find the leader's optimal mixed strategy given that all followers know the leader's strategy when optimizing their rewards. The leader's pure strategies consist of a set of generalized linear learning models $\langle \phi(x), w \rangle$ and the followers' pure strategies include a set of vectors performing data transformation $x \rightarrow x + \Delta x$.*

The defined Stackelberg game can be solved as a Mixed-Integer-Quadratic-Programming (MIQP) problem [17]. For a game with a single leader and m followers with n possible types where the m followers are independent of each other and their actions have no impact on each other's decisions, we reduce the problem to solving m instances of the *single leader single follower* game.

4   Experiments

In our experiments, we use three types of adversaries as discussed in Sect. 3.3. The first type $Adversary^{*1}$ can modify both positive and negative data, and the second type $Adversary^{*2}$ is only allowed to modify positive data as normally seen in spam filtering. The third type of adversary $Adversary^{*3}$ can transform data freely in the given domain. The prior distribution of the three adversary types is randomly set. Let p be the probability that the adversary modifies negative data. Then for each negative instance x^- in the test set, with probability p, x^- is modified as follows:

$$x^- = x^- + f_a \cdot (x^+ - x^-) + \epsilon$$

where ϵ is local random noise, and x^+ is a random positive data point in the test set. The intensity of attacks is controlled by the attack factor $f_a \in (0, 1)$. The greater f_a is, the more aggressive the attacks are. Similarly, for each positive instance x^+ we modify x^+ as follows:

$$x^+ = x^+ + f_a \cdot (x^- - x^+) + \epsilon$$

where x^- is a random negative data point in the test set. For the third type of attack, x^+ and x^- can be freely transformed in the data domain as follows:

$$x^{\pm} = \{ \begin{matrix} min(x^{max}, x^{\pm} + f_a \cdot \delta \cdot (x^{max} - x^{min})) & \delta > 0 \\ max(x^{min}, x^{\pm} + f_a \cdot \delta \cdot (x^{max} - x^{min})) & \delta \leq 0 \end{matrix}$$

where δ is randomly set and $\delta \in (-1, 1)$, x^{max} and x^{min} is the maximum and minimum values an instance can take. The learner's pure strategy set contains three learning models as discussed in Sect. 3.3: (1) Stackelberg equilibrium predictor Equi*; and (2) two SVM models SVM*1 and SVM*2 trained on equilibrium data transformations. Note that SVM*1 and SVM*2 are optimal only when the SVM learner knows the adversary's strategy ahead of time. Therefore, SVM*s alone are not robust solutions to the adversarial learning problem. When solving the prediction games, we assume the adversary can modify data in both classes. Note that even though we have two Stackelberg equilibrium predictors Equi*1 and Equi*2 corresponding to an aggressive and a mild adversary, we use only one in the learner's strategy set and refer to it as Equi* because equilibrium predictors rarely disagree. However, we list the classification errors of both Equi*1 and Equi*2 for comparison. SVM*1 and SVM*2 are trained on the two equilibrium data transformations when λ_f is set to 0.01 and 0.02. The two SVM models are essentially optimal strategies against the adversaries' equilibrium strategies. The learner will choose which learning model to play according to the probability distribution determined in the mixed strategy. The results are displayed as *Mixed* in the following sections. We also compare our results to the invariant SVM [6] and the standard SVM methods. We test our learning strategy on two artificial datasets and two real datasets. In all of our experiments, we modify the test sets to simulate the three types of adversaries.

4.1   Artificial Datasets

We generate two artificial datasets to test our learning framework. Each dataset is generated using a bivariate normal distribution with specified means of the two classes, and a covariance matrix. We purposefully generate the datasets so that positive data and negative data are clearly separated in the first dataset and largely fused in the second dataset. Training and test data are generated separately. Adversarial attacks are simulated in the test data as discussed above. In all experiments, when $p = 0.1$, the adversary has 10 % chance of modifying the negative data in the test set; when $p = 0.5$, the adversary is allowed to modify the negative data half of the time. Positive data are always modified. Detailed results are shown in Tables 2. All the results on the artificial data are averaged over 10 random runs. We summarize our observations as follows: (1) Equilibrium predictors have similar predictive errors as the invariant SVM and the standard SVM methods; (2) SVM*[1] outperforms SVM*[2] as expected when attacks are moderate and aggressive—both SVM*[1] and SVM*[2] outperform the equilibrium predictors, the invariant and the standard SVMs; (3) The mixed strategy lies in between the best and the worst predictors, however, consistently in line with the best results under all circumstances. Our proposed mixed strategy is most

Table 2. Error rates of component classifiers and the mixed strategy, as well as invariant SVM, and standard SVM on artificial datasets. $\lambda_f \in \{0.01, 0.02\}$ and $\lambda_\ell = 1$. The probability of modifying negative data is set to $p \in \{0.1, 0.5\}$. The best results are marked with *, and the mixed strategy is bolded for easy comparisons to others.

Dataset I						
p=0.1		$f_a = 0.1$	$f_a = 0.3$	$f_a = 0.5$	$f_a = 0.7$	$f_a = 0.9$
Stackelberg	Equi*[1]	0.0250 ± 0.0088	0.1560 ± 0.1109	0.3195 ± 0.0962	0.4460 ± 0.0222	0.5190 ± 0.0126
	SVM*[1]	0.0305 ± 0.0140	0.0815 ± 0.0266*	0.2055 ± 0.1504*	0.3005 ± 0.0370*	0.4650 ± 0.0373*
	Equi*[2]	0.0250 ± 0.0088	0.1560 ± 0.1109	0.3195 ± 0.0962	0.4460 ± 0.0222	0.5190 ± 0.0126
	SVM*[2]	0.0160 ± 0.0077*	0.1165 ± 0.0884	0.2735 ± 0.1204	0.3865 ± 0.0286	0.5045 ± 0.0117
	Mixed	**0.0273 ± 0.0149**	**0.0845 ± 0.0272**	**0.2132 ± 0.1500**	**0.3563 ± 0.0655**	**0.4860 ± 0.0341**
Invariant SVM		0.0255 ± 0.0086	0.1560 ± 0.1122	0.3195 ± 0.0962	0.4455 ± 0.0224	0.5190 ± 0.0126
SVM		0.0265 ± 0.0067	0.1595 ± 0.1207	0.3280 ± 0.0952	0.4360 ± 0.0313	0.5205 ± 0.0140
p=0.5		$f_a = 0.1$	$f_a = 0.3$	$f_a = 0.5$	$f_a = 0.7$	$f_a = 0.9$
Stackelberg	Equi*[1]	0.0350 ± 0.0151*	0.1520 ± 0.0643	0.3905 ± 0.0480	0.5805 ± 0.0472	0.6905 ± 0.0682
	SVM*[1]	0.0915 ± 0.0176	0.1515 ± 0.0297	0.3055 ± 0.0370*	0.4815 ± 0.0316*	0.6150 ± 0.0361*
	Equi*[2]	0.0350 ± 0.0151*	0.1520 ± 0.0643	0.3905 ± 0.0480	0.5805 ± 0.0472	0.6905 ± 0.0682
	SVM*[2]	0.0385 ± 0.0147	0.1365 ± 0.0385*	0.3725 ± 0.0505	0.5520 ± 0.0348	0.6805 ± 0.0652
	Mixed	**0.0631 ± 0.0328**	**0.1429 ± 0.0243**	**0.3264 ± 0.0430**	**0.5042 ± 0.0478**	**0.6367 ± 0.0588**
Invariant SVM		0.0355 ± 0.0157	0.1530 ± 0.0656	0.3910 ± 0.0482	0.5810 ± 0.0478	0.6900 ± 0.0680
SVM		0.0370 ± 0.0151	0.1545 ± 0.0724	0.3955 ± 0.0474	0.5770 ± 0.0458	0.6845 ± 0.0669
Dataset II						
p=0.1		$f_a = 0.1$	$f_a = 0.3$	$f_a = 0.5$	$f_a = 0.7$	$f_a = 0.9$
Stackelberg	Equi*[1]	0.1895 ± 0.0273	0.3050 ± 0.1146	0.3570 ± 0.0747	0.4300 ± 0.0255	0.5020 ± 0.0385
	SVM*[1]	0.1955 ± 0.0231	0.2920 ± 0.1232	0.3465 ± 0.1232	0.3985 ± 0.0342*	0.5090 ± 0.0731
	Equi*[2]	0.1895 ± 0.0273	0.3050 ± 0.1146	0.3570 ± 0.0747	0.4300 ± 0.0255	0.5020 ± 0.0385
	SVM*[2]	0.1810 ± 0.0240*	0.2915 ± 0.1172*	0.3530 ± 0.0985	0.4240 ± 0.0133	0.5045 ± 0.0524
	Mixed	**0.1929 ± 0.0229**	**0.2993 ± 0.1212**	**0.3444 ± 0.0841***	**0.4099 ± 0.0268**	**0.5096 ± 0.0664**
Invariant SVM		0.1835 ± 0.0247	0.3010 ± 0.1167	0.3570 ± 0.0775	0.4350 ± 0.0203	0.4995 ± 0.0403
SVM		0.1840 ± 0.0271	0.2970 ± 0.1169	0.3605 ± 0.0825	0.4330 ± 0.0121	0.4970 ± 0.0429*
p=0.5		$f_a = 0.1$	$f_a = 0.3$	$f_a = 0.5$	$f_a = 0.7$	$f_a = 0.9$
Stackelberg	Equi*[1]	0.2000 ± 0.0350	0.2985 ± 0.0232	0.4065 ± 0.0207	0.5340 ± 0.0181	0.6190 ± 0.0357
	SVM*[1]	0.2140 ± 0.0422	0.2955 ± 0.0201	0.3940 ± 0.0094*	0.5240 ± 0.0459*	0.6090 ± 0.0208*
	Equi*[2]	0.2000 ± 0.0350	0.2985 ± 0.0232	0.4070 ± 0.0211	0.5340 ± 0.0181	0.6195 ± 0.0344
	SVM*[2]	0.2010 ± 0.0330	0.2875 ± 0.0148*	0.4025 ± 0.0262	0.5365 ± 0.0184	0.6170 ± 0.0309
	Mixed	**0.2074 ± 0.0461**	**0.2900 ± 0.0167**	**0.4021 ± 0.0245**	**0.5284 ± 0.0394**	**0.6150 ± 0.0234**
Invariant SVM		0.1970 ± 0.0342*	0.3010 ± 0.0242	0.4045 ± 0.0211	0.5370 ± 0.0196	0.6210 ± 0.0353
SVM		0.2015 ± 0.0323	0.2960 ± 0.0221	0.4075 ± 0.0204	0.5370 ± 0.0153	0.6220 ± 0.0396

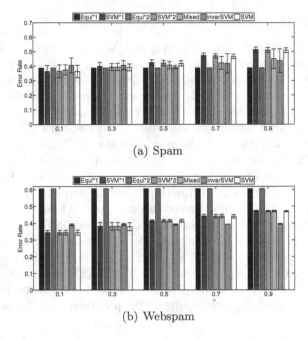

(a) Spam

(b) Webspam

Fig. 1. Classification error rates (with error bars) of $Equi^{*1}$, SVM^{*1}, $Equi^{*2}$, SVM^{*2}, $Mixed$, $invariant\ SVM$, and SVM on the `spambase` and `webspam` datasets.

reliable in that it is *either the best predictor or very competitive with the best in all cases*. We would like to emphasize the importance of consistency in adversarial learning tasks. When there are many adversaries of various types, the forms of attacks are inexhaustible. Therefore, there is no single best learning model to handle all possible scenarios in all problem domains. Our objective is for the mixed strategy to be always comparable to the best component predictor.

4.2   Real Datasets

The two real datasets we use in our experiments are: *spambase* from the UCI data repository [18] and *web spam* from the LibSVM website [19]. The learning tasks are binary classification problems, differentiating spam from legitimate e-mail or websites. We make the learning tasks more complicated by making the attack factor $f_a \in (0, 1)$ completely random under uniform distribution for each attacked sample in the test set. We assume the positive data is always modified by the adversary. In addition, we allow the probability of negative data being attacked to increase gradually from 0.1 to 0.9. The advantage of our mixed strategy is more obvious on these two datasets as illustrated in Fig. 1. The equilibrium predictors Equi^{*1,2} are better than the SVM^{*1,2} predictors on the *spambase* data, but significantly worse on the *web spam* data. Our mixed strategy consistently outperforms SVM^{*1,2} on the *spambase* data, and outperforms Equi^{*1,2} on the *web spam* data.

5 Conclusions

We tackle the challenges data mining applications face when there are adversaries of different types. Existing adversarial learning techniques focus on learning a robust learning model against a single type of adversary. In this paper, we present a nested Stackelberg game framework in which the lower level learners provide robust solutions to adversarial data transformations, while at the upper level, the learner plays a *single leader multiple followers* game and searches for an optimal mixed strategy to play. We demonstrate with empirical evidence that the mixed strategy is more reliable in general without knowing what types of adversaries the data mining applications are facing in the real world.

References

1. Lowd, D.: Good word attacks on statistical spam filters. In: Proceedings of the Second Conference on Email and Anti-Spam (CEAS) (2005)
2. Globerson, A., Roweis, S.: Nightmare at test time: robust learning by feature deletion. In: ICML, pp. 353–360. ACM (2006)
3. El Ghaoui, L., Lanckriet, G.R.G., Natsoulis, G.: Robust classification with interval data. Technical report UCB/CSD-03-1279, EECS Department, University of California, Berkeley, October 2003
4. Zhou, Y., Kantarcioglu, M., Thuraisingham, B., Xi, B.: Adversarial support vector machine learning. In: SIGKDD, pp. 1059–1067. ACM (2012)
5. Lanckriet, G.R.G., Ghaoui, L.E., Bhattacharyya, C., Jordan, M.I.: A robust minimax approach to classification. J. Mach. Learn. Res. **3**, 555–582 (2002)
6. Teo, C.H., Globerson, A., Roweis, S.T., Smola, A.J.: Convex learning with invariances. In: Advances in Neural Information Processing Systems (2007)
7. Dekel, O., Shamir, O.: Learning to classify with missing and corrupted features. In: ICML, pp. 216–223. ACM (2008)
8. Dekel, O., Shamir, O., Xiao, L.: Learning to classify with missing and corrupted features. Mach. Learn. **81**(2), 149–178 (2010)
9. Kantarcioglu, M., Xi, B., Clifton, C.: Classifier evaluation and attribute selection against active adversaries. Data Min. Knowl. Discov. **22**, 291–335 (2011)
10. Liu, W., Chawla, S.: A game theoretical model for adversarial learning. In: Proceedings of the 2009 IEEE International Conference on Data Mining Workshops. ICDMW 2009, pp. 25–30, Washington, DC, USA. IEEE Computer Society (2009)
11. Bruckner, M., Scheffer, T.: Nash equilibria of static prediction games. In: Advances in Neural Information Processing Systems, MIT Press, Cambridge (2009)
12. Brückner, M., Scheffer, T.: Stackelberg games for adversarial prediction problems. In: KDD, pp. 547–555, New York (2011)
13. Biggio, B., Nelson, B., Laskov, P.: Poisoning attacks against support vector machines. In: ICML, pp. 1807–1814 (2012)
14. Zhou, Y., Kantarcioglu, M., Thuraisingham, B.M.: Sparse bayesian adversarial learning using relevance vector machine ensembles. In: ICDM, pp. 1206–1211 (2012)
15. Zhou, Y., Kantarcioglu, M.: Adversarial learning with bayesian hierarchical mixtures of experts. In: SDM, pp. 929–937 (2014)
16. Basar, T., Olsder, G.J.: Dynamic Noncooperative Game Theory. Society for Industrial and Applied Mathematics, Classics in Applied Mathematics (1999)

17. Paruchuri, P.: Playing games for security: an efficient exact algorithm for solving bayesian stackelberg games. In: AAMAS (2008)
18. UCI:UCI Machine Learning Repository (2014). http://archive.ics.uci.edu/ml/
19. LIBSVM:LIBSVM Data: Classification, Regression, and Multi-label (2014). http://www.csie.ntu.edu.tw/~cjlin/libsvmtools/datasets/

Fast and Semantic Measurements on Collaborative Tagging Quality

Yuqing Sun[1,2(✉)], Haiqi Sun[3], and Reynold Cheng[4]

[1] School of Computer Science and Technology, Shandong University, Jinan, China
sun_yuqing@sdu.edu.cn
[2] Engineering Research Center of Digital Media Technology, MOE, Jinan, China
[3] Software College, Shandong University, Jinan, China
shqonline@yeah.net
[4] Department of Computer Science, The University of Hong Kong, Hong Kong, China
ckcheng@cs.hku.hk

Abstract. This paper focuses on the problem of tagging quality evaluation in collaborative tagging systems. By investigating the dynamics of tagging process, we find that high frequency tags almost cover the main aspects of a resource content and can be determined stable much earlier than a whole tag set. Motivated by this finding, we design the swapping index and smart moving index on tagging quality. We also study the correlations in tag usage and propose the semantic measurement on tagging quality. The proposed methods are evaluated against real datasets and the results show that they are more efficient than previous methods, which are appropriate for a large number of web resources. The effectiveness is justified by the results in tag based applications. The light weight metrics bring a little loss on the performance, while the semantic metric is better than current methods.

1 Introduction

Collaborative tagging, also known as crowdsourcing or folksonomy system, is widely adopted by web social applications, such as *Del.icio.us*, *Flickr* and *Movie-Lens*. It encourages users to annotate resources, such as URLs, images or movies, with bookmarks according to their understanding. Each bookmark contains a set of tags. After receiving a number of bookmarks, the tag frequency distribution of a resource would remain stable [14]. These tags and frequencies are regarded as the meta data of resources and are used for recommendation or information retrieval. This method provides an easy way to organize a large quantity of web resources, which is more efficient than traditional classification by specialists. Besides, collaborative tagging plays an important role in tag based applications, such as recommendation [4], web clustering [10], and search [1]. For example, tag-based retrieval is more efficient and effective than traditional full-text search [2].

This work is supported by NSF China (61173140), SAICT Experts Program, Independent Innovation & Achievements Transformation Program (2014ZZCX03301) and Science & Technology Development Program of Shandong Province (2014GGX101046).

J. Bailey et al. (Eds.): PAKDD 2016, Part II, LNAI 9652, pp. 363–375, 2016.
DOI: 10.1007/978-3-319-31750-2_29

Since collaborative tagging is a kind of user subjective action, users often choose their interested resources for tagging. So resources have different numbers of bookmarks. In practice only a small proportion of resources receive enough bookmarks and their tagging states reach stable. That is to say, the tag frequency distributions remain almost the same even if they continuously receive new bookmarks. Considering a large number of resources with few bookmark, their tag frequency distributions change with a new coming bookmark. The states of these resources are called *under-tagged* [16]. According to the findings in [13], a stable tag set is helpful for tag-based applications, such as retrieval or recommendation. But the tags of *under-tagged* resources often affect the correctness of results.

To verify whether a resource reaches tagging stable, the notion of tagging quality is introduced. One widely adopted criterion is the similarity score, which computes the tag distribution similarity in several consecutive tagging points [9,15]. Relative entropy is another way to gauge the stability, which computes the distance between tag frequency distributions on two tag sets [6]. Although these methods provide objective evaluation on tagging state, there are two shortcomings. One is lack of internal link evaluation on tag usage. For example, the phrases *iOS* and *Apple phone* together appear in many resources, which indicates their closeness in semantics and similar usages. However, these correlations have been overlooked in the current methods. Another is time consuming. It takes much computation on measuring two tag frequency distribution, which is not appropriate for a large quantity of web resources. So, it is necessary to take these characteristics into consideration when evaluating the tagging state.

By analyzing the tagging process, we find that noisy tags are the main cause on influencing a tag set stable. According to the previous study [7], most noisy tags are not related to a resource content, which may be caused by user misoperation or misunderstanding. Since the purpose of collaborative tagging is to find the semantics of a resource, the representative tags are often desired to describe a resource and the noisy tags can be neglected. By exploiting the evolution of tag set in collaborative tagging process, we find that some high frequency tags almost cover the main aspects of a resource content and can be determined stable much earlier than the whole tag set. This motivates us to design novel measurements, the swapping index and the smart moving index, to evaluate a resource tagging quality against these representative tags. We also study the inner links between tags and find that some tags often together appear in bookmarks. The notion *concept* is introduced to represent their relationships and semantics. Then a tag set can be reorganized semantically. Based on *concept*, we propose a semantic measurement on tagging quality. We perform a series of experiments on real datasets to justify our methods. The results show that the proposed metrics are more efficient than previous methods. We also adopt some tag based applications to verify their results against stable tag sets under different metrics. The results show the effectiveness of our methods.

The remainder of this paper is organized as follows: Sect. 2 presents the related works and basic notions. In Sect. 3 we describes the datasets and present our findings. Sections 4 and 5 introduce the efficient metrics and semantic measurement on tagging quality, respectively, as well as the experiments. Finally, conclusions and future works are discussed.

2 Related Works

Dynamics on Collaborative Tagging. Collaborative tagging systems have attracted much attention in recent years. Golder et al. analyze the characteristics of tagging systems and find that the tag frequency distribution of a resource remains almost unchanged after receiving enough bookmarks, named as the stable state [5]. Halpin et al. analyze several aspects of the dynamics in collaborative tagging, including why tag distribution follows the *Power Law*, the patterns of tag distribution for stable resources, and tag correlation or completeness etc. [6]. Trushkowsky et al. estimate the completeness of answers for enumerative questions in crowdsourced database [12]. Different from these works, we study the influence of tag correlations and representatives in tagging quality.

Tag Based Application. Bischoff et al. evaluate the effectiveness of tags obtained in collaborative tagging systems in information retrieval [2]. They find that not every tag well describes the content of a resource. There are vocabulary problems such as tag polysemy and tag synonymy in collaborative systems. Heymann et al. compare the tag based search to the full text based search so as to improve retrieval results [7]. Chi et al. also investigate the efficiency of collaborative tagging systems in information retrieval by use of *information theory* and propose some methods to improve the tag-based search [3]. These motivate us to evaluate the quality of a stable tag set against tag-based applications.

Incentive System and Measurement on Tagging Quality. In collaborative tagging systems, only a few resources can get enough tags such that their tagging quality is good enough to describe resource contents [16]. To promote the tagging quality of under-tagged resources, Yang et al. propose an incentive-based tagging mechanism which rewards users on tagging unstable resources. They propose the Moving Average (MA) score to measure the tagging quality of resources [16]. Halpin et al. also evaluate the tagging quality by the relative entropy (KL divergence) between tag frequency distributions at several consecutive tagging steps [6]. Since these measurements highly rely on the whole tag set, the computation is time consuming and the results are affected by noisy tags.

Basic Notions. The basic notions are cited from the related work [16]. Let $\mathcal{R} = \{r_1, r_2, \ldots, r_n\}$ and $\mathcal{T} = \{t_1, t_2, \ldots, t_m\}, n, m \in N^+$, be the resource set and tag set in a collaborative tagging system, respectively. A bookmark is a finite nonempty set of tags annotated to a resource by a user in one tagging operation. The jth bookmark received by resource r_i is denoted as $b_i(j) = \{t_1, t_2, \ldots, t_l\} \subset \mathcal{T}, j \geq 1, l \in N^+$. Let π_i^n denote the time point of r_i receiving its nth bookmark. For r_i, the tag set at π_i^n is denoted by $T_i(n) = \bigcup_{1 \leq j \leq n} b_i(j)$. The frequency of tag t for r_i at π_i^n is the number of bookmarks containing t that r_i has received at π_i^n, denoted by $h_i(t, n) = |\{b_i(j)|1 \leq j \leq n, t \in b_i(j)\}|$. The relative frequency is the normalized frequency $f_i(t,n) = \frac{h_i(t,n)}{\sum_{t' \in T_i(n)} h_i(t',n)}$. The **relative tag frequency distribution (rfd)** of r_i at π_i^n is a vector $\boldsymbol{F_i}(n)$, where $\boldsymbol{F_i}(n)[j] = f_i(t_j,n)$.

When a resource receives enough bookmarks, its rfd changes less and reaches stable, called **stable rfd** and denoted by $\varphi_i = \lim_{n \to \infty} \boldsymbol{F_i}(n)$. The **tagging quality** of r_i at π_i^n is the similarity between its current rfd $\boldsymbol{F_i}(n)$ and its stable rfd φ_i, denoted by $q_i(n) = sim(\boldsymbol{F_i}(n), \varphi_i)$. Actually, the ideal stable state is impossible to get to since collaborative tagging is an infinite process. So a practical evaluation on tagging quality **Moving Average score(MA)** is used to quantify the changes of tag frequency distribution in several consecutive steps. Given a parameter $\omega \geq 2$, the MA score of r_i at $\pi_i^n (n \geq \omega)$ is $m_i(n,\omega) = \frac{1}{\omega-1} \sum_{j=n-\omega+2}^{n} sim(\boldsymbol{F_i}(j-1), \boldsymbol{F_i}(j))$. For a given parameter τ (close to 1) as the threshold, r_i is defined tagging stable when $m_i(n,\omega) \geq \tau$.

3   Dataset and the Dynamics

In this section, we introduce the datasets adopted in this paper and investigate their dynamics from two aspect: which are the representative tags of a resources content and their evolution in a collaborative tagging process.

We adopt three real datasets. The *del.icio.us-* 2007 dataset contains the resources from web application *del.icio.us*. There are 5000 stable resources, 562,048 bookmarks, and 2,027,747 tags. On average each resource received 112 bookmarks, which contain 83 distinct tags. The second dataset *Last.fm* is about a music website, which contains the data from August 2005 to May 2011. There are 33 stable resources and each receives 178 bookmarks on average. The third dataset *MovieLens* contains the rating data for movies from Dec 2005 to March 2015 selected from website *MovieLens*. We select 256 resources, where each receives 413 bookmarks on average. The resources we select have reached their stable states against the MA score with $\tau = 0.9999$ and $\omega = 20$.

To understand the tag set of a resource, we analyze tag distribution and representatives tag. We firstly compute all tag frequency distributions, which follow the *Power Law* as the previous work [6]. To further evaluate a high frequency tag, we count the ratio of users on each tag, namely $\frac{f_i(t,n)}{n}, t \in T$, and find that the *top-1* tag is used by approximately 62 % users and even the 10th popular tag is used by more than 10 % users. High frequency tags indicate the consensus of a large population on resource content, which are helpful for tag-based applications. On the contrary, a low frequency tag is considered as someone's personal understanding and is often regarded noisy. This motivates us to employ tag frequency as the criterion on representative tags.

Then we study when representative tags can be determined in a tagging process. For each resource, we select four time points of its tagging process. The first point N_s records the number of bookmarks when a resource reaches stable. The average N_s of resources in datasets *delicious*, *Last.fm* and *MovieLens* are 84, 130 and 246, respectively. The other three points are selected $\frac{1}{4}N_s$, $\frac{2}{4}N_s$ and $\frac{3}{4}N_s$, respectively. For each point, we count the frequencies of popular tags of $r_i \in R$, shown in Fig. 1(a), where x axis gives a tag rank (top 1 to 60), y lists

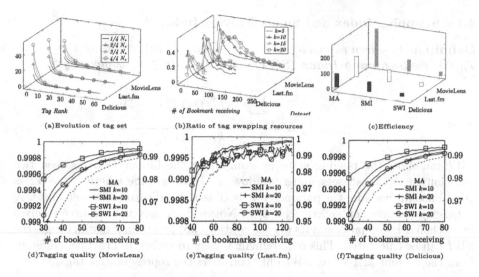

Fig. 1. The dynamics of collaborative tagging

the datasets and z gives the average tag frequency of resources in a dataset. The results show that all tag frequency distributions follow the *Power Law* on different time points. The more bookmarks, the more obvious the trend. For example, the top-1 tag in *Dilicious* is covered by approximate 50 bookmarks at N_s, while this number is about 24 at $1/2N_s$.

A coming question is whether the representative tags of a resource remain unchanged in a tagging process. We consider the top k tags and introduce the concept *tag swapping* to denote the change of top k tags at two consecutive points. For a fix k, we compute the ratio of tag-swapping resources at each point, shown in Fig. 1(b), where x gives time points of tagging process, y lists the datasets and z gives the ratio. We can see as the increase of bookmarks, the ratio decreases, namely the stability of a tag set gets better. Comparatively, there are more *tag swapping* under a larger k. This is caused by low frequency tags included in a top-k tag set, whose frequencies are very close and a new bookmark may change their rankings. But if k is set too small, only a few aspects of a resource can be captured. So, a decent k is desired.

4   The Light Weight Metrics on Tagging Quality

To reduce the influence of noisy tags and improve the efficiency of measuring tagging quality, we introduce two novel measurements *Swapping Index* and *Smart Moving Index*, which are *light weight* in computation.

4.1   Swapping Index and Smart Moving Index

Definition 1. *Given parameters $\omega \geq 2$, $k \geq 1$ and $T_i^k(n) = \{t|rank(t) \leq k \cap t \in T_i(n)\}$, the **Swapping Index (SWI)** of r_i at π_i^n ($n \geq \omega$) is given by*

$$swi_i(n, \omega, k) = \frac{\sum\limits_{j=n-\omega+2}^{n} |T_i^k(j) \cap T_i^k(j-1)|}{k(\omega - 1)} \tag{1}$$

SWI measures the tagging state by computing the intersection between two consecutive *top-k* tag sets under window ω. To determine whether a *SWI* is good enough, we adopt a parameter $\tau > 0$ as the indicator of stable state. r_i is tagging stable when $swi_i(n, \omega, k) > \tau$. Notice that, only if a resource receives more than ω bookmarks and its tag set contains not less than k distinct tags, the *SWI* can be calculated. This requirement is easy to satisfy in practice. Different from the current *MA* score, *SWI* only considers the representative tags.

Definition 2. *Given parameters $\omega \geq 2$ and $k \geq 1$, the **Smart Moving Index (SMI)** of resource r_i at π_i^n ($n \geq \omega$) is given in Eq. 2, where $F_i^k(n)$ is the top-k relative tag frequency distribution whose members $F_i^k(n)[j] = f_i(t_j, n), t_j \in T_i^k(n)$, and $F_i(j-1)_{T_i^k(j)}[m] = f_i(t_m, j-1), t_m \in T_i^k(j-1) \cap T_i^k(j)$. sim is a metric to quantify the similarity between two adjacent tag vectors.*

$$smi_i(n, \omega, k) = \frac{1}{\omega - 1} \sum\limits_{j=n-\omega+2}^{n} sim(F_i^k(j-1)_{T_i^k(j)}, F_i^k(j)) \tag{2}$$

SMI evaluates a tag set by the similarity between adjacent tag frequency distributions. Given a parameter $\tau > 0$ as a threshold, r_i is called tagging stable when $smi_i(n, \omega, k) > \tau$. Notice that only if a resource has received more than ω bookmarks and the size of its tag set is not less than k, *SMI* can be defined.

4.2   Experimental Study

The programs in this paper are written in C++ and experiments are executed on a machine with 4G memory, Intel i3 CPU, installed with 32-bit Linux system. As comparison, we adopt the widely adopted moving score (*MA*).

Tagging State Evaluation. We first verify the trend of each index in a tagging process. The results for three datasets are shown in Fig. 1(d),(e) and (f), where x axis gives the number of bookmarks a resource receives, y gives the score of *MA*, *SMI* and *SWI*. Since *MA* and *SMI* have the same domain, they follow the left vertical axis, while *SWI* is against the right axis. The results are the average value of all resources. The parameters here are $\omega = 5, k = 10, 20$. The results show that there are similar trends for three methods. With the increase of bookmarks, the tagging quality of resources get better.

Table 1. Number of Bookmarks Required vs k

	SMI k=5	SMI k=10	SMI k=15	SMI k=20	SWI k=5	SWI k=10	SWI k=15	SWI k=20
MovieLens	38.7188	78.695	113.234	137.87	22.2812	29.5156	37.2031	45.8125
Lastfm	49.0909	75.8182	83.9394	89.5758	15.3636	24.4242	30.9091	37.4242
Delicious	51.0962	62.7162	67.5676	70.7408	13.9108	19.1994	24.52	30.306

Efficiency Measurement. The efficiency is evaluated on two sides, where in experiments $\omega = 5$, $\tau = 0.9999$, and $k = 10$. One is to compute the required number of bookmarks for a resource to reach stable under different metrics, shown in Fig. 1(c), where x-axis lists the three stability indexes, y lists the datasets and the z-axis gives the average number of required bookmarks. The results show that SWI requires the least bookmarks for a resource to be stable and MA requires the most bookmarks. We further compare the results under different k. From the results in Table 1, we can see that the required number of bookmarks scales with the increasing k for both SMI and SWI. This is because a larger k takes into account more low frequency tags, whose rankings are less stable than high frequency tags.

Another is to evaluate the runtime. The first experiment tests how the performance scales with the number of resources, which helps us understand a system workload. The average runtime on 5000 resources in *delicious* dataset are shown in Fig. 2(a), where x-axis gives the quantity of resources and y-axis gives the runtime. The results show that our methods are more efficient than MA, which illustrates that they are appropriate for a large scale system. Comparatively, SWI is faster than SMI. We further analyze the effect of k setting on runtime. The results in Fig. 2(b) show that SWI and SMI are much more efficient than MA, since MA computes the similarity for all tag frequencies, which is time-consuming. Besides, MA stays relative stable since it is not affected by k. The runtime for both SMI and SWI gets longer as the increasing k. The reason is that a larger k means more tags involved in measurement. SWI is always more efficient than SMI.

Comparison on Stable Tag Set Usage. In order to understand how our proposed methods work in practical applications, we investigate the usage of a stable tag set in tag related applications. Specially, we consider the tag-based recommendation, which is one of the most web popular applications. Three representative recommendation algorithms are adopted in our experiments, namely *Item-based Hierarchical Clustering*, *Item-based k-Means Clustering* etc. The benchmark is the recommendation results returned by the final tag set, denoted by Rec_{final}. When resources are evaluated stable by SMI, the recommendation results are computed against the current tag sets, denoted as Rec_{SMI}. Similarly, the recommendation results for SWI and MA are denoted as Rec_{SWI} and Rec_{MA}, respectively.

We consider two criteria on the accuracy of recommendation results. One is the similarity between two recommended sets, denoted as $Accuracy_{set} = sim(Rec^*, Rec_{final})$, where $sim()$ is the similarity evaluation function and Rec^*

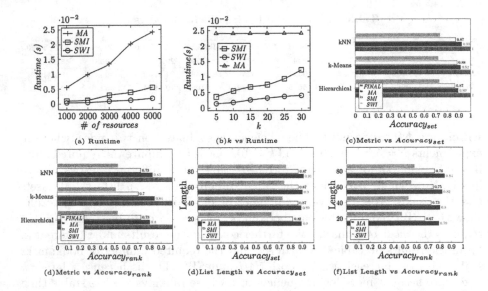

Fig. 2. Tag-based Recommendation Accuracy

indicates an alternative recommendation result based on different metric, such as Rec_{SMI}, Rec_{SWI} or Rec_{MA}. Another criterion evaluates the ranking difference between two recommended lists. We adopt the *Kendall tau distance*, which is widely used to evaluate the number of pairs in opposite order in two rankings [8]. Let $K(Rec^*, Rec_{final})$ denote the *Kendall tau distance*. This accuracy is defined as $Accuracy_{rank} = \frac{|Rec^* \cap Rec_{final}|}{|Rec^*|} * (1 - K(Rec^*, Rec_{final}))$.

For each recommendation algorithm, we compare the results on different stable tag sets against *MA*, *SWI* and *SMI*, respectively. This experiment is performed on the *delicious* dataset and the parameters are $\omega = 5$, $\tau = 0.9999$ and $k = 10$. Each value is computed as the average recommendation accuracy for all resources, i.e. the 5000 resources in the dataset. The results under different algorithms are shown as different colored bars. We first investigate the accuracy on different metrics and show the results in Fig. 2(c) and (d), where x gives the accuracy and y lists the algorithms. Here we choose the top 20 elements in the recommendation results. The result of *SWI* is lower than others. This is because *SWI* only considers the components of representative tags and ignores their frequency. The accuracy of *SMI* is close to *MA*. Comparing two accuracy evaluation methods, $Accuracy_{set}$ is higher than $Accuracy_{rank}$. This is because the $Accuracy_{set}$ metric only focuses on the members in a recommendation result, while the *Kendall tau distance* method considers the ranking of each element. Then we investigate how accuracy scales with the recommendation list length, shown in Fig. 2(e) and (f), where y gives the length of recommendation results. The accuracy increases with the length.

Summary. Taking into account the above evaluations, we can see that SWI and SMI are more efficient than the current method, which resides on two sides: the direct computation time and the required period for a resource becoming stable. For an expected stable tagging state, fewer bookmarks are required against our proposed metrics. For example, SWI needs about 36 % bookmarks against MA. This is because they compute the representative tags, which reflect the consent by a large population and can reach stable much earlier than some unpopular tags. Consiering the usage of stable tag set, SMI is much closer to MA. As shown in Fig. 2(b), when k is set 25, the result of SMI is very close to MA. But it saves about 60 % computation time than MA. So the proposed metrics on tagging quality are appropriate for a large scale system.

5   Semantic Measurement on Tagging Quality

Tags reflect user understanding on a resource content. Different tags reflect different facets of a resource, while two users may adopt different tags to describe the same aspect of a resource. So, multiple tags may share the identical sense and reflect the similar facet of resource content. For example, the word *iphone* is highly related to *Apple* than *tomato*. This motivates us to investigate the intrinsic associations between tags and take into account tag semantics in tagging quality evaluation.

5.1   The Semantic Metric on Tagging Quality

In this paper, we adopt the notion *concept* to model a resource content, which can be hierarchically organized. A *concept* is defined as a set of tags associated with the semantics in a system. Considering a collaborative tagging system, a concept is calculated by the tags together used in a bookmark. Formally, a tag vector is defined as $t = < f_1(t), f_2(t), \ldots, f_{|R|}(t) >, t \in \mathcal{T}$, where $f_i(t)$ is the relative frequency of tag t for resource r_i. For two tags t_i and t_j, its similarity $sim(t_i, t_j)$ can be calculated against the tag vector. Given a threshold τ and a distance increment δ, concepts can be iteratively clustered as a hierarchical tree. Details on the hierarchical clustering can be got in [11]. Each concept maps to a node in the hierarchies. Let h denote the height of the tree, and $\eta \in [1, h]$ denote a specific level of tree, which reflects the semantic closeness of tags. Each concept of level 1 (leaf node) maps to a concrete tag. A concept in a higher level has a broader semantics, which is iteratively generated by combining the semantically close concepts. Formally,

Definition 3. *A hierarchical tree consists of* **concepts***. The concept on level $\eta \in [1, h]$ with m children is given below, where $c_j^{\eta-1}$ is the jth child of c^η.*

$$c^\eta = \begin{cases} \{t\} \subset \mathcal{T} & \eta = 1 \\ \bigcup_{1 \leq j \leq m} c_j^{\eta-1} & \eta > 1 \end{cases} \tag{3}$$

Actually, the hierarchies reflect the crowd intelligence. Compared to the plain folksonmy, it takes the advantage of the taxonomy systems to manage tags in a systematic way. Besides, it can locate the related resources quickly when processing a query [5]. Base on the notion *concept*, we introduce the semantic measurement of tagging quality.

Definition 4. *Give a parameter η, the **resource concept distribution function (rcf)** of resource r_i at π_i^k is a vector $\boldsymbol{F_i}^\eta(k)$, s.t. the jth component is the frequency of concept c_j^η for r_i at π_i^k, $\boldsymbol{F_i}^\eta(k)[j] = f_i(c_j^\eta, k) = \frac{\sum_{t \in T_i(k) \cap c_j^\eta} h_i(t,k)}{\sum_{t \in T_i(k)} h_i(t,k)}$.*

Definition 5. *Given parameters $\omega \geq 2$, η and τ, the **Semantics Index(SI)** of $r_i \in \mathcal{T}$ at $\pi_i^k(k \geq \omega)$ is given by Eq. 4. r_i is called stable if $sem_i^\eta(k, \omega) > \tau$.*

$$sem_i^\eta(k, \omega) = \frac{1}{\omega - 1} \sum_{j=k-\omega+2}^{k} sim(\boldsymbol{F_i}^\eta(j-1), \boldsymbol{F_i}^\eta(j)) \tag{4}$$

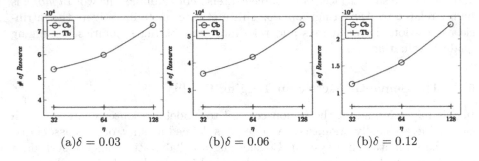

(a)$\delta = 0.03$ (b)$\delta = 0.06$ (b)$\delta = 0.12$

Fig. 3. Comparison on Semantic Index and Tag based Metrics

5.2   Experiment Analysis

The Tag Based Application. To evaluate the effectiveness of the semantic index, we select the tag based retrieval, one of the most popular applications, as the evaluation tool, which answers a tag based query with related resources. We justify the tagging quality metrics by the output of tag based retrieval against each tag set that is evaluated stable against a metric. The experiments are performed on the *delicious* dataset. In a collaborative tagging system, a tag frequency reflects how much people have a consensus on a resource by the tag, which is also admitted by the tag based retrieval. Let parameter $\delta \in [0..1]$ denote the relatedness threshold between a tag and a resource. For once query with tag t, resource r_i is selected as the result if $f_i(t, n) > \delta$. In practice, this parameter can be learned from user multiple retrievals. Similarly, the concept based evaluation is $f_i(c, n) > \delta$ or $\frac{f_i(c^\eta, n)}{|\{t|t \in c^\eta \cap T_i(n)\}|} > \delta$, where $t \in c^\eta$.

Effectiveness Comparison. We compare the concept-based measurement (Cb for short) and Semantic Index (SI) with the *Tag-based* measurement (Tb for short) MA. The parameters in this experiment are $\omega = 5, \tau = 0.9999$. Figure 3 shows the results for the retrieval application under different constraints, where x axis gives η settings and y gives the average number of returned resources on tag based queries. Overall, Cb is better than Tb since it can find more related resources which are not in an obvious mode. A larger δ means a stricter relatedness on tag and resource and results in fewer satisfied resources. A concept in a higher level η contains more semantics such that more resources satisfy a query.

We further evaluate the returned results on retrieval application against the final tag set in a collaborative tagging system. Let tp denote the number of resources in the *true positive* case, i.e. the resources appear in the retrieval results by both the SI stable tag set and the final tag set. Similarly, tn, fp and fn denote the cases *true negative*, *false positive*, and *false negative*, respectively. Thus, $precision = \frac{tp}{tp+fp}$, and $recall = \frac{tp}{tp+fn}$. The parameters are $\omega = 5$, $\tau = 0.9999$ and $\delta = 0.05$ for SI and MA. The comparison results are listed in Table 2, which show that the *precision* and *recall* are very close under SI and MA. A higher η brings a slight lower *precision* and *recall* since more general content are extracted and fewer bookmarks are desired for a resource being tagging stable. In practice, the parameters is set by an administrator at first so as to solve the *cool start* problem. Then it can be learned in the process of tag based applications.

Efficiency Measurement. It is necessary to consider both the required bookmarks for a resource be stable and the runtime for computing metrics. Comparing SI and MA, the complexity on each metric computation is the same. So, their difference depends on the number of required bookmarks for each resource reaching its stable state. From the results in Table 2, we can see the required number of bookmarks for a resource to be tagging stable is smaller under SI than MA. With an increasing η, fewer bookmarks are desired for a resource being tagging stable. At the first glance, the reduced number of bookmarks is not very large, such as the average number of saved bookmarks is about 20 when $\eta = 128$. However, if we take into account the whole tagging process, each resource receives 100 bookmarks on average around a whole year. It is easy to conclude that 20 bookmarks are expected for approximately one month. So, the improvement on evaluation efficiency is meaningful in practice.

Table 2. Comparison of SI and MA on Retrieval Results

	SI η=32	MA η=32	SI η=64	MA η=64	SI η=128	MA η=128
Precision	0.962848	0.967474	0.960242	0.968447	0.956816	0.97073
Recall	0.958032	0.963424	0.952392	0.963933	0.945782	0.96517
Required Bookmark	77.247	83.03	70.949	83.03	63.1314	83.03

6   Conclusion

In this paper, we propose several metrics on collaborative tagging quality evaluation, which are light weight on computation and effective than the current method. The SWI and SMI metrics take the representative tags for measurement to get rid of the influence of noisy tags in making a resource tagging stable. The semantic measurement SI takes tag intrinsic associations into consideration, which makes tagging quality evaluation more effective. A series of experiments are performed against several real datasets and results show the efficiency and effectiveness of our methods. This illustrates that the proposed methods are especially appropriate for a large quantity of web resources. In the future, we will take user personal preferences into account for quality study. Another direction is to investigate how to apply our methods into other crowdsourcing applications.

References

1. Bi, B., Lee, S.D., Kao, B., Cheng, R.: Cubelsi: an effective and efficient method for searching resources in social tagging systems. In: Proceedings of the 27th IEEE International Conference on Data Engineering (ICDE), pp. 27–38. IEEE (2011)
2. Bischoff, K., Firan, C.S., Nejdl, W., Paiu, R.: Can all tags be used for search? In: Proceedings of the 17th ACM Conference on Information and knowledge Management, pp. 193–202. ACM (2008)
3. Chi, E.H., Mytkowicz, T.: Understanding the efficiency of social tagging systems using information theory. In: Proceedings of the 19th ACM Conference on Hypertext and Hypermedia, pp. 81–88. ACM (2008)
4. Durao, F., Dolog, P.: A personalized tag-based recommendation in social web systems. In: Adaptation and Personalization for Web 2.0, p. 40 (2009)
5. Golder, S.A., Huberman, B.A.: Usage patterns of collaborative tagging systems. J. Inf. Sci. **32**(2), 198–208 (2006)
6. Halpin, H., Robu, V., Shepherd, H.: The complex dynamics of collaborative tagging. In: Proceedings of the 16th International Conference on World Wide Web, pp. 211–220. ACM (2007)
7. Heymann, P., Koutrika, G., Garcia-Molina, H.: Can social bookmarking improve web search? In: Proceedings of the 2008 International Conference on Web Search and Data Mining, pp. 195–206. ACM (2008)
8. Kendall, M.G.: Rank Correlation Methods. Griffin, London (1948)
9. Lei, S., Yang, X.S., Mo, L., Maniu, S., Cheng, R.: Itag: incentive-based tagging. In: Proceedings of 30th IEEE International Conference on Data Engineering (ICDE), pp. 1186–1189. IEEE (2014)
10. Ramage, D., Heymann, P., Manning, C.D., Garcia-Molina, H.: Clustering the tagged web. In: Proceedings of the Second ACM International Conference on Web Search and Data Mining, pp. 54–63. ACM (2009)
11. Shepitsen, A., Gemmell, J., Mobasher, B., Burke, R.: Personalized recommendation in social tagging systems using hierarchical clustering. In: Proceedings of the 2008 ACM Conference on Recommender Systems, pp. 259–266. ACM (2008)
12. Trushkowsky, B., Kraska, T., Franklin, M.J., Sarkar, P.: Crowdsourced enumeration queries. In: Proceedings of the 29th IEEE International Conference on Data Engineering (ICDE), pp. 673–684. IEEE (2013)

13. Van Damme, C., Hepp, M., Coenen, T.: Quality metrics for tags of broad folksonomies. In: Proceedings of International Conference on Semantic Systems (I-SEMANTICS), pp. 118–125 (2008)
14. Wagner, C., Singer, P., Strohmaier, M., Huberman, B.A.: Semantic stability in social tagging streams. In: Proceedings of the 23rd International Conference on World Wide Web, pp. 735–746. ACM (2014)
15. Xu, H., Zhou, D., Sun, Y., Sun, H.: Quality based dynamic incentive tagging. Distrib. Parallel Databases **33**(1), 69–93 (2015)
16. Yang, X.S., Cheng, R., Mo, L., Kao, B., Cheung, D.W.-l.: On incentive-based tagging. In: Proceedings of the 29th IEEE International Conference on Data Engineering (ICDE), pp. 685–696. IEEE (2013)

Matrices, Compression, Learning Curves: Formulation, and the GROUPNTEACH Algorithms

Bryan Hooi[1(✉)], Hyun Ah Song[1], Evangelos Papalexakis[1],
Rakesh Agrawal[2], and Christos Faloutsos[1]

[1] Carnegie Mellon University, Pittsburgh, PA 15213, USA
bhooi@andrew.cmu.edu, {hyunahs,epapalex,christos}@cs.cmu.edu
[2] Data Insights Laboratories, San Francisco, USA
ragrawal@acm.org

Abstract. Suppose you are a teacher, and have to convey a set of object-property pairs ('lions eat meat'). A good teacher will convey a lot of information, with little effort on the student side. What is the best and most intuitive way to convey this information to the student, without the student being overwhelmed? A related, harder problem is: how can we assign a numerical score to each lesson plan (i.e., way of conveying information)? Here, we give a **formal definition** of this problem of forming learning units and we provide a **metric** for comparing different approaches based on information theory. We also design an **algorithm**, GROUPNTEACH, for this problem. Our proposed GROUPNTEACH is *scalable* (near-linear in the dataset size); it is *effective*, achieving excellent results on real data, both with respect to our proposed metric, but also with respect to encoding length; and it is *intuitive*, conforming to well-known educational principles. Experiments on real and synthetic datasets demonstrate the effectiveness of GROUPNTEACH.

1 Introduction

If you were given Fig. 1 (c) and (d) to memorize, which would you find easier to memorize? If you were a zoology teacher, how would you come up with a lesson plan to teach the facts in Table 1, containing animals and their properties?

In our formulation, our facts consist of simple *(object, property)* pairs (Table 1a). Informally, our problem can be stated as follows:

Informal Problem 1 (Transmission/Teaching Rate Problem). *Given a large, sparse binary matrix whose rows represent objects, columns represent properties, in which ones represent facts, how do we measure how good a particular encoding of the matrix is for student learning, and how do we optimize this metric?*

Table 1 illustrates our intuition behind the solution: most people would agree that randomly stating facts ('salmons have fins') would be painful for the student. A good teacher would group animals and properties, as in Table 1b, and use analogies and comparison, such as 'tigers are like lions, but have stripes.'

Our contributions are as follows:

© Springer International Publishing Switzerland 2016
J. Bailey et al. (Eds.): PAKDD 2016, Part II, LNAI 9652, pp. 376–387, 2016.
DOI: 10.1007/978-3-319-31750-2_30

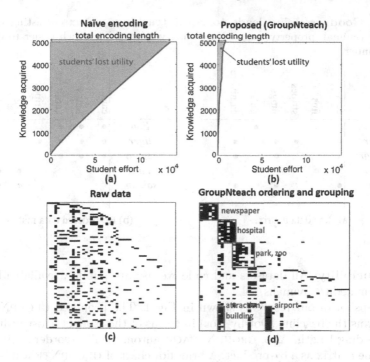

Fig. 1. GROUPNTEACH agrees with intuition. GROUPNTEACH (b) encodes facts with much lower total encoding length and students' lost utility than the (a) naive encoding, which encodes the nonzero entries of the matrix one by one. As a side effect, given randomly ordered data (c), GROUPNTEACH finds a (d) reordering and groupings of the facts along with labels which are intuitive.

- **Problem Formulation**: we formulate the *Transmission Rate Problem* formally in the context of matrix compression, and define how in the context of an educational setting, our goals differ from those of the standard matrix compression problem.
- **Optimization Goal**: we formulate a new metric which prioritizes consistent learning, rather than purely maximum compression, which maximizes student utility. This enables us to design *parameter-free* algorithm that picks optimal parameters for the defined criterion.
- **Algorithm**: we propose GROUPNTEACH, an algorithm for encoding and ordering a set of facts for student learning. GROUPNTEACH has the following properties:
  1. Scalable: it scales near-linearly in the data size, allowing it to scale to datasets with millions of facts.
  2. Effective: it encodes real datasets more efficiently than standard approaches for encoding sparse data, under both compression length and our metric.
  3. Intuitive: it follows educational principles such as grouping related concepts.

Table 1. Good grouping leads to good teaching: matrix consisting of facts, which are (animal, property) pairs. Note that version (b) is much easier to describe and remember.

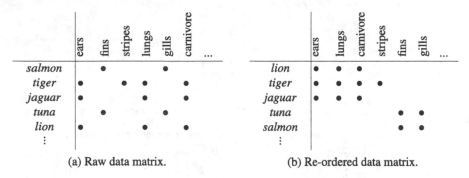

	ears	fins	stripes	lungs	gills	carnivore	...
salmon		•			•		
tiger	•		•	•		•	
jaguar	•			•		•	
tuna		•			•		
lion	•			•		•	
⋮							

(a) Raw data matrix.

	ears	lungs	carnivore	stripes	fins	gills	...
lion	•	•	•				
tiger	•	•	•	•			
jaguar	•	•	•				
tuna					•	•	
salmon					•	•	
⋮							

(b) Re-ordered data matrix.

Reproducibility: All datasets and code we use are publicly available http://www.cs.cmu.edu/~hyunahs/tol.

A snapshot of our results is shown in Fig. 1. The proposed GROUPNTEACH outperforms the baseline (Dot-by-Dot) in terms of the student's lost utility, and total encoding length. Also, GROUPNTEACH automatically reorders and groups facts of the matrix as a by-product. A brief flowchart of GROUPNTEACH is shown in Fig. 2, which will be discussed in more detail in later sections.

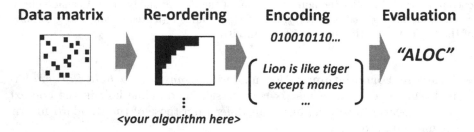

Fig. 2. Flowchart of GROUPNTEACH. GROUPNTEACH reorders the input data, encodes the information, and finally evaluates each plugin in GROUPNTEACH and returns the winning result. GROUPNTEACH tries several plugins to reorder the data, but any alternate reordering method can be plugged in 'your algorithm here.'

2   Background and Related Work

Support from Learning Theory. Improving student learning has been a great interest in various domains including psychology, education [7,10], as well as data mining [11,14]. In our study, we find that algorithms that do well under our metrics indeed agree with educational instructional principles [10], which are explained in Table 2. In the last column, we show the keyword of GROUPNTEACH (Table 4) that reflects the principles.

Table 2. Instructional principles found in [10], examples of how GROUPNTEACH conforms to these principles, and the corresponding language in GROUPNTEACH.

	details	example	GROUPNTEACH
P1 **(Linking)**	link multiple objects together	'Tigers, lions, and jaguars all have teeth'	'ranges'
P2 **(Pre-conceptions)**	refer and relate to prior knowledge	'Tigers have teeth' 'So do lions'	'like'
P3 **(Comparison)**	compare and contrast multiple instances	'Tigers are like lions except that they have stripes'	'except'

Table 3. GROUPNTEACH qualitatively outperforms competitors. Block-based methods in general link multiple objects (**P1**), but only GROUPNTEACH teaches new concepts based on students' existing concepts (**P2**) and communicates based on similarity and difference. (**P3**) The Shneiderman mantra refers to communicating a high-level summary first, followed by finer details.

	Functions	Matrix Factorization	Matrix Compression[4, 2]	FCA[6]	GROUPNTEACH
	Ordering				✓
	Compression	✓	✓		✓
	Grouping	✓	✓	✓	✓
Teachability	P1: *11th* Linking	✓	✓	✓	✓
	P2: *13th* Pre-conceptions				✓
	P3: *17th* Comparison				✓
HCI	"Shneiderman mantra"		✓	✓	✓

Matrix Compression. There are several methods for efficiently compressing a matrix [16]. Our goal is different, particularly due to the educational setting: (a) we want metrics that prioritize consistent learning rather than pure encoding length; (b) we want to encode new information based on a learner's existing knowledge.

Bipartite Clustering and Binary Matrix Reordering. Various bipartite clustering and matrix reordering algorithms [5,17] can be plugged in to our reordering stage as shown in Fig. 2. In Table 3, we compare GROUPNTEACH to related algorithms in terms of the functions they perform such as ordering, compression, and grouping, as well as their relation to educational and human computer interaction (HCI) principles.

Minimum Spanning Trees. In one of our proposed plugins, GROUPNTEACH-Tree, we adopt EUCLIDEAN-MST [13].

3   Proposed Metric

How can we assign a numerical score for comparing different ways of teaching a collection of facts to students?

Definition 1 *Performance-for-Price curve* $p(n)$. Given an encoding algorithm, define $p(n)$ as the number of nonzero entries of the matrix that are decodable based on the first n bits output by the encoding algorithm (e.g. see Fig. 1; top).

Definition 2 *Area Left of Curve (ALOC)*. The *ALOC* metric is the area left of the curve $p(n)$. Lower *ALOC* is better.

Utility interpretation. Assuming the students gain utility at each time step according to how much they know, the total utility gained by students is the area under the curve $p(n)$. Then *ALOC* corresponds to the utility lost by a student over time compared to having known all the information in advance.

ALOC uses the number of bits transmitted as the units along the x-axis when plotting $p(n)$ since this represents the amount of attention that students need to understand the lesson content. Hence transmitting the message efficiently lowers the amount of effort students need to understand the material.

4   Proposed Method: GROUPNTEACH

In this section, we describe GROUPNTEACH; it finds efficient and interpretable encodings for a collection of facts, and as a by-product, it sequences the facts so as to find groupings of related facts, and a good teaching order for the groups.

All-engulfing approach. Since real-world datasets differ greatly in their underlying patterns, we expect different algorithms to perform well on different datasets. Hence, we propose four plugins, each designed to perform well on a particular type of dataset. Our *all-engulfing* approach tries each plugin to encode a dataset, and chooses the encoding with lowest *ALOC*. The complete algorithm is given in Algorithm 1.

Algorithm 1. GROUPNTEACH encoding method that also returns grouped and ordered data

---

Data: Data matrix M (a binary matrix of objects by properties)
Result: Encoding for M and re-ordering and groupings of rows and columns of M
Plugins = {*Block, Tree, Chain, Fishbone*}   *(any other heuristics can be added)*;
$i^* = \arg\min_i ALOC(\text{Plugins}[i](M))$ *(find best performing component)*;
Output binary representation of i^* *(index of the winning method)*;
Output *encoding* of M using method Plugins[i^*];
Reorder rows and columns of M using the orderings induced by GROUPNTEACH-Chain (see Sect. 6);
Group rows and columns according to Algorithm 5;

Table 4 summarizes the encoding structure and keywords each plugin uses to encode information.

Table 4. Encoding structure used for each method. *r-id* and *c-id* refer to *row index* and *column index*, respectively. The **except** keyword communicates exceptions, e.g. for the Block plugin, exceptions are the zeroes within the current block. For Tree and Chain, exceptions are differences between the current and compared rows. **end-statement** terminates the list of exceptions.

Method	Encoding structure	Keywords	Encoding
Block	*(r-id, c-id)* + *(row-length)* + *(column-length)* ***except*** + *(r-id, c-id)* + ***except*** + *(r-id, c-id)* + ... + ***end-statement***		
		except	1
		end-statement	0
Tree / Chain	*(r-id)* + *(comparison r-id)* + **except** + *(c-id)* + **except** + *(c-id)* + ... + **end-statement**		
Fishbone	*(length of block)* + **except** + *(c-id)* + **except** + *(c-id)* + ... + **end-statement**		

5   GROUPNTEACH-plugins

In this section, we give detailed explanations on each of the plugins.

5.1   GROUPNTEACH-Block

GROUPNTEACH-Block, explained in Algorithm 2, is designed to encode block-structured data highly efficiently. It does this by clustering the rows and columns to produce dense blocks, then re-orders the block regions, and encodes the block information (starting point, row and column length of the block), along with the missing elements in the block, and the additional elements outside the defined blocks.

5.2   GROUPNTEACH-Tree

GROUPNTEACH-Tree encodes a row by describing its differences from a similar row. For example, to describe *tigers*, we say that *tigers* are like *lions* except that they have stripes. Since real-world datasets typically have many similar items, most rows end up having very short encodings.

The first row encoded is the row with the most ones, encoded directly as a binary string. All subsequent rows are encoded using statements like 'row i is like row j except in positions $k, l, \ldots$'. This means row i can be obtained by starting with row j and flipping the bits in positions $k, l, \ldots$. To construct this encoding, GROUPNTEACH-Tree first constructs a distance function $d(i, j)$, equal to the number of differences between row i and row j. Then finding an encoding is equivalent to constructing a spanning tree: for example, if we encoded row i based on similarity to row j, then (i, j) is an edge of weight $d(i, j)$ in the corresponding tree. It is a tree because each row has exactly one ancestor (the row used to encode it), except the root. Then, we can minimize the number of

---

Algorithm 2. GROUPNTEACH-BLOCK.

Data: Data matrix M
Result: Encoding of M
Step 1: Partition rows and columns into groups;
    Input: M, k;
    Cluster rows and columns of matrix M into k groups using NMF;
    Output: row and column indices for each cluster regions;
Step 2: Get the best permuted M_{pm^*};
    Input: cluster information;
    Compute density of each block;
    Output: M_{pm^*}, The best permuted matrix with highest density in the
diagonal blocks;
Step 3: Encode M_{pm^*};
    Using ENCODE-BLOCK(M_{pm^*}), encode the top-left corner of the block, row
and column lengths of the block.;
    Output: Encoding of M;

Algorithm 3. GROUPNTEACH-TREE: fast approximate minimum spanning tree-based encoding method

---

Data: Data matrix M
Result: Row-wise encoding for M
Let $(M_i)_{i=1}^n$ be the rows of M;
Generate random vectors $f_1, \ldots, f_p \in \mathbb{R}^n$;
for $i=1,\ldots,m$ **do**
    $x_i = (M_i \cdot f_1, \ldots, M_i \cdot f_p)$ *(Construct feature vectors)*
$T = $ EUCLIDEAN-MST($x_1, \ldots, x_m$);
Choose row index r with largest row sum;
Output M_r;
Let O be a BFS traversal of T with root r;
for *each edge (i, j) in O* **do**
    Let $D(i, j)$ be the set of column indices at which M_i differs from M_j;
    Output $\langle i, j, D(i, j) \rangle$ *(output that row j is like row i except in columns $D(i, j)$)*

---

differences we need to encode by minimizing the weight of the spanning tree. We could do this by constructing a distance matrix $d(i, j)$ between the rows, then finding the MST using e.g. Kruskal's algorithm. Since Kruskal's algorithm would require quadratic time, however, we instead use the Euclidean MST algorithm which takes $O(n \log n)$ time, as given in Algorithm 3.

5.3 GROUPNTEACH-Chain

GROUPNTEACH-Chain is similar to GROUPNTEACH-Tree: we also encode each row based on a similar row. However, unlike the tree pattern of GROUPNTEACH-Tree, here each row is encoded based on comparison to the last encoded row.

For example, we may encode *lions* based on *tigers*, then *jaguars* based on *lions*, and so on, forming a chain. This allows the encoding of *lions* to not encode its parent *tigers*, since its parent *tigers* can be deduced from being the last animal encoded. This encodes each row more cheaply and is more efficient for sparse data.

To find a good ordering of the rows, we first use the same random projections method as GROUPNTEACH-Tree to obtain feature representations $x_1, \ldots, x_m$ of the rows. We then use the Euclidean distance between x_i and x_j as a proxy for the number of differences between rows i and j. Starting from the row with the most ones, we repeatedly find the next row to encode as follows: randomly sample a fixed k of the remaining unused rows; choose the closest of these k rows as the next row to encode; then continue this until all rows are encoded.

5.4   GROUPNTEACH-Fishbone

GROUPNTEACH-Fishbone aims to efficiently encode data with uneven number of ones in each row, such as power-law degree distributions which are common in online communities [2]. GROUPNTEACH-Fishbone rearranges as many ones as possible to the top-left of the matrix, then takes advantage of the density of that region to encode the data efficiently. To do this, GROUPNTEACH-Fishbone first reorders the rows and columns in descending order of their row or column sum. Then, it encodes the top row by encoding a number k followed by a list of exceptions $p, q, r, \ldots$. This indicates that except at positions $p, q, r, \ldots$, the row contains k ones, then $n - k$ zeroes. k is chosen by trying all possible k and using the shortest encoding. For efficiency, we terminate the search for k early if the current encoding is some fixed constant C bits worse than the best found encoding. Having encoded the top row, we then encode the first column of the remaining matrix in the same way, and so on, as shown in Algorithm 4.

Algorithm 4. GROUPNTEACH-FISHBONE.

Data: Data matrix M
Requires: ENCODE-ROW, a function that encodes a vector of length n by comparing it to k ones followed by $n - k$ zeroes, and listing all exceptions to this pattern;
Result: Encoding of M
Reorder rows and columns of M in descending order of row and column sums;
while M *is non-empty* **do**
  Let $(M_i)_{i=1}^n$ be the rows of M;
  $k^* = \arg\min_k \text{LENGTH}(\text{ENCODE-ROW}(M_1, k)))$;
  Output ENCODE-ROW(M_1, k^*);
  Remove M_1 from M and transpose M;

5.5   Extensibility

GROUPNTEACH can be broken down into two parts: *reorganization (reordering of rows and columns)*, and *encoding of the reorganized matrix*. GROUPNTEACH can be easily extended by plugging in any matrix reorganization method, such as Cross Association [4] or METIS [8].

6   GROUPNTEACH-post processing: ordering, grouping and curriculum development

As a by-product, GROUPNTEACH produces an intuitive ordering and grouping of the objects in the dataset, as was shown in Fig. 1.

The process of grouping is described in Algorithm 5.

Algorithm 5. GROUPINGCODE: groupings of the related facts on the reordered matrix

---

Data: Reordered data matrix $\tilde{M}$ recovered as output of GROUPNTEACH, threshold C

Result: A set of groups $\mathcal{G}$ for the reordered data matrix $\tilde{M}$

Group data in $\tilde{M}$ into rectangles by combining nearby entries if they form rectangular blocks;

while *not converged* **do**

    for $i = 1, \ldots, |\mathcal{G}|$ **do**

        for $j = 1, \ldots, |\mathcal{G}|$ **do**

            Let R be the smallest bounding box covering R_i and R_j;

            if $\frac{number\ of\ 1s\ in R}{area\ of R} \geq C$ **then**

                remove R_i and R_j from $\mathcal{G}$, and add R to $\mathcal{G}$;

    Output $\mathcal{G}$;

---

7   Experiments

In this section we demonstrate the efficiency and effectiveness of GROUPNTEACH using real and synthetic datasets. We implemented GROUPNTEACH in MATLAB; all experiments were carried out on a 2.4 GHz Intel Core i5 Macbook Pro, 16 GB RAM, running OS X 10.9.5. Our code and all our datasets are publicly available at http://www.cs.cmu.edu/~hyunahs/tol. We used 100 features ($p = 100$) for GROUPNTEACH-Chain and GROUPNTEACH-Tree, and threshold $C = 50$ for GROUPNTEACH-Fishbone. The real datasets used are shown in Table 5. The synthetic datasets used are: 1.KRONECKER: a 256×256 Kronecker graph [12], 2.BLOCKS: two 50×70 blocks of ones in a matrix, 3.HYPERBOLIC: a 20×20

Table 5. Real datasets used.

	Size	Number of nonzeros	Content
ANIMAL [3]	34 by 13	136	animal-property
NELL [1]	212 985 by 217	1.1 million	object-category
DRUG-BANK [9]	1581 by 16 883	109 339	drug-property
QUESTIONS [15]	60 by 218	5252	question-answer

matrix containing 3 overlapping communities of sizes 20, 8 and 4, each resembling a scale-free network.

We conducted several experiments to answer the following questions: **Q1. Scalability, Q2. Effectiveness, Q3. Discoveries.**

Q1. Scalability: Figure 3 (a) shows the linear or near-linear performance of our algorithms. The algorithms are run on random matrices of varying number of rows, fixed to 1000 columns and an average of 10 ones per row.

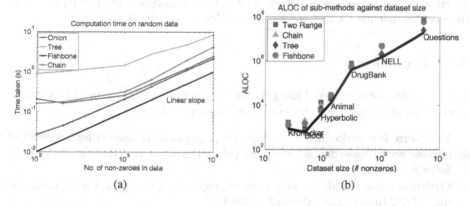

(a)   (b)

Fig. 3. (a) GROUPNTEACH **scales linearly.** GROUPNTEACH scales linearly with the input size (The line $y = cx$ for $c = \frac{1}{10000}$ is added for comparison). (b) GROUPNTEACH **needs all plugins**: different plugins win on different datasets (lower is better). The black lower bound is the final result by GROUPNTEACH.

Q2. Effectiveness: We demonstrate that the multiple plugins of GROUPN-TEACH allow it to do well on diverse types of data. Figure 3 (b) shows that the various plugins of GROUPNTEACH do well on different types on data: GROUP-NTEACH-Block for block-wise, GROUPNTEACH-Fishbone HYPERBOLIC, GROUP-NTEACH-Chain and GROUPNTEACH-Tree datasets with similar rows or columns (which is the case for the real datasets). No one method dominates the others.

Q3. Discoveries: As a by-product, GROUPNTEACH automatically reorders and groups the data. We analyze this property using the DRUG-BANK dataset; GROUPNTEACH finds a teaching order with several desirable characteristics, as shown in Fig. 4.

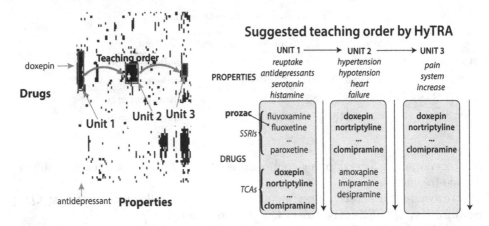

Fig. 4. GROUPNTEACH **leads to curriculum discovery. Left:** re-ordered and grouped DRUG-BANK drug-property data. GROUPNTEACH constructs a teaching order. **Right:** curriculum constructed by GROUPNTEACH. The units ('blobs') obtained are intuitive: e.g. drugs in the TCA family are known to have several groups of effects, which our algorithm groups as follows. Unit 1: their antidepressant properties; Unit 2: their cardiac side-effects; and, some of them, Unit 3: treating chronic pain.

8   Conclusion

In this paper, we considered the problem of teaching a collection of facts while minimizing student effort. Our contributions are as follows:

- **Problem Formulation**: we define the problem of transmitting a matrix of objects and properties adhering to principles from the theory of (human) learning.
- **Optimization Goal**: We define an appropriate optimization goal; minimizing $ALOC$ (maximizing student utility).
- **Algorithm**: We propose GROUPNTEACH, an *all-engulfing* method that encodes the data while reordering and grouping the data. We evaluate GROUP-NTEACH on synthetic and real datasets, showing that it encodes data more efficiently than a naive encoding approach, measured using both $ALOC$ and total encoding length.
- **Ordering of Groups**: When applying GROUPNTEACH on real datasets, we find that the orderings and groupings it produces are meaningful.

Acknowledgments. This material is based upon work supported by the National Science Foundation under Grant No. IIS-1247489

Research was sponsored by the Army Research Laboratory and was accomplished under Cooperative Agreement Number W911NF-09-2-0053.

This work is also partially supported by an IBM Faculty Award and a Google Focused Research Award. Any opinions, findings, and conclusions or recommendations expressed in this material are those of the author(s) and do not necessarily reflect the views of the National Science Foundation, or other funding parties. The U.S. Government is authorized to reproduce and distribute reprints for Government purposes notwithstanding any copyright notation here on.

References

1. Read the web. http://rtw.ml.cmu.edu/rtw/
2. Araujo, M., Günnemann, S., Mateos, G., Faloutsos, C.: Beyond blocks: hyperbolic community detection. In: Calders, T., Esposito, F., Hüllermeier, E., Meo, R. (eds.) ECML PKDD 2014, Part I. LNCS, vol. 8724, pp. 50–65. Springer, Heidelberg (2014)
3. Bro, R., Papalexakis, E.E., Acar, E., Sidiropoulos, N.D.: Coclustering - a useful tool for chemometrics. J. Chemom. **26**(6), 256–263 (2012)
4. Chakrabarti, D., Papadimitriou, S., Modha, D.S., Faloutsos, C.: Fully automatic cross-associations. In: ACM KDD, pp. 79–88 (2004)
5. Dhillon, I.S.: Co-clustering documents and words using bipartite spectral graph partitioning. In: 7th ACM SIGKDD, pp. 269–274. ACM (2001)
6. Ganter, B., Stumme, G., Wille, R.: Formal Concept Analysis: Foundations and Applications, vol. 3626. Springer Science & Business Media, New York (2005)
7. Gobet, F., Lane, P.C., Croker, S., Cheng, P.C., Jones, G., Oliver, I., Pine, J.M.: Chunking mechanisms in human learning. Trends Cogn. Sci. **5**(6), 236–243 (2001)
8. Karypis, G., Kumar, V.: METIS-unstructured graph partitioning and sparse matrix ordering system, version 2.0 (1995)
9. Knox, C., Law, V., Jewison, T., Liu, P., Ly, S., Frolkis, A., Pon, A., Banco, K., Mak, C., Neveu, V., et al.: Drugbank 3.0: a comprehensive resource for omics research on drugs. Nucleic Acids Res. **39**(suppl 1), D1035–D1041 (2011)
10. Koedinger, K.R., Booth, J.L., Klahr, D.: Instructional complexity and the science to constrain it. Science **342**(6161), 935–937 (2013)
11. Koedinger, K.R., Brunskill, E., de Baker, R.S.J., McLaughlin, E.A., Stamper, J.C.: New potentials for data-driven intelligent tutoring system development and optimization. AI Mag. **34**(3), 27–41 (2013)
12. Leskovec, J., Chakrabarti, D., Kleinberg, J., Faloutsos, C., Ghahramani, Z.: Kronecker graphs: an approach to modeling networks. JMLR **11**, 985–1042 (2010)
13. March, W.B., Ram, P., Gray, A.G.: Fast Euclidean minimum spanning tree: algorithm, analysis, and applications. In: ACM KDD, pp. 603–612 (2010)
14. Matsuda, N., Cohen, W.W., Koedinger, K.R.: Teaching the teacher: tutoring SimStudent leads to more effective cognitive tutor authoring. IJAIED **25**, 1–34 (2014)
15. Murphy, B., Talukdar, P., Mitchell, T.: Selecting corpus-semantic models for neurolinguistic decoding. In: ACL *SEM, pp. 114–123. Association for Computational Linguistics (2012)
16. Tarjan, R.E., Yao, A.C.-C.: Storing a sparse table. CACM **22**(11), 606–611 (1979)
17. Zha, H., He, X., Ding, C., Simon, H., Gu, M.: Bipartite graph partitioning and data clustering. In: 10th CIKM, pp. 25–32. ACM (2001)

Privacy Aware K-Means Clustering
with High Utility

Thanh Dai Nguyen$^{(\boxtimes)}$, Sunil Gupta, Santu Rana, and Svetha Venkatesh

Center for Pattern Recognition and Data Analytics,
Deakin University, Geelong 3216, Australia
{thanh,sunil.gupta,santu.rana,svetha.venkatesh}@deakin.edu.au

Abstract. Privacy-preserving data mining aims to keep data safe, yet useful. But algorithms providing strong guarantees often end up with low utility. We propose a novel privacy preserving framework that thwarts an adversary from inferring an unknown data point by ensuring that the estimation error is almost invariant to the inclusion/exclusion of the data point. By focusing directly on the estimation error of the data point, our framework is able to significantly lower the perturbation required. We use this framework to propose a new privacy aware K-means clustering algorithm. Using both synthetic and real datasets, we demonstrate that the utility of this algorithm is almost equal to that of the unperturbed K-means, and at strict privacy levels, almost twice as good as compared to the differential privacy counterpart.

1 Introduction

Data mining is transforming the world. The scope is enormous. Not only do institutions collect data, people too "exhume" data - from black-boxes in their cars, to Fitbits they wear, to posts on Facebook. Personal data, however, cannot be accessed freely. But we could quickly learn about dangerous road conditions if we could utilize the data from each car. In another scenario, we could give early warnings of a heart attack if we could access and integrate data from various sources such as Fitbit and hospital Electronic Medical Records data. For decades, we have protected sensitive data by barricading it. This has choked potential benefits available from data utilization. Privacy preserving data mining offers a way to be able *to utilize all the data safely*.

Privacy preserving data mining has become an active research area. There are different ways to achieve privacy. For example, Agrawal and Srikant [1] developed a privacy preserving decision tree by *perturbing data*. Another way to protect privacy is anonymization. In this approach, sensitive information like name, date of birth, social security number are removed from data. However, Sweeney showed that if an adversary has access to auxiliary information, these frameworks may be revealing [2]. She also proposed a k-anonymity framework [2] where some attributes of the data are removed such that if a record is in the database, there are at least k-1 identical records in the database. This reduces the risk of

© Springer International Publishing Switzerland 2016
J. Bailey et al. (Eds.): PAKDD 2016, Part II, LNAI 9652, pp. 388–400, 2016.
DOI: 10.1007/978-3-319-31750-2_31

revealing up to k records. This framework gave rise to a number of privacy preserving methods (see survey in [3]). Privacy can also be achieved using additive noise, data swapping or synthetic data [4]. These methods aim to retain useful statistical information about data while changing individual records.

An important task in data mining is *clustering*, where similar records are grouped together. Clustering has enormous applications for data explorations [5], data organization [6] and retrieval [7]. One of the most popular clustering algorithm is K-means. The need to perform clustering in a privacy aware manner has prompted researchers to develop privacy preserving K-means algorithms. Vaidya and Clifton [8] propose one such algorithm for vertically partitioned data using secure multiparty computation. A similar algorithm for horizontally partitioned data was proposed by Inan et al. [9]. In a more general work, Jagannathan and Wright extended these works for both the horizontally and vertically partitioned data [10]. All these works assume that the adversary does not have access to auxiliary information. In real world, when such assumptions do not hold, these methods may not protect privacy leading to distrust among the users about the system.

Recently, differential privacy [11] has emerged as a strong privacy preserving framework. It protects the data privacy even when an adversary has access to auxiliary information. Several machine learning and data mining models using this framework have been explored such as logistic regression [12], decision tree learning [13] and matrix factorization [14]. Differentially private K-means clustering algorithms are proposed in [15–17]. In [15], Blum et al. proposed SuLQ framework, which releases noisy answer for a query. They used K-means algorithm as an example to demonstrate the SuLQ framework. Similarly, PINQ system was proposed by McSherry [16], who provided a programming interface for privacy preserving analysis. K-means clustering has been implemented using PINQ as an example of data analysis algorithm. Su et al. [17] proposed a differentially private K-means under different settings where the learner is distrusted. Although differential privacy provides a strong guarantee on privacy, it often perturbs the output of algorithms so much that their utility drops to unacceptable levels. *The problem of developing a privacy framework that provides high utility under strong privacy guarantees is therefore still open.*

Inspired from a recent private random forest model [18], we propose a new privacy preserving framework that provides strong guarantee on privacy of each data point in the database ensuring high utility. This framework can handle arbitrary amounts of auxiliary knowledge about the database, that is, even if an adversary has access to all but a one data point, the framework still thwarts the adversary from inferring the unknown data point. We achieve this by randomizing the output of the algorithm using a well known statistical estimation technique known as bootstrap aggregation. Exploiting the randomness offered by bootstrap, our framework ensures that the variance of the error in the adversary's estimation does not reduce significantly due to the participation of a data point in the database. By ensuring that the error in estimation by the adversary is almost invariant to the inclusion/exclusion of the data point in the database, the adversary is defeated. Our framework significantly departs from differential privacy in the manner that in presence/absence of a data point, differential privacy preserves the *likelihood* of algorithm output while our framework preserves the *error variance*. By focusing directly on the estimation error for the data point, our framework is able to use significantly *smaller perturbation* in the algorithm output compared to the differential privacy.

Using our new privacy framework, we construct a novel, privacy preserving K-means algorithm. The key idea is to perturb the cluster centroids before their release. We do this by using bootstrap aggregation to compute the cluster centroids. We analyze our method theoretically, and derive bounds on the size of bootstrap ensemble to ensure the stipulated privacy. We consider two cases - when the cluster a data point belongs to is either *known* or *unknown* to the adversary. Using both synthetic and real datasets, we compare our algorithm against baselines - the conventional, non-private K-means and differentially private K-means. The results are remarkable - *at high levels of privacy, the utility of our method is almost the same as the non-private K-means*, and *at least twice as good as the differential privacy counterpart*. This is because for the same privacy level, we need to add significantly lower levels of noise compared to differential privacy - as example, the noise in our framework is almost 20 times lower for high privacy stipulated by leakage parameter ϵ less than 0.1.

In summary, our contributions are:

- A new privacy preserving framework;
- A novel privacy preserving K-means algorithm with high utility using the proposed privacy framework;
- Theoretical analysis of the proposed K-means algorithm and a derivation of the upper bound on the size of bootstrap ensemble to guarantee the requisite privacy;
- Illustration and validation of the usefulness of the proposed K-means through experiments on both synthetic and real datasets.

2   The Proposed Solution

In this section, we present a new privacy framework where our goal is to provide strong guarantee on privacy of every data point in the database while ensuring that utility of algorithms remain high. The proposed framework is capable of handling the arbitrary amount of auxiliary knowledge about the database in the sense that even if an adversary has access to all but one data point, the framework still thwarts an adversary from inferring the unknown data point. We use this new framework of privacy to develop a privacy preserving K-means clustering algorithm that has high clustering performance.

2.1   A New Privacy Framework

Let us denote by $D_N = \{x_1, x_2, ..., x_N\}, x_i \in R^d$ a dataset with N data points. Further denote by $D_{N \setminus r}$ a dataset that all the data points of D_N except a data point x_r. Next assume that $f(D_N)$ and $f(D_{N \setminus r})$ are the *randomized* answers of a system for a statistical query about the dataset D_N and $D_{N \setminus r}$ respectively. Inspired by the strong guarantees of differential privacy framework [11,19], we demand our framework to protect the privacy of a data point x_r even when an adversary has access to data points in $D_{N \setminus r}$. Specifically, our proposed framework controls the level of privacy leakage for the data point x_r based on a pre-specified leakage parameter ϵ. In particular, the adversary's estimation of x_r derived using $f(D_N)$ is guaranteed to be only "ϵ-fraction better" than a estimate that is derived using $f(D_{N \setminus r})$. Thus the presence of the data point x_r in the database brings only *negligible* risk on its privacy for a small value of ϵ. Assume that the variance of the error in the adversary's estimate of j-th attribute of x_r using $f(D_N)$, which is computed with data points including x_r, is denoted as $\mathcal{E}_{\text{inc}}(\hat{x}_{rj})$. Similarly, assume that the variance of the error in the adversary's estimate of j-th attribute of x_r using $f(D_{N \setminus r})$, which is computed using all data points except x_r, is denoted as $\mathcal{E}_{\text{exc}}(\hat{x}_{rj})$. Formally, our proposed framework ensures the inequality

$$\frac{\mathcal{E}_{\text{inc}}(\hat{x}_{rj})}{\mathcal{E}_{\text{exc}}(\hat{x}_{rj})} \geq \exp(-\epsilon). \tag{1}$$

In the above inequality, when the value of ϵ is 0, the *strongest level of privacy* is offered. In other words, adversary can not estimate x_r any better than an estimate that is obtained without x_r's participation in the database. As the value of ϵ is increased, the level of privacy drops. We refer to this framework as *Error Preserving Privacy* (**EPP**).

2.2   Privacy Preserving K-Means Clustering

Given the dataset D_N, the K-means clustering algorithm aims to partition D_N into K disjoint sets $\{C_1, C_2, ..., C_K\}$ by minimizing the following cost function:

$$\min_{C_1, ..., C_K} \sum_{k=1}^{K} \sum_{x_i \in C_k} \|x_i - m_k\|^2 \tag{2}$$

where m_k is the centroid of cluster C_k. The most popular algorithm for K-means clustering is due to Lloyd [20]. This algorithm first randomly picks K data points and uses them to initialize the centroids $m_1, m_2, ..., m_K$. Using these centroids, the algorithm assigns a data point x_i to cluster C_k if m_k is the nearest centroid. After this assignment, each centroid m_k is re-computed by averaging all data points that belong to cluster C_k. The algorithm is iterated between these two steps until it converges or exceeds the maximum number of iterations.

We propose a new privacy preserving K-means algorithm that can cluster the data while maintaining the data privacy under our proposed privacy framework in (1). The key to achieving privacy is to use a randomization in the answer of the query such that the inequality in (1) is satisfied. In doing so, our effort should be to use a mechanism for the randomization that does not degrade the utility of the answer for intended tasks. Motivated by this idea, we use a mechanism that is based on bootstrap sampling [21] of data points. The proposed mechanism not only offers the desired randomness but also retains the high utility of the original algorithm.

Similar to the Lloyd's algorithm, our algorithm iterates between the two steps of data assignment to cluster centroids and centroid re-computation until no improvement can be made. However, in the *last iteration* of our algorithm, the centroids are estimated using bootstrap aggregation (bagging) [21]. For each cluster, it generates a bag of data points through bootstrap sampling, *i.e.* uniformly randomly sampling of data points with replacement. The number of data points in each bag remains same as that in the original cluster. For each bag, the centroid is estimated by averaging the data points. A total of B such bags are generated and the aggregate centroid is computed by averaging the centroid estimates of all B bags. A step-by-step summary of our proposed algorithm is provided in Algorithm 1.

In the following analysis, we present a theoretical analysis of our algorithm showing that as long as the number of bags B in the bootstrap aggregation are smaller than a certain upper bound, the privacy of the algorithm is maintained under the framework of (1). This means given the bootstrap-perturbed cluster centroids and the data points except x_r, the adversary can not estimate x_r significantly better than an estimate made by using the centroids that were computed without x_r. We refer to this model as **Error Preserving Private K-means (EPP-KM)**.

2.3   The Analysis of Privacy Preserving K-Means Algorithm

Due to the randomness of bootstrapping, the adversary's estimate of unknown data point x_r is perturbed. In this section, we theoretically analyze the proposed model in the light of the adversary estimation of the unknown point. In general, we have the *two possible cases*: 'the adversary knows which cluster the unknown data point belongs to' *or* 'otherwise'.

Case-1 (The adversary knows which cluster x_r belongs to): Let us assume that the adversary knows that $x_r \in C_k$. Let us denote by N_k the number of data points in the cluster C_k and let x_{ij} be the j-th attribute value of data point $x_i \in C_k$. Using the centroid m_k and other data points of C_k, the best estimate of x_{rj} is given by:

$$\hat{x}_{rj} = N_k \times m_{kj} - \sum_{x_i \in C_k \setminus x_r} x_{ij}. \tag{3}$$

where m_{kj} is the j-th attribute of the centroid m_k. When the m_{kj} is estimated using bagging, it is a random variable. We will show that this randomness is used to preserve the privacy of x_{rj}. In (3), N_k and the sum of attributes are already known. Thus, the variance of the estimation error of $\hat{x}_{rj}$ is given by:

$$\mathcal{E}_{\mathrm{inc}}(\hat{x}_{rj}|D_{N \setminus r}, m_k, z_r = k) = N_k^2 var(m_{kj}|D_{N \setminus r}), \tag{4}$$

where the cluster indicator variable $z_r = k$ encodes the knowledge $x_r \in C_k$. Because of the bagging ensemble used in our privacy preserving algorithm, m_{kj} is given by:

$$m_{kj} = \frac{1}{B} \times \frac{1}{N_k} \times \sum_{x_r \in C_k} \alpha_r x_{rj},$$

where α_r denotes the number of times x_r is sampled in B bags of bootstrap during the computation of m_k. Clearly, α_r is a random variable following a binomial distribution with mean B and variance $B(1 - \frac{1}{N_k})$. Therefore, the conditional variance of m_k is:

$$var(m_{kj}|D_{N\backslash r}) = \frac{var(\alpha_r)}{B^2 N_k^2} \left(\sum_{x_r \in C_k} x_{rj}^2 \right) = \frac{1}{BN_k^2}(1 - \frac{1}{N_k}) \left(\sum_{x_r \in C_k} x_{rj}^2 \right). \quad (5)$$

Plugging (5) in (4), we have $\mathcal{E}_{\text{inc}}(\hat{x}_{rj}|D_{N\backslash r}, m_k, z_r = k) = \frac{1}{B}(1 - \frac{1}{N_k})(\sum_{x_r \in C_k} x_{rj}^2)$. To ensure that this estimation error variance follows the privacy framework in 1, the number of bootstrap bags B has to satisfy

$$B \leq \frac{(1 - \frac{1}{N_k}) \times (\sum_{x_r \in C_k} x_{rj}^2)}{\mathcal{E}_{\text{exc}}(\hat{x}_{rj}) \times \exp(-\epsilon)}. \quad (6)$$

The above bound is applicable to protect the j-th attribute of the data point x_r. Since the framework is required to protect all the attributes of all the data points in the cluster, the following needs to be satisfied

$$B \leq \min_j \frac{(1 - \frac{1}{N_k}) \times (\sum_{x_r \in C_k} x_{rj}^2)}{\mathcal{E}_{\text{exc}}(\hat{x}_{rj}) \times \exp(-\epsilon)} \quad (7)$$

We refer to this case as **EPP-KM (1)**.

Case-2 (The adversary doesn't know which cluster x_r belongs to): In this case, the adversary does not have the information of the cluster membership of x_r. The unavailability of this information creates a bias in his estimation. To see this, consider the adversary model in (3). Assuming that x_r truly belongs to cluster k', the expectation of the adversary estimate is given as

$$E(\hat{x}_{rj}) = E_{z_r}(E(\hat{x}_{rj} \mid z_r)) = \pi_{k'} x_{rj}$$

where z_r is a random variable and $z_r = k$ implies that x_r belongs to cluster C_k. We use $\pi_{k'}$ to denote the probability that x_r belongs to the cluster $C_{k'}$. The probability $\pi_{k'}$ can be approximately estimated using the partition of data $D_{N\backslash r}$. Clearly, the estimate $\hat{x}_{rj}$, in this case, is biased as $E(\hat{x}_{rj}) \neq x_{rj}$. The variance of the error 2 in the estimation can be derived by *the law of total variance* as below

$$\mathcal{E}_{\text{inc}}(\hat{x}_{rj}|D_{N\backslash r}, m_{1:K})$$
$$= E_{z_r}[var(\hat{x}_{rj}|z_r, D_{N\backslash r}, m_{1:K})] + var_{z_r}[E(\hat{x}_{rj}|z_r, D_{N\backslash r}, m_{1:K})]$$
$$= \sum_{k=1}^{K} \left[\frac{\pi_k}{B}(1 - \frac{1}{N_k}) \left(\sum_{x_r \in C_k} x_{rj}^2 \right) \right] + \pi_{k'}(1 - \pi_{k'}) x_{rj}^2$$

To satisfy the privacy framework in 1, the number of bootstrap bags B has to satisfy

$$B \leq \frac{\sum_{k=1}^{K} \pi_k \left(1 - \frac{1}{N_k}\right) \left(\sum_{x_r \in C_k} x_{rj}^2\right)}{\mathcal{E}_{\text{exc}}(\hat{x}_{rj}) \times \exp(-\epsilon) - \pi_{k'}(1 - \pi_{k'}) x_{rj}^2} \tag{8}$$

Once again, since the above bound should be applicable to protect all the attributes of all the data points in the cluster, the following needs to be satisfied

$$B \leq \min_{j,r} \frac{\sum_{k=1}^{K} \pi_k \left(1 - \frac{1}{N_k}\right) \left(\sum_{x_r \in C_k} x_{rj}^2\right)}{\mathcal{E}_{\text{exc}}(\hat{x}_{rj}) \times \exp(-\epsilon) - \pi_{k'}(1 - \pi_{k'}) x_{rj}^2} \tag{9}$$

We refer to this case as **EPP-KM (2).**

Algorithm 1. Error Privacy Preserving K-means algorithm

Input: Dataset $D = \{x_1, ..., x_N\}, x_i \in R^d$, number of clusters K.
Output: The bootstrap estimated cluster centroids: $m_1, ..., m_K$.
Initialization: Randomly initialize the cluster centroids $m_1, ..., m_K$.
1: **repeat**
2:      **for** each point x_i **do**
3:           **if** x_i is the closest to m_k out of all centroids $m_1, ..., m_K$ **then**
4:                Assign x_i to C_k
5:           **end if**
6:      **end for**
7:      **for** $k = 1$ to K **do**
8:           Compute m_k by averaging all $x_i \in C_k$
9:      **end for**
10: **until** clustering converges
11: **for** $k = 1$ to K **do**
12:      Calculate the value of B using (7) or (9) depending on if the adversary knows the cluster membership of data points or not.
13:      Compute m_k using aggregation of B bootstrap samples.
14: **end for**

---

3   Experiments

We experiment with a total of *three* clustering datasets: one synthetic and two real datasets. Experiments with the synthetic data illustrate the behavior of our proposed model in a controlled setting. Experiments with the real datasets show the effectiveness of our model for clustering under privacy constraints.

Baselines Methods. To evaluate the efficacy of our model, we compare its performance with the following baseline methods:

- **The Original K-means (Non-Private)**: This algorithm is the standard K-means algorithm. We note that this method does not protect privacy of database. We refer to this method as **KM**.
- **Differentially Private K-means**: This algorithm is a variant of K-means that protects the privacy of database under the framework of differential privacy [22]. In this algorithm, the j-th element of k-th K-means centroid is made ϵ-differential private by adding to it a noise η_{kj} that follows a Laplacian distribution with mean zero and standard deviation S_{kj}/ϵ where S_{kj} is the sensitivity of the j-th element of the k-th centroid. The sensitivity S_{kj} with respect to the presence/absence of any data point is approximately $\frac{1}{N_k} \max_r x_{rj}$, where N_k is the number of data points in the k-th cluster. We refer to this method as **DP-KM**.

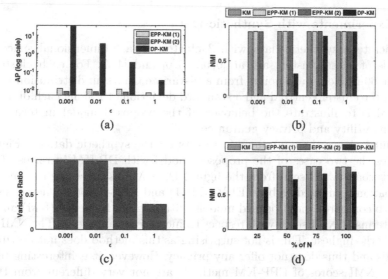

Fig. 1. Results using Synthetic dataset with $N = 180, K = 3$. (a) Average perturbation in cluster centroid with respect to ϵ, (b) NMI with respect to ϵ, (c) Ratio of variance for estimation errors $\mathcal{E}_{inc}$ and $\mathcal{E}_{exc}$, (d) NMI for varying number of data points at $\epsilon = 0.1$.

Performance Measures. We use four different metrics for performance evaluation: Normalized Mutual Information (NMI) [23], Rand Index [23] and Purity [23] to evaluate the clustering performance, and Average Perturbation (AP) of privacy-preserving models to evaluate how much noise a model adds to the cluster centroids before releasing them for end use. The first three measures are widely used in clustering literature. The last evaluation measure is a normalized version of mean absolute error (MAE). Given K clusters with the original centroids $\{\mathbf{m}_k\}_{k=1}^{K}$ and the perturbed centroids $\{\mathbf{m}'_k\}_{k=1}^{K}$, the average perturbation is calculated as $AP = \frac{1}{K} \sum_k \frac{\|\mathbf{m}_k - \mathbf{m}'_k\|}{\|\mathbf{m}_k\|}$.

Experimental Setting. For both synthetic and real data experiments, the clustering performance of each algorithm is studied with respect to varying privacy levels (ϵ) and the number of data points in the database. For the experiments showing clustering performance with respect to ϵ, we average the performance of each algorithm for 30 random centroid initializations for each value of ϵ. For the experiments showing clustering performance with respect to varying number of data points (N), we vary N from 25% to 100% of the data set size at a step of 25%. The average performance is reported over 40 different random subsamples of size N and 20 random centroid initializations. To demonstrate the privacy guarantee of the proposed model, we estimate every data point in the database using the perturbed means and the adversary model in Eq. (3). We report the ratio of the estimation errors made by the adversary under presence/absence of the data points in the database as per our EPP framework (see Eq. (1)).

3.1   Experiments with Synthetic Data

We generate a synthetic data with 3 clusters in a 2-dimensional space. The centroids of these clusters are at $[0, 0]$, $[5, 0]$ and $[4, 4]$. For each cluster, we generate 60 random data points from a bi-variate Gaussian distribution with its mean at the cluster centroid and a standard deviation of 1 along each dimension. Our goal is to illustrate the behavior of the proposed model in terms of its clustering utility and privacy guarantees.

Figure 1 shows the experimental results for the synthetic dataset. Figure 1a compares the two cases of the proposed model with DP-KM in terms of average perturbation. As seen from the figure, DP-KM has much higher amount of perturbation compared to both EPP-KM (1) and EPP-KM (2) when ϵ is small. Figure 1b compares the proposed models with original K-means (KM) and DP-KM in terms of NMI score with respect to increasing values of ϵ. The NMI score of KM is the highest. This is not surprising as this method does not perturb the centroids and thus does not offer any privacy. However, it is interesting to note that the NMI scores of EPP-KM methods are not very different from that of KM in spite of the strong privacy guarantees offered by EPP-KM. On the other hand, DP-KM performs poorly as its NMI scores are significantly lower compared to the other methods. This poor performance of DP-KM is evident from the high levels of perturbations made by this algorithm to the cluster centroids. In Fig. 1c, we demonstrate the privacy guarantee offered by EPP-KM models. As seen from the figure, the variance of the error in an adversary's estimation for any data point changes by a factor of only $\exp(-\epsilon)$ due to its participation in the database. We can see that for low values of ϵ, e.g. when $\epsilon = 0.001$, the ratio of the error variance in the adversary's estimation is around 1, meaning that no extra reduction in uncertainty is achieved by the adversary. At the other values of ϵ, the plot follows the EPP framework of Eq. (1). We also study the effect of the number of data points in the database on the clustering performance. Figure 1d compares the NMI score of the proposed models with KM and DP-KM. For this experiment, the privacy parameter ϵ is fixed at 0.1. The performance of all the algorithms improve with the number of data points due to reduction in the perturbation. The NMI scores of EPP-KM variants are close to that of KM. Once again the performance of DP-KM is poor in the beginning as it needs high perturbations due to small cluster size.

3.2   Experiment with Real Data

We use the following datasets from UCI machine learning repository[1]:

- **Seeds dataset:** This dataset consists of 210 data points of three wheat types: *Kama*, *Rosa* and *Canadian*. Each data point has 7 geometric attributes of wheat kernels: area, perimeter, compactness, length of kernel, width of kernel, asymmetry coefficient, length of kernel groove. Our task is to use these attributes to cluster the data points in 3 different categories.
- **User Knowledge Modeling dataset (UKM):** The dataset is about student's knowledge level about a subject of Electrical DC Machines. There are 4 levels of knowledge: Very Low, Low, Middle, High. The UKM dataset has 258 data points and each data point has 5 attributes: STG, SCG, STR, LPR, PEG. Our task is to use these attributes to cluster the data points in 4 different categories.

Experimental Results. The experimental results with the Seeds dataset and the UKM dataset are shown in Figs. 2 and 3 respectively. The results follow similar patterns as in the Synthetic dataset. As seen from Figs. 2a and 3a, the average perturbations used in the centroids by both the proposed EPP-KM variants are quite small. In contrast, the average perturbation by DP-KM is extremely high for small values of ϵ. The NMI performance of the proposed EPP-KM models with respect to ϵ is approximately 0.7 and 0.3, which is close to that of KM (see Figs. 2b and 3b) while the performance of DP-KM is extremely poor at small values of ϵ and only improves at higher values of ϵ. Similar to the Synthetic dataset, Figs. 2c and 3c demonstrate that the adversary gains almost no extra information about any data point at small values of ϵ (at strict privacy).

We also study the effect of the number of data points in the database on the clustering performance. From Figs. 2d and 3d we can see that the NMI score of both EPP-KM variants are almost same as that of KM. On the contrary, the performance of DP-KM is quite poor as when using 25 % fraction of data points, NMI score of DP-KM drops to as low as 0.54 and 0.17 for Seeds and UKM dataset respectively.

A more complete set of results showing other clustering measures, in particular, Purity and Rand Index are reported in Table 1. As seen from the Table, both EPP-KM variants consistently achieve high level of clustering performance in terms of all three evaluation metrics. At times, we observed that the performances of EPP-KM (2) were slightly better than even KM. After further investigation, we found that this happens due to the robustness of bootstrap sampling to outliers [24].

---

[1] available at URL https://archive.ics.uci.edu/ml/datasets.html.

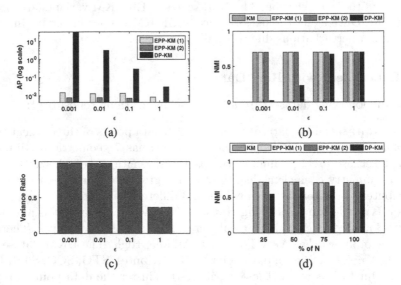

Fig. 2. Results using Seeds dataset with $N = 210, K = 3$, (a) Average perturbation in cluster centroid with respect to ϵ, (b) NMI with respect to ϵ, (c) Ratio of variance for estimation errors $\mathcal{E}_{\text{inc}}$ and $\mathcal{E}_{\text{exc}}$, (d) NMI for varying number of data points at $\epsilon = 0.1$.

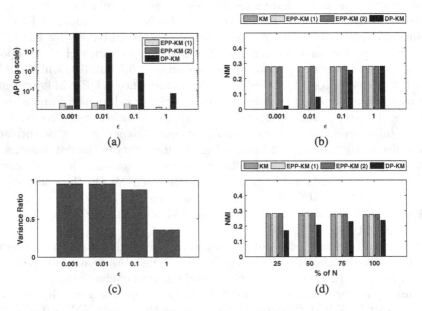

Fig. 3. Results using UKM dataset with $N = 258, K = 4$, (a) Average perturbation in cluster centroid with respect to ϵ, (b) NMI with respect to ϵ, (c) Ratio of variance for estimation errors $\mathcal{E}_{\text{inc}}$ and $\mathcal{E}_{\text{exc}}$, (d) NMI for varying number of data points at $\epsilon = 0.1$.

Table 1. Comparison with the baselines in terms of various metrics at $\epsilon = 0.1$. Average results over 30 random centroid initializations are reported with the standard errors in parenthesis. The bold face indicates the best results among private algorithms.

		Synthetic	Seeds	UKM
NMI	KM	0.9152 (0.0128)	0.7010 (0.0014)	0.2778 (0.0107)
	EPP-KM (1)	0.9160 (0.0121)	**0.7017** (0.0016)	0.2781 (0.0109)
	EPP-KM(2)	**0.9162** (0.0128)	0.7010 (0.0014)	**0.2790** (0.0106)
	DP-KM	0.8514 (0.0192)	0.6709 (0.0064)	0.2534 (0.0108)
Purity	KM	0.9707 (0.0112)	0.8933 (0.0004)	0.5683 (0.0078)
	EPP-KM (1)	0.9709 (0.0110)	**0.8938** (0.0004)	0.5686 (0.0080)
	EPP-KM (2)	**0.9711** (0.0112)	0.8933 (0.0004)	**0.5691** (0.0078)
	DP-KM	0.9446 (0.0128)	0.8759 (0.0047)	0.5536 (0.0062)
Rand index	KM	0.9669 (0.0093)	0.8732 (0.0003)	0.6819 (0.0033)
	EPP-KM (1)	0.9672 (0.0090)	**0.8736** (0.0003)	0.6820 (0.0033)
	EPP-KM (2)	**0.9674** (0.0093)	0.8732 (0.0003)	**0.6823** (0.0033)
	DP-KM	0.9368 (0.0115)	0.8559 (0.0041)	0.6642 (0.0046)
Average perturbation	EPP-KM (1)	0.0108 (0.0007)	0.0131 (0.0008)	0.0210 (0.0008)
	EPP-KM (2)	**0.0075** (0.0004)	**0.0071** (0.0004)	**0.0183** (0.0007)
	DP-KM	0.3275 (0.0227)	0.2811 (0.0176)	0.7059 (0.0433)

4 Conclusion

We proposed a novel framework for privacy preserving data mining and developed a K-means clustering algorithm under this framework. The proposed framework provides strong privacy guarantees even when an adversary has access to auxiliary knowledge about the database. Our private K-means algorithm calculates cluster centroids using bootstrap aggregation, which introduces just enough perturbation to ensure that privacy of every data point is maintained. We theoretically analyze our method and derive bounds on the size of bootstrap ensemble, which ensures the privacy under the proposed framework. The experimental results clearly show that our algorithm has high utility with strong privacy guarantees.

References

1. Agrawal, R., Srikant, R.: Privacy-preserving data mining. ACM SIGMOD Rec. **29**(2), 439–450 (2000). ACM
2. Sweeney, L.: k-anonymity: a model for protecting privacy. Int. J. Uncertainty Fuzziness Knowl. Based Syst. **10**(05), 557–570 (2002)
3. Ciriani, V., di Vimercati, S.D.C., Foresti, S., Samarati, P.: k-anonymous data mining: a survey. In: Aggarwal, C.C., Yu, P.S. (eds.) Privacy-Preserving Data Mining. Advances in Database Systems, vol. 34, pp. 105–136. Springer, US (2008)
4. Malik, M.B., Ghazi, M.A., Ali, R.: Privacy preserving data mining techniques: current scenario and future prospects. In: ICCCT 2012, pp. 26–32. IEEE (2012)

5. Begelman, G., Keller, P., Smadja, F., et al.: Automated tag clustering: improving search and exploration in the tag space. In: Collaborative Web Tagging Workshop at WWW2006, pp. 15–33 (2006)
6. Fred, A.L., Jain, A.K.: Data clustering using evidence accumulation. In: ICPR 2002, vol. 4, pp. 276–280. IEEE (2002)
7. Zeng, H.-J., He, Q.-C., Chen, Z., Ma, W.-Y., Ma, J.: Learning to cluster web search results. In: ACM SIGIR 2004, pp. 210–217 (2004)
8. Vaidya, J., Clifton, C.: Privacy-preserving k-means clustering over vertically partitioned data. In: KDD 2003, pp. 206–215. ACM (2003)
9. Inan, A., Kaya, S.V., Saygın, Y., Savaş, E., Hintoğlu, A.A., Levi, A.: Privacy preserving clustering on horizontally partitioned data. Data Knowl. Eng. 63(3), 646–666 (2007)
10. Jagannathan, G., Wright, R.N.: Privacy-preserving distributed k-means clustering over arbitrarily partitioned data. In: KDD 2005, pp. 593–599. ACM (2005)
11. Dwork, C.: Differential privacy. In: Bugliesi, M., Preneel, B., Sassone, V., Wegener, I. (eds.) ICALP 2006. LNCS, vol. 4052, pp. 1–12. Springer, Heidelberg (2006)
12. Chaudhuri, K., Monteleoni, C.: Privacy-preserving logistic regression. In: NIPS 2009, pp. 289–296 (2009)
13. Jagannathan, G., Pillaipakkamnatt, K., Wright, R.N.: A practical differentially private random decision tree classifier. In: ICDMW 2009, pp. 114–121. IEEE (2009)
14. Hua, J., Xia, C., Zhong, S.: Differentially private matrix factorization. In: IJCAI (2015)
15. Blum, A., Dwork, C., McSherry, F., Nissim, K.: Practical privacy: the sulq framework. In: PODS 2005, pp. 128–138. ACM (2005)
16. McSherry, F.D.: Privacy integrated queries: an extensible platform for privacy-preserving data analysis. In: ACM SIGMOD International Conference on Management of Data (2009)
17. Su, D., Cao, J., Li, N., Bertino, E., Jin, H.: Differentially private k-means clustering. CoRR, abs/1504.05998 (2015)
18. Rana, S., Gupta, S., Venkatesh, S.: Differentially private random forest with high utility. In: IEEE International Conference on Data Mining (2015)
19. Dwork, C.: Differential privacy: a survey of results. In: Agrawal, M., Du, D.-Z., Duan, Z., Li, A. (eds.) TAMC 2008. LNCS, vol. 4978, pp. 1–19. Springer, Heidelberg (2008)
20. Lloyd, S.P.: Least squares quantization in PCM. IEEE Trans. Inf. Theor. 28(2), 129–137 (1982)
21. Efron, B., Tibshirani, R.J.: An Introduction to the Bootstrap. CRC Press, Boca Raton (1994)
22. Dwork, C., McSherry, F., Nissim, K., Smith, A.: Calibrating noise to sensitivity in private data analysis. In: Halevi, S., Rabin, T. (eds.) TCC 2006. LNCS, vol. 3876, pp. 265–284. Springer, Heidelberg (2006)
23. Manning, C.D., Raghavan, P., Schütze, H., et al.: Introduction to Information Retrieval, vol. 1. Cambridge University Press, Cambridge (2008)
24. Salibian-Barrera, M., Zamar, R.H.: Bootstrapping robust estimates of regression. Ann. Stat. 30, 556–582 (2002)

Secure k-NN Query on Encrypted Cloud Data with Limited Key-Disclosure and Offline Data Owner

Youwen Zhu[1,2]($\boxtimes$), Zhikuan Wang[1], and Yue Zhang[1]

[1] Department of Computer,
Nanjing University of Aeronautics and Astronautics, Nanjing, China
[2] Collaborative Innovation Center of Novel Software Technology and
Industrialization, Nanjing 210016, China
zhuyw@nuaa.edu.cn

Abstract. Recently, many schemes have been proposed to support k-nearest neighbors (k-NN) query on encrypted cloud data. However, existing approaches either assume query users are fully-trusted, or require data owner to be online all the time. Query users in fully-trusted assumption can access the key to encrypt/decrypt outsourced data, thus, untrusted cloud server can completely break the data upon obtaining the key from any untrustworthy query user. The online requirement introduces much cost to data owner. This paper presents a new scheme to support k-NN query on encrypted cloud database while preserving the privacy of database and query points. Our proposed approach only discloses limited information about the key to query users, and does not require an online data owner. Theoretical analysis and extensive experiments confirm the security and efficiency of our scheme.

Keywords: Cloud computing · Privacy · k-nearest neighbors · Query

1 Introduction

Nowadays, many enterprises are increasingly interested in outsourcing their database storage and management services to cloud service provider for enjoying the attractive economical and technological benefits of cloud. Usually, the database has to been encrypted by its owner before being outsourced so that its privacy, such as costumer privacy and business secrets, cannot be breached in cloud computing system. Since it is of much difficulty to execute query and analysis operations on the encrypted data in traditional cryptosystem which focus on protecting the plaintext from being revealed, lots of schemes [1–11] with new properties have been recently proposed to efficiently support the query over encrypted cloud dataset.

The objective of k nearest neighbors (k-NN) query on database is to find out the k nearest points for a given query point. Because of its fundamental significance, several works [9–14, 16, 17] have addressed the security and privacy

© Springer International Publishing Switzerland 2016
J. Bailey et al. (Eds.): PAKDD 2016, Part II, LNAI 9652, pp. 401–414, 2016.
DOI: 10.1007/978-3-319-31750-2_32

problems of k-NN query over encrypted database. Hu *et al.* [10] present a secure protocol for processing k-NN queries on R-tree index, where encrypted dataset and the decryption are sent to query user and cloud server respectively. Namely, the work [10] does not outsource data storage to cloud, and the privacy of data owner will be seriously breached in case cloud server discloses the decryption key to any query user. Besides, query users and cloud server have to conduct complicated computation and communication for securely searching k-NN nodes. The schemes [11,12] can answer k-NN query on encrypted cloud data, but they only return approximate results in cloud. The work [9] puts forward an asymmetric scalar-product-preserving encryption (ASPE) scheme to securely support database outsourcing and k-NN query service on encrypted dataset. Nevertheless, query users have the access to the full key for encrypting and decrypting outsourced data in ASPE, and cloud server can completely breaks the outsourced database once he obtains the key from any untrustworthy query user. The work [9] also gives an enhanced ASPE schme to resist the chosen-plaintext attack. However, [12] has proved that the enhanced ASPE in [9] cannot achieve the declared security under chosen-plaintext attack, and no accurate k-NN query scheme can resist the chosen-plaintext attack. Elmehdwi et al.'s scheme [17] does not disclose data owner's key to query users, but reveals the key to a cloud server. Namely, data owner in [17] still shares his key with other party. Recently, [13,14] improves ASPE to address the problems owing to the key-sharing. However, the improved scheme in [13,14] requires data owner to be online all the time to encrypt each query point in a collaborative manner. It will introduce not only lots of computation burden and IT device cost to data owner, but also many communication overheads for each query. While a big number of queries are submitted for encrypting in a short period, data owner may become a bottleneck of the query system. Therefore, the online data owner is still undesirably strong requirement in real applications. The work in [16] lately proposes a secure k-NN query scheme on encrypted cloud data which can support both data privacy, key confidentiality, query privacy and query controllability. Nevertheless, the state-of-the-art scheme in [16] still requires data owner to be online for each query.

In this paper, we focus on the data security and privacy problems of outsourced database storage and k-NN query in cloud computing. Through enhancing the scheme in [13,14], we propose a new secure scheme, which does not need an online data owner for query encryption and computation, but can still prevent query users accessing the full key to encrypt/decrypt outsourced database. The main contributions of this work can be summarized as follows:

- We present a new approach to encrypt the outsourced database and query points. It can effectively support k-NN computation over encrypted data, and strongly preserve the privacy of database and query points.
- The new scheme reveals limited information about data owner's key for encrypting/decrypting the outsourced database to query users, but does not require an online data owner.

- Through analysis and extensive experiments, we evaluate the computation and communication overheads of our new approach and compare it with existing schemes. It shows that the new scheme is of efficiency and practicality.

The rest of the paper is organized as follows. In Sect. 2, we introduce the system model, design goals, some notations. Section 3 proposes our new secure k-NN query scheme on encrypted cloud data. Section 4 analyzes the overheads and security of the new approach, and compares it with existing work. Section 5 illustrates the efficiency of our scheme through extensive experiments. At last, Sect. 6 concludes the paper.

2 System Model and Design Goals

2.1 System Model

We consider the cloud computing model including three types of entities: one cloud server, one data owner and several query users. The cloud server has huge but bounded storage space and computation capability to provide outsourcing data storage and query service. Data owner utilizes the cloud service to store his private database. For the data privacy, only encrypted data is outsourced to cloud server. Let $\mathbf{D}$ and $\mathbf{D}$' denote data owner's original database and encrypted database, respectively. In this work, we assume that there are m points $\{\boldsymbol{p}_1, \boldsymbol{p}_2, \cdots, \boldsymbol{p}_m\}$ in $\mathbf{D}$, and each point is a d-dimensional vector, i.e., $\boldsymbol{p}_i = (p_{i1}, p_{i2}, \cdots, p_{id})$, for all $i = 1, 2, \cdots, m$. Each query user holds some private d-dimensional query points. For the query point $\boldsymbol{q} = (q_1, q_2, \cdots, q_d)$, query user would like to search its k-NN points in $\mathbf{D}$ according to the Euclidean distance, that is, $\|\boldsymbol{p}_i, \boldsymbol{q}\| = \sqrt{\sum_{j=1}^{d}(p_{ij} - q_j)^2}$. In our system model, query users will locally encrypt her query points, and send the encrypted values to cloud server, who completes the k-NN computation and returns the index of k-NN points to the corresponding query user. In this paper, a tuple denotes a point, and we use them interchangeably.

2.2 Threat Model and Design Goals

Data Privacy. Data privacy requires that the plain dataset $\mathbf{D}$ is not revealed to anybody else except for data owner himself while outsourcing storage and k-NN computation. Cloud server is the potential attacker against data privacy, as it can access the outsourced database $\mathbf{D}$' in the system model. To prevent the attacker from learning the private plain data, only encrypted data will be outsourced and stored in cloud. We consider three security levels for data privacy.

Level-1 attacker only knows the encrypted database and perturbed query points. This corresponds to the ciphertext-only attack in cryptography.

Level-2 attacker is supposed to also know some original tuples in $\mathbf{D}$, except the encrypted points. The attack is the same as known-sample attack in [15].

Level-3 attacker is supposed to also know all the numbers that data owner releases to query users, except a level-1 attacker learns. Considering the level-3 attack is to address the problems of key-sharing pointed out by the work [13,14].

It is easy to say the exsiting work in [9,12] cannot resist the level-3 attacker, since data owner in [9,12] will share his decryption key with query users.

Query Privacy. The query privacy requires that each query point is privately kept to the corresponding query user throughout the k-NN computation.

Efficiency. Additionally, our scheme will preserve the security and privacy in an efficient and practical manner. Concretely speaking, (1) the key generation and database point encryption in data owner should be of high efficiency in computation, (2) the encryption time of query points should be practically low, (3) computing k-NN on the encrypted database should not notably increase the cost of cloud server comparing to that on plain dataset. Additionally, for the goals of outsourcing and data owner's IT cost reduction, the query encryption and computation should not involve data owner, i.e., it will support offline data owner.

2.3   Notations

- $\mathbf{D}$ – data owner's database $\mathbf{D}=\{\boldsymbol{p}_1, \boldsymbol{p}_2, \cdots, \boldsymbol{p}_m\}$, where $\boldsymbol{p}_i = (p_{i1}, p_{i2}, \cdots, p_{id})$.
- $||\boldsymbol{p}_i||$ – Euclidean norm of $\boldsymbol{p}_i$, i.e., $||\boldsymbol{p}_i|| = \sqrt{\sum_{j=1}^{d} p_{ij}^2}$.
- $\mathbf{D}'$ – encrypted $\mathbf{D}$, and $\mathbf{D}'=\{\boldsymbol{p}_1', \boldsymbol{p}_2', \cdots, \boldsymbol{p}_m'\}$ where $\boldsymbol{p}_i'$ is the encrypted result of $\boldsymbol{p}_i$.
- $\boldsymbol{q}$ – a query point, and q_j denotes its j-th dimension.
- $\boldsymbol{q}'$ – the encrypted result of $\boldsymbol{q}$.
- $\boldsymbol{M}_{i*}$ – the i-th row of matrix $\boldsymbol{M}$.
- $\mathcal{I}_q$ – index set of k-NN of query point $\boldsymbol{q}$.
- $[X]$ – the set $\{1, 2, \cdots, X\}$, for any positive integer X.
- π – a random permutation of several numbers
- π^{-1} – the permutation inverse to π
- $\dot{\boldsymbol{A}}$ and $\boldsymbol{U}$ – $\dot{\boldsymbol{A}}$ is a $(2d+2) \times d$ matrix, and $\boldsymbol{U}$ is a $(2d+2)$-dimensional vector. They are the numbers that data owner discloses to query user, and will be used to generate encrypted query points.

3   Our Proposed Scheme

3.1   Highlight of Our Scheme

Let the $(d+1)$-dimensional vectors $\hat{\boldsymbol{p}}_i = (-2\boldsymbol{p}_i, ||\boldsymbol{p}_i||^2)$ and $\hat{\boldsymbol{q}} = (\boldsymbol{q}, 1)$. The work [9] has shown that it will never change the k-NN query result if we use the scalar $\hat{\boldsymbol{p}}_i \cdot \hat{\boldsymbol{q}}$, instead of Euclidean distance of $\boldsymbol{p}_i$ and $\boldsymbol{q}$, as the distance of points $\boldsymbol{p}_i$ and $\boldsymbol{q}$.

In our secure scheme, data owner randomly generates a $(d+1)$-dimensional vector $\boldsymbol{r} \in \mathbb{R}^{d+1}$ and a permutation π of $(d+1)$ numbers. Then, $\hat{\boldsymbol{p}}_i =$

$(-2\boldsymbol{p}_i, \|\boldsymbol{p}_i\|^2)$ and $\hat{\boldsymbol{q}} = (\boldsymbol{q}, 1)$ are permuted into $\dot{\boldsymbol{p}}_i = \boldsymbol{\pi}(\boldsymbol{r} + \hat{\boldsymbol{p}}_i)$ and $\dot{\boldsymbol{q}} = \boldsymbol{\pi}(\hat{\boldsymbol{q}})$, respectively. It is easy to say $\dot{\boldsymbol{p}}_i \cdot \dot{\boldsymbol{q}} = \hat{\boldsymbol{p}}_i \cdot \hat{\boldsymbol{q}} + \boldsymbol{r} \cdot \hat{\boldsymbol{q}}$. We further extend $\dot{\boldsymbol{p}}_i$ and $\dot{\boldsymbol{q}}$ to $(2d + 2)$-dimensional vectors $\ddot{\boldsymbol{p}}_i$ and $\ddot{\boldsymbol{q}}$ with two $(d + 1)$-dimensional random vectors $\boldsymbol{S}$ and $\boldsymbol{\eta}^{(q)}$ such that for all $j \in [d+1]$, $\ddot{p}_{i,2j-1} = \dot{p}_{i,j}$, $\ddot{p}_{i,2j} = S_j$, $\ddot{q}_{2j-1} = \dot{q}_j$ and $\ddot{q}_{2j} = \eta_j^{(q)}$. Here, $\ddot{p}_{i,j}$, $\dot{p}_{i,j}$, $\ddot{q}_j$, $\dot{q}_j$, S_j and $\eta_j^{(q)}$ denote the j-th dimension of vectors $\ddot{\boldsymbol{p}}_i$, $\dot{\boldsymbol{p}}_i$, $\ddot{\boldsymbol{q}}$, $\dot{\boldsymbol{q}}$, $\boldsymbol{S}$ and $\boldsymbol{\eta}^{(q)}$, respectively. After the extension, we have

$$\ddot{\boldsymbol{p}}_i \cdot \ddot{\boldsymbol{q}} = \dot{\boldsymbol{p}}_i \cdot \dot{\boldsymbol{q}} + \boldsymbol{S} \cdot \boldsymbol{\eta}^{(q)} = \hat{\boldsymbol{p}}_i \cdot \hat{\boldsymbol{q}} + \boldsymbol{r} \cdot \hat{\boldsymbol{q}} + \boldsymbol{S} \cdot \boldsymbol{\eta}^{(q)}. \tag{1}$$

Then, for all $i, j \in [m]$, there is $\ddot{\boldsymbol{p}}_i \cdot \ddot{\boldsymbol{q}} - \ddot{\boldsymbol{p}}_j \cdot \ddot{\boldsymbol{q}} = \hat{\boldsymbol{p}}_i \cdot \hat{\boldsymbol{q}} - \hat{\boldsymbol{p}}_j \cdot \hat{\boldsymbol{q}}$. Thus, the k-NN points of $\boldsymbol{q}$ can be exactly found out by adopting the new distance $\ddot{\boldsymbol{p}}_i \cdot \ddot{\boldsymbol{q}}$. It is remarkable that $\boldsymbol{\pi}$, $\boldsymbol{r}$ and $\boldsymbol{S}$ all are private information of data owner. The vector $\boldsymbol{\eta}^{(q)}$ is selected by a special way so that nobody learns the values of it, which will be later explained in detail.

In our scheme, the encrypted result of $\boldsymbol{p}_i$ is $\boldsymbol{p}'_i = \ddot{\boldsymbol{p}}_i \boldsymbol{M}^{-1}$ which can be locally completed by data owner. Correspondingly, $\boldsymbol{q}$ is encrypted into $\boldsymbol{q}' = \theta \boldsymbol{M} \ddot{\boldsymbol{q}}^T$. Here, θ is a positive random real number, $\boldsymbol{M}$ is a $(2d + 2) \times (2d + 2)$ random invertible matrix, and they both are independently selected and privately kept by data owner. While computing the encrypted query point $\boldsymbol{q}'$, query user holds the private plain query point $\boldsymbol{q}$, but the other inputs, such as θ, $\boldsymbol{M}$, $\boldsymbol{\pi}$, $\boldsymbol{r}$ and $\boldsymbol{S}$, all are private to data owner. In the existing scheme [9], data owner's private key is directly and completely disclosed to query users such that the latter can locally encrypt the query points. To deal with the problems of key-disclosure, the work [13,14] computes $\boldsymbol{q}'$ by a secure protocol which contains several rounds of communication between query user and data owner, thus it only reveals partial information about the key to query users, but data owner is required to be online all the time for encrypting each query point. In this paper, we propose a new secure k-NN query scheme on encrypted cloud data. Comparing to the previous solution [13,14], an important advantage of our scheme is that query users can locally encrypt their query points without an online data owner while the information disclosure about data owner's key is similar to that in [13,14].

Let $\boldsymbol{M}_{i*}$ denote the i-th row of matrix $\boldsymbol{M}$. Next, we will introduce the details for computing $(\theta \boldsymbol{M}_{1*} \cdot \ddot{\boldsymbol{q}})$ which is the first dimension of $\boldsymbol{q}'$, i.e., $q'_1 = \theta \boldsymbol{M}_{1*} \cdot \ddot{\boldsymbol{q}}$. The other dimensions of $\boldsymbol{q}'$ can be obtained through similar steps.

According to the aforementioned transformation and extension, we have

$$\ddot{\boldsymbol{q}} = (\dot{q}_1, \eta_1^{(q)}, \dot{q}_2, \eta_2^{(q)}, \cdots, \dot{q}_{d+1}, \eta_{d+1}^{(q)}).$$

Here, we use two $(d + 1)$-dimensional vectors $\boldsymbol{M}_{1*}^{od} = (M_{11}, M_{13}, \cdots, M_{1,2d+1})$ and $\boldsymbol{M}_{1*}^{ev} = (M_{12}, M_{14}, \cdots, M_{1,2d+2})$. Then,

$$\theta \boldsymbol{M}_{1*} \cdot \ddot{\boldsymbol{q}} = \theta \boldsymbol{M}_{1*}^{od} \cdot \dot{\boldsymbol{q}} + \theta \boldsymbol{M}_{1*}^{ev} \cdot \boldsymbol{\eta}^{(q)} = \theta \boldsymbol{M}_{1*}^{od} \cdot \boldsymbol{\pi}(\hat{\boldsymbol{q}}) + \theta \boldsymbol{M}_{1*}^{ev} \cdot \boldsymbol{\eta}^{(q)}$$

$$= \boldsymbol{\pi}^{-1}(\theta \boldsymbol{M}_{1*}^{od}) \cdot \hat{\boldsymbol{q}} + \boldsymbol{\pi}^{-1}(\theta \boldsymbol{M}_{1*}^{ev}) \cdot \boldsymbol{\pi}^{-1}(\boldsymbol{\eta}^{(q)}).$$

We definite ζ as a stagger function, which takes as inputs two vectors of the same length x (x is a positive integer) and returns a $2x$-dimensional vector by sequentially and alternately arranging each dimension of input vectors.

For example, if $a = (a_1, a_2, \cdots, a_x)$ and $b = (b_1, b_2, \cdots, b_x)$, then $\zeta(a, b) = (a_1, b_1, a_2, b_2, \cdots, a_x, b_x)$. For ease of description, we set two $(2d+2)$-dimensional vectors $\tilde{M}_{1*} = \zeta(\pi^{-1}(\theta M_{1*}^{\text{od}}), \pi^{-1}(\theta M_{1*}^{\text{ev}}))$ and $\tilde{q} = \zeta(\hat{q}, \pi^{-1}(\eta^{(q)}))$. Then, $q_1' = \theta M_{1*} \cdot \ddot{q} = \tilde{M}_{1*} \cdot \tilde{q}$. To support both limited key disclosure and offline data owner, we generate the random vector $\eta^{(q)}$ as follows.

Data owner selects a random $(d + 1)$-dimensional vector $X = (X_1, X_2, \cdots, X_{d+1})$, and $\eta^{(q)}$ inherently equals to $\pi(X + \hat{q})$. In fact, the value of $\eta^{(q)}$ is not visibly computed in our scheme and no one knows it. Further, there is $\tilde{q} = \zeta(\hat{q}, \pi^{-1}(\eta^{(q)})) = \zeta(\hat{q}, X + \hat{q})$. As data owner holds X, θ, M and the permutation π, he can calculate the sum $\alpha_j = \tilde{M}_{1,2j-1} + \tilde{M}_{1,2j}$ (for all $j \in [d + 1]$) and $u = \sum_{j=1}^{d+1} X_j \tilde{M}_{1,2j}$ on his own. If data owner sends $\{\alpha_1, \alpha_2, \cdots, \alpha_{d+1}, u\}$ to query user, the latter party can locally compute $q_1' = \theta M_{1*} \cdot \ddot{q} = \tilde{M}_{1*} \cdot \tilde{q} = u + \sum_{j=1}^{d+1} \alpha_j \hat{q}_j$, because

$$q_1' = \theta M_{1*} \cdot \ddot{q} = \tilde{M}_{1*} \cdot \tilde{q} = \sum_{i=1}^{2d+2} \tilde{M}_{1i} \tilde{q}_i = \sum_{j=1}^{d+1} \tilde{M}_{1,2j-1} \tilde{q}_{2j-1} + \sum_{j=1}^{d+1} \tilde{M}_{1,2j} \tilde{q}_{2j}$$

$$= \sum_{j=1}^{d+1} \tilde{M}_{1,2j-1} \hat{q}_j + \sum_{j=1}^{d+1} \tilde{M}_{1,2j}(X_j + \hat{q}_j) = \sum_{j=1}^{d+1}(\tilde{M}_{1,2j-1} + \tilde{M}_{1,2j})\hat{q}_j + \sum_{j=1}^{d+1} \tilde{M}_{1,2j} X_j$$

$$= u + \sum_{j=1}^{d+1} \alpha_j \tilde{q}_{2j-1} = u + \sum_{j=1}^{d+1} \alpha_j \hat{q}_j.$$

Since the last dimension of $\hat{q}$ is a constant 1 (i.e. $\hat{q}_{d+1} \equiv 1$), the data being disclosed to query users can be further revised to the set $\{A_{11}, A_{12}, \cdots, A_{1d}, U_1\}$ in which $A_{1j} = \tilde{M}_{1,2j-1} + \tilde{M}_{1,2j}$ (for all $j \in [d]$) and $U_1 = (\tilde{M}_{1,2d+1} + \tilde{M}_{1,2d+2}) + \sum_{j=1}^{d+1} X_j \tilde{M}_{1,2j}$. Then, query user can obtain $q_1' = \tilde{M}_{1*} \cdot \tilde{q} = U_1 + \sum_{j=1}^{d} \hat{q}_j A_{1j} = U_1 + \sum_{j=1}^{d} q_j A_{1j}$.

Similarly, for $i = 2$ to $2d + 2$, query user can access the data set $\{A_{i1}, A_{i2}, \cdots, A_{id}, U_i\}$ based on which the query user can compute the encrypted dimension q_i'. Thus, if data owner reveals the matrix $A = [A_{ij}]_{(2d+2) \times d}$ and vector $U = (U_1, U_2, \cdots, U_{2d+2})$ to query users, then q' can be computed at query user by $q' = U^T + Aq^T$.

At last, we further adjust our scheme to enhance the security. While registering with the system, each query user independently generates a random $d \times d$ invertible matrix W and submits it to data owner, then data owner releases matrix $\dot{A}$ and vector U to the corresponding query user, where $\dot{A} = AW^{-1}$. Then, her query encryption can be completed by computing $q' = U^T + \dot{A}Wq^T$.

As $\dot{A}$ and U can be figured out based on matrix M, permutation π, T, random parameters θ and X, then, each number in $\dot{A}$ and U does not depend on the values of any query point. Hence, data owner can compute $\dot{A}$ and U during key generation, and distribute them to query users. After the distribution and outsourcing the encrypted database to cloud, query user can complete the encryption of her query points by herself, and cloud server can find out the exact index set of k-NN for the encrypted query points, therefore data owner is not required to be online any more. Section 4.2 will in detail discuss the security of our scheme.

3.2   Formal Steps of Our Scheme

Our solution consists of four stages, which will be detailedly introduced in this section.

- KeyGen($\cdot$) $\mapsto$ $\{Key\}\&\{\dot{A}, U\}$(at data owner), $\{W\}$(at query user): In this stage, data owner generates his private key for encrypting the outsourced database, then he will distribute the random matrix $\dot{A}$ and vector U to the registered query user for encrypting query points. W is a $d \times d$ random invertible matrix independently selected by the query user. Here, U is shared by each query user, but $\dot{A}$ and W is randomly generated by individual users, and no query user knows the matrices $\dot{A}$ and W of other query users. After this stage, the registered query user can locally complete query points encryption without an online data owner. The steps of this stage are as follows.

Data owner randomly selects an invertible real matrix $M \in \mathbb{R}^{(2d+2) \times (2d+2)}$, two $(d+1)$-dimensional vectors $r = (r_1, r_2, \cdots, r_{d+1})$ and $S = (S_1, S_2, \cdots, S_{d+1}) \in \mathbb{R}^{d+1}$, and a permutation π of $(d+1)$ numbers. Data owner further generates a random positive number $\theta \in \mathbb{R}^+$ and a random $(d+1)$-dimension vector $X = (X_1, X_2, \cdots, X_{d+1}) \in \mathbb{R}^{d+1}$, and for all $i \in [2d+2]$, sets the vector

$$\dot{M}_{i*} = \pi^{-1}(M_{i1} + M_{i2}, M_{i3} + M_{i4}, \cdots, M_{i,2d+1} + M_{i,2d+2}).$$

Then, data owner computes a $(2d+2) \times d$ matrix A and a $(2d+2)$-dimensional vector $U = (U_1, U_2, \cdots, U_{2d+2})$ by the following way,

$$A_{i*} = \theta \cdot (\dot{M}_{i1}, \dot{M}_{i2}, \cdots, \dot{M}_{id}), \tag{2}$$

$$U_i = \theta \cdot \dot{M}_{i,d+1} + \sum_{j=1}^{d+1} X_j M_{i,2j}. \tag{3}$$

Here, M_{ij} is the value in row i, column j of matrix M, $\dot{M}_{ij}$ denotes the j-th dimension of $\dot{M}_{i*}$, and A_{i*} denotes the i-th row of matrix A. The matrix M is required to make the rank of A be less than d.

At last, data owner sets $Key = \{r, S, M, \pi\}$, and keeps Key in private. While each query user registers in the system, the query user independently selects a $d \times d$ random matrix W, and sends the matrix to data owner, then the latter distributes matrix $\dot{A}$ and vector U to the corresponding query user for her local computation of encrypted query points, where $\dot{A} = AW^{-1}$.

- DbEnc($\mathbf{D}$, Key) $\mapsto$ $\{\mathbf{D}'\}$: This stage will encrypt each point in $\mathbf{D}$, and is completed by only the data owner. Every point is encrypted as a single unit. The output is the encrypted database $\mathbf{D}' = \{p'_i \mid p_i \in \mathbf{D}\}$. The steps are

Let $r_{-(d+1)}$ denote $r \setminus \{r_{d+1}\}$, i.e., $r_{-(d+1)}$ is the d-dimensional vector $(r_1, r_2, \cdots, r_d)$. For each tuple $p_i \in \mathbf{D}$, data owner computes

$$\dot{p}_i = \pi\left(r_{-(d+1)} - 2p_i, r_{d+1} + \| p_i \|^2\right), \tag{4}$$

and extends $\dot{p}_i$ to the $(2d+2)$-dimensional vector $\ddot{p}_i = (\ddot{p}_{i1}, \ddot{p}_{i2}, \cdots, \ddot{p}_{i,2d+2})$ in which $\ddot{p}_{i,2j-1} = \dot{p}_{ij}$, $\ddot{p}_{i,2j} = S_j$ for all $j \in [d+1]$. Here, $\dot{p}_{ij}$ is the j-th dimension of $\dot{p}_i$. Then, data owner obtains the encrypted result

$$p_i' = \ddot{p}_i M^{-1}, \tag{5}$$

and uploads p_i' to cloud server.

• QEnc($q, U, \dot{A}, W$) $\mapsto \{q'\}$: Taking the private point q and $\dot{A}, U, W$ as inputs, query user locally encrypt the query point q into q' by the following steps.

After obtaining vector U and matrix $\dot{A}$, query user can locally encrypt each query point $q = (q_1, q_2, \cdots, q_d)$ into $q' = (q_1', q_2', \cdots, q_{2d+2}')$ in which

$$q_i' = U_i + \sum_{j=1}^{d} A_{ij} \sum_{t=1}^{d} W_{jt} q_t, \quad \text{for all } i \in [2d+2]. \tag{6}$$

Here, A_{ij} is the element in row i, column j of matrix A, U_i denotes the i-th dimension of U, and W_{jt} is the value in row j, column t of matrix W.

• KNNComp($\mathbf{D}', q', k$) $\mapsto \{\mathcal{I}_q\}$:

Query user uploads q' to cloud server, then cloud server computes the index set of k-NN in $\mathbf{D}'$ of the encrypted query point q' according to the distance $p_i' \cdot q'$, and sends the index set $\mathcal{I}_q$ of the k-NN in $\mathbf{D}'$ to the corresponding query user.

4   Analysis and Comparison

4.1   Cost Analysis

Computation Complexity. During key generation, it takes data owner $O(d^2)$ time to generate $Key = \{r, S, M, \pi\}$ and $O(d^2)$ time to compute $\{U, \dot{A}\}$.

In database encryption stage, data owner spends $O(d^2)$ computation time on encrypting each point in $\mathbf{D}$, thus it cost $O(d^2 m)$ time to complete the encryption of the total database.

While encrypting query points, our approach takes $O(d^2)$ addition/multiplication operations to compute one encrypted query point, which is completed by query user.

In outsourcing k-NN computation stage, cloud server can compute the k-NN by linearly scanning. The computation of distance $p_i' q'$ requires $O(d)$ multiplications and additions. Besides, it takes $O(m \log k)$ comparisons to find out the k-NN points where m is the number of points in $\mathbf{D}$. The computation complexity of KNNComp is $O(md \log k)$.

Communication Overheads. We analyze the communication overheads from four aspects: (a) uploading encrypted database to cloud, (b) distributing full/partial information about key to query users (or communication in encrypting query points), (c) submitting encrypted query points to cloud, (d) returning the index of k-NN to query users. Suppose it needs ω bits to transport one dimension of each point or one index of k-NN. Since one encrypted database point and query point has $(2d + 2)$ dimensions, it takes $m(2d + 2)\omega$ bits to upload $\mathbf{D}$' to cloud and $(2d + 2)\omega$ bits to submit an encrypted query point. During key distribution, a $d \times d$ matrix, a $(2d + 2) \times d$ matrix and a $(2d + 2)$-dimensional vector are transferred between data owner and each query user, and no communication occurs in the stage QEnc of our scheme, thus, communication overheads of aspect (b) are $(3d^2 + 4d + 2)\omega$ bits. For a k-NN query, cloud server will return k indexes to query user, which cost $k\omega$ bits.

4.2 Security Analysis

Data Privacy. We guarantee the data privacy of our scheme by Theorem 1.

Theorem 1. *The encrypted database $\mathbf{D}'$ in our scheme is secure against level-2 attack.*

Proof. In our scheme, $\mathbf{D}'$ is obtained by the manner of Eqs. (4) and (5). It is obvious that our encryption is more secure than the simple matrix encryption $p'_i = \hat{p}M$ in ASPE. Since [9] has shown that the simple matrix encryption is secure against level-2 attack (see Sect. 2.2) when M is a random invertible matrix, our scheme can also securely resist against the level-2 attackers, which completes the proof of the theorem.

Since a level-2 attacker is supposed to obtain some plain points in addition to all data that the level-1 attacker knows, our scheme can preserve data privacy under level-1 attack based on its security against the level-2 attacker. Next, we will discuss data privacy in level-3 attack, i.e., the attacker obtains matrix $\dot{A}$ and vector U except all the data a level-1 attacher knows.

Security against Level-3 Attacker. While obtaining matrix $\dot{A}$ and vector U from some untrustworthy query users, the attacker cannot infer the matrix A, since A is perturbed by a random invertible matrix W. Besides, the work [13,14] has shown that even if attaining $\beta(M_{i,2j} + M_{i,2j-1})$ (for $i \in [2d+2]$ and $j \in [d]$) where β is a random coefficient, the adversary still cannot recover the database $\mathbf{D}$. According to Eq. (2), the matrix A in our scheme just consists of $\theta(M_{i,2j} + M_{i,2j-1})$ (for $i \in [2d + 2]$ and $j \in [d]$), and moreover the values in each row is randomly permuted. Here, the coefficient θ is also randomly selected by data owner. Therefore, our scheme can resist level-3 attacker as well.

Query Privacy. Since query user only sends encrypted query points to cloud server and data owner learns nothing about the query points, we will analyze the view of cloud server from different levels to prove the query privacy. In our scheme, cloud server can legally learn the encrypted value q' about the query point q, where $q' = U^T + \dot{A}W q^T$. We discuss the query privacy from two levels as follows.

(i) Cloud server knows encrypted query points, encrypted database $\mathbf{D}'$ and several plain points in $\mathbf{D}$. Since θ, M, $\dot{A}$, U and W are random according to the view of cloud server, it cannot infer q from q'.

(ii) Except the data in above first level, cloud server also receives $\dot{A}$ and U from some untrustworthy query users. Then, the cloud can obtain an equation

$$\dot{A}W q^T = q' - U^T. \tag{7}$$

In this situation, the cloud server still cannot figure out the query point q due to two reasons: (ii-1) Because $\dot{A} = AW^{-1}$ and the rank of matrix A is less than d, the rank of $\dot{A}$ must be less than d as well. Thus, the d-dimensional vector $(W q^T)$ cannot be inferred while the attacker learns $\dot{A}$ and $q' - U^T$ in Eq. (7). (ii-2) The random matrix W is kept private to data owner and the corresponding query user, therefore, query privacy can also be well preserved even if cloud server works out an accurate estimation of $(W q^T)$ from Eq. (7).

Table 1. Computation complexity comparison

Schemes	Key generation		Database encryption	Query encryption[a]		k-NN computation[b]
	Data owner	Query user	Data owner	Data owner	Query user	Cloud server
ASPE [9]	$O(d^2)$	No	$O(md^2)$	No	$O(d^2)$	$O(md \log k)$
Scheme in [13, 14]	$O(d^2)$	No	$O(md^2)$	$O(d^2)$	$O(d^2)$	$O(md \log k)$
Our scheme	$O(d^2)$	$O(d^2)$	$O(md^2)$	No	$O(d^2)$	$O(md \log k)$

[a] Here is the computation complexity for encrypting one query point.
[b] The computation cost that cloud server spends on computing k-NN points for one query.

4.3   Comparison

Computation Complexity Comparison. In Table 1, we compare the computation complexity of our approach with that of two efficient existing schemes. The computation cost in key generation is used to create random key for encrypting database and query points. In our scheme, each query user will submit a random $d \times d$ invertible matrix while registering with the system. Thus, it introduces $O(d^2)$ expense to query user, but each query user only does this once regardless of the number of her query points. During encrypting the database for outsourcing, each solution takes almost the same computation cost. While encrypting each query points, ASPE and our approach do not need an online data owner, and query user can locally finish it, nevertheless, query encryption of previous scheme in [13,14] is an interactive protocol between query users and data owner in which both parties spend $O(d^2)$ computation cost on encrypting one query point. After query user submits encrypted query point, cloud server in all the three approaches will compute its k-NN points by linearly scanning. For a database consisting of m points, the linearly scanning will require $O(m \log k)$ distance comparisons, and each distance computation consumes $O(d)$, thus, each scheme costs $O(md \log k)$ in the stage KNNComp.

In general, our scheme has higher computation complexity only in KeyGen stage comparing to ASPE – the most efficient one, but our approach is secure against level-3 attack in which situation the outsouced database in ASPE will be entirely breached. Since it only needs to compute the distributed key once per query user, the KeyGen of our scheme is also practical, which will be further confirmed by extensive experiments in Sect. 5. The scheme in [13, 14] can achieve similar security with our approach, however, the previous solution [13, 14] requires that data owner must be online for encrypting each query points, thus its query encryption is much more time-consuming than that of ASPE and our new scheme, especially when each query user wants to encrypt a big number of query points.

Table 2. Comparison of security and other properties

Properties	ASPE [9]	Scheme in [13, 14]	Our scheme
Security	level-2	level-3	level-3
Need online data owner?	No	Yes	No

Comparison of Security and other Properties. Table 2 further compares our new solution with ASPE and the scheme in [13, 14]. It shows that the scheme in [13, 14] and our solution can achieve better security than ASPE, but our scheme has an obvious advantage that it does not require an online data owner. The property, that need no online data owner, denotes the query users can securely complete each k-NN query through directly communicating with cloud server after the outsourced database is uploaded to cloud and the query user has gained necessary data (such as the key) while registering with the system in KeyGen stage. Comparing with the scheme in [13, 14], our main advantage is that an online data owner is not requisite. The feature can not only reduce the communication and computation overheads and the IT expense of data owner, but also improve the feasibility and scale of our outsourcing storage and query system.

5   Performance Evaluation

In this section, we evaluate and compare the performance of ASPE, the approach in [13, 14] and our scheme by simulation. All experiments are performed on the Windows XP operating system with Intel(R) Core(TM) i5-2430M 2.40 GHz CPU and 2.0 GB memory. Every scheme is implemented using Microsoft Visual C++.

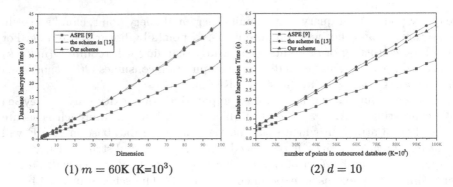

(1) $m = 60\mathrm{K}$ (K=10^3)   (2) $d = 10$

Fig. 1. Average time of database encryption

The experiment results show that key generation in our scheme cost a little more time for generating the key than that in the existing solutions, but our scheme still takes less than 1 ms to complete key generation even when d is as high as 100.

We further evaluate the encryption time for $m = 10\,\mathrm{K}$ to $100\,\mathrm{K}$ (here, K=10^3) and $d = 2$ to 100. Figure 1(1) shows the running time while m equals to $60\,\mathrm{K}$ and d ranges from 2 to 100. Figure 1(2) shows the encryption time for various m with $d = 10$. As we can see from the figures, our new approach expends almost the same time as the scheme in [13,14]. Within 6.5 s, our approach can complete encrypting the database of $100\,\mathrm{K}$ 10-dimensional points. While $m = 60\,\mathrm{K}$ and $d = 100$, the database encryption time of our scheme is still less than 43 s. Therefore, the database encryption costs are relatively practical.

For query encryption, our running time is similar to that of ASPE, which is less than that of the scheme in [13,14], since there is no communication cost during the query encryption of our solution and ASPE. While query point is 100-dimensional, our scheme can encrypt one point within 1 ms.

Same as the two existing schemes, cloud server in our scheme also uses linearly scanning to compute the k-NN points. For the encrypted database with the same number of points in the same dimension space, the query time is exactly the same. In our approach and the scheme in [13,14], the dimension of encrypted points is $(2d + 2)$. Therefore, the k-NN searching time of both solutions is the same as each other.

6   Conclusion

In this paper, we proposed a new secure outsourcing storage and k-NN query scheme, which can well preserve the privacy of database and query points in cloud, and efficiently support k-NN query on encrypted data. We also consider the security against the attacker who learns the data that data owner releases to query user except the encrypted dataset. After outsourcing his database, the data owner is not required to be online for the query. Theoretical analysis and extensive experiments confirm the above properties of our approach.

In future work, we will consider some stronger attack models, such as the collusion attack of cloud server and malicious query users.

Acknowledgments. This work is partly supported by the Fundamental Research Funds for the Central Universities (NZ2015108), Natural Science Foundation of Jiangsu province (BK20150760), the China Postdoctoral Science Foundation funded project (2015M571752), Jiangsu province postdoctoral research funds (1402033C), and NSFC (61472470, 61370224).

References

1. Hore, B., Mehrotra, S., Canim, M., Kantarcioglu, M.: Secure multidimensional range queries over outsourced data. VLDB J. **21**(3), 333–358 (2012)
2. Wang, C., Cao, N., Li, J., Ren, K., Lou, W.J.: Secure ranked keyword search over encrypted cloud data. In: Proceedings of the 30th IEEE ICDCS, pp. 253–262 (2010)
3. Cao, N., Yang, Z., Wang, C., Ren, K., Lou, W.J.: Privacy-preserving query over encrypted graph-structured data in cloud computing. In: Proceedings of the 31st IEEE ICDCS, pp. 393–402 (2011)
4. Li, M., Yu, S.C., Cao, N., Lou, W.J.: Authorized private keyword search over encrypted data in cloud computing. In: Proceedings of the 31st IEEE ICDCS, pp. 383–392 (2011)
5. Mykletun, E., Tsudik, G.: Aggregation queries in the database-as-a-service model. In: Proceedings of DBSEC, pp. 89–103 (2006)
6. Shi, E., Bethencourt, J., Chan, T.H.H., Song, D., Perrig, A.: Multi-dimensional range query over encrypted data. In: IEEE Symposium on Security and Privacy, pp. 350–364 (2007)
7. Yang, Z., Zhong, S., Wright, R.N.: Privacy-preserving queries on encrypted data. In: Gollmann, D., Meier, J., Sabelfeld, A. (eds.) ESORICS 2006. LNCS, vol. 4189, pp. 479–495. Springer, Heidelberg (2006)
8. Yiu, M.L., Assent, I., Jensen, C.S., Kalnis, P.: Outsourced similarity search on metric data assets. IEEE Trans. Knowl. Data Eng. **24**(2), 338–352 (2012)
9. Wong, W.J., Cheung, D.W., Kao, B., Mamoulis, N.: Secure kNN computation on encrypted databases. In: Proceedings of the 35th SIGMOD, pp. 139–152 (2009)
10. Hu, H.B., Xu, J.L., Ren, C.S., Choi, B.: Processing private queries over untrusted data cloud through privacy homomorphism. In: Proceedings of the 27th IEEE ICDE, pp. 601–612 (2011)
11. Xu, H.Q., Guo, S.M., Chen, K.K.: Building confidential and efficient query services in the cloud with rasp data perturbation. IEEE Trans. Knowl. Data Eng. **26**(2), 322 (2014)
12. Yao, B., Li, F.F., Xiao, X.K.: Secure nearest neighbor revisited. In: Proceedings of the 29th IEEE ICDE, pp. 733–744 (2013)
13. Zhu, Y.W., Xu, R., Takagi, T.: Secure k-NN computation on encrypted cloud data without sharing key with query users. In: ACM Workshop on Security in Cloud Computing, pp. 55–60 (2013)
14. Zhu, Y.W., Xu, R., Takagi, T.: Secure k-NN query on encrypted cloud database without key-sharing. Int. J. Electron. Secur. Digit. Forensics **5**(3/4), 201–217 (2013)

15. Liu, K., Giannella, C.M., Kargupta, H.: An attacker's view of distance preserving maps for privacy preserving data mining. In: Fürnkranz, J., Scheffer, T., Spiliopoulou, M. (eds.) PKDD 2006. LNCS (LNAI), vol. 4213, pp. 297–308. Springer, Heidelberg (2006)
16. Zhu, Y.W., Huang, Z.Q., Takagi, T.: Secure and controllable k-NN query over encrypted cloud data with key confidentiality. J. Parallel Distrib. Comput. **89**, 1 (2016). doi:10.1016/j.jpdc.2015.11.004
17. Elmehdwi, Y., Samanthula, B.K., Jiang, W.: Secure k-nearest neighbor query over encrypted data in outsourced environments. In: Proceedings of IEEE 30th International Conference on Data Engineering, pp. 664–675 (2014)

Hashing-Based Distributed Multi-party Blocking for Privacy-Preserving Record Linkage

Thilina Ranbaduge[1]([⊠]), Dinusha Vatsalan[1],
Peter Christen[1], and Vassilios Verykios[2]

[1] Research School of Computer Science,
The Australian National University, Canberra, ACT 0200, Australia
{thilina.ranbaduge,dinusha.vatsalan,peter.christen}@anu.edu.au
[2] School of Science and Technology, Hellenic Open University, Patras, Greece
verykios@eap.gr

Abstract. In many application domains organizations require information from multiple sources to be integrated. Due to privacy and confidentiality concerns often these organizations are not willing or allowed to reveal their sensitive and personal data to other database owners, and to any external party. This has led to the emerging research discipline of privacy-preserving record linkage (PPRL). We propose a novel blocking approach for multi-party PPRL to efficiently and effectively prune the record sets that are unlikely to match. Our approach allows each database owner to perform blocking independently except for the initial agreement of parameter settings and a final central hashing-based clustering. We provide an analysis of our technique in terms of complexity, quality, and privacy, and conduct an empirical study with large datasets. The results show that our approach is scalable with the size of the datasets and the number of parties, while providing better quality and privacy than previous multi-party private blocking approaches.

Keywords: Locality sensitive hashing · Clustering · Bloom filters

1 Introduction

As the world is moving into the Big Data era, it is becoming increasingly challenging to integrate and combine records that correspond to the same real-world entities across multiple heterogeneous data sources for efficient and effective decision making. The process of matching and aggregating records that relate to the same real-world entities from different data sources is known as 'record linkage', 'data matching' or 'entity resolution' [3].

In general, not all the databases to be linked have unique entity identifiers. In record linkage, it has become a common practice to use quasi-identifiers (QIDs) such as first and last name, address details, age, etc. [16] for accurate linkage of records. However, when linking personal or confidential data, organizations usually do not want their sensitive information to be revealed to other data

© Springer International Publishing Switzerland 2016
J. Bailey et al. (Eds.): PAKDD 2016, Part II, LNAI 9652, pp. 415–427, 2016.
DOI: 10.1007/978-3-319-31750-2_33

sources due to privacy and confidentiality concerns [3]. This has led to a new research area known as 'privacy-preserving record linkage' (PPRL) [18].

Considering the large volume of data in many databases and increasing number of participating parties, achieving scalability has become a challenging task in both classical record linkage and PPRL. Many blocking techniques have recently been developed [3] to overcome scalability issues. They reduce the comparison space by grouping similar record sets into blocks that likely corresponded to true matches, while excluding as many unlikely matching record sets as possible. As a result, expensive similarity comparisons are only required on a smaller number of candidate record sets. In the PPRL context, blocking techniques need to be performed with the additional requirement of preserving privacy [16].

In this paper, we propose a novel distributed blocking mechanism for multi-party PPRL which allows each database owner to generate their set of blocks by clustering their own databases independently without revealing any private information across parties, and identify the candidate cluster sets to be compared by hashing these blocks in a lower dimensional space. Besides an initial exchange of parameter settings and the final central clustering, our approach does not require any communication among the parties. This provides flexibility and control over the efficiency and privacy of blocking for the users on their datasets. Our approach also provides a mechanism to identify the most similar blocks that need to be compared in sub-groups of participating parties. To the best of our knowledge, no such distributed and independent multi-party private blocking technique has been developed in the literature so far.

Our contributions in this paper are: (1) a novel scalable distributed and independent multi-party PPRL blocking protocol based on clustering and hashing, (2) a theoretical analysis of the proposed technique in terms of complexity, blocking quality, and privacy, and (3) an empirical evaluation of our technique using large datasets. We compare our approach with two state-of-the-art multi-party private blocking techniques [12,13] which shows that our approach outperforms earlier approaches in terms of efficiency, effectiveness, and privacy.

2   Related Work

Blocking techniques have been employed in record linkage for decades [3]. Some of the developed techniques have been adapted for PPRL [1,5,12,13]. Al-Lawati et al. [1] were the first to introduce blocking for PPRL using standard blocking [3]. Mapping-based blocking [14], clustering [13], locality sensitive hashing (LSH) [5,8], and reference values-based blocking [7] are some other techniques that have been adopted for PPRL. However, performing scalable record linkage that provides better linkage quality while preserving privacy is still an open research question that needs further investigation [16].

The use of Bloom filters in PPRL [5,8,12,13,15,17] has become popular due to their capability of preserving the distances (similarities) between the original strings in the databases. Durham [5] was the first to suggest a private blocking technique using Bloom filters in the three-party context. This approach uses

Hamming-based LSH functions to efficiently generate candidate record pairs. Recently, Karapiperis and Verykios [8] have suggested a three-party PPRL blocking framework which also uses LSH functions for blocking.

Blocking techniques for multi-party PPRL have not been addressed until Ranbaduge et al. [12] recently proposed a multi-party blocking approach based on Bloom filters. The suggested technique uses a single-bit tree [9] data structure which requires all the parties to communicate to create each node in the tree. In this approach the blocks generated might miss some true matches due to the recursive splitting of Bloom filter sets. This has been addressed by the same authors using a clustering-based blocking approach [13], where the Bloom filters are first split into small mini-blocks, which are then merged into large blocks based on their similarity using a canopy clustering technique. However, these suggested approaches require all the participating parties to communicate frequently, and they lack flexibility and control over the block generation as all the parties need to agree on the same parameter settings. In contrast to these existing blocking protocols, our approach aims to provide users with control over the block generation process in terms of sizes and number of blocks, which leads to an efficient distributed blocking approach.

3   Private Multi-party Blocking Protocol

The aim of our protocol is to allow each database owner to cluster their databases completely independently. Importantly, it allows users to block their data according to their computational resources and privacy requirements.

Our protocol is designed to operate with multiple ($P \geq 2$) parties. For ease of presentation we assume that the databases held by the parties have the same schema. If the databases have different schemas, a private schema matching approach can be used to align the different schemas [14]. The main steps of our protocol are:

1. *Bloom Filter Generation:* The parties agree on the parameter settings for Bloom filter encodings and the blocking key attributes (BKA). Each party encodes their records into Bloom filters.
2. *Local Cluster Generation:* Each party performs a local clustering over their databases until the size of the final clusters (blocks) falls within a specific lower and upper range.
3. *Cluster Representative Generation:* For each local cluster a Min-Hash based representative is generated. Each party sends these representatives together with a cluster identifier to the *linkage unit* (LU).
4. *Candidate Cluster Sets Generation:* The LU uses a LSH-based blocking on the received cluster representatives to generate candidate cluster sets.

In step 4, we employ a *linkage unit* (LU) to determine candidate cluster sets to be compared from sets of local clusters generated by the different parties. A LU is commonly employed in PPRL [1,5,8] to conduct the linkage of the data sent to it by the database owners. An illustrative example of our protocol is shown in Fig. 1 and each step is discussed further in the following subsections.

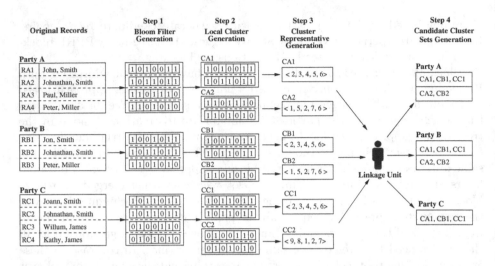

Fig. 1. Overview of our blocking protocol.

3.1 Bloom Filter Generation

As the first step in the protocol, each party independently encodes their database records into Bloom filters (BFs), where each BF is a bit vector of length m [15]. First, a set of quasi-identifier (QID) values is selected as blocking key attribute (BKA) values. These selected BKA values are then converted into a set $S = \{s_1, s_2, \cdots, s_n\}$ of sub-strings of length q (known as q-grams) [3]. Each $s_i \in S$ is encoded into a given BF by using k independent hash functions $h_1, h_2, \cdots, h_k$ and all bits having index positions $h_j(s_i)$ for $1 \leq j \leq k$ in the BF are set to 1. All P parties need to agree upon the parameters required for Bloom filter generation, including m, q (in characters), the k hash functions, and the BKAs.

3.2 Local Cluster Generation

The second step is to perform clustering over the Bloom filters independently by each party to generate a set of blocks. The distributed nature of our protocol allows each party to select an appropriate clustering technique depending upon their computational resources. A key point is that the local cluster generation can be considered as a *black box* where the clustering technique does not need to be the same across the parties as long as the minimum (s_{min}) and maximum (s_{max}) size of a cluster can be controlled. Each party can set s_{min} and s_{max} independently without making any agreements with each other, which provides more control over the block generation for the database owners.

The output of this local clustering step is a set of clusters each containing a set of Bloom filters. For our experimental study, we will use a threshold-based hierarchical clustering technique, which is explained in detail in Sect. 5.

3.3   Cluster Representative Generation

As the third step in the protocol, each party independently generates a cluster representative pair (CRP) for each local cluster. A CRP consists of a cluster ID (CID) and a Min-Hash signature (MhS) [2] in the form of $\langle CID, MhS \rangle$.

Prior to the CRP generation, for each cluster a ratio vector v_R is computed that defines the densities of all m bit positions of the BFs in a cluster, which allows each party to select the most dense bit positions in a cluster for signature generation, and hide information about less dense bit positions which can potentially reveal sensitive information to other parties [5,10]. The density of a bit position i is calculated as $\mu_i = \frac{o_i}{n}, 1 \le i \le m$, where o_i is the number of 1's in i and n is the number of BFs in a given cluster. Then, the bit positions in v_R are ranked according to the density values in descending order and the top-most dense bit positions $d < m$ are selected. Each v_R is then converted into a bit vector v_B, where all the selected d bits are set to 1 and the other $(m - d)$ positions are set to 0. Finally, a v_B is used to generate the MhS for its cluster.

To compute the MhS, we select l independent hash values for each index in v_B by using a random seed value R_{seed}, which defines l random permutations [2]. Using a single pass over the set of v_B, the corresponding l Min-Hash values $g_1(v_B), \cdots, g_l(v_B)$ are determined for each v_B under each of the l random permutations. Finally, the generated MhS is used in the CRP for each cluster. Before generating the CRPs independently, all parties need to agree on the parameters d, l, and R_{seed}, and the importance of each is described next.

Number of Top-Most Dense Bit Positions in MhS Generation (d): In order to calculate a suitable value for d, the database owners can individually simulate a linkage attack on their databases to compute an interval of the minimum and maximum d that can be selected [18], as will be described in Sect. 4. All parties need to use the same d, since different values can generate different signatures for similar clusters in different parties. This would result in these clusters not being considered as candidates in step 4 of our protocol.

The Length of the Min-Hash Signature (l): The length l provides a trade-off between accuracy and computational cost in candidate cluster generation. In general, the probability that two bit vectors v_B' and v_B'' generate the same Min-Hash for a hash function g_i is computed as $Pr[g_i(v_B') = g_i(v_B'')] = sim_J(v_B', v_B'')$, where $sim_J(v_B', v_B'')$ is the Jaccard similarity between v_B' and v_B'' [2]. For length l MhSs, the $sim_J(v_B', v_B'')$ can be estimated as $sim_J(v_B', v_B'') \approx (y/l) = |\{i|1 \le i \le l \text{ and } g_i(v_B') = g_i(v_B'')\}|/l$, where y is the number of hash functions for which $g_i(v_B') = g_i(v_B'')$. If $sim_J(v_B', v_B'')$ deviates from (y/l) with an error ε, then the probability that $y = (1 - \varepsilon) \cdot sim_J(v_B', v_B'') \cdot l$ Min-Hash values of v_B' and v_B'' are equal can be defined as:

$$Pr[y, l, sim_J(v_B', v_B'')] = \sum_{j=0}^{y} \binom{l}{y} sim_J(v_B', v_B'')^j \cdot (1 - sim_J(v_B', v_B''))^{l-j}$$

Algorithm 1. Generate candidate cluster sets (Protocol step 4)

Input:
- **SCRP$_j$**:   Set of $\langle CID_j, MhS_j \rangle$ belonging to parties $P_j, 1 \leq j \leq P$
- b :   Number of bands considered in a MhS
- l : MhS length
- L :   Number of candidate cluster sets need to be matched
- s_g: Sub-group size, $2 \leq s_g \leq P$

Output:
- **SCCS**:   Set of candidate cluster ID tuples

```
1: B ← []; SCCS ← ∅; CS' ← []; λ ← l/b        // Initialize the variables
2: foreach i ∈ 1, 2, . . . , b do:              // Each MhS is hashed for all the bands
3:    B[i] ← {}                                  // Create an inverted index for band i
4:    foreach j ∈ 1, 2, . . . , P do:            // Hash MhSs from all the parties
5:       foreach⟨CID_j, MhS_j⟩ ∈ SCRP_j do:
6:          u ← getMinHashValues(MhS_j, i, λ)    // Get λ Min-hash values for band i from MhS_j
7:          if (u ∈ B[i]) then:                  // If a bucket is available with hash key u
8:             B[i][u].add(CID_j)                 // Add the CID_j to the bucket
9:          else:
10:            B[i][u] ← [CID_j]                  // Create a new bucket with λ Min-Hash values
11:   foreach u ∈ B[i] do:
12:      C' ← createCandidateSets(B[i][u],s_g)   // Compute CCSs according to the sub-group size
13:      foreach CCS ∈ C' do:
14:         CS'[CCS] ← incrementCounter(CS', CCS)
15: SCCS ← getCandidateSets(CS',L)        // Generate L CCS sets
```

By applying the Chernoff Bound [11] to this equation, the probability that $sim_J(v'_B, v''_B)$ deviates from (y/l) up to an error bound ε is given by:

$$Pr[|sim_J(v'_B, v''_B) - (y/l)| \leq \varepsilon] \leq 2e^{-\frac{\varepsilon^2}{2+\varepsilon}l} < \delta, \tag{1}$$

where $0 < \varepsilon \leq 1$. Equation (1) must satisfy $l \geq (2+\varepsilon)/\varepsilon^2 \cdot ln(2/\delta) = O(1/\varepsilon^2)$ with a probability $\delta > 0$ [4]. To this end, for any constant $l = O(1/\varepsilon^2)$ the expected error of the similarity estimate is at most ε. Based on this estimation the database owners can agree upon an appropriate value for l (with $l < m$).

Random Seed Value for Permutation Generation (R_{seed}): The value R_{seed} needs to be shared among the parties to generate the same set of l permutations for the MhS generation using a pseudo-random generator.

3.4   Candidate Cluster Set Generation

In order to identify the candidate cluster sets (CCSs), as a naive approach the linkage unit (LU) can compute the similarities between all MhSs and based on these similarity values group the CIDs into different CCS. This would require a complexity of $O(C^P)$, if each of the P parties generates C clusters, which can become infeasible in terms of the number of computations. To improve the efficiency of generating CCSs, a Locality Sensitive Hashing (LSH)-based blocking approach is used in step 4 of our protocol, as shown in Algorithm 1. LSH [6] is commonly used for searching nearest neighbors in high dimensional data.

   LSH involves the use of locality sensitive hash function families, which in our context can be defined as follows. The MhSs of each party are divided into b bands each consisting of λ values where $l = b \cdot \lambda$. If two signatures MhS' and

MhS'' have a Jaccard similarity s, then the probability that the signatures agree in all λ Min-Hash values of one particular band is s^λ. Thus, the probability that MhS' and MhS'' agree in all λ hash values of at least one band to become a candidate pair is $1 - (1 - s^\lambda)^b$ (see also [2]). This defines the LSH family to be $(s_1, s_2, 1 - (1 - s_1^\lambda)^b, 1 - (1 - s_2^\lambda)^b)$-sensitive if,

- $sim_J(MhS', MhS'') \geq s_1$, then $Pr_{i \in 1,...,b}[i|MhS' = i|MhS''] \geq p_1$, and
- $sim_J(MhS', MhS'') \leq s_2$, then $Pr_{i \in 1,...,b}[i|MhS' = i|MhS''] \leq p_2$,

where p_1 and p_2 are two probabilities such that $p_1 > p_2$, i is a band in a MhS, and s_1, s_2 (with $s_1 > s_2$) are two Jaccard similarity thresholds.

In Algorithm 1, the MhSs of all the parties are hashed for each band $i \in b$ into a list of inverted indexes B (lines 2 to 10). For a given band i, the λ values of a MhS are used as the hashing key of a bucket $u \in B[i]$ by concatenating all λ values and the corresponding CID is added to u (lines 7 to 10). If MhSs of different parties have the same λ values for band i then the corresponding CIDs are added to the same bucket in $B[i]$ which become a CCS to be compared. In line 12, the LU can generate CCSs for different groups of parties by specifying the parameter s_g in the function $createCandidateSets()$, which defines the sub-group size of participating parties. This allows the LU to identify the most similar clusters that need to be compared between sub-groups of parties. With a default value of $s_g = P$, the algorithm provides the set of CCSs that need to be compared among all the parties, while $2 \leq s_g < P$ produces sets of CCSs for different group combinations of parties according to the specified size.

Depending on the computational capability, the LU computes L candidate cluster sets that need to be compared in the private comparison and classification step (lines 13 to 15). The parameter L can be calculated as an approximation of the reduction ratio (RR) based on the cluster comparisons by the protocol where $RR = 1 - \frac{number\ of\ candidate\ record\ sets}{total\ number\ of\ record\ sets}$ [3]. By assuming each party generates C_i clusters (each of the same size), where $1 \leq i \leq P$, then the total number of cluster comparisons for P parties is equal to $\prod_{i=1}^{P} C_i$. The approximate RR can be considered as the fraction of cluster comparisons reduced from $\prod_{i=1}^{P} C_i$, computed as $RR_{approx} \approx 1 - (L / \prod_{i=1}^{P} C_i)$.

The selection of L CCSs allows the LU to identify the cluster sets which are more likely to be similar among all the CCSs generated. In order to identify the L most similar CCSs, the CCSs are ranked based on how often a given CCS is generated for a bucket $u \in B$ (line 14), where a CCS generated for all $u \in B$ gets the highest rank. The top L CCSs are added to the set **SCCS**. Finally, **SCCS** is sent to the relevant parties (line 15) to be compared in the private comparison step [16,17] as shown in Fig. 1 - step 4. The time and space requirement for Algorithm 1 is proportional to b which has a trade-off between quality and efficiency as we discuss in more detail in the following section.

4   Analysis of the Protocol

In this section we analyze the steps of our protocol in terms of complexity, quality, and privacy.

Complexity: By assuming there are N records in a dataset with each having an average of n q-grams in the attributes used for blocking, we analyze the computational and communication complexities in terms of a single party. In step 1 all records are encoded into m length Bloom filters using k hash functions. The Bloom filter generation for a single party is of $O(k \cdot n \cdot N)$ complexity. In step 2, the complexity depends on the clustering technique used by a party. We assume each party generates $C = N/s_{max}$ clusters and calculates the v_R for each cluster in step 3. This requires a complexity of $O(m \cdot N/s_{max})$, where s_{max} defines the maximum number of records in a block. Consequently, generating l length MhSs for all clusters requires a complexity of $O(l \cdot N/s_{max})$.

In step 4 of the protocol, the LU needs to hash MhSs of all the P parties into b bands, which requires $O(b \cdot P \cdot N/s_{max})$ complexity. By excluding the initial parameter agreements in the protocol, which has a constant complexity, the parties communicate only with the LU in step 4 of our protocol. The LU receives P messages each containing $C = N/s_{max}$ CRPs which leads to an overall protocol communication complexity of $O(l \cdot P \cdot N/s_{max})$.

Quality: We analyze the quality of our protocol in terms of effectiveness and efficiency [3]. Since each party performs the clustering mechanism independently, the overall blocking quality depends on these clustering mechanisms used. The lower (s_{min}) and upper (s_{max}) size bound of the clusters generated by each party indirectly determine the number of candidate record set comparisons required in the private comparison step [16].

In step 3 of our protocol, the length l of a MhS provides a trade-off between effectiveness and efficiency of the candidate cluster set generation in step 4. An increase of l decreases the error ε of similarity estimation by $\varepsilon = O(1/\sqrt{l})$, which would decrease the *false negative* rate [3], but would increase the number of *false positives* [3] in the candidate cluster generation, and vice versa. For a given signature length l, the running time and quality of the candidate cluster set generation in step 4 depend on the number of bands b considered by the LU. As b decreases the probability that MhSs are mapped to the same bucket decreases, which leads to the number of false positives to decrease, but would increase the number of false negatives, and vice versa. On the other hand, incrementing b requires more computational time as each MhS is hashed more often which leads to an increase in the number of candidate cluster sets.

Privacy: We assume that each party follows the honest-but-curious adversary model [18] with no collusion between parties. In steps 1 to 3 of the protocol, each party performs their computations independently without any communication across the parties except for the parameter agreements. Step 2 allows each party to select the values of s_{min} and s_{max} according to their privacy requirements, where s_{min} guarantees that every resulting cluster contains at least s_{min} records, which guarantees k-anonymous mapping ($k = s_{min}$) privacy [16,18].

In step 3 of the protocol, before generating MhSs, each party can individually simulate a linkage attack, based on the *probability of suspicion* (P_s) [18], on their databases to compute an interval $[d_{min}, d_{max}]$ for selecting the number of top-most dense bit positions d in BFs to be used for the signature generation. P_s is defined for a value in an encoded dataset as $1/n_g$, where n_g is the number of values in a global dataset (**G**) that match with the corresponding value in the encoded dataset **D**. Each party performs the attack by assuming the worst case scenario of **D** = **G**. d_{min} and d_{max} provide the minimum and maximum d required to avoid exact matching of a record in the database and a unique q-gram of a record, respectively. All parties agree upon an appropriate value for d by considering these intervals.

The LU is not capable of deducing anything about the parties' data as the LU only receives a set of CRPs from each party. The parameters used in the Bloom filter generation, the values for d, R_{seed}, and the sizes of the clusters, all are unknown to the LU. Therefore, the LU cannot learn the frequency distribution of the clusters generated to conduct a frequency attack [18].

5 Experimental Evaluation

We evaluated our protocol, which we named DCH (for Distributed Clustering and Hashing), using the datasets used in and provided by [12], which were extracted from a North Carolina Voter Registration (NCVR) database[1]. These datasets contain 5,000 to 1,000,000 records for 3, 5, 7 and 10 parties, where in each of these sub-sets, 50 % of records were matches. Some of these datasets included corrupted records, where the corruption levels were set to 20 % and 40 %. Experiments with these corrupted datasets allowed us to evaluate how our approach works with 'dirty' data. We used *Given name*, *Surname*, *Suburb*, and *Postcode* attributes as QIDs, as these are commonly used for record linkage.

We used a threshold-based hierarchical clustering for all the parties, which splits the Bloom filters into a set of mini-blocks, as adapted from [13], and merges these mini-blocks using hierarchical clustering until all clusters are at least of size s_{min}. For comparative evaluation purposes we used two state-of-the-art multi-party private blocking techniques, the single-bit tree (SBT) multi-party PPRL blocking approach [12] and the hierarchical canopy clustering-based (HCC) multi-party PPRL blocking approach [13] by Ranbaduge et al. as to our knowledge there are no other blocking approaches for multi-party PPRL available. All experiments were run on a server with 64-bit Intel Xeon (2.4 GHz) CPUs, 128 GBytes of main memory, and running Ubuntu 14.04. We implemented all the approaches using Python (version 2.7.3), and these programs and test datasets are available from the authors.

Following earlier Bloom filter work in PPRL [5,12,13,15] we set the Bloom filter parameters as $m = 1000$, $k = 30$, and $q = 2$. Based on a set of parameter evaluation experiments we set the parameters $d = 500$, $l = 500$, $b = 100$, and $s_{min} = 1$ % of the dataset size. We set the parameters of the SBT and HCC approaches according to the settings provided by the authors [12,13]. Complexity is evaluated using the average total runtime for the protocol. We evaluate the blocking quality using the standard measures RR (as explained in Sect. 3.4) and

---

[1] Available from: ftp://alt.ncsbe.gov/data/.

pairs completeness (PC), which is the fraction of true matching record sets that are included in the candidate record sets generated by a blocking technique [3]. To evaluate the privacy against frequency attacks we use the measure *probability of suspicion* (P_s) [18] for a single party by assuming all parties contain similar block structures.

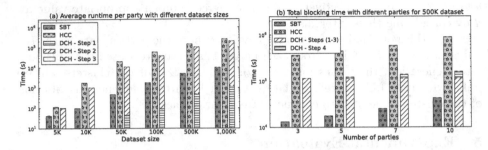

Fig. 2. (a) Average runtime per party and (b) total blocking time for the 500 K dataset. The time required for step 3 of DCH is below 1 % of the total time and therefore not visible. Note that the y-axis is in log scale and K = 1,000 records.

Figure 2 illustrates the scalability of our approach in terms of the average time required with different dataset sizes and the number of parties. Our approach shows a linear scalability with the size of the datasets and the number of parties. As shown in Fig. 2(a), the runtime required for a given dataset depends mostly on the local clustering step, which suggests that the efficiency of the protocol improves with faster clustering algorithms. According to Fig. 2(b), total blocking time increases with the number of parties because more hashing is required by the LU. For the dataset with 1,000,000 records, SBT and HCC require each party to send 2,010 and 16,128 messages with floating-point vectors, respectively. In our approach the parties require to communicate only once with the LU in step 3 which makes the protocol more communication efficient.

Figure 3(a) to (c) illustrate PC against RR for the dataset with 500,000 records for different number of parties and different corruption levels when all parties have the same block sizes. This shows our approach can produce better blocking quality with more cluster comparisons. According to the results illustrated in Fig. 3(*b*) and (*c*), our approach DCH can provide better blocking quality than SBT and HCC even with 'dirty' data. However, with the increment in the number of parties, a lower RR value would generate more candidate cluster sets which in-turn requires more candidate record sets to be compared in the private comparison and classification step in PPRL. Figure 3(d) illustrates PC against RR for the dataset with 500,000 records with different block sizes for 3 parties. This shows our approach provides high blocking quality for lower block sizes and even when the parties use different block sizes.

As shown in Fig. 4(a), DCH provides significantly better privacy against frequency attacks compared to SBT and HCC. According to Fig. 4(b), the

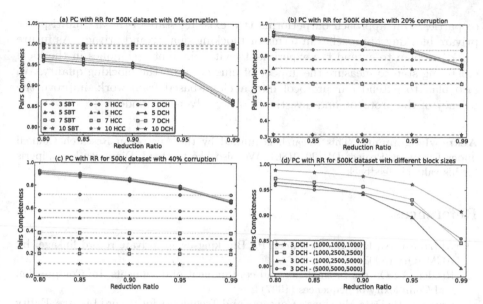

Fig. 3. Pairs completeness (PC) and Reduction ratio (RR) for 500 K dataset with (a) 0 %, (b) 20 %, and (c) 40 % corruption with same block sizes for all parties, and (d) 0 % corruption with different block sizes. Note the different y-axis scales.

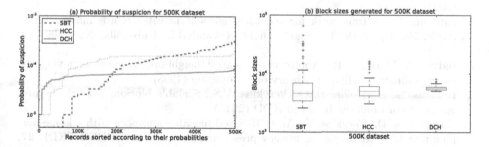

Fig. 4. (a) Probability of suspicion (P_s) against frequency attacks and (b) block sizes generated by the different approaches using the 500K dataset.

variance between cluster sizes generated by DCH is lower compared to the other approaches. DCH provides clusters within the acceptable size limit of $[s_{min}, s_{max}]$. This illustrates that our technique provides better privacy than the other approaches while achieving higher RR and PC for dirty data.

6 Conclusions

We proposed a novel communication efficient distributed blocking protocol for multi-party PPRL based on local clustering followed by a global candidate cluster set identification. We evaluated our approach with large datasets which indicates our approach is scalable with both the size of the datasets and the number of parties. Our approach outperforms two existing state-of-the-art multi-party private blocking approaches in terms of blocking quality and privacy. As future work, we plan to extend the experiments with different techniques in the local clustering step to measure the impact of clustering on the blocking quality. We also plan to extend our protocol into a cloud-based framework, improve the privacy of the protocol, and investigate other adversary models.

Acknowledgements. This research is funded by the Australian Research Council under Discovery Project DP130101801. We also like to thank Dimitrios Karapiperis for his valuable feedback.

References

1. Al-Lawati, A., Lee, D., McDaniel, P.: Blocking aware private record linkage. In: ACM IQIS (2005)
2. Broder, A.: On the resemblance and containment of documents. In: Compression and Complexity of Sequences (1997)
3. Christen, P.: Data Matching - Concepts and Techniques for Record Linkage, Entity Resolution, and Duplicate Detection. Springer, Berlin (2012)
4. Cohen, E., Datar, M., Fujiwara, S., Gionis, A., Indyk, P., Motwani, R., et al.: Finding interesting associations without support pruning. IEEE TKDE **13**, 64–78 (2001)
5. Durham, E.: A framework for accurate, efficient private record linkage. Ph.D. thesis, Faculty of the Graduate School of Vanderbilt University, Nashville, TN (2012)
6. Indyk, P., Motwani, R.: Approximate nearest neighbors: Towards removing the curse of dimensionality. In: Theory of Computing (1998)
7. Karakasidis, A., Koloniari, G., Verykios, V.S.: Scalable blocking for privacy preserving record linkage. In: ACM KDD (2015)
8. Karapiperis, D., Verykios, V.: An LSH-based blocking approach with a homomorphic matching technique for privacy-preserving record linkage. IEEE TKDE **27**, 909–921 (2015)
9. Kristensen, T.G., Nielsen, J., Pedersen, C.N.: A tree-based method for the rapid screening of chemical fingerprints. Algorithms Mol. Biol. **5**, 9 (2010)
10. Kuzu, M., Kantarcioglu, M., Durham, E., Malin, B.: A constraint satisfaction cryptanalysis of Bloom filters in private record linkage. In: PETS (2011)
11. Motwani, R., Raghavan, P.: Randomized Algorithms. Cambridge University Press, Cambridge (1995)
12. Ranbaduge, T., Vatsalan, D., Christen, P.: Tree based scalable indexing for multiparty privacy- preserving record linkage. In: AusDM, CRPIT (2014)
13. Ranbaduge, T., Vatsalan, D., Christen, P.: Clustering-based scalable indexing for multi-party privacy- preserving record linkage. In: PAKDD (2015)
14. Scannapieco, M., Figotin, I., Bertino, E., Elmagarmid, A.: Privacy preserving schema and data matching. In: ACM SIGMOD (2007)
15. Schnell, R., Bachteler, T., Reiher, J.: Privacy preserving record linkage using Bloom filters. BMC Med. Inform. Decis. Mak. **9**, 1–11 (2009)
16. Vatsalan, D., Christen, P., Verykios, V.S.: A taxonomy of privacy-preserving record linkage techniques. Elsevier JIS (2013)

17. Vatsalan, D., Christen, P.: Scalable privacy-preserving record linkage for multiple databases. In: ACM CIKM (2014)
18. Vatsalan, D., Christen, P., O'Keefe, C.M., Verykios, V.S.: An evaluation framework for privacy-preserving record linkage. JPC **6**, 13 (2014)

Text Mining and Recommender Systems

Enabling Hierarchical Dirichlet Processes to Work Better for Short Texts at Large Scale

Khai Mai, Sang Mai, Anh Nguyen, Ngo Van Linh, and Khoat Than[✉]

Hanoi University of Science and Technology,
No. 1, Dai Co Viet Road, Hanoi, Vietnam
khaimaitien@gmail.com, magicmasterno1@gmail.com,
{anhnk,linhnv,khoattq}@soict.hust.edu.vn

Abstract. Analyzing texts from social media often encounters many challenges, including shortness, dynamic, and huge size. Short texts do not provide enough information so that statistical models often fail to work. In this paper, we present a very simple approach (namely, *bag-of-biterms*) that helps statistical models such as Hierarchical Dirichlet Processes (HDP) to work well with short texts. By using both terms (words) and biterms to represent documents, bag-of-biterms (BoB) provides significant benefits: (1) it naturally lengthens representation and thus helps us reduce bad effects of shortness; (2) it enables the posterior inference in a large class of probabilistic models including HDP to be less intractable; (3) no modification of existing models/methods is necessary, and thus BoB can be easily employed in a wide class of statistical models. To evaluate those benefits of BoB, we take Online HDP into account in that it can deal with dynamic and massive text collections, and we do experiments on three large corpora of short texts which are crawled from Twitter, Yahoo Q&A, and New York Times. Extensive experiments show that BoB can help HDP work significantly better in both predictiveness and quality.

Keywords: Online HDP · Bag of biterms · Document representation · Short texts

1 Introduction

Recently, with the explosion of social networks, microblogs, instant messages, Question & Answer forums, the number of texts that users generate is extremely massive. For example, according to the statistics from http://www.internetlivestats.com/twitter-statistics/, the number of tweets posted per day is about 500 millions. With the increase in the popularization of internet, this number will be even bigger in the future. These sources of information are really attractive to data scientists because analyzing them would provide a lot of insights into users. From this, companies can come up with many strategies in doing business.

© Springer International Publishing Switzerland 2016
J. Bailey et al. (Eds.): PAKDD 2016, Part II, LNAI 9652, pp. 431–442, 2016.
DOI: 10.1007/978-3-319-31750-2_34

Unfortunately, unlike formal or official documents (academic papers, news articles, etc.), data from these sources are often short texts and most are originated from social sources. Those kinds of data feature three challenges:

- **Short**: Short texts have extremely short length. In fact, in some popular social networks such as Twitter, users are restricted to write a status (tweet) with no more than 140 characters.
- **Massive**: The number of short texts is big and increases tremendously. For example, the number of tweets generated per day is more than 500 milions.
- **Dynamic**: The topics of short texts are highly dynamic and reflect the trends of the society. A topic might emerge and disappear over time.

The limited length of short texts poses various difficulties which have been revealed by existing studies [2,4,6,7,9–13,19]. Statistical models such as Latent Dirichlet Allocation (LDA) [3] and Hierarchical Dirichlet Processes (HDP) [15] often do not work well on short texts. Those models often base on the co-occurrence of terms. However, a short text often contains few co-occurrences and does not provide a clear context. In statistical perspectives [14], we can never recover/learn correctly a model from very short texts even though we may have arbitrarily large collections.

In this paper, we show a very simple approach to dealing with short texts. Instead of using bag-of-words to represent documents, we propose to use both terms (words) and biterms which lead to *bag-of-biterms* (BoB). By this way, BoB brings us many significant benefits:

(1) BoB naturally lengthens documents and thus helps us reduce bad effects of shortness. Therefore, a statistical model might be recovered better [14].
(2) BoB enables the posterior inference in a large class of probabilistic models including HDP to be less intractable [16]. This suggests that BoB helps learn a model better.
(3) No modification of existing models/methods is necessary, and thus BoB can be easily employed in a wide class of statistical models.

To evaluate the benefits of BoB, we take HDP into account as it is a popular model and is the base for many other models. HDP [15] is a nonparametric model which can automatically grow as more data come in. Therefore, HDP can deal with the dynamic of short texts. In combination with an online learning algorithm in [17], HDP can deal with two challenges (massive and dynamic) mentioned above. Data for evaluation are three big short text collections which have been crawled from Twitter, Yahoo Q&A, and New York Times.

From extensive experiments we find that in most cases BoB helps Online HDP [17] work significantly better than bag-of-words. Both predictiveness and quality of the learned models are significantly improved. We also find that BoB helps Online HDP less sensitive to learning rate paramaters, which is an important property in pratice. This suggests that BoB really helps us recover HDP better.

The rest of paper is organzied as follows. In Sect. 2, we present a short review of previous approaches to dealing with short texts. In Sect. 3, we present general idea about Hierarchical Dirichlet Processes and the online learning algorithm [17]. In Sect. 4, we present BoB and related definitions. In Sect. 5, we describe our experiments and the evaluation of the new representation. In Sect. 6, we draw some conclusions for the paper.

2 Related Work

Besides simply applying traditional models such as LDA, HDP to short texts directly, there have been many other approaches to dealing with analyzing short texts. Here we summarize some of them to have a general view about previous approaches.

The first approach is to find additional context of the short texts by using powerful search engines (Google, yahoo, etc.) [4,12,19]. In this way, top results from searching keywords in each short document are used to evaluate its semantic meaning. One disadvantage of this method is the dependence on external tools, we cannot guarantee the quality of search engine for any keyword, especially those from social networks. Therefore it is hard to use this approach for a general short text corpus.

In another attempt, some researchers utilize external sources to deal with short texts [2,11,13]. More concretely, in both [2] and [13], the authors use Wikipedia as a source of additional information for short texts. For a short document, the authors search the closest articles from Wikipedia and take advantage of this extra information. On the other hand, in [11], the authors use Wikipedia as a universal knowledge to build a model by applying LDA. From built model, the authors do inference for each document and integrate the result into its vector. The weak point of this approach is there would be nothing sure about the correlation between the universal corpus and the short text corpora as they are almost generated from social networks with informal content and noises.

Grouping documents in rational ways is also a technique for dealing with short texts, as described in [6,7,9,10]. For example, the authors group tweets by hashtags or by the authors of tweets, the result is surely better when applying LDA to grouped texts. However, by grouping, one must know the metadata of the corpus, not all corpora of short texts have metadata as tweets. Therefore this technique is not a general approach to dealing with any corpus of short texts.

In [5,18] the authors directly model the word co-occurence pattern instead of a single word. More specifically, they generate all pairs of words (called biterm) for each document and aggregate all the biterms in all documents into one collection, and next they model the generative process for this collection. The good point of this approach is that it requires no additional information or any sources of knowledge. However, this approach does not model the generative process for each document, the authors use a heuristics to infer the topic proportions for each document. This does not guarantee the consistency between training phase and inferring phase.

In our work, we try to devise an approach that does not require any additional metadata or external sources of knowledge and able to take advantage of the flexibility of HDP. More concretely, we deal with the problem of analyzing short texts by coming up with a new representation of document and applying it to Online HDP. By this way, we can fully deal with three challenges of short texts mentioned above.

3   Hierarchical Dirichlet Processes

In topic modeling, HDP [15] is considered as the nonparametric model of LDA [3], the generative process is

$$G_0 \sim DP(\gamma, Dir(\eta)), \quad G_d \sim DP(\alpha_0, G_0), \quad \phi_{z_{di}} \sim G_d, \quad w_{di} \sim Mult(\phi_{z_{di}})$$

where Dir is Dirichlet distribution, DP is Dirichlet process and $Mult$ is multinomial distribution. γ, η, α_0 are hyperparameters. ϕ_k (for $k \in \{1, 2, \ldots\}$) are corpus-level atoms. z_{di} is topic index of w_{di} and w_{di} is the ith word in document d,

We follow [17] to use stick-breaking construction [15] to have a closed-form coordinate ascent variational inference:

$$G_0 = \sum_{k=1}^{\infty} \beta_k \delta_{\phi_k}, \quad G_d = \sum_{t=1}^{\infty} \pi_{dt} \delta_{\phi_{c_{dt}}}, \quad c_{dt} \sim Mult(\boldsymbol{\beta})$$

$$\beta'_k \sim Beta(1, \gamma), \beta_k = \beta'_k \prod_{l=1}^{k-1} (1 - \beta'_l), \quad \pi'_{dk} \sim Beta(1, \alpha_0), \pi_{dt} = \pi'_{dt} \prod_{l=1}^{t-1} (1 - \pi'_{dt})$$

where c_{dt} are document-level topic indices. $Beta$ is Beta distribution.

Approximating the posterior distribution of its latent variables bases on the idea of mean-field variational inference [8]. The general idea of this technique is considering a family of distribution over latent variables, which is defined by free parameters and then finding the closest member of this family to the true posterior. Following [17], the variational distribution has the following form:

$$q(\mathbf{c}, \mathbf{z}, \boldsymbol{\phi}, \boldsymbol{\beta}', \boldsymbol{\pi}') = \prod_d \prod_t q(c_{dt} \mid \varphi_{dt}) \prod_d \prod_n q(z_{dn} \mid \zeta_{dn}) \prod_k q(\phi_k \mid \lambda_k)$$

$$\prod_{k=1}^{K-1} q(\beta'_k \mid u_k, v_k) \prod_d \prod_{t=1}^{T-1} q(\pi'_{dt} \mid a_{dt}, b_{dt})$$

where φ_{dt}, ζ_{dn} are the variational parameters of multinomial distribution, λ_k is the variational parameter of Dirichlet distribution, (u_k, v_k) and (a_k, b_k) are variational parameters of Beta distribution, K is topic-level truncation, T is document-level truncation. Using Jensen inequality for the log likelihood of observed data we obtain a lower bound called ELBO. Taking derivatives with respect to each variational parameter, we have the coordinate ascent updates, more details at [17].

Online learning for HDP bases on stochastic optimization [8]. The idea of the technique is based on computing noisy estimates of the gradient of ELBO instead of the true gradient which requires iterating through the whole dataset. At time t the fast noisy estimates of gradient is computed by subsampling a small set of document (called minibatch). Based on noisy estimates, the intermediate global parameters $(\hat{\boldsymbol{\lambda}}, \hat{\mathbf{u}}, \hat{\mathbf{v}})$ are calculated and then update global parameters with a decreasing learning rate schedule: $\boldsymbol{\lambda}^{(t)} \leftarrow (1 - \rho_t)\boldsymbol{\lambda}^{(t-1)} + \rho_t\hat{\boldsymbol{\lambda}}, \quad \mathbf{u}^{(t)} \leftarrow (1 - \rho_t)\mathbf{u}^{(t-1)} + \rho_t\hat{\mathbf{u}}, \quad \mathbf{v}^{(t)} \leftarrow (1 - \rho_t)\mathbf{v}^{(t-1)} + \rho_t\hat{\mathbf{v}}$ where $\rho_t \leftarrow (\tau + t)^{-\kappa}$ (κ is *forgetting rate*, τ is *delay*, see [8]). The procedure will converge to an optimum [8].

4   Bag-of-biterms and Representation of Documents

4.1   Bag-of-Words (BoW)

Bag-of-words (BoW) is the conventional representation of documents. In this representation, a document is considered as a container and words as items regardless of the information about the order of words. BoW has been widely used in common models for topic modeling such as LDA and HDP. Though being widely used in practice, BoW exhibits many limitations in modeling short texts [6]. The reason is that statistical models such as HDP often base on the statistical information about the co-occurence of words; when the document is short, the co-occurence is not enough to provide a clear context. This fact inspired us to try to find out a new representation that is applicable to HDP and more suitable than BoW for modeling short documents.

4.2   Bag-of-biterms (BoB)

First, we define a "biterm" is created by a pair of words that co-occur in some document. For example, given a document containing two words "character" and "story", (character, story) is a created biterm and (story, character) is also a created biterm.

Consider a document d with its set of distinct words $\{w_1, w_2, \ldots, w_n\}$ and associated frequency set $\{f_1, f_2, \ldots, f_n\}$. In BoB, this document is represented by set of created biterms $\{b_{ij}\}$ from words (w_i, w_j) where $i = 1, \ldots n$ and $j = 1 \ldots n$ and associated frequencies defined by:

- Frequency of b_{ii} from (w_i, w_i) is f_i
- Frequency of b_{ij} from (w_i, w_j) where $i \neq j$ is 1

For example, a document d with set of words $\{w_1, w_2, w_3\}$ and set of frequencies $\{2, 2, 4\}$. In BoB, this document is represented by set of biterms $\{b_{11}, b_{22}, b_{33}, b_{12}, b_{21}, b_{13}, b_{31}, b_{23}, b_{32}\}$ and set of frequencies $\{2, 2, 4, 1, 1, 1, 1, 1, 1\}$. About the frequency of biterms, we have the observation that short documents only have several distinct words, and each often appears 1 time, therefore, to guarantee that the influence of additional biterms would not surpass original words, we set 1 for the frequency of each biterm.

Note that our approach is different from that in [18]. About the definition, biterm in our perspective can be created from two identical words while biterm in [18] is only created by two different ones. Moreover, the purpose of creating biterms in our approach is to find a new representation of documents and apply it to a wide class of statistical models, while in [18], the authors aggregate all biterms derived from all documents to form a collection of biterms and model this collection as a mixture of topics, there is no concern about documents in the course of modeling process, only in inferring step, the authors use a heuristic to infer the topic proportion for each document. This does not guarantee the consistency between modeling phase and infering phase.

Based on the definition presented above, we see that BoB has the same form like BoW, therefore applying BoB in HDP is the same as BoW. Moreover, we can see several properties of BoB:

- The document representations in BoB are improved thanks to additional biterms b_{ij} where $i \neq j$.
- Size of vocabulary in BoB (the number of distinct biterms) is higher than that of BoW. The fact is that each biterm is created from a pair of words.
- BoB is an expansion of BoW: BoB can be viewed as BoW adding additional "ingredients". In fact, in BoB, biterms b_{ii} can be considered as original words w_i and biterms $b_{ij}, i \neq j$ are additional ingredients.

These properties are the foundations for the explanation for the superiority of BoB in comparison with BoW which is presented in Sect. 4.3.

The Size of Vocabulary and Biterm Threshold in BoB. Note that in BoB, because of the symmetry of biterms b_{ij} and $b_{ji}(i \neq j)$ in every document, in practice, we can reduce the memory for storage by merging b_{ij} and b_{ji}, we only use b_{ij} $(i < j)$ with frequency 2 instead of 1.

The biggest challenge of BoB is the dilation of vocabulary set. In theory, the number of possible biterms might be up to $V_b = V(V-1)/2 + V = V(V+1)/2$ where V is the number of distinct words in corpus. This number is quite large that requires a lot of memory to store and the training time might last so long that makes the algorithm become infeasible. Fortunately, owing to the shortness of short texts, the practical number of distinct biterms is often not so large (see the description of datasets in Sect. 5.1) that storing and running are possible in a regular personal computer.

However, besides the purpose of improving the speed of learning the model and reducing the memory needed during training process, removing low-frequency biterms is also considered as the preprocessing step like removing words with low frequency (noises) in preprocessing text. Therefore, we need to have a threshold called "biterm threshold". The biterms with document frequency lower than this threshold are eliminated from the representation of document. The size of vocabulary in BoB can be adjusted by changing biterm threshold. In our experiments presented at Sect. 5, we also evaluate the influence of V_b on the quality of the model. Note that when biterm threshold is ∞, all biterms are removed, therefore BoB becomes BoW.

4.3   Explanation for the Potential Superiority of BoB for Short Texts

There are several reasons for believing in a better performance of BoB.

Firstly, BoB deals with the shortness of short documents. In BoB, the length of a document is much more longer than that in BoW. Namely, in BoB each document is $d(d-1)$ longer than document in BoW, where d is the number of distinct words in this document. In [14], the authors showed that document length plays an important role in topic modeling, when documents are extremely short, the learned model is expected to have poor performance. The idea of lengthening short documents has also been carried out in previous researches [6, 7,9,10]. In these researches, the authors aggregate short documents in a rational way to make them longer, consequently bring about better results.

Secondly, BoB increases the vocabulary size considerably. Recent research in topic modeling has indicated that the MAP inference for topic mixtures of each document can reach a global optimum when the length of document and the size of vocabulary are large enough [16]. Therefore, BoB can help us do posterior inference in topic models better than BoW. As a result, the learned model from BoB is expectedly better than that in BoW.

Finally, BoB is also considered as adding extra words to the original document in a rational way. In fact, biterms $b_{ii} = (w_i, w_i)$ can be treated as the original word w_i and biterms $b_{ij} = (w_i, w_j)$ with $i \neq j$ are the additional "ingredients". In a short document, due to the lack of information, the context becomes ambiguous and unclear. The ingredients being added would reinforce the context, make it clearer.

5   Experimental Results

In this section, we conduct extensive experiments on three different large short text collections to evaluate the effectiveness of our approach for dealing with short texts. We run Online HDP (as described in Sect. 3) in both BoW and BoB and compare their two performance measures and their sensitivity to learning rate parameters. We also investigate the influence of the size of vocabulary in BoB on the quality of the model. We use the source code of Online HDP from: http://www.cs.cmu.edu/~chongw/software/onlinehdp.tar.gz.

5.1   Datasets

To prepare for our experiments, we had crawled three large collections of short texts, as described below:

- *Yahoo Questions*: This dataset is crawled from https://answers.yahoo.com/, which is a forum that users post questions and wait for the answers from other users. Each document is a question.
- *Tweets*: This dataset is a set of tweets crawled from Twitter (http://twitter. com/) with 69 hashtags containing various kinds of topic. Each document is the text content of a tweet.

– *Nytimes Titles*: This dataset is a set of titles of articles from The New York Times (http://www.nytimes.com/) from 01/01/1980 to 29/11/2014. Each document is the title of an article.

These datasets went through a preprocessing procedure including tokenizing, stemming, removing stopwords, removing low-frequency words (appear in less than 3 documents) and removing extremely short documents (less than 3 words). About BoB, for each corpus we set a different biterm threshold (definition of biterm threshold in Sect. 4.2) to make sure that the number of created biterms is not so large. The detailed description for each dataset is on Table 1.

Table 1. Description of three datasets used in experiments

	Corpus size	Average length per doc	V	Biterm threshold	V_b
Yahoo Questions	537,770	4.73	24,420	2	722,238
Tweets	1,485,068	10.14	89,474	10	764,385
Nytimes Titles	1,684,127	5.15	55,488	5	756,700

5.2 Performance Measures

To compare the two representations, we use two very common performance measures in topic modeling: the LPP and NPMI.

Log Predictive Probability (LPP). LPP measures the predictiveness and generalization of a learned model to new data. The procedure for this measure is similar to [8]. For each dataset, we randomly separate into training set and test set. In test set, we only pick out documents with length greater than 4. For each chosen document, we divide it randomly into two part $(\mathbf{w}_1, \mathbf{w}_2)$ with ratio 4:1, we did inference for $\mathbf{w}_1$ and estimated the distribution of $\mathbf{w}_2$ given $\mathbf{w}_1$. In BoB, when estimating the distribution of $\mathbf{w}_2$ given $\mathbf{w}_1$, we need to convert the topic-over-biterms (distribution over biterms) to topic-over-words (distribution over words), the conversion fomula is described in Appendix.

Normalized Pointwise Mutual Information (NPMI). NPMI [1] measures the coherence of topics of a model. This measure indicates the quality of the topics learned from dataset by evaluating words with highest probability of each topic (called top words). Here we compute this measure by considering 10 top words for each topic. In case of BoB, we first need to convert the topic-over-biterms (distribution over biterms) to topic-over-words (distribution over words) the conversion fomula is described in Appendix.

5.3 Result

Performance Comparison. To evaluate thoroughly the performance of the new representation, we run Online HDP in various settings for each representation in each dataset. More concretely, The settings of learning rate parameters

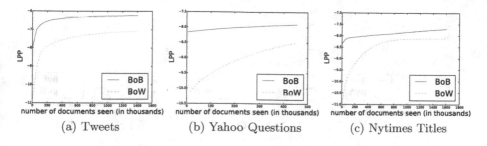

(a) Tweets   (b) Yahoo Questions   (c) Nytimes Titles

Fig. 1. Predictiveness on three datasets in both BoW and BoB.

(τ, κ) form a grid: $\tau \in \{1, 20, 40, 60, 80, 100\}$, $\kappa \in \{0.6, 0.7, 0.8, 0.9\}$. For each setting of (κ, τ), we fix the minibatch size $= 5000$, truncation for corpus $K = 100$, truncation for document $T = 20$ and $\alpha_0 = 1.0$, $\gamma = 1.0$.

We first compare the predictive capability of two representations. Figure 1 shows the median of 24 LPPs associated with 24 settings of (κ, τ) on three datasets in both representations. As we can see, When the number of documents seen increases, both Bow and BoB attain a better predictive capability regardless of learning rate parameters. However, in all three datasets, BoB outperforms BoW significantly. Moreover, BoB has good starting points while BoW always has poor ones and needs much time to be stable. This superiority of model learned from BoB can be explained by better inference for each document. As mentioned in Sect. 4.3, the goodness of this inference may originate from the longer length in each document, larger vocabulary size and clearer context by additional ingredients. As the inference for each document is more precise, the model is recovered better.

About the NPMI score, as we can see in Fig. 2 showing the median of 24 NPMIs associated with 24 settings of (κ, τ) in both BoB and BoW , there is also a huge distinction between BoB and BoW like LPP, especially in *Tweets*. Apart from outperforming BoW significantly, BoB also has much better starting points. All of these observations indicates the superiority of BoB to BoW in the quality of learned topics in Online HDP. Besides the reasons explained above, the goodness of BoB in terms of NPMI can also be explained by the conversion from topic-over-biterms to topic-over-words (presented in appendix). Biterm b_{ii} in BoB can be considered as w_i in BoW. However in BoB there are additional ingredients $b_{ij}, i \neq j$, these items would contribute their probability to w_i in BoB. Therefore, a word with more biterms containing it would be stressed more than a word with less biterms containing. By this way, the probability of words in each topic would converge to the true one more quickly and more precisely.

Sensitivity of Learning Rate Parameters. To compare the sensitivity of τ in BoB and BoW, we fix $\kappa = 0.9$ and explore different values of $\tau \in \{1, 20, 40, 60, 80, 100\}$, the result is described in Fig. 3. Similarly, to compare the sensitivity of κ in BoB and BoW, we fix $\tau = 20$ and explore different values of $\kappa \in \{0.6, 0.7, 0.8, 0.9\}$, the result is described in Fig. 4. Observing two figures, we

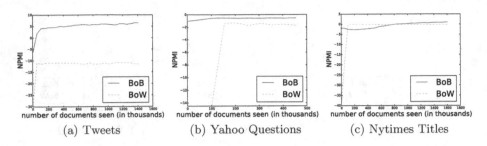

(a) Tweets   (b) Yahoo Questions   (c) Nytimes Titles

Fig. 2. Quality of the topics learned from BoW and BoB on three datasets. The higher is the better

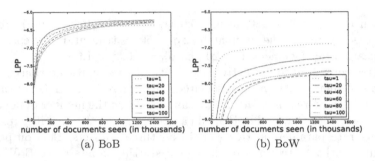

(a) BoB   (b) BoW

Fig. 3. Sensitivity of κ in BoB and BoW when fixing $\kappa = 0.9$ in *Tweets*

see that BoB is much less sensitive to both κ and τ than BoW. Less dependence on learning rate parameters is a good property of BoB in practice, one does not have to consider exhaustive model selection for learning rate parameters.

The Influence of Size of Vocabulary in BoB (V_b). To evaluate the influence of size of vocabulary in BoB (V_b), we fix $\kappa = 0.8, \tau = 80$ (this is the best setting of learning rate in BoB with dataset *Tweets*) and carry out experiments with varying V_b by changing biterm threshold in set $\{10, 15, 20, 25, 30, \infty\}$. Remember

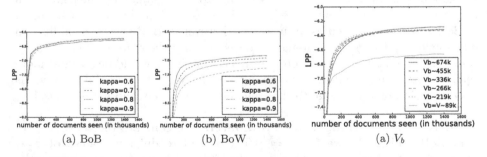

(a) BoB   (b) BoW   (a) V_b

Fig. 4. Sensitivity of κ in BoB and BoW when fixing $\tau = 20$ in *Tweets*

Fig. 5. LPP with varying sizes of V_b in BoB in *Tweets*

that as biterm threshold is ∞, BoB becomes BoW. The experimental result is described in Fig. 5. We observe that when $V_b = V$ (BoB becomes BoW), the model has the poorest predictive capability. The model attains best predictive performance when V_b is not the smallest or the biggest, it seems that there is a tradeoff in V_b. Namely when biterm threshold is large, there are many biterms being removed from document in BoB, as a result, the documents become shorter and we return to the problem of short text. On the other hand, if biterm threshold is small, there are many biterms being added to documents even they only appear in a small number of documents. These biterms can be considered as noises that might degrade the model quality.

6   Conclusion

We investigated a new representation which is called bag-of-biterms (BoB) for documents. We found that BoB has many interesting properties, which overcome some severe limitations of bag-of-words representation. It can help us learn significantly better statistical models such as HDP from short texts. It can be easily employed in a large class of statistical models and methods without any need of modification. It also compliments to existing approaches to deal with short texts. Therefore, we believe that BoB is a potential approach, and can be used in a wide context.

Acknowledgments. This work was partially supported by Vietnam National Foundation for Science and Technology Development (NAFOSTED Project No. 102.05-2014.28), and by AOARD (US Air Force) and ITC-PAC (US Army) under agreement number FA2386-15-1-4011.

Appendix: Conversion of topic-over-biterms (distribution over biterms) to topic-over-words (distribution over words)

In BoB, after we finish training the model, we obtain topics, each is the multinomial distribution over biterms and we want to convert to the topics with distribution over words. Assume that ϕ_k is the distribution over biterms of topic k. According to probability:

$p(w_i \mid z = k) = \sum_{j=1}^{V} p(w_i, w_j \mid z = k) = \sum_{j=1}^{V} p(b_{ij} \mid z = k) = \sum_{j=1}^{V} \phi_{kb_{ij}},$

As discussed in Sect. 4.2, in implementing BoB, we can merge b_{ij} and b_{ji} into b_{ij} with $i < j$. Because of identical occurence in every document, after finishing training process, the value of $p(b_{ij} \mid z = k)$ will be expectedly the same as $p(b_{ji} \mid z = k)$. Therefore, in grouping these biterms into one, the conversion version of this implementation is: $p(w_i \mid z = k) = \sum_{j=1}^{V} p(b_{ij} \mid z = k) = \phi_{kb_{ii}} + \frac{1}{2} \sum_{b:\ biterms\ contain\ w_i} \phi_{kb}.$

References

1. Aletras, N., Stevenson, M.: Evaluating topic coherence using distributional semantics. In: Proceedings of the 10th International Conference on Computational Semantics (IWCS 2013), pp. 13–22 (2013)
2. Banerjee, S., Ramanathan, K., Gupta, A.: Clustering short texts using wikipedia. In: Proceedings of the 30th Annual International ACM SIGIR Conference on Research and Development in Information Retrieval, pp. 787–788. ACM (2007)
3. Blei, D.M., Ng, A.Y., Jordan, M.I.: Latent dirichlet allocation. J. Mach. Learn. Res. **3**, 993–1022 (2003)
4. Bollegala, D., Matsuo, Y., Ishizuka, M.: Measuring semantic similarity between words using web search engines. WWW **7**, 757–766 (2007)
5. Cheng, X., Yan, X., Lan, Y., Guo, J.: BTM: topic modeling over short texts. IEEE Trans. Knowl. Data Eng. **26**(12), 2928–2941 (2014)
6. Ye, C., Wen Wushao, P.Y.: TM-HDP: an effective nonparametric topic model for tibetan messages. J. Comput. Inf. Syst. **10**, 10433–10444 (2014)
7. Grant, C.E., George, C.P., Jenneisch, C., Wilson, J.N.: Online topic modeling for real-time twitter search. In: TREC (2011)
8. Hoffman, M.D., Blei, D.M., Wang, C., Paisley, J.: Stochastic variational inference. J. Mach. Learn. Res. **14**(1), 1303–1347 (2013)
9. Hong, L., Davison, B.D.: Empirical study of topic modeling in twitter. In: Proceedings of the First Workshop on Social Media Analytics, pp. 80–88. ACM (2010)
10. Mehrotra, R., Sanner, S., Buntine, W., Xie, L.: Improving lda topic models for microblogs via tweet pooling and automatic labeling. In: Proceedings of the 36th International ACM SIGIR Conference on Research and Development in Information Retrieval, pp. 889–892. ACM (2013)
11. Phan, X.H., Nguyen, L.M., Horiguchi, S.: Learning to classify short and sparse text & web with hidden topics from large-scale data collections. In: Proceedings of the 17th International Conference on World Wide Web, pp. 91–100. ACM (2008)
12. Sahami, M., Heilman, T.D.: A web-based kernel function for measuring the similarity of short text snippets. In: Proceedings of the 15th International Conference On World Wide Web, pp. 377–386. ACM (2006)
13. Schönhofen, P.: Identifying document topics using the wikipedia category network. Web Intell. Agent Syst. **7**(2), 195–207 (2009)
14. Tang, J., Meng, Z., Nguyen, X., Mei, Q., Zhang, M.: Understanding the limiting factors of topic modeling via posterior contraction analysis. In: Proceedings of The 31st International Conference on Machine Learning, pp. 190–198 (2014)
15. Teh, Y.W., Jordan, M.I., Beal, M.J., Blei, D.M.: Hierarchical dirichlet processes. J. Am. Stat. Assoc. **101**(476), 1566–1581 (2006)
16. Than, K., Doan, T.: Dual online inference for latent dirichlet allocation. In: Proceedings of the Sixth Asian Conference on Machine Learning (ACML), pp. 80–95 (2014)
17. Wang, C., Paisley, J.W., Blei, D.M.: Online variational inference for the hierarchical dirichlet process. In: International Conference on Artificial Intelligence and Statistics, pp. 752–760 (2011)
18. Yan, X., Guo, J., Lan, Y., Cheng, X.: A biterm topic model for short texts. In: Proceedings of the 22nd International Conference on World Wide Web, International World Wide Web Conferences Steering Committee, pp. 1445–1456 (2013)
19. Yih, W.T., Meek, C.: Improving similarity measures for short segments of text. AAAI **7**, 1489–1494 (2007)

Query-Focused Multi-document Summarization Based on Concept Importance

Hai-Tao Zheng, Ji-Min Guo(✉), Yong Jiang, and Shu-Tao Xia

Tsinghua-Southampton Web Science Laboratory,
Graduate School at Shenzhen, Tsinghua University, Shenzhen, China
{zheng.haitao,jiangy,xiast}@sz.tsinghua.edu.cn,
guojm14@mails.tsinghua.edu.cn

Abstract. With the exponential growth of the web documents and the requirement of limited bandwidth for mobile devices, it becomes more and more difficult for users to get information they look forward to from the vast amount of information. Query-focused summarization gets more attention from both the research and engineering area in recent years. However, existing query-focused summarization methods don't consider the conceptual relation and the concept importance that make up the sentences, a concept is the title of a wikipedia article and can express an entity or action. In this article. We propose a novel method called Query-focused Multi-document Summarization based on Concept Importance (QMSCI). We first map sentence to concepts and get ranked weighted concepts by reinforcement between the concepts of sentences and concepts of the query in a bipartite graph, then we use the ranked weighted concepts to help to rank the sentences in a hyper-graph model, sentences that contain important concepts, related with the query and also central among sentences are ranked higher and comprise the summary. We experiment on the DUC datasets, the experimental result demonstrates the effectiveness of our proposed method compared to the state-of-art methods.

Keywords: Query-focused summarization · Bipartite-graph model · Hyper-graph model

1 Introduction

Text Summarization starts off from the 1950s' when people use simple statistical and linguistic knowledge to generate summary of the documents [12]. Till now, many summarization methods have been proposed, they can be categorized as extractive summarization [5] and abstractive summarization. The former ranks the sentences in the document and extracts the top-ranked sentences to comprise the summary, the latter usually needs sentence compression [8], information fusion [1] and re-formulation [13], it requires natural language understanding and representation knowledge, it's more complicated and not mature even now.

© Springer International Publishing Switzerland 2016
J. Bailey et al. (Eds.): PAKDD 2016, Part II, LNAI 9652, pp. 443–453, 2016.
DOI: 10.1007/978-3-319-31750-2_35

So most summarization methods belong to extractive summarization. Summarization can be also categorized as generic summarization and query-focused summarization from different purposes of the summary, generic summarization generates summary in all aspects about the document, while query-focused summarization generates summary mainly considering the user's preference or the query. Our approach is for extractive query-focused multi-document summarization.

The purpose of the query-focused summarization is to extract sentences that are highly relevant to the query and representative of the documents, so measuring the relevance between the sentence and query accurately is very important. The paper [14] first proposed to use wikipedia concepts to help to generate query-focused summarization, as Wikipedia has grown to become the largest encyclopedia with over 2 million articles [18], using wikipedia concepts can greatly enrich the literal meaning of the sentences. The method proposed in [14] is to map sentences to concept-vector and use the concept-vector and concept-relatedness matrix to get sentence-query relatedness, sentences that are most correlated with the query are selected to comprise the summary. This method generates more accurate summary compared to the previous term-vector method. There also exist summarization methods that exploit the conceptual relation of sentences in generic multi-document summarization [20], because concept is the basic element of language that can express an entity or action, by further exploring the conceptual relations we can get more informative summary. While in query-focused summarization, in order to generate more accurate summary, we are the first to further explore the conceptual relation in the effective graph model and use the ranked weighted concepts to help to rank the sentences. In this paper, we first map the query and sentences in the documents to the concept set, construct a bipartite-graph between the concepts of the query and concepts of the sentences, by mutual reinforcement between the concepts, we can get the ranked weighted concepts. Then we use a hyper-graph model to represent the complicated relations between sentences and concepts and get top-ranked sentences to comprise the summary, sentences that are more correlated with the query, contain more important concepts and also correlated with other sentences in the documents ranked higher. Finally remove redundancy and reorder the sentences comprising the summary according to the original sentence position.

The contribution of this paper is as follows:

(1) We explore the implicit semantic information of the sentence using Wikipedia concept and take the concept as the basic element of the documents which is more close to human's comprehension.

(2) We get the ranked weighted concepts by mutual reinforcement between the concepts of sentences and concepts of query in a bipartite-graph model, which help to rank the sentences comprising the summary.

(3) We utilize hyper-graph model to represent the complicated relation between sentences and concepts, and get sentences that are central in the graph, contain important concepts and correlate with the query to comprise the summary.

The rest of the paper is organized as follows: Sect. 2 introduces the related work about the query-focused summarization, Sect. 3 introduces our proposed method in detail, Sect. 4 presents the experiment of our method, Sect. 5 concludes this paper.

2 Related Work

Query-focused summarization is to extract sentences that can satisfy the query's need to comprise the summary. The most direct way is to calculate the similarity between sentence and query, sentences that are more similar to the query are ranked higher and comprise the summary. To calculate the similarity between sentences, Miao [14] proposed to use Wikipedia concept vector and concept-relatedness matrix to get the similarity between sentence and query, compared to the previous TFIDF vector method, Miao [14] gets more accurate summary because it takes into account the implicit semantic meaning of the sentences. The most query-focused summarization methods used graph-based model [3,4, 16,17]. It's an extension to the generic graph-based model. In a graph-based model, take the sentences as vertex, the similarity between sentences as the edge weight between vertexes. By random walk on the graph, get sentences that are more central in the graph to comprise the summary. while the previous graph-based method only consider the relation between two sentences and ignore the implicit semantic meaning of sentences, the paper [2] proposed to use hyper-graph to represent textual content to get generic text summarization, hyper-graph can represent the relation between more than two sentences as in real-world condition, but this method takes word as the basic language element and don't consider the semantic meaning of sentences. There are also some methods concerning the expansion of query through external corpus such as Word Net [19] or selecting important terms in the documents [21], these methods can enrich the query and get more information about the query but it can also result in noise in the query, so many unnecessary terms may add to the query and result in unwanted result. There also exist some other query-focused summarization methods extended from the classical generic summarization methods like MMR which also takes into account the query-related feature and the previous features to decide the weight of sentences [11].

3 Query-Focused Multi-document Summarization Based on Concept Importance

We propose a method to generate query-focused multi-document summarization based on concept importance(QMSCI). Our method is composed of three components: (1) map the query and sentences in the document to the concept set, construct bipartite-graph between the concepts of the query and concepts of the sentences, the weight between two concept is calculated using WLVM(Wikipedia Link Vector based Measure) metric and is calculated based on the Wikipedia hyperlink structure. After running the random walk algorithm like HITS [7] on the bipartite-graph, we get the ranked weighted concepts; (2) construct hyper-graph using sentence as hyper-edge and concepts as node, initiate the sentence weight according to the concepts importance of the sentence and the similarity between the sentence and the query, random walk on the hyper-graph to get the ranked weighted sentences; (3) select the top-ranked sentences, remove redundancy and reorder the sentences according to the original position in the document and generate the summary. The main framework of the method is shown as follows in Fig. 1.

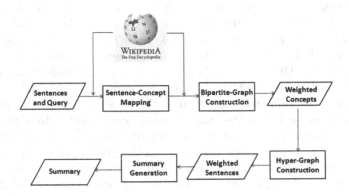

Fig. 1. Framework of the proposed method

3.1   Bipartite-Graph Construction

There exist two methods to map the sentences to concepts [6]. We use the related-match method to map the query and sentences in the documents to the concepts. The mapping process is shown as follows in Fig. 2.

s_i represent the ith sentence in the documents, and c_j are the concepts mapped from the sentences, w_{ij} are the relatedness between the sentence s_i and concept c_j, for convenient processing, we only select the top-N concepts that most correlated with the sentence as the sentence-mapped concept vector. After mapping the query and sentences to the concept vector, we construct a bipartite-graph between all concepts of the sentences and concepts of the query. The bipartite-graph is shown as follows in Fig. 3.

Q_i represent the concepts of the query, C_j represents the concepts of the sentences, the relatedness between Q_i and C_j is measured by WLVM metric, a sophisticated measure to calculate the relatedness between two concepts based on the hyperlink structure in Wikipedia [15]. The principle behind this measure is that more common concepts the two concepts point to, while less other concepts point to the target concepts, the more correlated is between the two concepts. The process of calculating the relatedness between two concepts by WLVM is as follows: c_i, c_j are two concepts and represented as a vector of the weight pointing to the target concepts:

$$c_i = \{w(c_i \rightarrow L_1), w(c_i \rightarrow L_2), \cdots, w(c_i \rightarrow L_n)\}$$
$$c_j = \{w(c_j \rightarrow L_1), w(c_j \rightarrow L_2), \cdots, w(c_j \rightarrow L_n)\}$$

$L_i(i = 1, 2, \cdots, n)$ are all the target concepts that c_i and c_j point to.

$w(c_i \rightarrow L_i) = |c_i \rightarrow L_i| * log(\sum_{x=1}^{t} \frac{t}{|x \rightarrow L_i|})$, $|c_i \rightarrow L_i|$ is the number of links point from concept c_i to L_i, t is the total number of concepts in Wikipedia, the relatedness between two concepts c_i and c_j is then calculated as the cosine similarity between the two weight vectors.

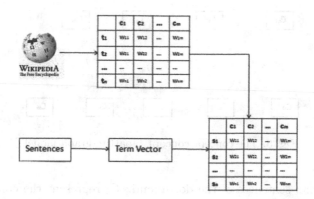

Fig. 2. The sentence-concept mapping process

After constructing the bipartite-graph, we take the concepts of the query as the authority nodes and concepts of the sentences as the hub nodes, run the random walk algorithm like hits on the bipartite-graph until convergence to get the ranked-weighted concepts. We initiate the hub node weight as $\frac{1}{\sqrt{m}}$, m is the total number of hub nodes, and initiate the authority node weight as $\frac{1}{\sqrt{n}}$, n is the total number of authority nodes. In the random process, every iteration step,the score of every authority node is the total score of the hub node that point to it, and the score of every hub node is the total score of the authority node that it point to. To guarantee the random walk process is convergent, every iteration step the weight of the vertexes is normalized.

3.2 Hyper-graph Construction

Hyper-graph is a generalization of graphs [2], the previous graph models the pairwise-relation between two vertexes as an edge between the two vertexes, it can only represents one-to-one relationship between two vertexes, while in real world, there may exist many complicated relations among more than two vertexes, so hyper-graph can better represent the real-world problem and express more complicated relations between more than two vertexes. The random walk process in the hyper-graph can favor the sentences that are more important and informative, while in the previous graph model, a sentence which is uninformative but have some unimportant common words with many other sentences can be ranked higher by random walk on the graph. A hyper-graph can be defined as $G = \{E, V, W_e, W_v\}$, where V is the vertex set, E is hyper-edge set, a hyper-edge is a subset of the total vertex set, v is incident to e when v belong to e, and W_e is the hyper-edge weight, W_v is the vertex weight. We construct a hyper-graph using the sentences as hyper-edge and the concepts as vertexes, our constructed hyper-graph is shown as follows in Fig. 4.

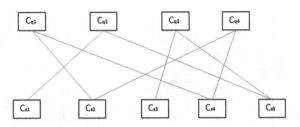

Fig. 3. The concept-concept bipartite-graph model

S_i represents sentences of the documents; C_i represent the concepts, C_i is incident to S_j if C_i belong to the concept set of S_j. In the above hyper-graph, S_1 contains concepts C_1, C_4, C_5 and S_2 contains concepts C_2, C_5 and so on. The vertex weight and hyper-edge weight is the weight of the corresponding sentences and concepts . From the above section we have get the concept weight, and we initiate the sentence weight as

$$w(S) = \sum_{i=1}^{k} w(C_i) + \lambda * sim(S, q) \qquad (1)$$

C_i is the concept of sentence S, $w(C_i)$ is the weight of the concept C_i, we initiate the sentence weight because in the random walk process, we want to favor sentences that have high initial score. The initial score of a sentence depends on two factors: the average score of the concepts the sentence mapped to and the similarity between the sentence and the query, the similarity is measured by the term overlap measure. The parameter λ is used to balance the proposition of the two factors influencing the initial sentence score. In the experiment we set λ to 0.8. Sentences that are more similar to the query and contain more important concepts get higher initial weight. Then random walk among the hyper-edges, the process of random walk from a hyper-edge S_i to another hyper-edge is as follows: choose a vertex that is incident to S_i proportional to the vertex weight, choose another hyper-edge the vertex incident to proportional to the hyper-edge weight, the probability of random walk from hyper-edge S_i to S_j is as follows:

$$p(S_i, S_j) = \frac{\sum_{C_i \in S_i \cap S_j} w(C_i) * w(S_j)}{\sum_{S_k \in S} \{\sum_{C_j \in S_i \cap S_k} [w(C_j) * w(S_k)]\}} \qquad (2)$$

where C_i is the vertexes that incident to both S_i and S_j, more common vertexes that incident to the two hyper-edges, and more important the vertexes are, it is more probable to random walk from S_i to S_j, and the random walk process favors the high score initiated sentences. Run page rank algorithm to get the ranked weighted sentences. Sentences that is more central in the graph and have more initial scores will rank higher at last.

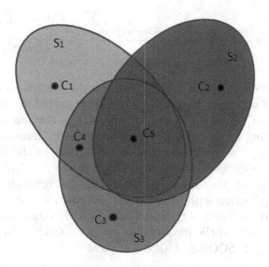

Fig. 4. The sentence-concept hyper-graph model

3.3 Summary Generation

Since sentences are ranked, reorder the sentences by weight in descending order. Use list A to store the reordered sentences and list B to store the top ranked sentences, initiate list B as an empty list. Every time we choose the first element in list A and calculate the similarity with the sentences in B, if the similarity is beyond a given threshold, the sentence is discarded, else the sentence is removed from list A and added to list B. Repeat this procedure until the total sentence length in B is up to the summary length. Reorder sentence in list B according to the original position of the sentence in the document to comprise the summary.

4   Experiment

4.1 Experiment Setup

We use DUC2006 and DUC2007 datasets, DUC2006 contains 50 topics, every topic includes 25 documents, DUC2007 contains 45 topics and every topic also includes 25 documents. For every topic of the documents, we use the sentences inserted between the ⟨title⟩ and ⟨narr⟩ tag as the query sentences of the corresponding documents set. We used the ROUGE (Recall-Oriented Understanding for Gisting Evaluation) as evaluation method, ROUGE measure the quality of the system-generated summary by comparing the overlap between the system-generated summary and human-generated summary [9]. The length of the summary is 250 words. The ROUGE-N metrics is defined as:

$$ROUGE - N = \frac{\sum\limits_{S \in \{ReferenceSummaries\}} \sum\limits_{gram_n \in S} Count_{match}(gram_n)}{\sum\limits_{S \in \{ReferenceSummaries\}} \sum\limits_{gram_n \in S} Count(gram_n)} \qquad (3)$$

Rouge-N is based on n-gram overlap between the system-generated summary and a set of model summary that is the human generated summary [10]. N represent the length of n-gram, $Count_{match}(gram_n)$ is the maximum number of n-grams co-occurring in a candidate summary and the human-generated summary, $Count(gram_n)$ is the total number of n-gram in the human-generated summary. We used F-SCORE as the final result of the ROUGE metrics. We get the F-SCORE of Rouge-1(unigram-based), Rouge-2 (bigram-based) and Rouge-SU4(based on skip bigram with a maximum skip distance of 4), where Rouge-1 and Rouge-SU4 are regarded as the most correlated results with the human evaluation. For every topic of the dataset, we get a F-SCORE value, and the final result is the average F-SCORE of all the topics.

4.2 Experiment Results

In our experiment, we choose the extended graph-based method used for query-focused summarization named as Graph-based, query expansion method [19] named as Query-expansion and improved sentence-query similarity method which used concept-vector and concept-relatedness matrix to rank the sentences [14] named as Word+Matrix, the experimental results on duc2006 and duc2007 are shown in Tables 1 and 2.

Table 1. Comparison results on DUC2006

Systems	ROUGE-1	ROUGE-2	ROUGE-SU4
Graph-based	0.35839	0.06688	0.1249
Query-expansion	0.34971	0.05805	0.10797
Word+Matrix	0.35053	0.05544	0.11114
QMSCI	**0.38029**	**0.06773**	**0.12676**

The experimental result demonstrates the effectiveness of our proposed method, in duc2006 and duc2007, our method performs better than other methods. Because we not only consider the implicit semantic information of sentences, but also utilize bipartite graph model and hyper-graph model to get the ranked

Table 2. Comparison results on DUC2007

Systems	ROUGE-1	ROUGE-2	ROUGE-SU4
graph-based	0.36168	0.07132	0.12383
query-expansion	0.37604	0.07718	0.12527
Word+Matrix	0.38118	0.07627	0.13177
QMSCI	**0.40016**	**0.0865**	**0.13878**

weighted concepts and ranked weighted sentences, we rank sentences according to the concept importance of the sentence, the sentence-query similarity and sentence centrality in the graph, so the generated summary is more correlated with the query's need. We also study the influence of different summary length to the summary quality, we define the summary size as 100,150,200,250,300 respectively, because the ROUGE-1 and ROUGE-SU4 are considered as the most correlated metrics compared with our human comprehension, we use ROUGE-1 and ROUGE-SU4 metrics to test the influence of the summary length using the DUC 2006 dataset. The experimental result is as follows in Fig. 5.

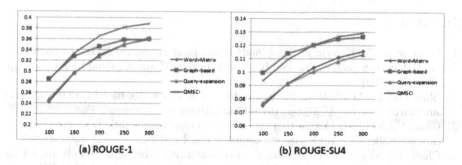

(a) ROUGE-1

(b) ROUGE-SU4

Fig. 5. Comparison result of different summary length using ROUGE-1 and ROUGE-SU4 metrics in duc2006

It's obvious that with the increase of the summary length, the quality of the summary is increased. Because with the increase of the summary length, the summary contains more important information, and the quality of the summary is increased. Other phenomenon is that with the length of the summary increasing, our method has more advantages. It may because the basic language element we consider is concept, with the increase of the summary length, more important concepts are added to the summary, while other methods may add more redundancy. There is another phenomenon, in the ROUGE-SU4 metric, when the summary length is less 200, the extended graph-based method outperforms our method, when the summary size increases, the advantage of our method can merge.

5   Conclusion and Future Work

In this article, we propose a new query-focused multi-document summarization method based on concept importance. First we use a bipartite-graph model to get ranked weighted concepts, then we use the weighted concepts to help to rank sentences in a hyper-graph model. We are the first to utilize the concept-concept and concept-sentence relation and reinforcement to get query-focused summarization. The experimental results demonstrate the effectiveness of our proposed method. In the future, we want to do some research on how to reduce the spatial complexity and time complexity of this algorithm. And we want to improve the sentence-concept mapping method by considering the whole document content and not only the sentence to adjust the mapping process more tailored to document summarization and help to generate more accurate summary.

Acknowledgments. This research is supported by the 863 project of China (2013AA013300), National Natural Science Foundation of China (Grant No. 61375054 and 61402045), Tsinghua University Initiative Scientific Research Program Grant No. 20131089256, and Cross fund of Graduate School at Shenzhen, Tsinghua University (Grant No. JC20140001).

References

1. Barzilay, R., McKeown, K. R., Elhadad, M.: Information fusion in the context of multi-document summarization. In: Association for Computational Linguistics, pp. 550–557 (1999)
2. Ellaachia, A., Al-Dhelaan, M.: Multi-document hyperedge-based ranking for text summarization. In: ACM, pp. 1919–1922 (2014)
3. Chali, Y., Joty, S.R.: Exploiting syntactic and shallow semantic kernels to improve random walks for complex question answering. In: IEEE, pp. 123–130 (2008)
4. Chali, Y., Joty, S.R.: Improving the performance of the random walk model for answering complex questions. In: Association for Computational Linguistics, pp. 9–12 (2008)
5. Goldstein, J., Mittal, V., Carbonell, J., Kantrowitz, M.: Multi-document summarization by sentence extraction. In: Association for Computational Linguistics, pp. 40–48 (2000)
6. Hu, X., Zhang, X., Lu, C., Park, E.K., Zhou, X.: Exploiting wikipedia as external knowledge for document clustering. In: ACM, pp. 389–396 (2009)
7. Kleinberg, J.M.: Authoritative sources in a hyperlinked environment. JACM **3**(2), 604–632 (1999)
8. Knight, K., Marcu, D.: Summarization beyond sentence extraction: a probabilistic approach to sentence compression. In: Artificial Intelligence, pp. 91–107 (2002)
9. Lin, C.-Y.: Rouge: a package for automatic evaluation of summaries. In: Proceedings of the ACL-2004 workshop (2004)
10. Lin, C.-Y., Hovy, E.: Automatic evaluation of summaries using n-gram co-occurrence statistics. In: Association for Computational Linguistics, pp. 71–78 (2003)

11. Lin, J., Madnani, N., Dorr, B.J.: Putting the user in the loop: interactive maximal marginal relevance for query-focused summarization. In: Association for Computational Linguistics, pp. 305–308 (2010)
12. Luhn, H.P.: The automatic creation of literature abstracts. IBM J. Res. Dev. **2**(2), 159–165 (1958)
13. McKeown, K., Klavans, J., Hatzivassiloglou, V., Barzilay, R., Eskin, E.: Mining topic-level influence in heterogeneous networks. In: AAAI, pp. 453–460 (1999)
14. Miao, Y., Li, C.: Enhancing query-oriented summarization based on sentence wikification. In: Workshop of the 33rd Annual International (2010)
15. Milne, D.: Computing semantic relatedness using wikipedia link structure. In: Citeseer (2007)
16. Otterbacher, J., Erkan, G., Radev, D.R.: Using random walks for question-focused sentence retrieval. In: Association for Computational Linguistics, pp. 915–922 (2005)
17. Otterbacher, J., Erkan, G., Radev, D.R.: Cross-domain collaboration recommendation. In: Information Processing and Management, pp. 42–54 (2009)
18. Ramanathan, K., Sankarasubramaniam, Y., Mathur, N., Gupta, A.: Document summarization using wikipedia. Proceedings of the First International Conference on Intelligent Human Computer Interaction, pp. 254–260. Springer, New Delhi (2009)
19. Zhao, L., Wu, L., Huang, X.: Using query expansion in graph-based approach for query-focused multi-document summarization. In: Information Processing and Management, pp. 35–41 (2009)
20. Zheng, H.-T., Gong, S.-Q., Guo, J.-M., Wu, W.-Z.: Exploiting conceptual relations of sentences for multi-document summarization. In: Li, J., Dong, X.L., Dong, X.L., Yu, X., Sun, Y., Sun, Y. (eds.) WAIM 2015. LNCS, vol. 9098, pp. 506–510. Springer, Heidelberg (2015). doi:10.1007/978-3-319-21042-1_51
21. Zhou, L., Lin, C. Y., Hovy, E.: A be-based multi-document summarizer with query interpretation (2005)

Mirror on the Wall: Finding Similar Questions with Deep Structured Topic Modeling

Arpita Das[1(✉)], Manish Shrivastava[1], and Manoj Chinnakotla[2]

[1] International Institute of Information Technology, Hyderabad, India
`arpita.das@research.iiit.ac.in, m.shrivastava@iiit.ac.in`
[2] Microsoft, Hyderabad, India
`manojc@microsoft.com`

Abstract. Internet users today prefer getting precise answers to their questions rather than sifting through a bunch of relevant documents provided by search engines. This has led to the huge popularity of Community Question Answering (cQA) services like Yahoo! Answers, Baidu Zhidao, Quora, StackOverflow *etc.*, where forum users respond to questions with precise answers. Over time, such cQA archives become rich repositories of knowledge encoded in the form of questions and user generated answers. In cQA archives, retrieval of similar questions, which have already been answered in some form, is important for improving the effectiveness of such forums. The main challenge while retrieving similar questions is the "lexico-syntactic" gap between the user query and the questions already present in the forum. In this paper, we propose a novel approach called "Deep Structured Topic Model (DSTM)" to bridge the lexico-syntactic gap between the question posed by the user and forum questions. DSTM employs a two-step process consisting of initially retrieving similar questions that lie in the vicinity of the query and latent topic vector space and then re-ranking them using a deep layered semantic model. Experiments on large scale real-life cQA dataset show that our approach outperforms the state-of-the-art translation and topic based baseline approaches.

Keywords: Community question answering · Machine learning · Deep neural networks

1 Introduction

Community Question Answering (cQA) services such as Yahoo! Answers, Baidu Zhidao, Quora, StackOverflow *etc.* build a community of "experts" on various topics of interest. They provide a platform where any internet user can pose a question on specific topics and get answers from other users or experts. Compared to search engines, cQA forums offer the following advantages - (a) they provide the exact answer to user questions in stead of having to sift through a list of web pages (b) they provide answers to opinion based questions which are usually answered by other users based on their personal experience. Though,

© Springer International Publishing Switzerland 2016
J. Bailey et al. (Eds.): PAKDD 2016, Part II, LNAI 9652, pp. 454–465, 2016.
DOI: 10.1007/978-3-319-31750-2_36

the answers provided by cQA users are usually of good quality, sometimes they can be noisy and redundant too. However, the noise in the data is offset by other forum meta-data associated with each question such as user provided ratings, categories, best answers ratings *etc*. Over time, cQA forums accumulate an archive of rich knowledge encoded in the form of user question and answers on various topics of interest.

The effectiveness of a cQA forum, depends on its ability to retrieve similar questions, which have already been answered, in response to new questions posed by users. This helps in - (a) Saving the time and attention of experts/users by allowing them to focus on new unanswered questions in stead of repeated questions expressed in a different language. (b) Minimizing the time lag of the user waiting for an answer (c) Preventing question starvation where if a question goes unanswered for a long time, answers to similar questions could be offered to the user and (d) Generating automated answers from the answers provided to similar questions in various cQA forums. The major challenge associated with this problem is bridging the lexico-syntactic gap. Two questions can convey the same meaning even though they are syntactically and lexically different. For example: "Can caffeine contents be harmful to health?" and "Does coffee have adverse effects on health?" are semantically similar.

In this paper, we address the above problem of retrieval of similar questions from cQA archives. Previous approaches in literature have relied on - (a) learning word or phrase level translation models [4, 7, 9] from question-answer pairs in parallel corpora of same language. The similarity function between questions is then defined as the probability of translating a given question into another or (b) learning topic models from question-answer pairs [2, 5, 8]. Here, the similarity between questions, is defined in the latent topic space discovered by the topic model. However, they suffer from the following short-comings - (a) Large parallel sentence-aligned question corpora is usually a scarce resource. This limits the effectiveness of translation-based approaches. (b) Topic based approaches usually perform well in improving recall but sacrifice on precision. Recently, Deep Structured Semantic Models (DSSM) [3] have shown state-of-the-art performance when compared to translation models and topic models for learning latent semantic representations. DSSM leverages large volumes of click-through data available to commercial search engines to learn a low-dimensional representation of the queries and documents. This is done by discriminatively training a neural network which maximizes the conditional likelihood of the clicked documents given a query. Further, Convolutional DSSM (CDSSM) adds a convolutional layer to the DSSM structure that projects each word within a context window to a local contextual feature vector. Semantically similar words within a context are projected to vectors that are close to each other in the contextual feature space.

In this paper, we propose a novel approach called "Deep Structured Topic Model (DSTM)" which leverages the strengths of both deep learning based methods and topic models. In the training phase, DSTM learns - (a) A topic model, Latent Dirichlet Allocation (LDA) [1], for discovering the latent topics from the question-answer archive and (b) CDSSM semantic model for modeling the hidden semantic similarities between question-answer pairs. Given a new user question, DSTM employs a two-step process consisting of initially retrieving similar questions that lie in the vicinity of the query in the latent topic vector space discovered by LDA and then re-ranking them using the CDSSM model. Since, topic models such as LDA are expected to have better recall, the initial step will ensure that most relevant questions are filtered from the entire corpus. Later, this filtered set is re-ranked using CDSSM such that the most semantically similar questions are ranked higher in the order. Thus, we combine the strengths of both topic models and deep semantic models. Our contributions are (1) We enhance the topic modeling approach using deep neural networks. We employ deep semantic models to capture semantic similarity between the question pairs with same topic distribution. (2) Qualitative and quantitative analysis of our results using large scale Yahoo! dataset and comparison with other state-of-the-art methods.

The rest of the paper is organized as follows. Section 2 presents our work in the context of related work. Section 3 describes the process of filtering relevant questions from corpus using LDA. Section 4 gives more details on Deep Semantic Modeling. Section 5 gives details of our system such as architecture and training methodology. Section 6 describes the experimental set-up, details of the evaluation datasets, evaluation metrics and also presents the quantitative and qualitative results. Finally, Sect. 7 concludes the paper.

2   Related Work

[4] visualized question retrieval problem as word-based translation problem. They utilized the similarity between answers in the archives to estimate translation probabilities. [7] incorporated query likelihood language model and word based translation model to enhance the performance. In order to capture contextual information, [9] proposed phrase based translation model for similar question retrieval. These models use question answer pairs as parallel corpus.

Another way of approaching this retrieval problem is topic modeling. [2] exploited latent topic information of the question content. They infused "category"(every question in cQA data have a category and a set of sub-categories) information in the topic space to find the questions semantically nearer to the query. [5] proposed Question-Answer Topic Model (QATM) to learn the latent topics aligned across the question-answer. They also introduced QATM with posterior regularization (QATM-PR) to prevent the topic vector representation of question-answer pair to be dominated by the answers which are much longer in length than the questions. [8] used supervised question-answer topic model. They infused language model with topic model for matching the questions on both term and topic levels.

[10] learned word embeddings of questions with "category" information. They employed a fisher kernel to convert the variable length question vectors to a fixed size representation. Our work is aligned towards the topic modeling and deep semantic models [3,6].

3   LDA Based Selection of Relevant Questions

Traditionally, LDA [1] has given good results for related question selection in cQA forums [2,5,8]. LDA assumes that a document exhibits multiple topics. In this problem setting each document is a question and its best answer pair. Assume, there are M documents in the corpus and every document is composed of words W. N is the total number of words in a particular document and K is the total number of topics to be considered. LDA is a probabilistic model where observed variables are the bag of words per document and the hidden random variables are θ, ϕ and $Z(i, j)$ where θ is the topic distribution per document, ϕ is the distribution of the vocabulary per topic and $Z(i, j)$ is the topic of the jth word in the ith document. It can be considered as a unsupervised model where documents are unlabeled and no information of topic distribution is known beforehand. Our goal is to calculate the posterior probability of the hidden variables given the observations. The graphical model of LDA can be seen in Fig. 1. Documents are represented as random mixtures over latent topics, where each topic is characterized by a distribution over words. The generative process of LDA is used to find the joint probability of the observed variables and hidden variables. It is carried out as follows:

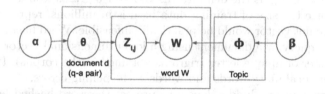

Fig. 1. Graphical model of generative LDA.

1. Choose $\theta_i \sim \text{Dir}(\alpha)$ where $i \in \{1,\ldots,M\}$ and $\text{Dir}(\alpha)$ is the Dirichlet distribution for parameter α
2. Choose $\phi_k \sim \text{Dir}(\beta)$ where $k \in \{1,\ldots,K\}$
3. For each word position (i, j) where $j \in \{1,\ldots,N_i\}$, and $i \in \{1,\ldots,M\}$
   (a) Choose a topic $z_{i,j} \sim \text{Multinomial}(\theta_i)$.
   (b) Choose a word $w_{i,j} \sim \text{Multinomial}(\phi_{z_{i,j}})$.

After finding the joint probability of hidden and observed variables we can find the posterior or conditional probability of the hidden variable given the observed one by Eq. 1.

$$P(\beta_{1:K}, \theta_{1:K}, Z_{1:K} | w_{1:M}) = \frac{P(\beta_{1:K}, \theta_{1:K}, Z_{1:K}, w_{1:M})}{P(w_{1:M})} \qquad (1)$$

The posterior probability is used to obtain the topic vector distribution of each document. Thus, each document can be represented in latent topic space.

But, considering the noisy and redundant nature of data found in cQA sites it would not be a good idea to assume that semantically identical questions will have similar topic distribution.

As shown in Table 1, few questions retrieved for the query "How the human species evolved?" like "Who is the most mature species in planet?" or "How old is human being?" are not semantically identical to the query in spite of having similar topic distribution. This means that topic models may not be capturing the relevant semantic content of the query entirely. To rectify this problem we need to employ a model which takes deeper semantic similarity into account. This leads us to develop DSTM.

4   Deep Semantic Modeling

DSSM [3] and CDSSM [6] are the deep neural networks that map raw textual features into vectors in semantic space. [3] have shown that these models work with click-through data consisting of query and set of clicked documents. It is interesting to note that the input to these networks is not high dimensional term vectors of documents. Instead, word hashing involving letter n-gram is used to reduce the dimensionality of term vectors. For a word, say, "book" represented as (#book#) where # is the delimiter used, letter 3-grams would be #bo, boo, ook, ok#. Since the size of training data used is in millions, representing every word with one hot vector would be practically infeasible. Word hashing allows us to represent a query or a document using a lower dimensional vector (dimension equal to number of unique letter trigrams obtained in the corpus). It also takes care of out-of-vocabulary words and words with spelling errors.

DSSM uses multiple hidden layers to project the word-hashed features into the semantic space. The semantic relatedness between query and document is scored on the basis of cosine similarity between them. DSSM proposes a supervised training method to learn the weight matrices and the bias vectors of the deep neural network (DNN). The objective of DSSM is to maximize the conditional likelihood of the clicked documents given the queries or to minimize the loss function in Eq. 2.

$$L(\Lambda) = -log \prod_{(Q,D^+)} P(D^+|Q) \qquad (2)$$

where Λ denotes the set of parameters of the DNN, Q is the query, D^+ is the set of documents clicked for query Q and $P(D^+|Q)$ is the posterior probability of a clicked document given a query from the semantic relevance score between them using a softmax function.

CDSSM [6] introduces a convolutional layer in addition to DSSM. The advantage of CDSSM is that it considers words at a contextual level and projects each word within a context to a local contextual feature vector. Further, CDSSM uses a max pooling layer to extract the crucial local features to form a fixed-length global feature vector. Affine transformations and element-wise non-linear functions is then applied to the global feature vectors to extract highly non-linear and effective features. Both DSSM and CDSSM captures the non-linear semantic structures between a query and a document. These models significantly outperformed the existing Web retrieval models.

5 Deep Structured Topic Modeling (DSTM)

Inspired by the success of DSSM and CDSSM in the fields of web retrieval, we propose DSTM or Deep Structured Topic Model which aims to capture non linear semantic relatedness and topic information between questions in cQA forums. The system architecture is showed in Fig. 2.

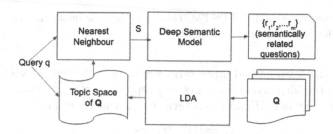

Fig. 2. System architecture of DSTM model.

We use LDA to calculate the latent topic distribution of each document (a question and its best answer) in our dataset. Each question Q is represented with a topic vector of the question-best answer pair $(Q\text{-}BA)$ obtained using LDA. The question Q in the document d has a topic space representation: $[t_1, t_2, \ldots, t_K]$ where $t_j \in [0,1]$ is the indicator variable denoting whether the topic t_j is assigned to document or not, $j = \{1,2,\ldots,K\}$ and K is the total number of topics. Given a query q we find its n nearest neighbors in the topic space using cosine similarity. The candidate set is $C = \{r_1, r_2, \ldots, r_n\}$, where each r_i is a relevant candidate question retrieved. Set S is the set of pairs of queried question and the retrieved candidates r_i present in C.

$$S = \{(q, r_1), (q, r_2), \ldots, (q, r_n)\} \tag{3}$$

As explained earlier, questions with same topic distribution may not be semantically similar thus we employ DNN to capture non linear semantic relations between them.

The term vector of question pairs in S are word hashed [3] and fed to the DNN. It maps the word hashed vectors to their corresponding semantic concept vectors. For testing, the term vector of retrieved question r_i is calculated using both the question and its answer (best answer and most voted answers) pair. The similarity between the query q and a candidate question r_i is calculated using cosine similarity of their corresponding semantic concept vectors.

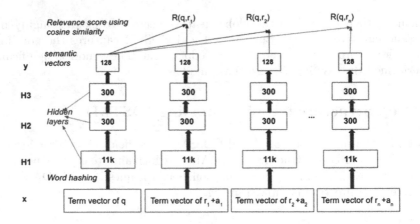

Fig. 3. Deep semantic model of DSTM. Here q is the new query and $r_i + a_i$ are the nearest question-answer pairs given by LDA.

Let x be the word hashed input term vector, y is the output vector and h is the number of hidden layers used. Let, H_j represents the jth intermediate layer whose weight matrix is W_j and bias term is b_j, where $j = \{1, 2, \ldots, h\}$.

$$l_j = f(W_j H_{j-1} + b_j) \tag{4}$$

where $j = \{2,3,...,h\}$ and $H_1 = W_1 x$

$$y = f(W_h H_{h-1} + b_h) \tag{5}$$

where we use tanh as the activation function.

$$f(x) = \frac{1 - e^{-2x}}{1 + e^{-2x}} \tag{6}$$

The relevance score between query and candidate questions is:

$$R(q, r_i) = \frac{y_q^T y_{r_i}}{|y_q||y_{r_i}|} \tag{7}$$

The training of neural network is explained in the next section. After the training is complete the questions pairs in S are fed to the network. We select the questions with the highest relevance scores with the query. For a query we output those questions that are semantically similar in the topic space. The testing phase of deep neural network in DSTM is shown in Fig. 3.

For Convolutional Deep Structured Topic Model (CDSTM) we use a convolutional layer that captures local context of word within a sliding window. Every word of a document is represented by a count vector of its letter-trigrams. The word hashed vectors of the words in a window are concatenated to form a context window vector. Global feature vector of each query or document is obtained by max pooling which retains only the important features from each context window. Non linear projections of global feature gives the semantic vector. The remaining operations are the same as in DSTM.

5.1   Training the Network

Large-scale standard dataset of similar questions is not available. Therefore, we cannot train the network to maximize the conditional likelihood of a similar question given the question. However, richer information is available to us in cQA data in terms of questions, associated answers and metadata associated with them. We train the network for a question while looking for semantic similarity with their best answers. We train our deep neural network with the questions and their answer pairs (we consider only best answer and most voted answers as valid answers to the question). We learn the parameters of our network W_i and b_i to maximize the conditional likelihood of the corresponding answers given the question. The posterior probability of the answer given the question is:

$$P(A|Q) = \frac{\exp(\gamma R(Q, A))}{\sum_{A' \in A} \exp(\gamma R(Q, A'))} \tag{8}$$

where γ is the smoothing factor of the softmax function, $\mathbf{A}$ is the set of user-given answers to question Q.

Ideally all the answers provided to the query must be considered, but since the answers are human generated there is a lot of noise involved with them. For example, a query "Whats the best way to delete a computer virus?" (in Yahoo! Answers) has answers like "setup your computer.", "throw it out your window.", "Throw the computer in the garbage and buy a new one." etc. These answers are irrelevant and do not give any valid solution to the problem. So, in order to filter out noise and redundancy, we approximate $\mathbf{A}$ using A^+ which contains the best answer and the most voted answers for a query Q and A^- which contains any five answers from a random question other than Q. Finally, the problem boils down to the minimization of the loss function in Eq. 9.

$$L(\Lambda) = -log \prod_{(Q, A^+)} P(A^+|Q) \tag{9}$$

6   Experiments

We collected Yahoo! Answers dataset from Yahoo! Labs Webscope. It has about 5 million questions where each question contains title, description, best answer, most voted answers and meta-data like categories, sub categories etc. The questions in this dataset are distributed across a number of categories. Each question was pre-processed by lower-casing, stemming, stopword and special character removal.

We randomly selected a set of thousand questions from the above Yahoo! dataset as our test set. In order to obtain the true labels for the test set, we split the test set into 10 batches containing 100 questions each. For each question in a batch, we retrieved the top 20 similar questions as options using BM25 ranking algorithm in Lucene scoring function. Each batch of questions was distributed to three different human judges. Each judge was asked to vote for the most relevant question expressing the same intent from the provided options. Since each batch was distributed to three different annotators, majority of their votes was taken to judge relevance of a question given a query.

On this gold data, we evaluate the performance of all the models using four evaluation criteria: Mean Average Precision (MAP), Mean Reciprocal Rank (MRR), R-Precision (R-Prec) and Precision at K (P@K).

6.1 Results

In this section, a comparative study of the previous methods is done with respect to our method. The baseline methods considered are query likelihood language model (LM) and BM25. To show how translation based methods perform we implemented the papers [4,7,9]. [4] deals with word based translation, [7] enhanced the first by adding language model to it and [9] implements phrase based translation method. As seen in Table 2, the translation based methods outperform the baseline significantly. The models are trained using GIZA++ tool with the question and best answer pair as the parallel corpus. We also implemented the papers using unsupervised and supervised topic modeling approaches presented in [5,8] respectively. The topic-based models form the current state-of-the-art. Again it is visible from the Table 2 topic based approaches shows comparable performance with translation based methods but they show significant improvement over baseline. We have used Sent2Vec to evaluate the results shown by DSSM and CDSSM [3,6]. These models outperforms the traditional methods but DSTM and CDSTM outperforms them. This proves that latent topic information enhances the deep semantic models in [3,6].

We also evaluated DSTM and CDSTM models in the dataset released by [8]. The comparison of performance are shown in Table 3. The better performance of DSTM and CDSTM on this new dataset proves their stability. Our method which combines topic modeling and deep semantic pruning outperforms the other approaches significantly. The improvement could be attributed to two factors - (1) We assume that semantically similar questions may not have similar topic distribution and (2) We model the non-linear semantic dependencies between the question and best answer pair and utilize it to output the most relevant similar questions given a query.

6.2 Qualitative Analysis

In Table 1, examples are shown to depict performance of supervised topic based method and DSTM. In Q1, the questions that DSTM outputs are semantically much closer, for example "How our society has evolved?" or "Does history of human evolution begin from two persons?". These are very relevant to the person who wants to know "How the human species evolved?". In Q2, the query is about the best way to boot a Windows CD. The questions retrieved by both the models are related to "computers", "booting" etc. but DSTM really understands that the question is about making a "Windows" CD bootable. DSTM is able to capture that Win98se, WinXP, WinProfessional are related to Windows.

Table 1. This table compares the performance of supervised topic model and DSTM. DSTM performs better in Q1–3 while topic model shows better performance in Q4.

Q1: How the human species evolved?	
Supervised Topic Model	1.Except human, what is the most evolved and intelligent species on this planet? 2.How old is the human species ? 3.Who is the most mature species on the planet?
Our Model	1.How our society has evolved? 2.Does history of human evolution begin from two persons? 3.Is it real that humans evolved from monkeys ?
Q2:What's the best way to create a bootable windows CD?	
Supervised Topic Model	1.How do I create a bootable winxp dvd? 2.How can I make a cd from the C-drive? 3.How to make a bootable OS CD?
Our Model	1.How to make windows bootable cd for win98se? 2.How do I create a bootable winxp dvd? 3.How can I create a bootable winxppro cddisk?
Q3: I am looking for a new bike... what's the best bike for on-road exercise?	
Supervised Topic Model	1.What should be my new bike? 2.What is the maintenance for my new bike? 3.What is the best website to buy a new bike?
Our Model	1.What should I buy for a mountain bike to get into sports? 2.Buying a bike, any ideas? 3.Looking for a two seater?
Q4: What's the least expensive way to make a movie using local talent ?	
Supervised Topic Model	1.What was the most expensive movie to make ? 2.How can I make a movie with a camcorder and a bunch of friends ? 3.Hi, I need to make a movie of my times in college. Please help..
Our Model	1.What was the most expensive movie to make ? 2.What is the basic thing to make a movie ? 3.Will they ever make a dark tower movie prolly someday ?

In Q3, the user want suggestions for buying a new bike, the supervised topic model retrieves question that deals with a bike in general i.e., queries dealing with maintenance and website of new bikes are retrieved. DSTM model output questions that are exclusive about buying a new bike, also it is interesting to note that it can capture that "two seater" and bike are semantically similar.

Q4 is an instance where the performance of supervised topic model is better than DSTM. Both the models understand that the query is about making movies but most of the questions retrieved by topic model are more aligned towards

Table 2. Results on Yahoo dataset. The best results obtained by CDSTM are bold faced. The difference between the results of DSTM,CDSTM and other methods are statistically significant with p < 0.05.

Method	MAP	MRR	R-Prec	P@5
LM	0.355	0.392	0.238	0.204
BM25	0.360	0.399	0.275	0.262
Translation(word) [4]	0.385	0.431	0.301	0.292
Translation+LM [7]	0.399	0.464	0.296	0.343
Translation(phrase) [9]	0.479	0.486	0.380	0.361
Q-A topic model [5]	0.463	0.477	0.365	0.351
Q-A topic model(s) [8]	0.474	0.482	0.368	0.359
DSSM [3]	0.513	0.659	0.388	0.386
CDSSM [6]	0.515	0.661	0.387	0.392
DSTM	0.531	0.574	0.392	0.401
CDSTM	**0.532**	**0.574**	**0.393**	**0.420**

Table 3. Results on dataset released by [8]. $TBLM_{SQATM}$ is the model introduced by [8]. The best results obtained by CDSTM are bold faced.

Method	MAP	MRR	R-Prec	P@5
$TBLM_{SQATM}$[8]	0.805	0.889	0.718	0.831
DSTM	0.834	0.903	0.759	0.870
CDSTM	**0.836**	**0.912**	**0.759**	**0.877**

how movies can be made in the least expensive way (using camrecorder, casting friends, making college life movie etc.) than DSTM.

7   Conclusion

In this paper, we proposed a novel approach called "Deep Structured Topic Model (DSTM)" to bridge the lexico-syntactic gap between the question posed by the user and forum questions. DSTM employs a two-step process consisting of initially retrieving similar questions that lie in the vicinity of the query and latent topic vector space and then re-ranking them using a deep layered semantic model. We showed that questions with same topic distribution may not be semantically identical and hence deep semantic models help in enhancing the performance of topic models. We evaluated our approach using large scale datasets from real-world cQA forums and showed that DSTM and CDSTM performs better than other state-of-the-art baseline approaches. Qualitative analysis of our results reveals that DSTM is indeed able to leverage the advantages of both topic modeling approaches such as LDA and also deep semantic models such as CDSSM. As part of future work, we would like to combine the metadata information and the topic vector representation of the questions to train DSTM and CDSTM to find out if metadata can enhance our existing models.

References

1. Blei, D.M., Ng, A.Y., Jordan, M.I.: Latent dirichlet allocation. J. Mach. Learn. Res. **3**, 993–1022 (2003)
2. Cai, L., Zhou, G., Liu, K., Zhao, J.: Learning the latent topics for question retrieval in community QA. In: Fifth International Joint Conference on Natural Language Processing, IJCNLP 2011, pp. 273–281 (2011)
3. Huang, P., He, X., Gao, J., Deng, L., Acero, A., Heck, L.P.: Learning deep structured semantic models for web search using clickthrough data. In: 22nd ACM International Conference on Information and Knowledge Management, CIKM 2013, pp. 2333–2338 (2013)
4. Jeon, J., Croft, W.B., Lee, J.H.: Finding similar questions in large question and answer archives. In: Proceedings of the 2005 ACM CIKM International Conference on Information and Knowledge Management, pp. 84–90. ACM (2005)
5. Ji, Z., Xu, F., Wang, B., He, B.: Question-answer topic model for question retrieval in community question answering. In: 21st ACM International Conference on Information and Knowledge Management, CIKM 2012, pp. 2471–2474 (2012)
6. Shen, Y., He, X., Gao, J., Deng, L., Mesnil, G.: Learning semantic representations using convolutional neural networks for web search. In: 23rd International World Wide Web Conference, WWW, pp. 373–374 (2014)
7. Xue, X., Jeon, J., Croft, W.B.: Retrieval models for question and answer archives. In: Proceedings of the 31st Annual International ACM SIGIR Conference on Research and Development in Information Retrieval, SIGIR 2008, pp. 475–482 (2008)
8. Zhang, K., Wu, W., Wu, H., Li, Z., Zhou, M.: Question retrieval with high quality answers in community question answering. In: Proceedings of the 23rd ACM International Conference on Conference on Information and Knowledge Management, CIKM, pp. 371–380 (2014)
9. Zhou, G., Cai, L., Zhao, J., Liu, K.: Phrase-based translation model for question retrieval in community question answer archives. In: Proceedings of the 49th Annual Meeting of the Association for Computational Linguistics: Human Language Technologies, pp. 653–662 (2011)
10. Zhou, G., He, T., Zhao, J., Hu, P.: Learning continuous word embedding with metadata for question retrieval in community question answering. In: Proceedings of the 53rd Annual Meeting of the Association for Computational Linguistics and the 7th International Joint Conference on Natural Language Processing of the Asian Federation of Natural Language Processing, ACL 2015, pp. 250–259 (2015)

An Efficient Dynamic Programming Algorithm for STR-IC-STR-IC-LCS Problem

Daxin Zhu[1], Yingjie Wu[2], and Xiaodong Wang[3]([⊠])

[1] Quanzhou Normal University, Quanzhou 362000, China
dex@qztc.edu.cn
[2] Fuzhou University, Fuzhou 350002, China
yjwu@fzu.edu.cn
[3] Fujian University of Technology, Fuzhou 350108, China
wangxd135@139.com

Abstract. In this paper, we consider a generalized longest common subsequence problem, in which a constraining sequence of length s must be included as a substring and the other constraining sequence of length t must be included as a subsequence of two main sequences and the length of the result must be maximal. For the two input sequences X and Y of lengths n and m, and the given two constraining sequences of length s and t, we present an $O(nmst)$ time dynamic programming algorithm for solving the new generalized longest common subsequence problem. The time complexity can be reduced further to cubic time in a more detailed analysis. The correctness of the new algorithm is proved.

Keywords: Longest common subsequence problem · Dynamic programming · Substring inclusion constraints · Time complexity

1 Introduction

The longest common subsequence (LCS) problem is a well-known measurement for computing the similarity of two strings. It can be widely applied in diverse areas, such as file comparison, pattern matching and computational biology [2–6,8,9].

Given two sequences X and Y, the longest common subsequence (LCS) problem is to find a subsequence of X and Y whose length is the longest among all common subsequences of the two given sequences.

For some biological applications some constraints must be applied to the LCS problem. These kinds of variants of the LCS problem are called the constrained LCS (CLCS) problem. Recently, a new kind of generalized forms of the constrained LCS problem, the generalized constrained LCS (GC-LCS) problem was proposed by Chen and Chao [1]. For the two input sequences X and Y with their lengths n and m, respectively, and a constrained string P of length r, the generalized constrained LCS problem can be formulated as a set of four problems to find the longest common subsequence of X and Y including/excluding P as a subsequence or a substring, respectively.

© Springer International Publishing Switzerland 2016
J. Bailey et al. (Eds.): PAKDD 2016, Part II, LNAI 9652, pp. 466–477, 2016.
DOI: 10.1007/978-3-319-31750-2_37

In this paper, we consider a more general constrained longest common subsequence problem called STR-IC-STR-IC-LCS, in which two constraining sequence of length s and t must be included as substrings of two main sequences and the length of the result must be maximal. We will present the first efficient dynamic programming algorithm for solving this problem.

The organization of the paper is as follows.

In the following 4 sections, we describe our presented dynamic programming algorithm for the STR-IC-STR-IC-LCS problem.

In Sect. 2 the preliminary knowledge for presenting our algorithm for the STR-IC-STR-IC-LCS problem is discussed. In Sect. 3 we give a new dynamic programming solution for the STR-IC-STR-IC-LCS problem with time complexity $O(nmst)$, where n and m are the lengths of the two given input strings, and s and t the lengths of the two constraining sequences. In Sect. 4, the time complexity is further improved to $O(nms)$. Some concluding remarks are in Sect. 5.

2 Characterization of the Generalized LCS Problem

A sequence can be characterized as a string of characters over some alphabet $\sum$. A subsequence is obtained by deleting some characters (not necessarily contiguous), zero or more, from the original sequence X. A substring is a subsequence of X and its characters are successive within X.

For a given sequence $X = x_1 x_2 \cdots x_n$ with length n, its ith character can be denoted as $x_i \in \sum$ for any $i = 1, \cdots, n$. $X[i : j] = x_i x_{i+1} \cdots x_j$ is used to denote the substring of X from position i to j. The substring $X[i : j] = x_i x_{i+1} \cdots x_j$ is designated as a proper substring of X, if $i \neq 1$ or $j \neq n$. If $i = 1$ or $j = n$, then the substring $X[i : j] = x_i x_{i+1} \cdots x_j$ is referred to as a prefix of X or a suffix of X, respectively.

An appearance of sequence $X = x_1 x_2 \cdots x_n$ in sequence $Y = y_1 y_2 \cdots y_m$, for any X and Y, starting at position j is a sequence of strictly increasing indexes $i_1, i_2, \cdots, i_n$ such that $i_1 = j$, and $X = y_{i_1}, y_{i_2}, \cdots, y_{i_n}$. A compact appearance of X in Y starting at position j is the appearance of the smallest last index i_n. A match for sequences X and Y is a pair (i, j) such that $x_i = y_j$. The total number of matches for X and Y is denoted by δ. It is obvious that $\delta \leq nm$.

In the description of our new algorithm, a function σ will be mentioned frequently. For any string S and a fixed constraint string P, the length of the longest suffix of S that is also a prefix of P is denoted by function $\sigma(P, S)$.

The symbol $\oplus$ is also used to denote the string concatenation.

For example, if $P = aaba$ and $S = aabaaab$, then substring aab is the longest suffix of S that is also a prefix of P, and therefore $\sigma(P, S) = 3$.

It is readily seen that $S \oplus P = aabaaabaaba$.

For the two input sequences $X = x_1 x_2 \cdots x_n$ and $Y = y_1 y_2 \cdots y_m$ of lengths n and m, respectively, and two constrained sequences $P = p_1 p_2 \cdots p_s$ and $Q = q_1 q_2 \cdots q_t$ of lengths s and t, the STR-IC-STR-IC-LCS problem is to find a constrained LCS of X and Y including both P and Q as its substrings.

Definition 1. *Let* $Z(i, j, k, r)$ *denote the set of all LCSs of* $X[1 : i]$ *and* $Y[1 : j]$ *such that for each* $z \in Z(i, j, k, r)$, $\sigma(P, z) = k$ *and* $\sigma(Q, z) = r$, *where* $1 \leq i \leq n, 1 \leq j \leq m, 0 \leq k \leq s$, *and* $0 \leq r \leq t$. *The length of an LCS in* $Z(i, j, k, r)$ *is denoted as* $f(i, j, k, r)$.

Definition 2. *Let* $V(i, j, r)$ *denote the set of all LCSs of* $X[1 : i]$ *and* $Y[1 : j]$ *such that for each* $v \in V(i, j, r)$, $\sigma(P, v) = r$, *where* $1 \leq i \leq n, 1 \leq j \leq m, 0 \leq r \leq s$. *The length of an LCS in* $V(i, j, r)$ *is denoted as* $h(i, j, r)$.

Definition 3. *For each character* $x \in \sum$ *and any* $1 \leq i \leq n, 1 \leq j \leq m, 0 \leq k \leq s, 0 \leq r \leq t$, *let*

$$\alpha(i, j, k, r) = \max \left\{ f(i - 1, j - 1, c, d) | c \in \mu(P, i, k), d \in \mu(Q, i, r) \right\} \quad (1)$$

where, $\mu(P, i, k) = \{c | 0 \leq c \leq s, \sigma(P, P[1 : c] \oplus x_i) = k\}$ *and* $\mu(Q, i, r) = \{d | 0 \leq d \leq t, \sigma(Q, Q[1 : d] \oplus x_i) = r\}$.

The indices of $c, 0 \leq c \leq s$ *and* $d, 0 \leq d \leq t$, *achieving the maximum are denoted as* $\beta(P, i, j, k, r)$ *and* $\beta(Q, i, j, k, r)$ *respectively, i.e.,*

$$\alpha(i, j, k, r) = f(i - 1, j - 1, \beta(P, i, j, k, r), \beta(Q, i, j, k, r)) \quad (2)$$

The following theorem characterizes the structure of $f(i, j, k, r)$, the length of an LCS in $Z(i, j, k, r)$, for any $1 \leq i \leq n, 1 \leq j \leq m, 0 \leq k \leq s$, and $0 \leq r \leq t$.

Theorem 1. *If* $Z[1 : l] = z_1, z_2, \cdots, z_l \in Z(i, j, k, r)$, *then the following formula holds:*
For any $1 \leq i \leq n, 1 \leq j \leq m, 0 \leq k \leq s$, *and* $0 \leq r \leq t$, $f(i, j, k, r)$ *can be computed by the following recursive formula* (3).

$$f(i, j, k, r) = \begin{cases} \max \left\{ f(i - 1, j, k, r), f(i, j - 1, k, r) \right\} & if\ x_i \neq y_j \\ f(i - 1, j - 1, k, r) & if\ k > 0 \wedge x_i \neq p_k \\ & \vee r > 0 \wedge x_i \neq q_r \\ \max \left\{ f(i - 1, j - 1, k, r), 1 + \alpha(i, j, k, r) \right\} & otherwise \end{cases}$$
$$(3)$$

The boundary conditions of this recursive formula are
$f(i, 0, 0, r) = f(0, j, 0, r) = 0$ for any $0 \leq i \leq n, 0 \leq j \leq m$, and $0 \leq r \leq t$
and
$f(i, 0, k, r) = f(0, j, k, r) = -\infty$ for any $0 \leq i \leq n, 0 \leq j \leq m, 1 \leq k \leq s$, and $0 \leq r \leq t$.

Proof. 1. In the case of $x_i \neq y_j$, there are two subcases to be distinguished.
(1.1) If $z_l \neq x_i$, then $Z[1 : l]$ must be a common subsequence of $X[1 : i-1]$ and $Y[1 : j]$ including both $P[1 : k]$ and $Q[1 : r]$ as its suffixes. It is obvious that $Z[1 : l]$ is also an LCS of $X[1 : i - 1]$ and $Y[1 : j]$ including both $P[1 : k]$ and $Q[1 : r]$ as its suffixes, i.e., $Z[1 : l] \in Z(i - 1, j, k, r)$.

(1.2) If $z_l \neq y_j$, then $Z[1:l]$ must be a common subsequence of $X[1:i]$ and $Y[1:j-1]$ including both $P[1:k]$ and $Q[1:r]$ as its suffixes. It is obvious that $Z[1:l]$ is also an LCS of $X[1:i]$ and $Y[1:j-1]$ including both $P[1:k]$ and $Q[1:r]$ as its suffixes, i.e., $Z[1:l] \in Z(i,j-1,k,r)$. Therefore, in the case of $x_i \neq y_j$, we have,
$$f(i,j,k,r) = \max\{f(i-1,j,k,r), f(i,j-1,k,r)\}.$$

2. In the case of $k > 0$, $x_i = y_j$ and $x_i \neq p_k$, if $x_i = y_j = z_l$, then $z_l \neq p_k$, and thus $P[1:k]$ is not a suffix of $Z[1:l]$. Similarly, in the case of $r > 0$, $x_i = y_j$ and $x_i \neq q_r$, if $x_i = y_j = z_l$, then $z_l \neq q_r$, and thus $Q[1:r]$ is not a suffix of $Z[1:l]$. Therefore, we have $x_i = y_j \neq z_l$, and $Z[1:l]$ must be a common subsequence of $X[1:i-1]$ and $Y[1:j-1]$ including both $P[1:k]$ and $Q[1:r]$ as its suffixes, i.e., $Z[1:l] \in Z(i-1,j-1,k,r)$. Therefore, in this case we have, $f(i,j,k,r) = f(i-1,j-1,k,r)$.

3. In the remaining cases of $x_i = y_j$, There are also two subcases to be distinguished.

(3.1) If $x_i = y_j \neq z_l$, then $Z[1:l]$ must be a common subsequence of $X[1:i-1]$ and $Y[1:j-1]$ including both $P[1:k]$ and $Q[1:r]$ as its suffixes. It is obvious that $Z[1:l]$ is also an LCS of $X[1:i-1]$ and $Y[1:j-1]$ including both $P[1:k]$ and $Q[1:r]$ as its suffixes, i.e., $Z[1:l] \in Z(i-1,j-1,k,r)$, and thus $f(i,j,k,r) = f(i-1,j-1,k,r)$ in this subcase.

(3.2) If $x_i = y_j = z_l$, then $Z[1:l-1]$ must be a common subsequence of $X[1:i-1]$ and $Y[1:j-1]$ including both $P[1:c]$ and $Q[1:d]$ as its suffixes, where $c \in \mu(P,i,k), d \in \mu(Q,i,r)$. Therefore,

$$l-1 \leq \alpha(i,j,k,r) \qquad (4)$$

On the other hand, for any $z' \in Z(i-1,j-1,c,d)$ and $c \in \mu(P,i,k), d \in \mu(Q,i,r)$, $z' \oplus x_i$ is a common subsequence of $X[1:i]$ and $Y[1:j]$ including both $P[1:k]$ and $Q[1:r]$ as its suffixes and thus

$$l \geq 1 + \alpha(i,j,k,r) \qquad (5)$$

Combining (4) and (5) we have $l = 1 + \alpha(i,j,k,r)$. Therefore, we have
$$f(i,j,k,r) = \max\{f(i-1,j-1,k,r), 1+\alpha(i,j,k,r)\}.$$
The proof is completed. $\qquad\qquad\square$

The formula (3) can only be applied to solve the STR-IC-STR-IC-LCS Problem in the case when the given problem has a solution including both P and Q as its suffixes. If we extend the strings $P = p_1p_2\cdots p_s$ and $Q = q_1q_2\cdots q_t$ to the generalize strings $P = p_1p_2\cdots p_s p_{s+1}$ and $Q = q_1q_2\cdots q_t q_{t+1}$ respectively, and let $p_{s+1} = q_{t+1} = *$ be a wildcard character, which can match zero or any number of characters, then we can use formula (3) to find the solutions of the given problem. In this case, the length of the constrained LCS of X and Y containing both P and Q as its substrings must be

$$\max\{f(n,m,s,t), f(n,m,s+1,t), f(n,m,s,t+1), f(n,m,s+1,t+1)\}$$

3   A Simple Dynamic Programming Algorithm

Based on the recursive formula (3), our algorithm for computing $f(i, j, k, r)$ is a standard dynamic programming algorithm which can be implemented as the following Algorithm 1.

Algorithm 1. DP

Input: Strings $X = x_1 \cdots x_n$, $Y = y_1 \cdots y_m$ of lengths n and m, respectively, and two constrained sequences $P = p_1 p_2 \cdots p_s p_{s+1}$ and $Q = q_1 q_2 \cdots q_t q_{t+1}$ of lengths $s + 1$ and $t + 1$, where $p_{s+1} = q_{t+1} = *$ is a wildcard character, which can match zero or any number of characters.

Output: $f(i, j, k, r)$, the length of an LCS of $X[1 : i]$ and $Y[1 : j]$ including both $P[1 : k]$ and $Q[1 : r]$ as its suffixes, $1 \le i \le n, 1 \le j \le m, 0 \le k \le s + 1$, and $0 \le r \le t + 1$.

1: **for all** i, j, k, r , $0 \le i \le n, 0 \le j \le m, 0 \le k \le s + 1$ and $0 \le r \le t + 1$ **do**
2:     $f(i, 0, k, r), f(0, j, k, r) \leftarrow -\infty, g(i, 0, 0, 0), g(0, j, 0, 0) \leftarrow 0$ {boundary condition}
3: **end for**
4: **for all** i, j, k, r , $1 \le i \le n, 1 \le j \le m, 0 \le k \le s + 1$ and $0 \le r \le t + 1$ **do**
5:     **if** $x_i \ne y_j$ **then**
6:         $f(i, j, k, r) \leftarrow \max\{f(i - 1, j, k, r), f(i, j - 1, k, r)\}$
7:     **else if** $k > 0$ **and** $x_i \ne p_k$ **or** $r > 0$ **and** $x_i \ne q_r$ **then**
8:         $f(i, j, k, r) \leftarrow f(i - 1, j - 1, k, r)$
9:     **else**
10:        $f(i, j, k, r) \leftarrow \max\{f(i - 1, j - 1, k, r), 1 + \alpha(i, j, k, r)\}$
11:    **end if**
12: **end for**

To implement our new algorithm efficiently, the most important thing is to compte $\mu(P, i, k)$ and $\mu(Q, i, r)$ for each $x_i, 1 \le i \le n$ and $0 \le k \le s + 1$ and $0 \le r \le t + 1$ efficiently.

We consider how to compute $\sigma(P, P[1 : k] \oplus x_i)$ first. It is obvious that $\sigma(P[1 : k] \oplus x_i) = k + 1$ for the case of $x_i = p_{k+1}$. It will be more complex to compute $\sigma(P[1 : k] \oplus x_i)$ for the case of $x_i \ne p_{k+1}$. In this case the length of matched prefix of P has to be shortened to the largest $r < k$ such that $p_{k-r+1} \cdots p_k = p_1 \cdots p_r$ and $x_i = p_{r+1}$. Therefore, in this case, $\sigma(P[1 : k] \oplus x_i) = r + 1$.

This computation is very similar to the computation of the prefix function in KMP algorithm for solving the string matching problem [3, 7].

For the constraint string P of lengths s, its prefix function kmp can be pre-computed in $O(t)$ time as follows.

With this pre-computed prefix function kmp, the function $\sigma(P, P[1 : k] \oplus ch)$ for each character $ch \in \sum$ and $1 \le k \le s$ can be described as follows.

Algorithm 2. Prefix Function

[!t] **Input:** String $P = p_1 \cdots p_s$
Output: The prefix function kmp of P
1: $kmp(0) \leftarrow -1$
2: **for** $i = 2$ to s **do**
3:    $k \leftarrow 0$
4:    **while** $k \geq 0$ **and** $p_{k+1} \neq p_i$ **do**
5:       $k \leftarrow kmp(k)$
6:    **end while**
7:    $k \leftarrow k + 1$
8:    $kmp(i) \leftarrow k$
9: **end for**

If we pre-compute a table λ_P of the function $\sigma(P, P[1 : k] \oplus ch)$ for each character $ch \in \sum$ and $1 \leq k \leq s$, then we can speed up the computation of $\mu(P, i, k)$.

Algorithm 3. $\sigma(P, k, ch)$

Input: String $P = p_1 \cdots p_s$, integer k and character ch
Output: $\sigma(P, P[1 : k] \oplus ch)$
1: **while** $k \geq 0$ **and** $p_{k+1} \neq ch$ **do**
2:    $k \leftarrow kmp(k)$
3: **end while**
4: **return** $k + 1$

The time cost of above preprocessing algorithm is obviously $O(s|\Sigma|)$. By using this pre-computed table λ_P, the value of function $\sigma(P, P[1 : k] \oplus ch)$ for each character $ch \in \sum$ and $1 \leq k \leq s$ can be computed readily in $O(1)$ time.

Preprocessing computations of $\sigma(Q, Q[1 : r] \oplus x_i)$ and thus the table λ_Q is similar.

With these preprocessing, the function $\alpha(i, j, k, r)$ can be described as follows.

It is obvious that with these preprocessing, the algorithm requires $O(nmst)$ time and space. For each value of $f(i, j, k, r)$ computed by algorithm DP, the corresponding LCS z of $X[1 : i]$ and $Y[1 : j]$ including both $P[1 : k]$ and $Q[1 : r]$ as its suffixes, can be constructed by backtracking through the computation paths from (i, j, k, r) to $(0, 0, 0, 0)$.

4 Improvements of the Algorithm

S. Deorowicz [3] proposed the first quadratic-time algorithm for the STR-IC-LCS problem. A similar idea can be exploited to improve the time complexity of our dynamic programming algorithm for solving the STR-IC-STR-IC-LCS problem. The improved algorithm is also based on dynamic programming with some preprocessing. To show its correctness it is necessary to prove some more structural properties of the problem.

Algorithm 4. $\lambda_P(1:s, ch \in \Sigma)$

Input: String $P = p_1 \cdots p_s$, alphabet Σ
Output: A table λ_P

1: **for all** $a \in \Sigma$ and $a \neq p_1$ **do**
2:     $\lambda_P(0, a) \leftarrow 0$
3: **end for**
4: $\lambda_P(0, p_1) \leftarrow 1$
5: **for** $r = 1$ to $s - 1$ **do**
6:     **for all** $a \in \Sigma$ **do**
7:         **if** $a = p_{r+1}$ **then**
8:             $\lambda_P(r, a) \leftarrow r + 1$
9:         **else**
10:             $\lambda_P(r, a) \leftarrow \lambda_P(kmp(r), a)$
11:         **end if**
12:     **end for**
13: **end for**

Algorithm 5. $\alpha(i, j, k, r)$

Input: Integers i, j, k, r
Output: $\alpha(i, j, k, r)$

1: $\alpha \leftarrow -\infty$
2: **if** $k = 0$ and $r = 0$ **then**
3:     $\alpha \leftarrow f(i - 1, j - 1, 0, 0)$
4: **else if** $k = 0$ and $r > 0$ **then**
5:     **for** $d = 0$ to $s + 1$ **do**
6:         **if** $\lambda_P(d, x_i) = r$ and $f(i - 1, j - 1, k, d) > \alpha$ **then**
7:             $\alpha \leftarrow f(i - 1, j - 1, k, d)$
8:         **end if**
9:     **end for**
10: **else if** $k > 0$ and $r = 0$ **then**
11:     **for** $d = 0$ to $t + 1$ **do**
12:         **if** $\lambda_Q(d, x_i) = k$ and $f(i - 1, j - 1, d, r) > \alpha$ **then**
13:             $\alpha \leftarrow f(i - 1, j - 1, d, r)$
14:         **end if**
15:     **end for**
16: **else**
17:     **for** $c = 0$ to $s + 1$ **do**
18:         **for** $d = 0$ to $t + 1$ **do**
19:             **if** $\lambda_P(c, x_i) = k$ and $\lambda_Q(d, x_i) = r$ and $f(i - 1, j - 1, c, d) > \alpha$ **then**
20:                 $\alpha \leftarrow f(i - 1, j - 1, c, d)$
21:             **end if**
22:         **end for**
23:     **end for**
24: **end if**
25: **return** α

Let $Z[1:l] = z_1, z_2, \cdots, z_l$ be a constrained LCS of X and Y including both P and Q as its substring. Let also $I = (i_1, j_1), (i_2, j_2), \cdots, (i_l, j_l)$ be a sequence of indices of X and Y such that $Z[1:l] = x_{i_1}, x_{i_2}, \cdots, x_{i_l}$ and $Z[1:l] = y_{j_1}, y_{j_2}, \cdots, y_{j_l}$. From the problem statement, there must exist an index $d \in [1, l-t+1]$ such that $P = x_{i_d}, x_{i_{d+1}}, \cdots, x_{i_{d+t-1}}$ and $P = y_{j_d}, y_{j_{d+1}}, \cdots, y_{j_{d+t-1}}$.

Theorem 2. *Let $i'_d = i_d$ and for all $e \in [1, s-1]$, i'_{d+e} be the smallest possible, but larger than i'_{d+e-1}, index of X such that $x_{i_{d+e}} = x_{i'_{d+e}}$. The sequence of indices*

$I' = (i_1, j_1), (i_2, j_2), \cdots, (i_{d-1}, j_{d-1}), (i'_d, j_d), (i'_{d+1}, j_{d+1}), \cdots,$
$(i'_{d+t-1}, j_{d+t-1}), (i_{d+t}, j_{d+t}), \cdots, (i_l, j_l)$
defines the same constrained LCS as $Z[1:l]$.

Proof. From the definition of indices i'_{d+e}, it is obvious that they form an increasing sequence, since $i'_d = i_d$, and $i'_{d+t-1} \leq i_{d+t-1}$. The sequence $i'_d, \cdots, i'_{d+t-1}$ is of course a compact appearance of P in X starting at i_d. Therefore, both components of I' pairs form increasing sequences and for any (i'_u, j_u), $x_{i'_u} = y_{j_u}$. Therefore, I' defines the same constrained LCS as $Z[1:l]$.
The proof is completed.   □

The same property is also true for the jth components of the sequence I. Therefore, we can conclude that when finding a constrained LCS of the given STR-IC-STR-IC-LCS problem, instead of checking any common subsequences of X and Y it suffices to check only such common subsequences that contain compact appearances of P both in X and Y. The number of different compact appearances of P in X and Y will be denoted by δ_x and δ_y, respectively. It is obvious that $\delta_x \delta_y \leq \delta$, since a pair (i, j) defines a compact appearance of P in X starting at ith position and compact appearance of P in Y starting at jth position only for some matches.

Base on Theorem 2, we can reduce the time complexity of our dynamic programming from $O(nmst)$ to $O(nm(s+t))$. The improved algorithm consists of three main stages. In the first stage, both sequences X and Y are preprocessed to determine two corresponding arrays lx and ly. For each occurrence i of the first character p_1 of P in X, the index j of the last character p_s of a compact appearance of P in X is recorded as $lx_i = j$. A similar preprocessing is applied to the sequence Y.

In the algorithm Prep, function $left$ is used to find the index lx_i of the last character p_s of a compact appearance of P.

In the second stage of the improved algorithm, the DP matrices $h(i, j, k)$ for the STR-IC-LCS problem, defined by Definition 2, is computed. The formula (3) can be modified somewhat to solve this problem, since it is a special case of $k = 0$ in the formula (3).

$$h(i, j, r) = \begin{cases} \max\{h(i-1, j, r), h(i, j-1, r)\} & \text{if } x_i \neq y_j \\ \max\{h(i-1, j-1, r), 1 + \alpha(i, j, 0, r)\} & \text{if } x_i = y_j \end{cases} \quad (6)$$

In the last stage, two preprocessed arrays lx and ly are used to determine the final results. To this end for each match (i, j) for X and Y the ends (lx_i, ly_i)

Algorithm 6. Prep

Input: X, Y
Output: For each $1 \leq i \leq n$, the minimal index $r = lx_i$ such that $X[i : r]$ includes P as a subsequence
For each $1 \leq j \leq m$, the minimal index $r = ly_j$ such that $Y[j : r]$ includes P as a subsequence

1: **for** $i = 1$ to n **do**
2:    **if** $x_i = p_1$ **then**
3:       $lx_i \leftarrow left(X, n, i)$
4:    **else**
5:       $lx_i \leftarrow 0$
6:    **end if**
7: **end for**
8: **for** $j = 1$ to m **do**
9:    **if** $y_j = p_1$ **then**
10:       $ly_j \leftarrow left(Y, m, j)$
11:    **else**
12:       $ly_j \leftarrow 0$
13:    **end if**
14: **end for**

Algorithm 7. $left(X, n, i)$

Input: Integers n, i and $X[1 : n]$
Output: The minimal index r such that $X[i : r]$ includes P as a subsequence

1: $a \leftarrow i + 1, b \leftarrow 2$
2: **while** $a \leq n$ and $b \leq s$ **do**
3:    **if** $x_a = p_b$ **then**
4:       $b \leftarrow b + 1$
5:    **else**
6:       $a \leftarrow a + 1$
7:    **end if**
8: **end while**
9: **if** $b > s$ **then**
10:    **return** $a - 1$
11: **else**
12:    **return** 0
13: **end if**

---

of compact appearances of P starting at position i in X and j in Y are read. The length of an STR-IC-STR-IC-LCS, $g(n, m, s + 1, r)$ defined by Definition 2, containing these appearances of P is determined as a sum of three parts. The first part is, for some indices i, j, r', $h(i - 1, j - 1, r')$, the length of u, the constrained LCS which is a prefix of X and Y ending at positions $i - 1$ and $j - 1$, including $Q[1 : r']$ as its suffix. The second part is s, the length of the constrained string P. The third part is $h(lx_i + 1, ly_j + 1, r + 1)$, the length of v, the constrained

---

Algorithm 8. *modify*

Input: The DP matrices $h(i,j,r)$ computed in stage 2
Output: The modified DP matrices $h(i,j,r)$

1: **for** $i = 1$ to n **do**
2:    **for** $j = 1$ to m **do**
3:       **if** $lx_i > 0$ **and** $ly_j > 0$ **then**
4:          **for** $d = 0$ to $t + 1$ **do**
5:             $h(lx_i, ly_j, \zeta(d)) \leftarrow \max\{h(lx_i, ly_j, d), h(i-1, j-1, d) + s\}$
6:          **end for**
7:       **end if**
8:    **end for**
9: **end for**
10: **for** $d = 0$ to $t + 1$ **do**
11:    **for** $i = 1$ to n **do**
12:       **for** $j = 1$ to m **do**
13:          **if** $h(i,j,d) < \max\{h(i-1,j,d), h(i,j-1,d)\}$ **then**
14:             $h(i,j,d) \leftarrow \max\{h(i-1,j,d), h(i,j-1,d)\}$
15:          **end if**
16:          **if** $h(i,j,d) < 0$ **then**
17:             $h(i,j,d) \leftarrow -\infty$
18:          **end if**
19:       **end for**
20:    **end for**
21: **end for**

Algorithm 9. $\zeta(1:r)$

Input: Strings $P = p_1 \cdots p_s$ and $Q = q_1 \cdots q_t$
Output: A table ζ

1: **for** $i = 0$ to $t + 1$ **do**
2:    $\zeta(i) \leftarrow i$
3: **end for**
4: $\zeta(t) \leftarrow t + 1$
5: **for** $i = 1$ to s **do**
6:    **for** $j = 1$ to $t - 1$ **do**
7:       $\zeta(j) \leftarrow \lambda_Q(\zeta(j), p_i)$
8:    **end for**
9: **end for**

LCS which is a suffix of X and Y starting at positions $lx_i + 1$ and $ly_j + 1$, such that $z = u \bigoplus P \bigoplus v$ including $Q[1:r]$ as its suffix.

We will show how to compute the third part for all $0 \leq r \leq t + 1$.

For each $0 \leq r \leq t$, let

$$\zeta(r) = \sigma(Q, Q[1:r] \bigoplus P[1:s]) \tag{7}$$

For each LCS z of $X[1:i]$ and $Y[1:j]$ including $Q[1:r]$ as its suffix, computed in stage 2, if the ends (lx_i, ly_i) of compact appearances of P starting at position i in X and j in Y are valid, then $z' = z \oplus P$ is an LCS including P as a suffix, and including $Q[1 : \zeta(r)]$ as its suffix. Therefore, we can modify the DP matrix $h(i, j, r)$ computed in stage 2 in these ends (lx_i, ly_i) to $h(lx_i, ly_i, \zeta(r)) = h(i - 1, j - 1, r) + s$.

With these modified values as boundary values, a dynamic programming algorithm for the STR-IC-LCS problem can then be applied to find the third part of the value $h(n, m, r) = f(n, m, s + 1, r)$.

The following algorithm $modify$ will modify the DP matrix $h(i, j, r)$ computed in stage 2 to our purpose.

Then a dynamic programming algorithm for the STR-IC-LCS problem is applied to this modified DP matrix. In our improved algorithms, the two dynamic programming algorithms for the STR-IC-LCS problem require both $O(nmt)$ time and space. In the algorithm $modify$, the values of $\zeta(d)$ for all $0 \le r \le t$ can be pre-computed as follows.

The time complexity of this algorithm is obviously $O(st)$. With this pre-computed table, the value of $\zeta(d)$ for each $0 \le d \le t$ can then be computed readily in $O(1)$ time. Therefore, the time complexity of the algorithm $modify$ is also $O(nmt)$.

Since the roles of constrained strings P and Q are symmetric, we can use the same algorithm to the case of the positions of P and Q swapped. In this case, the time and space complexities are $O(nms)$, and thus the total time and space are both $O(nm(s + t))$.

Finally we can conclude that our improved algorithm for solving the STR-IC-STR-IC-LCS problem requires $O(nm(s+t))$ time and $O(nmnm(s+t))$ space in the worst case.

5   Concluding Remarks

We have suggested a new dynamic programming solution for the new generalized constrained longest common subsequence problem STR-IC-STR-IC-LCS. The first dynamic programming algorithm requires $O(nmst)$ in the worst case, where n, m, s, t are the lengths of the four input sequences respectively. The time complexity can be reduced further to cubic time in a more detailed analysis. Many other generalized constrained longest common subsequence (GC-LCS) problems have similar structures. It is not clear that whether the same technique of this paper can be applied to these problems to achieve efficient algorithms. We will investigate these problems further.

References

1. Chen, Y.C., Chao, K.M.: On the generalized constrained longest common subsequence problems. J. Comb. Optim. **21**(3), 383–392 (2011)
2. Crochemore, M., Hancart, C., Lecroq, T.: Algorithms on Strings. Cambridge University Press, Cambridge (2007)

3. Deorowicz, S.: Quadratic-time algorithm for a string constrained LCS problem. Inf. Process. Lett. **112**(11), 423–426 (2012)
4. Deorowicz, S., Obstoj, J.: Constrained longest common subsequence computing algorithms in practice. Comput. Inf. **29**(3), 427–445 (2010)
5. Gotthilf, Z., Hermelin, D., Lewenstein, M.: Constrained LCS: hardness and approximation. In: Ferragina, P., Landau, G.M. (eds.) CPM 2008. LNCS, vol. 5029, pp. 255–262. Springer, Heidelberg (2008)
6. Gotthilf, Z., Hermelin, D., Landau, G.M., Lewenstein, M.: Restricted LCS. In: Chavez, E., Lonardi, S. (eds.) SPIRE 2010. LNCS, vol. 6393, pp. 250–257. Springer, Heidelberg (2010)
7. Gusfield, D.: Algorithms on Strings, Trees, and Sequences: Computer Science and Computational Biology. Cambridge University Press, Cambridge (1997)
8. Peng, Y.H., Yang, C.B., Huang, K.S., Tseng, K.T.: An algorithm and applications to sequence alignment with weighted constraints. Int. J. Found. Comput. Sci. **21**(1), 51–59 (2010)
9. Tang, C.Y., Lu, C.L.: Constrained multiple sequence alignment tool development and its application to RNase family alignment. J. Bioinform. Comput. Biol. **1**, 267–287 (2003)
10. Tseng, C.T., Yang, C.B., Ann, H.Y.: Efficient algorithms for the longest common subsequence problem with sequential substring constraints. J. Complex. **29**, 44–52 (2013)

Efficient Page-Level Data Extraction
via Schema Induction and Verification

Chia-Hui Chang[✉], Tian-Sheng Chen, Ming-Chuan Chen, and Jhung-Li Ding

CSIE, National Central University, Zhongli District, Taiwan
chia@csie.ncu.edu.tw

Abstract. Page-level data extraction provides a complete solution for all kinds of information requirement, however very few researches focus on this task because of the difficulties and complexities in the problem. On the other hands, previous page-level systems focus on how to achieve unsupervised data extraction and pay less attention on schema/wrapper generation and verification. In this paper, we emphasize the importance of schema verification for large-scale extraction tasks. Given a large amount of web pages for data extraction, the system uses part of the input pages for training the schema without supervision, and then extracts data from the rest of the input pages through schema verification. To speed up the processing, we utilize leaf nodes of the DOM trees as the processing units and dynamically adjust the encoding for better alignment. The proposed system works better than other page-level extraction systems in terms of schema correctness and extraction efficiency. Overall, the extraction efficiency is 2.7 times faster than state-of-the-art unsupervised approaches that extract data page by page without schema verification.

1 Introduction

Search results from deep Web are information with high quality and are often useful for many information integration applications. Most researches pursue automated methods that use unannotated Web page as training examples. Therefore, many unsupervised web data extractions also assume a single data-rich section which contains search results such that intelligent mining techniques can be applied for record segmentation from a single page input. For example, [2, 9, 10, 13], and all receive a single page as input and conduct analysis for set detection. We call these approaches as record-level extraction systems.

Since these unsupervised approaches do not require annotated pages as input, many systems also omit the wrapper generation procedure as required by supervised approaches, which receive annotated pages for training. The problem with these wrapper-free approaches is the concern to handle large volume of Web pages. Suppose there are N pages to be processed, wrapper-free record-level extraction systems would have to be executed N times for each page as shown in Fig. 1(a). On the other hand, if a wrapper is generated from a subset of training pages for schema and template induction, the rest can be wrapped quickly using the induced schema and template as shown in Fig. 1(b). As extraction tasks based on known schema and template are usually more

© Springer International Publishing Switzerland 2016
J. Bailey et al. (Eds.): PAKDD 2016, Part II, LNAI 9652, pp. 478–490, 2016.
DOI: 10.1007/978-3-319-31750-2_38

efficient than the complicate analysis process for schema and template induction, potential gain can be expected for unsupervised extraction systems with wrapper generation.

In addition to the scalability concern, another drawback of these page-by-page approaches is that they do not make full use of the information provided by all training pages since the extraction system process page independently. The concern is that the output schema may be different for different input pages even if they are generated by the same Web CGI program. Therefore, an additional step to integrate all extracted data (schema matching) is required for these "wrapper-free" unsupervised extraction systems.

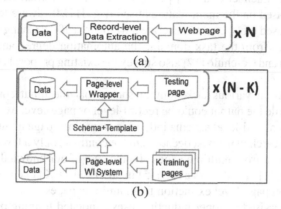

Fig. 1. Two architectures for annotation-free extraction systems without (top) and with (bottom) wrapper generation

On the contrast, when multiple pages are available for training, we are able to induce full schema and provide a total solution for various kinds of extraction needs. The benefit of varieties from multiple pages also allow the system to differentiate the roles of each element and decide whether they are template or data, set or tuple, option or fixed occurrence. That is to say, we have better chance to induce the complete schema and template from multiple pages.

However, inducing the full schema for page-level data extraction is a challenging task that has not been fully explored. In this paper, we propose a page-level schema induction system with wrapper verification procedure to speed up the extraction of large scale web pages. The system, called UWIDE (Unsupervised Wrapper Induction and Data Extraction), is composed of two subsystems. The first subsystem is for page-level schema induction, while the second subsystem is for data extraction and schema verification. The schema induction module conducts set region detection, record boundary segmentation, data alignment and schema generation. The verification module transforms the induced schema into a finite state machine and assigns labels to each leaf node of a test page to achieve data extraction. The experiments show UWIDE outperforms state-of-the-art page-level data extraction system TEX [11], ExAlg [1], and RoadRunner [4] in terms of schema correctness and extraction efficiency.

The rest of the paper is organized as follows: Section 2 introduces related work. Section 3 describes how to induce the full schema from unannotated input pages.

Section 4 explains the verification procedure for testing phase. Effectiveness of schema induction and efficiency of data extraction for test pages are presented in Sect. 5. Finally, Sect. 6 concludes this paper and discusses future work.

2   Related Work

Researches on web information extraction have been lasted for more than a decade. Several survey papers which classify these approaches from various aspects are available. For example, Laender et al. [7] propose taxonomy for characterizing Web data extraction tools based on the main features used (e.g. HTML-aware tools, NLP-based tools, Ontology-based tools, etc.) to generate a wrapper in 2002. Chang et al. [3] classify various approaches from the task domain, the automation degree and the technique. Recently, Sleiman and Corchulo [12] also survey the existing proposals regarding region extractors.

The analysis from various task domain shows that the input could be single or multiple pages while the output could be record-level or page-level extraction target. In fact, unsupervised record-level schema induction from singe page input has become the main stream of research in the past decade. On the contrary, only a few studies focus on page-level extraction from multiple pages. To the best of our knowledge, RoadRunner [4] in 2001, ExAlg [1] in 2003, FiVaTech [5] in 2010 and TEX [11] in 2013 are the only studies focusing on page-level extraction from multiple pages.

Note that, supervised wrapper induction uses annotated training pages to generate wrappers and conduct data extraction for test pages. On the other hand, unsupervised approaches often proceed without wrapper generation since they can extract data directly from any training pages (annotation-free). Thus, no maintenance is required.

Therefore, most wrapper maintenance researches focus on supervised record-level wrappers generated from annotated training pages. Their wrapper verification only checks the validity of extracted data to remedy the situations when the wrappers work normally but the extracted data are invalid. For example, RAPTURE [6] and DATA-PROG [8] are designed for two supervised wrapper induction systems WIEN and STALKER, respectively. For both systems, if the extraction process fails due to template change, the verification of the semantic of data is not necessary.

Reviewing state-of-the-art page-level data extraction systems, TEX does not output schema; ExAlg and FivaTech, despite the generation of schema, does not implement the wrapper module; the only page-level data extraction system with wrapper generation based on induced schema and template is RoadRunner which works as a wrapper to align tag token sequence of an input page with the existing wrapper.

In this paper, we argue that wrapper-free approaches, i.e. Fig. 1(a), may not be efficient for large-scale data extraction. This is because a generated wrapper guided by the induced schema and template, i.e. Fig. 1(b), is usually much faster than the complex analysis procedure for wrapper induction.

3 Schema Induction

The system, UWIDE, is composed of two parts: the first part is page-level schema induction from multiple pages, while the second part is the wrapper generation which is used to verify if a testing page complies with the schema and extract data accordingly as shown in Fig. 2.

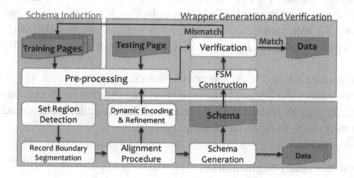

Fig. 2. UWIDE system architecture

3.1 Pre-processing

Our preprocessing step uses CyberNeko HTML Parser to transform the input pages into well-formed format and represent each page by a list of leaf nodes from the DOM (document object model) tree. We work on leaf nodes instead of tag tokens or word tokens since it can greatly reduce the data volume for multi-pages input. Each leaf node is composed of three attributes: path, text, and type. The benefit can be seen directly from the following statistics. For pages with an average of 23 KB, the numbers of word tokens and tag tokens are 3000 and 1130 respectively. In terms of leaf nodes, there are only 258 leaves per page, which is only 8.6 % word tokens.

3.2 Set Region Detection

To locate data-rich sections, many approaches have been proposed. Most approaches operate on token sequences that are abstracted from data records with various values.

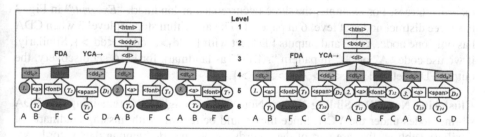

Fig. 3. Example pages P1 & P2

In this paper, we enumerate frequent leaf node patterns in every page for set region detection. For example, the *"Excerpt:"* leaf node with path "<html><body><dl><dd><font>" in page P1 and P2 of Fig. 3 has three and two occurrences respectively, representing a frequent pattern with length 1. For patterns with length longer than one, we encode each leaf node based on their path and content type to enable the discovery of repeat patterns. We categorize text content into five categories including *Number, Date, Time, URL*, and *Email*. If a text segment does not belong to any of these types, we denote it by *String* type. In Fig. 3, all String type leaf nodes T_2, T_5, T_7, T_9, and T_{12} with path "<html><body><dl><dd>" are encoded with code "C", while Date type leaf nodes D_1, D_2, and D_3 with the same path are encoded with a different code "D".

Given the encoded leaf node sequences, we choose the longest and the shortest sequences for pattern mining and remove smaller patterns that are dominated by larger patterns (i.e. the shorter pattern is a sub-string of a longer pattern and having the same occurrence frequency) to focus on maximal patterns.

3.3   Record Boundary Segmentation

Enumerating repeat patterns only detect possible set region. The challenging part is deciding how repeat pattern is used for segmenting record boundary in each data region. Since the record boundary can be somewhere between two adjacent occurrences of a pattern or somewhere inside the pattern leaf nodes, we need some heuristics to determine the record boundary. The idea is based on the following observations of data records in DOM tree structures: (1) records are mutually exclusive, i.e. a leaf node cannot belong to two records at the same time; (2) a record is usually composed of adjacent leaf nodes, forming either a subtree or a forest; (3) all subtrees or forests of these records are rooted at the same level, sharing common ancestors.

Given these observations, our approach for record boundary detection uses the youngest common ancestor (YCA) of all occurrences of a landmark to represent the set region and locates the largest subtree under YCA for each occurrence such that all record subtrees are not overlapped.

FDA Discovery: For each leaf node (called a landmark, L) in a frequent pattern p, we traverse up the path of every occurrence $a_1, a_2, ..., a_m$ of the landmark L in a page until the first common ancestor of two adjacent leaf nodes (not the common ancestor of all occurrences) is met and keep the farthest distinct ancestors (FDAs) at the same level as record subtrees. For example, the three occurrences of landmark *"Excerpt:"* in Fig. 3 has three distinct nodes at level 6 in page P1. The algorithm stops at level 3 when CDA has only one node, <dl>, and outputs FDA set with {<dd_1>,<dd_4>,<dd_6>}. Similarly, if we use code "A" or "B" in pattern "ABFC" as landmark for FDA set discovery, the output FDA set will be {<dt_0>,<dt_3>,<dt_5>}.

Clustering Non-FDA Sibling Nodes: Note that we might have remaining non-FDA sibling nodes under the YCA. Since they could be records with missing landmark, or auxiliary subtrees that are part of the records, or ads inserted among data records, we need to determine if the sibling node is similar to the discovered FDA set with the same tag name. We used the edit distance of the leaf code string in the sibling subtree and the longest common subsequence (LCS) of the FDAs' code strings to calculate the similarity. If the similarity is greater than 0.7, the sibling node is considered a record with optional landmark and is added to the FDA set.

For non-FDA sibling nodes with similarity less than 0.7, they are grouped first by their tag name and clustered to decide their subtree type. With this processing, we can assign these sibling nodes to its neighboring records based on their order in pages. For example, the non-FDA sibling nodes for landmark *"Excerpt:"* in P1 include $\{<dt_0>,<dd_2>,<dt_3>,<dt_5>\}$. Since $<dd_2>$ has the same tag name with FDAs, i.e. $\{<dd_1>,<dd_4>,<dd_6>\}$, we further compute the similarity of the code string *"GD"* with the LCS of the FDAs' code strings, i.e. *"FC"*, and assigned a new cluster to $<dd_2>$ since the similarity is 0. Meanwhile, the three siblings with tag name $<dt>$ all have the same code string *"AB"* and are clustered to the same group $\{<dt_0>,<dt_3>,<dt_5>\}$.

Data Record Assembling: When all subtrees under the YCA is properly clustered, the next step is to assemble the nodes from various clusters to form data records. By examining the order of these clusters in multiple input pages, we can assemble these nodes to form data records. In the above example, we obtain one FDA cluster with $<dd>$ subtrees, two non-FDA clusters with $<dt>$ and $<dd>$ subtrees. The $<dt>$ cluster comes before FDA node (with $<dd>$ cluster) in both P1 and P2, while non-FDA cluster with $<dd>$ subtree comes after all FDA nodes. Therefore, we can assemble $<dt>$ sibling node with the following FDA node and the $<dd>$ sibling node to form data records, generating three data records $\{<dt_0>,<dd_1>,<dd_2>\}$, $\{<dt_3>,<dd_4>\}$ and $\{<dt_5>,<dd_6>\}$ for P1.

Prioritizing FDA Sets: Since there could be more than one landmark or several repeat patterns discovered for a set region, we also need to decide how to handle interleaving FDA sets, i.e. the span of one FDA set overlaps with the span of the second FDA set. To evaluate whether an output FDA set represents a set region well, we define compression ratio (i.e. the number of aligned columns divided by the average number of leaf nodes in a page) and set density (i.e. the number of non-null cells divided by the number of all cells in the set region). If two FDA sets are interleaving, the FDA set with the larger density value divided by the compression ratio will be selected with higher priority.

3.4 Data Alignment

After the segmentation of records, we will be able to conduct alignment based on segmented records from each page for the same data region. Similarly, data alignment has to be applied to non-set region for full page alignment as well.

We apply **multiple string alignment** to find a common presentation for multiple records. We start from the record with the largest number of leaf nodes and define a match scoring function based on leaf nodes' path, type and content. Let w denote the aligned result of k records, $Cmode(w_i)$ denote the content with the highest frequency in

w_i and $Maxcount(w_i)$ denote the count of $Cmode(w_i)$ in w_i. The scoring function of matching the node at position i of w with a new record r at position j, denoted by r_j, is defined as the sum of the matching score for path, type, and content between w_i and r_j.

(1) Path score: If $path(w_i) = path(r_j)$ then return 0.5, or 0 otherwise.
(2) Type score: If $type(w_i) = type(r_j)$ then return 0.25; if $type(w_i)$ and $type(r_j)$ is a parent-child relationship then return 0.125, or 0 otherwise.
(3) Content score:
      a. $Cmode(w_i) = content(r_j)$: If $Maxcount(w_i) = k$ and $k > 1$ then return 1.25 (Full Match), else return 0.25 (Partial Match).
      b. $Cmode(w_i) \neq content(r_j)$: If $Maxcount(w_i) <= 2$ then return 0, else return -0.1 (Mismatch).

Note that we give extra score (1 point) for full match when there are more than one record and a small penalty (-0.1) for mismatch. This is to guide nodes with the same content to be aligned together but still allow the alignment of various values of categorical attributes. Therefore, a match score is either a value between 0 to 1, or a big value 2. With this scoring, leaf nodes with same path, same type, or same content will have larger value, while parent-child type relation is also considered to tolerate different type matching.

For each set region S_i with the aligned column set C_i and a total of $|S_i|$ records, we also have a $|S_i| \times |C_i|$ array. As for non-set region, we obtain a $|P| \times |C|$ array where P denotes the set of input pages and C denotes the set of aligned columns.

3.5   Dynamic Encoding and Refinement

While most of these techniques mentioned above have been applied in various scopes, the result is not always satisfactory. For example, product descriptions in commercial Web pages usually present more diversity and are hard to align since they may come from different vendors. As another example, several attribute value pairs can be incorrectly recognized as set regions. Meanwhile, repeat pattern mining for set region detection also relies on abstraction mechanism which needs to be tuned for input pages. In this paper, we propose a post-processing step to examine the alignment result and suggest dynamic encoding scheme for refinement.

The refinement procedure is triggered by the performance of alignment result. We define column density for a column c, $Cdens$ (c) as the percentage of non-null nodes in the column and set density $Sdens(S_i)$ as the percentage of non-null cells in the aligned array, i.e. $|S_i| \times |C_i|$.

Detecting false positive set: When a set density is lower than a threshold (0.6), it often implies a false positive set being detected. Therefore, we dismiss all segmented records of the set region and include them with non-set region for full page alignment.

Dynamic encoding for decorative tags: For deep Web pages, decorative tags are often used to emphasize query terms and result in lots of optional columns. Therefore, we count the number of low density columns between two template columns. If more than

30 % columns have density less than 0.6, we search for leaf nodes with one more decorative tag than adjacent text leaf nodes in this region. We then record these paths together with the data schema to guide the encoding for test pages.

Merging low-density basic nodes: If low density ratio between two template columns is higher than 0.3 but no decorative tags are found after dynamic encoding, we attribute the cause to inconsistent format from various product descriptions without uniform template. In this case, we will combine columns in this region and record the common path of these columns for testing.

Finally, we generate schema trees from the aligned results. Due to space limitation, we ignore the details and give an example schema tree as shown in Fig. 4.

Set 1					
B_1	B_2	T_1	B_3	B_4	B_5
1.	T_1	Excerpt:	T_2	T_3	D_1
2.	T_4	Excerpt:	T_5		
3.	T_6	Excerpt:	T_7		
1.	T_8	Excerpt:	T_9	T_{10}	D_2
2.	T_{11}	Excerpt:	T_{12}		
3.	T_{13}			T_{14}	D_3

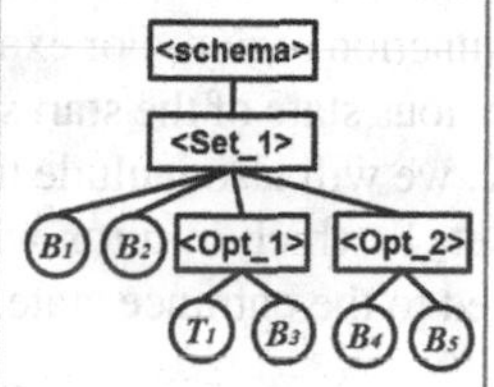

Fig. 4. Schema tree for Fig. 3

4 Verification and Data Extraction

In the extraction phase, the wrapper checks whether a testing page complies with the induced schema and template. The wrapper is a driver program which consists of two main modules: one is the transformation from the induced schema and template to finite state machine (FSM), while the other is the label assignment for each leaf node of an input page. If there exists a label sequence which complies with the FSM, the data could be extracted accordingly. Otherwise, the page can be included to train a better schema.

4.1 Finite State Machine Construction

Given the induced schema and template tree, we first build a state machine with basic transitions between adjacent leaf nodes. Two additional state "S" and "E" are added to denote the start and end state of the state machine (Fig. 5(a)). For each composite data type (i.e. internal nodes), we record the start position and end position of the region and find the entrance and exit states as described below.

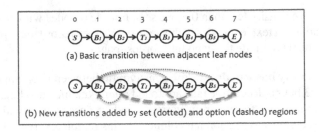

(a) Basic transition between adjacent leaf nodes

(b) New transitions added by set (dotted) and option (dashed) regions

Fig. 5. FSM construction

Finding Entrance Nodes: The major entrance node of a region comes from the start state's previous state, while set regions have additional entrance state from their region end state. However, we also need to consider if such an entrance state is included in an option region or disjunction region. For example in Table 1, the major entrance state for Set_1 include 0 (previous state of the start state) and 6 (end state). Since state 6 is inside option region Opt_2, we will also include the entrance state of Opt_2, i.e. state 4, in the entrance state for Set_1, which in turns will refer to Opt_1 and the entrance state for Opt_2, i.e. 2, is added to the entrance state.

Table 1. The span of each composite data

Region type	Tag ID	Start	End	Entrance	Exit
Set	Set_1	1	6	0, 6, 4, 2	7, 1
Option	Opt_1	3	4	2	5, 7
	Opt_2	5	6	4, 2	7

Finding Exit Nodes: Similarly, the major exit node of a region goes to the next state of its end state. For set regions, additional exit to the region start has to be included. If the added exit state is included in an option region or disjunction region, the exit state for the region has to be added. For example, the exit state of Set_1 in Table 1 is 7 (next state of the end state) and 1 (start state). Since state 1 is not inside any option region, the recursive process stops. For Opt_1, the exit state 5 is inside Opt_2; therefore, we also add the exit state of Opt_2, i.e. state 7, to the exit state of Opt_1.

Given the entrance and exit state set for each composite region, we add transitions for each region according to its data type. (1) If it is a **set region**, we add *entrance transitions* from its entrance nodes to the start state and *exit transitions* from the end state to its exit nodes. (2) If it is an **option region**, we add jump transitions from its entrance states to its exit states. (3) If it is a **disjunction region**, we remove transitions between adjacent states in the region, and add *entrance transitions* from its entrance states to every state in the region and *exit transitions* from every state to its exit states.

4.2   Label Sequence Assignment

We model the task of assigning labels to all leaf nodes as a constrained satisfaction problem, where each leaf node is a variable with a set of candidate labels as its domain and the transitions among states restrict possible state pairs for two consecutive leaf nodes. The steps are briefed below.

First, we list candidate states for each leaf node (after neglecting decorative tags as specified by training procedure) based on its tag path, type and content feature. Note that template states can only be a candidate for leaf nodes with the same tag path and content. For example, given the states specified by Fig. 5, a leaf node with tag path "<html><body><dl><dt>" and a number content can only have B_1 as its candidate state, while a leaf node with tag path "<html><body><dl><dd>" and date content could have B_5 as its candidate. For such leaf node with a single candidate, we can then apply constraint imposed by FSM to reduce impossible candidate states of its adjacent leaf node.

If there exists more than one label sequence that complies with the FSM, the system will choose a label sequence which best fit the input according to state transition probability and content similarity. In such cases, the verification is successful and we can output data for each attribute as shown in Fig. 5. Otherwise, if no label sequence complies with the FSM, then the verification fails.

5   Experiments

The following experiments are designed to show the effectiveness of schema induction as well as the efficiency gain with wrapper generation for page-level (full schema) data extraction.

5.1   Schema Correctness Evaluation

For schema correctness evaluation, we use 234 pages from 9 websites of ExAlg for the following experiments. The first 5 websites contain 82 list pages and the rest 4 websites contain 152 detail pages. We evaluate the effectiveness of schema induction by precision, recall, and F-measure in terms of the number of basic nodes (columns) for each website.

$$P = B_{sys} \cap B_{ans}/B_{sys}, \ R = B_{sys} \cap B_{ans}/B_{sys}, \ F = 2P^*R/(P+R)$$

where B_{ans} denotes the number of basic nodes in the golden dataset and B_{sys} denotes the number of basic nodes extracted by our system

As shown in Fig. 6, UWIDE improves the recall from 66.4 % to 92.0 % with the dynamic encoding and refinement procedure. However, the precision is less than ExAlg because the scoring tends to split a categorical data node into multiple columns. By combining these columns as a disjunctive node via GUI, the precision can be further improved. Figure 6 also shows the result of running RoadRunner and TEX. Since Road-Runner does not handle HTML tables, the performance is very limited. As for TEX, it outputs a lot of basic nodes for both list pages and detailed pages, showing the difficulties for full page alignment.

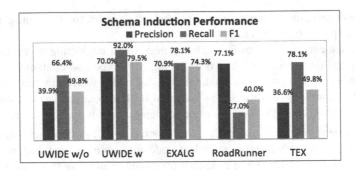

Fig. 6. Schema induction correctness

5.2 Extraction Accuracy Evaluation

Next, we compared our system with TEX using the data set of WEIR which contains 24,028 pages from 40 websites in four domains: soccer, stock, video games, and books. We evaluate the performance in terms of data cells:

$$\text{Accuracy} = ExtC/_{AnsS}$$

where AnsS denotes the number of cells to be extracted for a website in the golden answer provided by WEIR and $ExtC$ denotes the number of leaf nodes (cells) correctly extracted by our system for the website.

Figure 7 shows that UWIDE performs much better than TEX (81 %) with default threshold (94 %) or adjusted threshold (98 %). Adjusting threshold allows us to detect false positive set and trigger full page alignment with higher density. Meanwhile it also avoids unnecessary combination of low density basic nodes.

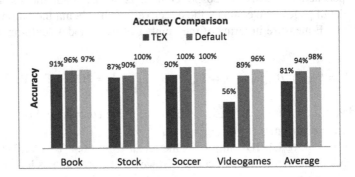

Fig. 7. Accuracy evaluation

5.3 Efficiency Gain with Active Training

In this section, we collected 2,519 pages from 12 websites to show the efficiency gain with generated wrappers. We start with two pages (the largest and smallest) to train a model for each website and test on the remaining pages using the generated wrapper. If some testing pages fail the extraction procedure, we randomly add a failure page for retraining. The process is repeated until all pages pass the extraction/verification procedure. In average, 12.5 pages (14.4 %) are required to train a complete model for a website.

The efficiency gain of the proposed method using the 2-phase (train+test) extraction framework over the 1-phase extraction using all training pages is shown in Fig. 8 where the websites are sorted by the number of input pages. In summary, the speedup is more significant for websites with more pages. However, the page size and the complexity of the schema also affect the speedup. For WEIR dataset, UWIDE needs 0.278 s to process a page when all pages are used for training, while TEX needs 0.408 s to process a page.

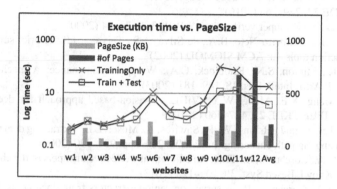

Fig. 8. Efficiency comparison of 1-phase vs 2-phase training on 12 websites

6   Conclusion

In this paper, we argue that combining wrapper induction and data extraction in one step may not be efficient when they have to be repeated many times for large-scale web data extraction. When a few pages are able to support page-level schema induction, we don't need to waste time to analyze all pages, but use the generated schema & wrapper for extracting data from the remaining pages. We confirm this conjecture by designing an unsupervised training module for page-level schema induction and a wrapper module for data extraction from testing pages. The wrapper induction module is an end-to-end solution to identify data records as well as data attributes for each record and page. The wrapper module assigns labels to the leaf node sequence of a test page to complete the extraction step according to the FSM generated from the page-level schema and template.

The experimental results show that our schema induction performs well for wrapper induction and data extraction. In terms of efficiency, the training phase is quite fast since we only need to process leaf nodes. The efficiency gain of the proposed framework is 2.7 times faster than the framework without wrapper generation.

References

1. Arasu, A., Garcia-Molina, H.: Extracting structured data from web pages. In: ACM SIGMOD 2003 San Diego, California, USA, pp. 337–348 (2003)
2. Bing, L., Lam, W., Gu, Y.: Towards a unified solution: Data record region detection and segmentation. In: CIKM 2011, Glasgow, Scotland, UK, pp. 1265–1274 (2011)
3. Chang, C.-H., Kayed, M., Girgis, M.R., Shaalan, K.F.: A survey of Web information extraction systems. IEEE TKDE **18**(10), 1411–1428 (2006)
4. Crescenzi, V., Mecca, G., Merialdo, P.: RoadRunner: Towards automatic data extraction from large web sites. In: VLDB 2001, Roma, Italy, pp. 109–118 (2001)
5. Kayed, M., Chang, C.H.: FiVaTech: Page-level web data extraction from template pages. IEEE TKDE **22**, 249–263 (2010)
6. Kushmerick, N.: Wrapper verification. WWW **3**(2), 79–94 (2000)
7. Laender, A.H.F., Ribeiro-Neto, B.A., de Silva, A.S., Teixeira, J.S.: A brief survey of Web data extraction tools. In: ACM SIGMOD (2002)
8. Lerman, K., Minton, S.N., Knoblock, C.A.: Wrapper maintenance: A machine learning approach. J. Artif. Intell. Res. **18**, 149–181 (2003)
9. Liu, W., Meng, X.F., Meng, W.Y.: ViDE: A vision-based approach for deep web data extraction. IEEE TKDE **22**, 447–460 (2010)
10. Miao, G., Tatemura, J., Hsiung, W.-P., Sawires, A., Moser, L.E.: Extracting data records from the web using tag path clustering. In: WWW 2009, pp. 981–990 (2009)
11. Sleiman, H.A., Corchuelo, R.: TEX: An efficient and effective unsupervised web information extractor. Knowl. Based Syst. **39**, 109–123 (2013)
12. Sleiman, H.A., Corchuelo, R.: A survey on region extractors from documents. IEEE Trans. Knowl. Data Eng. **25**(9), 1960–1981 (2013)
13. Zheng, S., Song, R., Wen, J.-R., Giles, C.L.: Efficient record-level wrapper induction. In: CIKM 2009, pp. 47–56 (2009)

Transfer-Learning Based Model for Reciprocal Recommendation

Chia-Hsin Ting, Hung-Yi Lo, and Shou-De Lin[✉]

Department of Computer Science and Information Engineering,
National Taiwan University, Taipei, Taiwan
{r00922092,d96023,sdlin}@csie.ntu.edu.tw

Abstract. This paper tackles the reciprocal recommendation task which has various applications such as online dating, employee recruitment and mentor-mentee matching. The major difference between traditional recommender systems and reciprocal recommender systems is that a reciprocal recommender has to satisfy the preference on both directions. This paper proposes a simple yet novel regularization term, the Mutual-Attraction Indicator, to model the mutual preferences of both parties. Given such indicator, we design a transfer-learning based CF model for reciprocal recommender. The experiments are based on two real world tasks, online dating and human resource matching, showing significantly improved performance over the original factorization model and state-of-the-art reciprocal recommenders.

Keywords: Reciprocal recommender system · Transfer learning

1 Introduction

1.1 Background

Recommender systems have been successfully used to provide each user with a list of recommended items on many e-commerce websites such as Amazon, Netflix, YouTube and iTunes. In general, the techniques for recommendation systems can be classified into three categories, collaborative filtering (CF), content-based filtering (CBF) and hybrid recommendation [1,2]. CF systems work by collecting explicit user feedbacks in the form of ratings for items and exploit similarities and differences among users in determining the unseen ratings. CBF approaches generate recommendations by matching content descriptions of items to users' interests. In the case of a hybrid recommendation, both information, CF and CBF, are used.

Sometimes due to the concern of the data privacy, users tend to provide the minimal information to these e-commerce services [3]. In this case, it is hard to use the CBF related approaches for the recommendation service. Under such circumstance we need to resort to a CF system, provide useful personal recommendations. CF systems use very little user or item information with the

© Springer International Publishing Switzerland 2016
J. Bailey et al. (Eds.): PAKDD 2016, Part II, LNAI 9652, pp. 491–502, 2016.
DOI: 10.1007/978-3-319-31750-2_39

exception of a partially observed rating matrix. The rating matrix contains ratings of items (columns) by users (rows) in numeric or binary form. For example, one to five stars in Netflix movie recommendation or like vs. dislike [4–6].

Nevertheless, most systems aim at recommending items to people. Recommending people to people has begun to emerge under the name of reciprocal recommender. In a reciprocal recommender, both sides can express their preferences and requirements. Therefore, a perfect match has to satisfy the preferences of both sides. Unfortunately, traditional recommenders only have to satisfy the preferences of one side, cannot be adopted directly for reciprocal recommender systems. Moreover, the opinions for matching are different between both sides (e.g., male-female, mentor-mentee and recruiters-job seekers) However, traditional recommenders did not distinguish user behaviors in different sides.

1.2   Contribution

In order to solve these challenges in reciprocal recommender system, we propose a simple and general reciprocal method to exploit a factorization-based and transfer learning based CF model, taking advantage of the efficient performance on the users' rating data of items in the traditional recommender [7–9]. Our contributions can be summarized as:

- We propose a novel reciprocal recommender model, and argue that such model should optimize the matching quality directly instead of optimizing the traditional metrics for a CF model based on the accuracy of the user-to-item ratings.
- We propose a transfer learning based CF method for transferring user opinions from two-sides in the reciprocal recommender task. To our knowledge, this is the first-ever model for reciprocal recommendation without the requirement of user profile or other meta information.
- In order to model the mutual preferences between two-sides of users of reciprocal online services, we propose a simple and meaningful indicator, the Mutual-Attraction Indicator. Using Mutual-Attraction Indicator, our major contribution lies in the connection between the transfer learning models to the reciprocal recommender task.
- We evaluate the model using two different types of real-world tasks, online dating and human resource job matching. The experiment results show that the proposed approaches and significantly outperform the state-of-the-art recommenders on AUC.

2   Evaluation Criteria Discussion

In previous works, CF models are widely used to predict user's ratings for items. The quality of a CF based recommendation system can be evaluated by using held out data to compare distance between the predicted values and held out values. Root Mean Squared Error (RMSE) is a commonly used metric for CF

(a)

	U1	U2	U3
U1	--	9	9
U2	9	--	5
U3	3	5	--

(b)

	U1	U2	U3
U1	--	4	9
U2	9	--	9
U3	3	9	--

(c)

	U1	U2	U3
U1	--	9	2
U2	9	--	1
U3	2	1	--

Fig. 1. A simple example to highlight the concern of the RMSE-based metrics. Note that (b) is better solution according to RMSE, but (c) is a better reciprocal solution since it recommend the right pair 'U2-U1'

[5,6,15,24] However, in the case of a reciprocal recommender, evaluating the quality of systems by RMSE is problematic. A good reciprocal recommendation should return matching pairs (i.e. pairs which provide high mutual rates to each other) and ignore pairs that none or only one side favors the other. The spirit of such metrics lies in that for pairs that like each other, the system needs to predict their high rating accurately; while for the other cases, it is ok for the system to not predict the ratings perfectly correct as long as it does not produce high ratings for both sides. Unfortunately, such properties cannot be reflected by an rating-based metrics such as RMSE. Take the case of Fig. 1, for example, there are three users (U1,U2,U3) and several observed ground truth ratings among these users for an online dating website. It is apparent that the best matching is U1 and U2 according to Fig. 1(a), since although U1 likes U3, U3 does not like U1, and U2 and U3 do not like each other very much. Assuming two recommender systems, S1 and S2, predict the preference ratings as Fig. 1(b) and Fig. 1(c), respectively. Obviously, S1 performs better than S2 on RMSE. However, according to S1, U2 and U3 should be matched, which is incorrect. On the other hand, S2 suggests U1 and U2 can be recommended to each other, which is indeed accurate despite that it does not predict the remaining ratings correctly. Therefore, in the reciprocal recommender, distinguishing user pairs is good match or not is more important than making accurate rating predictions. Furthermore, a good recommendation can help users find the match partner soon. For each user, a recommender system generates a candidate ranking list. This list, in general, contains some true positive matches and false positive ones. Comparing this list with the ground truth, we can then obtain a receiver operating characteristic (ROC) curve. Hence, in this work, we evaluate the reciprocal models with Area Under the ROC Curve (AUC).

3   Methodology

3.1   Problem Statement

In this section, we formally define our problem. We consider that there are two sets of users S and T. For example, in an online dating system, the two sets of users correspond to male and female; in a human resource system, they correspond to recruiters and job seekers. The sizes of two subsets are n and m respectively. Each user maintains a preference list towards a subset of users in the other set. Therefore, we can separate the feedbacks data into two subsets: the feedbacks from the users in S to the users in T, $S2T$ and the opposite referred as $T2S$. Note that the data have numerical ratings or binary click-through recordings (like/dislike). Given a user $s \in S$, our goal is to recommend S a list of user $Qs \subset T$ such that the feedback from s to t, r_{st}, and the feedback from any $t \in Qs$ to s, $\hat{r}_{st}$, are both high.

3.2   Mutual-Attraction Indicator

Differing from the traditional item to people recommenders, in the reciprocal recommender, both sides can express their likes and dislikes and a good match requires satisfying the preferences of both people. Therefore, we propose a reciprocal regularized term, $Mutual - Attraction\ Indicator\ (MAI)$, to measure whether each pair of people like each other or not. Furthermore, we focus on user's explicit feedbacks in the form of ratings or binary. Thus, if the feedbacks of both sides are high or non-zero, we can assume that there exists a higher chance that they yield a great match. According to this observation, the MAI between user $s \in S$ and user $t \in T$ can be determined from the rating or binary feedbacks of both sides, using the equations:

$$MAI = r_{st} \times \hat{r}_{st} \tag{1}$$

3.3   Reciprocal-Regularized Transfer by Collective Factorization

In this subsection, we will introduce simple and efficient approaches to modify the learning algorithms of Transfer by Collective Factorization (TCF). Previously we argue that a successful user-to-user recommendation should satisfy both sides' preference. That is, the MAI of this user-to-user pair will be high. Based on this objective function, we adjust the collectively learning functions of TCF with MAI, which is called $Reciprocal - Regularized\ TCF(RR - TCF)$.

Here we exploit the transfer learning framework to handle this problem. Transfer-based approaches such as [9] use the data in the auxiliary domain (e.g., click-through data 0/1) expressed in the binary form to help reduce the effect of data sparsity of numerical ratings (e.g., five-star rating data). They proposed a principled matrix-based transfer-learning framework, referred as Transfer by Collective Factorization (TCF), which tri-factorizes the data matrix in three parts: a user-specific latent feature matrix, an item-specific latent feature and two data-dependent matrices, to model user's heterogeneous feedbacks. Here we exploit this idea to transfer the reciprocal preference information into the original CF model to boost its performance. However, in the reciprocal recommendation case, we adjust several settings of TCF to match up the reciprocal recommender system scenario.

As mentioned in Sect. 3.1, we can divide the users into two subsets, S and T. In the feedback data of $S2T$, we have an incomplete matrix $R = [r_{st}]_{n \times m} \in \mathbb{R}^{n \times m}$ with q observed ratings, where $s \in \{1, 2, 3, \ldots, n\}$, $t \in \{1, 2, 3, \ldots, m\}$. Here we use a mask matrix $Y = [y_{st}]_{n \times m} \in \{0, 1\}^{n \times m}$ to denote whether the entry (s, t) is observed ($y_{st} = 1$) or not ($y_{st} = 0$). Similarly, in the $T2S$ data, we can build an incomplete matrix $\hat{R}$ with $\hat{q}$ observed ratings, where $\hat{R} = [\hat{r}_{st}]_{n \times m} \in \mathbb{R}^{n \times m}$. Similar to the $S2T$ data, we have a corresponding mask matrix $\hat{Y} = [\hat{y}_{st}]_{n \times m} \in \{0, 1\}^{n \times m}$. Note that there is one-to-one mapping between the users of R and $\hat{R}$.

Next, there is an assumption that the feedback of a user $s \in S$ on a user $t \in T$ in the $S2T$ data, r_{st}, is generated from the user-specific latent feature vectors, $U_s \in \mathbb{R}^{1 \times d_u}$ and $V_t \in \mathbb{R}^{1 \times d_v}$, and some data-dependency modeled as $B \in \mathbb{R}^{d_u \times d_v}$. For notation simplicity, we fix $d = d_u = d_v$. Therefore, we can denote the tri-factorization in the $S2T$ data as:

$$F(R \sim UBV^T) = \sum_{s=1}^{n} \sum_{t=1}^{m} y_{st} \left[\frac{1}{2}(r_{st} - U_s B V_t^T)^2 + \frac{\alpha_u}{2}\|U_s\|_F^2 + \frac{\alpha_v}{2}\|V_t\|_F^2 \right] + \frac{\beta}{2}\|B\|_F^2 \quad (2)$$

where regularization terms, $\|U_s\|$, $\|V_t\|$ and $\|B\|$, are used to avoid overfitting.

Similar in the $T2S$ data, we have $F(\hat{R} \sim U\hat{B}V)$. To factorize R and $\hat{R}$ collectively, we obtain the following optimization problem for TCF.

$$\min_{U,V,B,\hat{B}} F(R \sim UBV^T) + \lambda \times F(\hat{R} \sim U\hat{B}V) \quad (3)$$

where $\lambda > 0$ is a tradeoff parameter to balance the $S2T$ data and $T2S$ data, $U \in \mathbb{R}_{n \times d}$ and $V \in \mathbb{R}_{m \times d}$.

In order to solve the optimization problem, we can collectively factorize two data matrices of R and $\hat{R}$ to learn U and V, and estimate B and $\hat{B}$ separately [9]. The knowledge of latent features U and V is transferred by collective factorization of the rating matrices R and $\hat{R}$. Next, we will introduce the detail how to learn U and V, and how to estimate B and $\hat{B}$ in each data.

Learning U and V. Given B and V, we have gradient on the latent feature vector U_s of user $s \in S$ as function (4):

$$\frac{[\partial F(R \sim UBV^T) + \lambda \times F(\hat{R} \sim U\hat{B}V)]}{\partial U_s} = -b_s + U_s C_s \quad (4)$$

where

$$C_s = \sum_{t=1}^{m} \left(y_{st} B V_t^T V_t B^T + \lambda \times \hat{y}_{st} \hat{B} V_t^T V_t \hat{B}^T \right) + \alpha_u \sum_{t=1}^{m} \left(y_{st} + \hat{y}_{st} \right) \mathbf{I}$$

$$b_s = \sum_{t=1}^{m} \left(y_{st} B V_t^T V_t B^T + \lambda \times \hat{y}_{st} \hat{B} V_t^T V_t \hat{B}^T \right)$$

Therefore, there is an update rule similar to alternative least square (ALS) approach [23].

$$U_s = b_s C_s^{-1} \tag{5}$$

And we can obtain the update rule for the latent feature vector V_t of the user $t \in T$ similarly.

Learning B and $\hat{B}$. Given U and V, we can estimate B and $\hat{B}$ separately in each data, e.g. for the $S2T$ data,

$$F(R \sim UBV^T) \propto \frac{1}{2}\|Y \odot (R - UBV^T)\|_F^2 + \frac{\beta}{2}\|B\|_F^2 \tag{6}$$

where the data-dependent parameter B can estimated in the same way as that of estimating w in a corresponding least-square SVM problem, where $w = vec(B) = [B_1^T B_2^T \dots B_d^T]^T \in \mathbb{R}^{p \times 1}$ is a big vector concatenated from the columns of matrix B. The instances can be built as (x_{st}, r_{st}) with $y_{st} = 1$, where $x_{st} = vec(U_s^T V_t) \in \mathbb{R}^{d^2 \times 1}$. Therefore, the corresponding least-square SVM problem is obtained as below.

$$\min_w \frac{1}{2}\|r - Xw\|_F^2 + \frac{\beta}{2}\|w\|_F^2 \tag{7}$$

Note that, $X = [\dots x_{st} \dots]^T \in \mathbb{R}^{p \times d^2}$ with $y_{st} = 1$ is the data matrix, and $r \in \mathbb{R}^{p \times 1}$ is the corresponding observed ratings from R. By having gradient on w, we have:

$$w = (X^T X + \beta \mathbf{I})^{-1} X^T r \tag{8}$$

Then we can estimate $\hat{B}$ for the $T2S$ data similarly.

In order to model the reciprocal recommender problem, we modify the function (8), for estimating B in the $S2T$ data. Assume x_{st} is the p'-th row of X, which means user s has reviewed user t. If the user t also reviewed user s, on the other hand, $\hat{y}_{st} = 1$, we will adjust the corresponding element in r as below.

$$r_{p'} = \gamma \times MAI \tag{9}$$

Similarly, we can adjust the function for estimating $\hat{B}$ for the $T2S$ data by this approach.

The MAI term reflects the potential match pair from the observed data. When two users vote each other with a high rating or positive feedback in the binary form, we can adjust the latent factor vectors of both sides with MAI to capture this information. Therefore, the reciprocal-regularized factorization models will predict a higher rating than original CF models. Although doing such can increase the prediction errors of individual ratings, it leads to better chance to recommend matchings satisfying both sides.

3.4 Heuristic for Initialization

It is well-known that good initializations can improve the speed and accuracy of the solutions of many factorization-based models. Therefore, we run TCF on one-side data, and use the results to initialize the values of TCF and RR-TCF. According to function (3), we use the one-side data as rating matrix R. And then we relabel high ratings in $\hat{R}$ as 1 (like) and low ratings as 0 (dislike). Then we use the TCF framework to learn the latent matrices, U and V. Finally, we can use these results as the initialization values of TCF and RR-TCF.

4 Experiment

4.1 Dataset and Evaluation

Online-Dating: The data is provided by a public Czech online-dating website Libimseti.[1] It contains more than 1.5×10^6 ratings with values in $[1, 10]$. Among the users, there are $61,365$ male users, $76,441$ female users and $83,164$ unknown users. The data set used in the experiment is constructed as follows.

- We first divide the original data into two parts: ratings from male to female ($M2F$) and the opposite ($F2M$). Moreover, $M2F$ contains $3,232,064$ reviews of the whole data, and $F2M$ contains $7,991,238$ reviews respectively. We remove the users who have fewer than two reviews.
- We then randomly choose 8/10 of the original data as training set, 1/10 of them as validation and the rest as testing set. Note that the validation set is used to determine the parameters of the model (e.g. the threshold to determine a match).
- Since the data provide only the review ratings between users (no final matching outcome), in this experiment we define a positive pair of users as the male and female who receive the highest ratings from one another. That is, if a male rates a female 10 and vice versa then we consider them as a positive pair, otherwise they will be treated as a negative pair. In this case, there are $53,376$ positive pairs in the dataset and $41,507$ negative pairs.

Human Resource Matching: This data is collected from 104 Corp.,[2] the largest human resource company in Taiwan. The data set contains $6,482,165$ records of the users' applications for jobs (referred as $U2J$) and $958,240$ interview invitations for users from companies (referred as $J2U$). Furthermore, we are provided the ground truth about the outcome of the final match (i.e. whether the users were offered and accepted the jobs eventually). We randomly separate the original data into three subsets: 1/3 for training set, 1/3 for testing set and the rest for validation.

---

[1] http://libimseti.cz/.
[2] http://www.104.com.tw/.

Evaluation Criteria. As argued previously, the reciprocal recommender system should be evaluated based on the mutual-matching quality rather than the rating errors, which is significantly different from the traditional CF problems. We treated the task as a ranking problem. In other words, we want to evaluate whether a recommender system can rank the matched pairs higher. Table 1 shows the perfect-match ratio of these two data.

For each user, recommendation systems generate top-k candidate ranking list. Comparing this list with the ground truth, we can then obtain a receiver operating characteristic (ROC) curve. We exploit the area under curve (AUC) to evaluate the models.

Table 1. Perfect-matching ratio of Libimseti and 104

	#perfect-matching pairs	#reciprocal pair	#perfect-matching ratio
Libimseti	53,376	94,883	0.563
104	1,972	358,641	0.0055

4.2 Baselines

We compare our methods with a non-factorization method, the most popular recommender (MP) and a state-of-the-art factorization model, Matrix Factorization. For the MP method, we always recommend the most popular items to users. Such model has enjoyed its success previously for e-commerce applications. For the MF model, we first generate Matrix Factorization on each one-side data (referred as F2M-MF, M2F-MF, U2J-MF, J2U-MF) and combine the data of both sides to construct single Matrix Factorization model (MF).

4.3 Result and Discussion

In this subsection, we first discuss the effects of initialization values for factorization models and propose an efficient approach to solve this problem; we then demonstrate the effectiveness of our proposed reciprocal-regularized approaches on improving the performance of the original models.

Results. The experiment results on test data (unavailable during training) of Libimseti and 104 are reported in Tables 2 and 3 respectively. And we can make the observations as follows.

- In the experiments on two different test data, reciprocal regularized models perform significantly better than all other non-reciprocal regularized factorization models and baseline.
- For the reciprocal-regularized model, RR-TCF, is much better than the models without reciprocal-regularized on AUC. Conceivably its performance on RMSE is worse than non-reciprocal regularized models; nevertheless, what we

Table 2. Prediction performance on Libimseti

Evaluation metric\model	Baseline MP	Without reciprocal regularized				With reciprocal regularized
		F2M-MF	M2F-MF	MF	TCF	RR-TCF
RMSE	–	1.639	1.890	1.712	1.802	1.824
RMSE(F2M)	–	1.639	–	1.631	1.700	1.778
RMSE(M2F)	–	–	1.890	1.880	2.012	2.290
AUC	0.586	0.643	0.621	0.622	0.678	0.883

Table 3. Prediction performance on 104

Evaluation metric\model	Baseline MP	Without reciprocal regularized				With reciprocal regularized
		U2J-MF	J2U-MF	MF	TCF	RR-TCF
RMSE	–	0.208	0.385	0.384	0.502	0.697
RMSE(U2J)	–	0.208	–	0.436	0.395	0.486
RMSE(J2U)	–	–	0.385	0.278	0.643	0.785
AUC	0.579	0.780	0.825	0.811	0.803	0.871

really care about here is to make a high-perfect-match recommendation than an accurate-rating-prediction recommendation. Using regular MF or TCF yields better rating prediction but much worse matching quality.

Effects of Different Initialization. As mentioned in Sect. 3.4, we apply the initialization processes to improve performance. Table 4 shows the results on Libimseti data after using initialization values. And according to the experiment results, it does improve the performances of TCF and RR-TCF.

Table 4. Meta data initializations for Libimseti

	Random-initialization		Meta-data initialization	
	TCF	RR-TCF	TCF	RR-TCF
RMSE	1.834	2.014	1.802	1.824
RMSE(F2M)	1.733	1.965	1.700	1.778
RMSE(M2F)	2.037	2.296	2.012	2.290
AUC	0.624	0.823	0.678	0.883

5   Related Work

Recently, reciprocal recommendations have attracted various academic investigations. These researches have studied how to do reciprocal recommendations in many reciprocal applications: online dating, mentor-mentee matching, social networks and job recommendation. Therefore, in this section, we review the previous works.

5.1   Reciprocal Recommendation System

There are numerous papers investigated in reciprocal recommendations based on the content-based and hybrid approaches. Pizzato et al. [13] proposed a content based system which used both user profiles and user interactions. To generate recommendations for a given user, they extracted users' explicit preferences from the interactions with the other users and then matched them with the user profiles. Akehurst et al. [14] proposed a recommender system for online dating that combined content-based and collaborative filtering approaches and used both user profiles and user interactions. Cai et al. [10] developed a people recommender system for a social network website where users can reply positive or negative messages to other users. They propose a collaborative filtering approach, called SocialCollab, which considers the preference of both sides. SocialCollab distinguishes two users are similar in attractiveness if they are like by a common group of users, and these two users are similar in taste if they like a common group of users. Moreover, Cai et al. [11] improved the results by using gradient descent to learn the relative contribution of similar users in the ranking of the recommendations provided by SocialCollab. In the same task, the work of Kutty et al. [12] used a model based on tensor decomposition to generate recommendations and also reported improvements over Cai et al. [11]. As mentioned, due to the privacy issue, we do not use user profile in this paper. Therefore, it is hard for us to reproduce these works for comparison.

5.2   Transfer Learning

Furthermore, there is an emerging research topic in recommender systems. Aiming to alleviate the sparsity problem in individual CF domains, it can help us to transfer knowledge among related domains. In [16], Li presents a review of collaborative filtering approaches for cross-domain recommendation. And Li distinguish three types of domains: system domains, data domains, and temporal domains. System domains are the different datasets upon which the recommender systems are built, and in which some kind of transfer learning is performed [17–21]. Data domains are the different representations of user preferences, which can be implicit (e.g. clicks, purchases) or explicit (e.g. ratings) [9, 21]. Finally, temporal domains are subsets in which a dataset is split based on timestamps [22]. In our case, we focus on the rating data. Therefore, we will use the data domain based approach in our work.

6   Conclusion and Future Work

We propose a novel reciprocal recommender task and connect its solution with a transfer CF approach. We demonstrated the usefulness of the proposed models in the domain of online dating and job matching. We believe such algorithm has a wider range of application scenarios such as mentor-mentee matching, doctor-patient matching, etc.

To our knowledge, this is the first ever CF-based solution proposed for a reciprocal recommender task, and we hope this paper can inspire more works in this direction. Future works include exploiting more sophisticated transfer-based recommendation models [25] or to learn the matching directly.

References

1. Sarwar, B., Karypis, G., Konstan, J., Riedl, J.: Item-based collaborative filtering recommendation algorithms. In: World Wide Web, pp. 285–295 (2001)
2. Adomavicius, G., Tuzhilin, A.: Toward the next generation of recommender systems: a survey of the state-of-the-art and possible extensions. IEEE Trans. Knowl. Data Eng. 17(6), 734–749 (2005)
3. Parameswaran, R., Blough, D.M.: Privacy preserving collaborative filtering using data obfuscation. In: Granular Computing, pp. 380–387 (2007)
4. Bennett, J., Lanning, S.: The netflix prize. In: Proceedings of KDD Cup and Workshop (2007)
5. Piotte, M., Chabbert, M.: The pragmatic theory solution to the netflix grand prize. In: Netflix Prize Documentation (2009)
6. Koren, Y.: The bellkor solution to the netflix grand prize. Netflix Prize Documentation 81 (2009)
7. Koren, Y., Bell, R., Volinsky, C.: Matrix factorization techniques for recommender systems. IEEE Comput. Soc. Press 42(8), 30–37 (2009)
8. Kolda, T.G., Bader, B.W.: Tensor decompositions and applications. SIAM Rev. 51(3), 455–500 (2009)
9. Pan, W., Liu, N N., Xiang, E.W., Yang, Q.: Transfer learning to predict missing ratings via heterogeneous user feedbacks. In: IJCAI Proceedings-International Joint Conference on Artificial Intelligence, vol. 22, no. 3, p. 2318 (2011)
10. Cai, X., Bain, M., Krzywicki, A., Wobcke, W., Kim, Y.S., Compton, P., Mahidadia, A.: Collaborative filtering for people to people recommendation in social networks. In: Li, J. (ed.) AI 2010. LNCS, vol. 6464, pp. 476–485. Springer, Heidelberg (2010)
11. Cai, X., et al.: Learning collaborative filtering and its application to people to people recommendation in social networks. In: 2010 IEEE 10th International Conference on Data Mining (ICDM), pp. 743–748. IEEE (2010)
12. Kutty, S., Chen, L., Nayak, R.: A people-to-people recommendation system using tensor space models. In: Proceedings of the 27th Annual ACM Symposium on Applied Computing, pp. 187–192 (2012)
13. Pizzato, L., et al.: RECON: a reciprocal recommender for online dating. In: Proceedings of ACM Conference on Recommender Systems (RecSys). ACM, Barcelona (2010)
14. Akehurst, J., Koprinska, I., Yacef, K., Pizzato, L., Kay, J., Rej, T.: Explicit and implicit user preferences in online dating. In: Cao, L., Huang, J.Z., Bailey, J., Koh, Y.S., Luo, J. (eds.) PAKDD Workshops 2011. LNCS, vol. 7104, pp. 15–27. Springer, Heidelberg (2012)
15. Celma, O., Cano, P.: From hits to niches? or how popular artists can bias music recommendation and discovery. In: Proceedings of 2nd Netflix-KDD Workshop (2008)

16. Li, B.: Cross-domain collaborative filtering: a brief survey. In: 23rd IEEE International Conference on Tools with Artificial Intelligence, pp. 1085–1086 (2011)
17. Li, B., Yang, Q., Xue, X.: Transfer learning for collaborative filtering via a rating-matrix generative model. In: Proceedings of the 26th Annual International Conference on Machine Learning, pp. 617–624. ACM (2009)
18. Cao, B., Liu, N.N.: Transfer learning for collective link prediction in multiple heterogeneous domains. In: Proceedings of the 27th International Conference on Machine Learning, pp. 159–166 (2010)
19. Zhang, Y., Cao, B., Yeung, D.-Y.: Multi-domain collaborative filtering. In: Proceedings of the 26th Conference on Uncertainty in Artificial Intelligence (UAI), pp. 725–732 (2010)
20. Shi, Y., Larson, M., Hanjalic, A.: Tags as bridges between domains: improving recommendation with tag-induced cross-domain collaborative filtering. In: Konstan, J.A., Conejo, R., Marzo, J.L., Oliver, N. (eds.) UMAP 2011. LNCS, vol. 6787, pp. 305–316. Springer, Heidelberg (2011)
21. Pan, W., Xiang, E.W., Liu, N.N., Yang, Q.: Transfer learning in collaborative filtering for sparsity reduction. In: AAAI, pp. 230–235 (2010)
22. Li, B., Zhu, X., et al.: Cross-domain collaborative filtering over time. In: Proceedings of the Twenty-Second International Joint Conference on Artificial Intelligence, vol. 3, pp. 2293–2298. AAAI Press (2011)
23. Bell, R.M., Koren, Y.: Scalable collaborative filtering with jointly derived neighborhood interpolation weights. In: Seventh IEEE International Conference on Data Mining, ICDM 2007, pp. 43–52. IEEE (2007)
24. Yu, H.-F., et al.: Feature engineering and classifier ensemble for KDD cup 2010. In: Proceedings of the KDD Cup 2010 Workshop, pp. 1–16 (2010)
25. Jing, H., Liang, A.-C., Lin, S.-D., Tsao, Y.: A transfer probabilistic collective factorization model to handle sparse data in collaborative filtering. In: 2014 IEEE International Conference on Data Mining (ICDM), pp. 250–259, 14–17 December 2014

Enhanced SVD for Collaborative Filtering

Xin Guan[1](✉), Chang-Tsun Li[1], and Yu Guan[2]

[1] Department of Computer Science, University of Warwick, Coventry, UK
{x.guan,c-t.li}@warwick.ac.uk
[2] School of Computing Science, Newcastle University, Newcastle upon Tyne, UK
yu.guan@ncl.ac.uk

Abstract. Matrix factorization is one of the most popular techniques for prediction problems in the fields of intelligent systems and data mining. It has shown its effectiveness in many real-world applications such as recommender systems. As a collaborative filtering method, it gives users recommendations based on their previous preferences (or ratings). Due to the extreme sparseness of the ratings matrix, active learning is used for eliciting ratings for a user to get better recommendations. In this paper, we propose a new matrix factorization model called Enhanced SVD (ESVD) which combines the classic matrix factorization method with a specific rating elicitation strategy. We evaluate the proposed ESVD method on the Movielens data set, and the experimental results suggest its effectiveness in terms of both accuracy and efficiency, when compared with traditional matrix factorization methods and active learning methods.

Keywords: Matrix factorization · Recommender systems

1 Introduction

Generally speaking, recommender systems provide users with personalized suggestions by predicting the rating or preference that the users would give to an item. They have become increasingly common recently and are used by many internet leaders. Examples include movie recommendation by Netflix [1], web page ranking by Google [2], related product recommendation by Amazon [3], social recommendation by Facebook [4], etc.

Recommender systems are used for generating recommendations to users, usually in one of the two ways: (1) content-based filtering based on the characteristics it has and the item descriptions which could be automatically extracted or manually created, and (2) collaborative filtering that predicts other items the users might like based on the knowledge about preferences (usually expressed in ratings) of users for some items. Due to the limitations of content-based filtering algorithm (sometimes characteristics are unobvious and item descriptions are hard to extract), most recommender systems are based on collaborative filtering.

Collaborative filtering is a technique used to predict the preferences of users according to the same taste or common experiences from others and based on the assumption that people who agreed in the past will also agree in the future

© Springer International Publishing Switzerland 2016
J. Bailey et al. (Eds.): PAKDD 2016, Part II, LNAI 9652, pp. 503–514, 2016.
DOI: 10.1007/978-3-319-31750-2_40

over time. There are two primary strategies to deal with collaborative filtering: the neighborhood approaches and latent factor models. Neighborhood methods [5] concentrate on the relationship between items or users, so they are good at detecting localized relationships (e.g., someone who likes Superman also likes Batman), but fall short of power in detecting the user's overall preference. By transforming both items and users to the same latent space, latent factor models try to explain the ratings by items and users, aiming at making them directly comparable. Latent factor models are good at estimating the overall structure (e.g., a user likes comedy movies), but less effective in analyzing associations among small sets of closely related items.

Matrix factorization methods, such as Non-Negative Matrix Factorization (NMF) [6] and Singular Value Decomposition (SVD) [7], are widely used for constructing a feature matrix for users and for items, respectively. Generally, matrix factorization, as one of the most successful realizations of latent factor models, can produce better accuracy than classic nearest neighbor methods when dealing with product recommendations because of the incorporation of additional information such as implicit feedback and temporal effects [8].

In real life situations, when a new user comes in, most recommender systems would only ask the user to rate a certain number of items (which is a small proportion comparing with the whole set). Therefore the ratings matrices are often extremely sparse, which means there is not enough knowledge to form accurate recommendations for the user. To get precise recommendations for this user, active learning in collaborative filtering is often used to acquire more high-quality data in order to improve the precision of recommendations for the target user. However, traditional active learning methods [9–11] only evaluate each user independently and only consider the benefits of the elicitation to the 'new' user, but pay less attention to the effects of the system. In this work we propose a rating elicitation strategy that improves the accuracy of the whole system by eliciting more ratings for 'existing' users. In some previous works, ratings were elicited one by one per request [9] or user's by user's per request [11]. The consequence is that the model is trained at each request, which is significantly time-consuming. In this paper, we design a series of methods that elicit ratings simultaneously with matrix factorization algorithms. Through this special preprocessing step not only are the computational costs reduced, but also the performance of matrix factorization methods is greatly improved.

The rest of the paper is organized as follows. We start with preliminaries and related work in Sect. 2. Then in Sect. 3 we introduce a new way to apply active learning to recommender systems. A more accurate model is proposed in Sect. 4. Experimental results and analyses are provided in Sect. 5. Section 6 concludes the work.

2   Preliminaries

2.1   Regularized SVD

Normally, matrix factorization methods are used to deal with the issue of missing values in a given ratings matrix. However, ratings matrices are usually extremely sparse. For example, the density of the famous Netflix and Movielens data sets are 1.18 % and 4.61 %, respectively, which means that only a few elements are rated while most of them are unknown. Another challenge is that the data set we use in real world recommender systems is typically of high dimensionality. Due to high sparseness and computational complexity, directly applying SVD algorithms to rating matrices is not appropriate.

In [12], Funk proposed an effective method called regularized SVD algorithm for collaborative filtering which decomposes the ratings matrix into two lower rank matrices. Suppose $R \in \mathbb{R}^{m \times n}$ is the ratings matrix of m users and n items. The regularized SVD algorithm finds two matrices $U \in \mathbb{R}^{k \times m}$ and $V \in \mathbb{R}^{k \times n}$ as the feature matrix of users and items:

$$\tilde{R} = U^T V \tag{1}$$

It assumes that each user's rating is composed of the sum of preferences about the various latent factors of that movie. So each rating r_{ij} the ith user to the jth movie in the matrix R can be represented as:

$$\tilde{r}_{ij} = U_i^T V_j \tag{2}$$

where U_i, V_j are the feature vectors of the ith user and the jth movie, respectively. Once we get the best approximation of U and V, we can obtain the best prediction accordingly. The optimization of U and V can be performed by minimizing the sum of squared errors between the existing scores and prediction values [12]:

$$E = \frac{1}{2} \sum_{i,j \in \kappa} (r_{ij} - \tilde{r}_{ij})^2 + \frac{k_u}{2} \sum_{i=1}^{m} U_i^2 + \frac{k_v}{2} \sum_{j=1}^{n} V_j^2 \tag{3}$$

where κ is a set of elements in the ratings matrix R that have been assigned values, k_u and k_v are regularization coefficients to prevent over-fitting.

To solve the optimization problem like Equation (3), Stochastic Gradient Descent (SGD) is widely used and has shown to be effective for matrix factorization [12–14]. SGD loops through all ratings in the training set κ and for each rating it modifies the parameters U and V in the direction of the negative gradient:

$$U_i \leftarrow U_i - \alpha \frac{\partial E_{ij}}{\partial U_i} \tag{4}$$

$$V_j \leftarrow V_j - \alpha \frac{\partial E_{ij}}{\partial V_j} \tag{5}$$

where α is the learning rate.

Unlike traditional SVD, regularized SVD is just a tool for finding those two smaller matrices which minimize the resulting approximation error in the least square sense. By solving this optimization problem, the end result is the same as SVD which just gets the diagonal matrix arbitrarily rolled into the two side matrices, but could be easily extracted if needed.

2.2   SVD++

Since matrix factorization for recommender systems based on regularized SVD was first proposed, several variants have been exploited with extra information on the ratings matrix. For example, Paterek [13] proposed an improved regularized SVD algorithm by adding a user bias and an item bias in the prediction function. Koren [14] extended the model by considering more implicit information about rated items and proposed a SVD++ model with the prediction function:

$$\tilde{r}_{ij} = u + \beta_i + \gamma_j + V_j^T (U_i + |I(i)|^{(-1/2)} \sum_{k \in I(i)} y_k), \tag{6}$$

where u is the global mean, β_i is the bias of the ith user and γ_j is the bias of the jth item. U_i is learnt from the given explicit ratings, $I(i)$ is the set of items user i has provided implicit feedback for. $|I(i)|^{(-1/2)} \sum_{k \in I(i)} y_k$ represents the influence of implicit feedback. The implicit information enables SVD++ to produce better performance than regularized SVD model.

2.3   Dataset

Movielens is a classic recommender system data set that recommends films to users through collaborative filtering algorithms. It contains 100,000 ratings from 943 users on 1,682 movies and each rating is an integer ranging from 1 to 5 which represents the interests the user has to this movie. All users were selected at random for inclusion and each user selected had rated at least 20 movies. The whole data set is randomly partitioned into training set and test set with exactly 10 ratings per user in the test set. The quality of results is usually measured by their RMSE:

$$\text{RMSE} = \sqrt{\frac{\sum\limits_{(i,j \in TestSet)} (r_{ij} - \tilde{r}_{ij})^2}{T}} \tag{7}$$

where T is the total number of test samples.

RMSE is one of the common error metrics and often used to evaluate recommender systems especially in the official Netflix contest [1].

3   A System View of Active Learning for Recommendation Systems

It is important to note that the characteristics of the prediction algorithm may influence the prediction accuracy. *Matrix factorization methods learn the model by fitting a limited number of existing ratings, hence we could make an assumption that the more reliable ratings we have, the better accuracy the model can reach.* However, in most recommendation system data sets, the ratings matrices are extremely sparse because a user typically only rates a small proportion of movies while most ratings are unknown, which motivates us to add more high quality data for matrix factorization.

3.1   The Item-Oriented Approach

Classic active learning methods focus on different individual rating elicitation strategies for a single user when a new user comes in. These strategies include:

1. *Randomization:* It can be regarded as a baseline method (e.g., [11,15,16]).
2. *Popularity-based:* Items with the largest number of ratings are preferred. It is based on the assumption that the more popular the items are, the more likely that they are known by this user (e.g., [15,17]).
3. *Entropy-based:* Items with the largest entropy are selected [15].
4. *Highest predicted* Items with the highest predicted ratings are preferred [16].
5. *Lowest predicted:* Items with the lowest predicted ratings are chosen. The items with the lowest ratings are supposed to be the most disliked movies for this user, which also may influence the user to rate them [16].
6. *Hybrid:* This includes Log(popularity)*entropy [15], *Voting*, which considers the overall effect of previous methods [11,18].

Previous works (e.g., [11,15,16]) focused on the accuracy of the recommendations for 'a single user' and the model was trained in each iteration user by user, which incurs high computational costs. While implementing classic active learning strategies, the items selected for different users to elicit are always different. For example, the items with the highest predicted ratings for a user may not be the same as another user's since not all users have the exactly same tastes. Hence strategy has to be applied repeatedly for each user to elicit ratings which are corresponding to different items but with one exception: popularity-based strategy.

On the other hand, the movie popularity may vary significantly. Take Movielens data set for example, the maximal and minimum number of popularity is 495 and 0, respectively, which means that the most popular movie is rated by 495 users while for some movies there is no one has rated them before. Popularity is based on the number of ratings of each item only and it is irrelevant to users, therefore the popularity remain unchanged no matter which users we choose. Here we only select the top N most popular movies for all the users to form a new block matrix, based on the idea that users tend to rate world-famous movies than movies in the minority. Then the missing values in this block would be the 'desirable' movies in some sense for the users who missed before. Our strategy is to elicit these specific ratings for all the users at the same time in one iteration through matrix factorization on this block matrix. After adding these ratings to the original rating matrix, a more accurate matrix factorization model could be extracted. The overall procedure of this approach are shown in Algorithm 1.

In summary, Algorithm 1 simultaneously elicit ratings of popular movies for all users, in order to improve the performance of the whole system. Therefore, it reduces the training iteration for as high as the number of users to only 2, and evaluate the benefits of the system instead of each user.

Algorithm 1. The item-oriented approach

Input: Ratings matrix $R \in \mathbb{R}^{m \times n}$, where $Q_{j \in [1,n]} \in \mathbb{R}^{m \times 1}$ is the column vector, κ is a set of elements in the ratings matrix that have been assigned values; the top number of items selected in the block matrix based on popularity N;

Output: RMSE of the test set;

Step 1: Sort items based on the number of ratings each user rate this item (popularity) in descending order $j(1), j(2), ..., j(m)$;

Step 2: Create a block M_1 by selecting the top N of items (columns) in **Step 1** based on the popularity. Therefore $M_1 = [Q_{j(1)}, Q_{j(2)}, ..., Q_{j(N)}](N < m)$;

Step 3: Apply basic matrix factorization (regularized SVD) on matrix M_1 to obtain feature matrices U and V according to Equation (1);

Step 4: Predict every missing value in block M_1 to acquire a non-null matrix M'_1 according to Equation (2). Then a series of ratings L_1 is elicited, such that $L_1 = \{$

$r_{i_{k(1)}, j(1)}, r_{i_{k(2)}, j(1)}, ..., r_{i_{k(n)}, j(1)},$

$r_{i_{k(1)}, j(2)}, r_{i_{k(2)}, j(2)}, ..., r_{i_{k(n')}, j(2)},$

$......,$

$r_{i_{k(1)}, j(N)}, r_{i_{k(2)}, j(N)}, ..., r_{i_{k(n'')}, j(N)}\}$

where $r_{i_k, j} \notin \kappa$;

Step 5: Fill ratings in the original matrix R with every predicted value by **Step 4** to acquire a new ratings matrix R'. That means we add the elicited ratings into the set of existing ratings. $\kappa = \{\kappa, L_1\}$;

Step 6: Apply basic matrix factorization (regularized SVD) on matrix R' to obtain feature matrices U' and V' according to Equation (1);

Step 7: Predict the target ratings (test set) according to Equation (2) and calculate RMSE according to Euqation (7);

3.2 The User-Oriented Approach

Traditional active learning for collaborative filtering is a set of techniques to intelligently select a number of 'items' to rate so as to improve the rating prediction for the user, while Carenini et al. [17] proposed an item-focused method that elicits ratings by choosing some special users to rate a specific item in order to improve the rating prediction for the item. Because of the feasibility of this item-focused method we also propose a user-oriented approach as shown in Algorithm 2 to further explore its potential.

Generally, the number of movies each user has rated varies significantly (e.g., in Movielens data set the maximal and minimum number for different user's are 727 and 10, respectively). In fact, though active users who are enthusiastic about movies may watch far more than the ones who are not into movies, there still exist some movies the users have watched but not yet rated. For these reasons, it is easier to accept that they would give ratings to the movies they have not rated yet. In Algorithm 2 we select this kind of special users based on the number of movies they have rated. After these movie enthusiasts are chosen, ratings of the movies they never rate would be elicited by matrix factorization as the missing values in the new block. Then we add these new ratings to the original matrix for matrix factorization again to get the target values we want.

In brief, Algorithm 2 tries to improve the performance of the whole system by eliciting ratings of all movies for only active users simultaneously. Therefore it also has the benefits that Algorithm 1 has. However, both algorithms still may suffer from large computational cost because of the extensively selection of elicitations, especially when the number of popular movies or active users selected in the block matrix is large.

Algorithm 2. The user-oriented approach

Input: Ratings matrix $R \in \mathbb{R}^{m \times n}$, where $P_{i \in [1,m]} \in \mathbb{R}^{1 \times n}$ is the row vector, κ is a set of elements in the ratings matrix that have been signed values; the top number of users selected in the block matrix based on activity N';

Output: RMSE of the test set;

    Step 1: Sort users based on the number of ratings they rates (activity) in descending order $i(1), i(2), ..., i(n)$;

    Step 2: Create a block M_2 by selecting the top N of users(rows) in **Step 1** based on the activity. Therefore $M_2 = [P_{i(1)}, P_{i(2)}, ..., P_{i(N)}](N < n)$;

    Step 3: Apply basic matrix factorization (regularized SVD) on matrix M_2 to obtain feature matrices U and V according to Equation (1);

    Step 4: Predict every missing value in block M_2 to acquire a non-null matrix M'_2 according to Equation (2). Then a series of ratings L_2 is elicited, such that $L_2 = \{$

$r_{i(1),j_k(1)}, r_{i(1),j_k(2)}, ..., r_{i(1),j_k(n)},$

$r_{i(2),j_k(1)}, r_{i(2),j_k(2)}, ..., r_{i(2),j_k(n')}$

$r_{i(N'),j_k(1)}, r_{i(N'),j_k(2)}, ..., r_{i(N'),j_k(n'')}\}$

where $r_{i,j_k} \notin \kappa$;

    Step 5: Fill ratings in the original matrix R with every predicted value by **Step 4** to acquire a new ratings matrix R'. That means we add the elicited ratings into the set of existing ratings. $\kappa = \{\kappa, L_2\}$;

    Step 6: Apply basic matrix factorization (regularized SVD) on matrix R' to obtain feature matrices U' and V' according to Equation (1);

    Step 7: Predict the target ratings (test set) according to Equation (2) and calculate RMSE according to Euqation (7);

---

4 The Proposed ESVD (Density-Oriented Approach)

So far we have presented an item-oriented approach and a user-oriented approach, both based on the idea that eliciting a group of reliable and meaningful ratings simultaneously for matrix factorization model to learn. The reason why these new ratings from Algorithms 1 and 2 should be reliable is because they are predicted from a denser block which is formed from the top numbers of ratings in either item-view or user-view by a matrix factorization method. Typically the denser the matrix is, the better the matrix factorization model we can obtain. Take the Movielens data set as an example, the density of the original matrix is 5.71 %. However, if only the top 5 % of the most popular movies are chosen, we can obtain a block of density 29.47 % which consists of more ratings that have been already rated by the users. While selecting the top 5 % of the most active users, the density of the new block we obtain is 23.33 %. Based on this assumption we propose a density-oriented approach which combines previous item-oriented and user-oriented methods in Algorithm 3.

Algorithm 3. The proposed density-oriented approach

Input: Ratings matrix $R \in \mathbb{R}^{m \times n}$, where $P_{i \in [1,m]} \in \mathbb{R}^{1 \times n}$ is the row vector and $Q_{j \in [1,n]} \in \mathbb{R}^{m \times 1}$ is the column vector, κ is a set of elements in the ratings matrix that have been assigned values; The top number of items selected in the block matrix based on popularity N and the top number of users selected in the block matrix based on activity N';

Output: RMSE of the test set;

    Step 1: Sort both items and users in descending order based on popularity and activity respectively. $j(1), j(2), ..., j(m)$; $i(1), i(2), ..., i(n)$;

    Step 2: Create a block M_3 by selecting the intersection of top N of items (columns) and top N' of users (rows) in **Step 1** based on the popularity and activity. Therefore $M_3 = M_1 \bigcap M_2$;

    Step 3: Apply basic matrix factorization (regularized SVD) on matrix M_3 to obtain feature matrices U and V according to Equation (1);

    Step 4: Predict every missing value in block M_3 to acquire a non-null matrix M'_3 according to Equation (2). Then a series of ratings L_3 is elicited, such that $L_3 = \{r_{i_{k(1)},j_{t(1)}}, ..., r_{i_{k(n)},j_{t(n')}}\}$ where $r_{i_k,j_t} \in (M_3 \bigcap \neg \kappa)$;

    Step 5: Fill ratings in the original matrix R with every predicted value by **Step 4**. That means we add the elicited ratings into the set of existing ratings. $\kappa = \{\kappa, L_3\}$;

    Step 6: Apply basic matrix factorization (regularized SVD) on matrix R' to obtain feature matrices U' and V' according to Equation (1);

    Step 7: Predict the target ratings (test set) according to Equation (2) and calculate RMSE according to Euqation (7);

Because both the popularity of movies and the activity of users depend on the numbers of ratings each user rates or each movie is rated, by choosing the top N popular movies and active users we can obtain the densest block matrix. With the Movielens data set, if we choose the top 5 % of the most popular movies and most active users, the density of the newly-formed block matrix would be 77.28 %. The missing values in this block matrix can be explained as ratings of the most famous movies but have not been rated by a group of most active users. By applying matrix factorization on this block, a better model could be extracted and the missing values (elicited ratings) could be predicted with higher accuracy. Finally a more accurate matrix factorization model can be learnt by fitting the existing ratings and extra high quality elicited ratings.

5   Experimental Results

We conducted experiments of our proposed item-oriented, user-oriented and density-oriented approach (ESVD) on the classic recommender system data set: Movielens 100K. We also performed some experiments with the larger version and obtained similar results. However, it requires much longer time to perform our experiments since we train and test the models each time for different choice of the number of N. Therefore, we focused on the smaller data set to be able to run more experiments, in order to explore how this parameter N affects the results of our matrix factorization methods. We use RMSE as the default metric which is widely used in the Netfilx Competition and proved to be effective to measure recommender systems. The number of the latent factors (rank) are set to be 10 for training each matrix factorization model. Although increasing it does raise the performance, the computational cost is proportional to latent factors. For matrix factorization of the comparatively smaller block matrix M_1, M_2 and M_3, the coefficient of the regularization term is 0.01 for both k_u and k_v, and the learning rate α is 0.1 with a decrease by a factor of 0.9 each iteration. For matrix factorization of the ratings matrix (with elicited ratings) R', the coefficient of regularization term is 0.1 for both k_u and k_v, and the learning rate α is 0.01 with decrease by a factor of 0.9 each iteration.

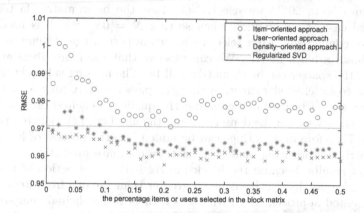

Fig. 1. RMSE comparisons of proposed methods based on SVD

Figure 1 shows the results of proposed methods based on different number of items or users N selected in the block matrix. We see that as the number of items or users N increases, the errors set keeps dropping since more ratings are added into the training matrix, afterwards the errors fluctuate in a small region. Specifically, the results of the item-oriented approach (Algorithm 1) are not promising. Because in the item-oriented approach we only elicit ratings based on the most popular movies, which may lead to a lot of bias and distort the latent factor model. For example, most people prefer happy endings, the consequence is that comedies are more popular than tragedies. Then a lot of comedy movies would be elicited for each user to give ratings which leads the latent factor model (regularized SVD in this case) to put more weights on the factor corresponding to comedies. Although the RMSE of user-oriented approach (Algorithm 2) fluctuates around the baseline method (the red line) at first, it gets lower as N goes up. The best result of Algorithm 2 we obtain is 0.9619 which reduces RMSE by 0.09 when compared with the regularized SVD 0.9709. It is apparent from the figure that our proposed ESVD consistently outperforms other methods including the baseline method: regularized SVD.

Table 1. RMSE of ESVD (the density-oriented approach)

Top items & users	0 %	5 %	10 %	15 %	20 %
Block density	Null	77.28 %	65.20 %	53.90 %	45.66 %
Elicited ratings	Null	897	5496	16381	34508
RMSE (Test)	0.9709	0.9677	0.9632	0.9630	0.9570

In Table 1, we illustrate the RMSE of proposed density-oriented (ESVD) method which incorporates both item-oriented and user-oriented approach on the same data set. We compare different RMSE based on how many items and users (from 0 % to 20 %) we select. Note that the basic matrix factorization is a special case of our method when setting $N = 0\%$, which is used as the baseline for comparison. After selecting a certain percentage of items and users, we can obtain a block matrix. We can observe that the more items and users we choose, the sparser the block matrix will be. The missing values in the block are chosen to be elicited ratings. Although sparser matrix may lead to a less accurate matrix factorization model and the quality of elicited ratings may not as good as the ones from the denser matrix, the number of the elicited ratings is increased. Therefore more ratings can be added into the process of learning the target matrix factorization model. At last we compute predictions on the test set and get results. Because the block matrix is the intersection of the top N items and N users, its density is much greater than the one from item-oriented or user-oriented approach. Even with less ratings to be elicited compared with item-oriented and user-oriented, the results are better. In our experiments, we observed that the performance fluctuates as the number of top projects increases (Table 1). The optimal point that balances the quality (density of block matrix) and the quantity (number of elicited ratings) depends on the distribution of ratings. The Algorithm 3 can reach to 0.9570 (when $N = 20\%$) which reduces the RMSE by 0.0139 compared with the regularized SVD 0.9709.

5.1   Extension: ESVD++

Broadly speaking, our proposed ESVD approach can be seen as a preprocessing step and it can be incorporated with other variants of SVD models, such as SVD++ [14]. We conduct ESVD++ by just changing the prediction algorithm from SVD to SVD++. Compared with SVD model, SVD++ improves the prediction accuracy by adding biases and implicit information $I(i)$, and the prediction function is shown in Eq. 6. Specifically, $I(i)$ contains all items for which the ith user has provided a rating, even if the value is unknown. Therefore, for prediction of elicited ratings as shown in Step 4 of Algorithm 3, $I(i)$ is set to be the number of existing ratings and the missing values in the block matrix that are also shown in the test set. For prediction of test set as shown in Step 7 of Algorithm 3, $I(i)$ is the same as the one in original matrix without considering the elicited ratings.

As the elicitation strategy is the same as ESVD, the ratings that need to be elicited would also be the same. The results show that the ESVD++ outperforms the state-of-art SVD++ model and greatly reduces the RMSE by 0.0214 (from the baseline SVD++ 0.9601 to 0.9387 when $N = 10\%$).

6 Conclusion

The lack of information is an acute challenge in most recommender systems. In this paper, we present a new matrix factorization method called ESVD which applies SVD with ratings elicitation that best approximates a given matrix with missing values. We conduct corresponding experiments on the Movielens data set. Instead of looking at the active learning from the individual user's point of view, we deal with the problem from the system's perspective. Although the proposed method cannot deal with the cold start problem where the database keeps growing as new users or items continue to be added, it does reduce the computational costs greatly (the iteration number from as high as the number of users to only 2) since all the ratings are elicited simultaneously. Compared with traditional matrix factorization models, it can be incorporated with different SVD-based algorithms and greatly improve the performance of the whole system. Because the model is learnt by extra high-quality-elicited ratings that are extracted from the densest block matrix we create based on item popularity and user activity. Even using basic SVD as a prediction algorithm, without considering the effect of implicit feedbacks, the performance is better than its counterpart SVD++.

References

1. Bennett, J., Lanning, S.: The netflix prize. In: Proceedings of KDD Cup and Workshop, p. 35 (2007)
2. Das, A.S., Datar, M., Garg, A., Rajaram, S.: Google news personalization: scalable online collaborative filtering. In: Proceedings of the 16th International Conference on World Wide Web, pp. 271–280. ACM (2007)
3. Linden, G., Smith, B., York, J.: Amazon.com recommendations: item-to-item collaborative filtering. IEEE Internet Comput. **7**(1), 76–80 (2003)
4. Baatarjav, E.-A., Phithakkitnukoon, S., Dantu, R.: Group recommendation system for facebook. In: Meersman, R., Tari, Z., Herrero, P. (eds.) OTM-WS 2008. LNCS, vol. 5333, pp. 211–219. Springer, Heidelberg (2008)
5. Bell, R.M., Koren, Y.: Scalable collaborative filtering with jointly derived neighborhood interpolation weights. In: Seventh IEEE International Conference on Data Mining, pp. 43–52. IEEE (2007)
6. Wang, Y.-X., Zhang, Y.-J.: Nonnegative matrix factorization: a comprehensive review. IEEE Trans. Knowl. Data Eng. **25**(6), 1336–1353 (2013)

7. Golub, G., Kahan, W.: Calculating the singular values and pseudo-inverse of a matrix. J. Soc. Ind. Appl. Math. Ser. B Numer. Anal. **2**(2), 205–224 (1965)
8. Koren, Y., Bell, R., Volinsky, C.: Matrix factorization techniques for recommender systems. Computer **8**, 30–37 (2009)
9. Harpale, A.S., Yang, Y.: Personalized active learning for collaborative filtering. In: Proceedings of the 31st Annual International ACM SIGIR Conference on Research and Development in Information Retrieval, pp. 91–98. ACM (2008)
10. Jin, R., Si, L.: A bayesian approach toward active learning for collaborative filtering. In: Proceedings of the 20th Conference on Uncertainty in Artificial Intelligence, pp. 278–285. AUAI Press (2004)
11. Elahi, M., Ricci, F., Rubens, N.: Active learning strategies for rating elicitation in collaborative filtering: a system-wide perspective. ACM Trans. Intell. Syst. Technol. (TIST) **5**(1), 13 (2013)
12. Funk, S.: Netflix update: try this at home (2006)
13. Paterek, A.: Improving regularized singular value decomposition for collaborative filtering. In: Proceedings of KDD Cup and Workshop, pp. 5–8 (2007)
14. Koren, Y.: Factorization meets the neighborhood: a multifaceted collaborative filtering model. In: Proceedings of the 14th ACM SIGKDD International Conference on Knowledge Discovery and Data Mining, pp. 426–434. ACM (2008)
15. Rashid, A.M., Albert, I., Cosley, D., Lam, S.K., McNee, S.M., Konstan, J.A., Riedl, J.: Getting to know you: learning new user preferences in recommender systems. In: Proceedings of the 7th International Conference on Intelligent User Interfaces, pp. 127–134. ACM (2002)
16. Marlin, B.M., Zemel, R.S., Roweis, S.T., Slaney, M.: Recommender systems, missing data and statistical model estimation. In: IJCAI, pp. 2686–2691 (2011)
17. Carenini, G., Smith, J., Poole, D.: Towards more conversational and collaborative recommender systems. In: Proceedings of the 8th International Conference on Intelligent User Interfaces, pp. 12–18. ACM (2003)
18. Golbandi, N., Koren, Y., Lempel, R.: On bootstrapping recommender systems. In: Proceedings of the 19th ACM International Conference on Information and Knowledge Management, pp. 1805–1808. ACM (2010)

Social Group Based Video Recommendation Addressing the Cold-Start Problem

Chunfeng Yang[1]([⊠]), Yipeng Zhou[2], Liang Chen[2], Xiaopeng Zhang[3], and Dah Ming Chiu[1]

[1] The Chinese University of Hong Kong, Hong Kong, China
{yc012,dmchiu}@ie.cuhk.edu.hk
[2] Shenzhen University, Shenzhen, China
{ypzhou,lchen}@szu.edu.cn
[3] Precise Recommendation Center, Tencent Inc., Shenzhen, China
xpzhang@tencent.com

Abstract. Video recommendation has become an essential part of online video services. Cold start, a problem relatively common in the practical online video recommendation service, occurs when the user who needs video recommendation has no viewing history (Cold start consists of the new-user problem and the new-item problem. In this paper, we discuss the new-user one). A promising approach to resolve this problem is to capitalize on information in online social networks (OSNs): Videos viewed by a user's friends may be good candidates for recommendation. However, in practice, this information is also quite limited, either because of insufficient friends or lack of abundant viewing history of friends. In this work, we utilize social groups with richer information to recommend videos. It is common that users may be affiliated with multiple groups in OSNs. Through members within the same group, we can reach a considerably larger set of users, hence more candidate videos for recommendation. In this paper, by collaborating with Tencent Video, we propose a social-group-based algorithm to produce personalized video recommendations by ranking candidate videos from the groups a user is affiliated with. This algorithm was implemented and tested in the Tencent Video service system. Compared with two state-of-the-art methods, the proposed algorithm not only improves the click-through rate, but also recommends more diverse videos.

1 Introduction

Video recommendation has become an integral part of today's online video services, such as those provided by YouTube and Netflix. Good recommendation not only increases user engagement, but also improves user loyalty. Although the problem of recommending videos on the basis of users' viewing history has been well studied, it is still challenging to recommend videos for those users with little or no viewing history. In the literature, this is known as the data sparsity

© Springer International Publishing Switzerland 2016
J. Bailey et al. (Eds.): PAKDD 2016, Part II, LNAI 9652, pp. 515–527, 2016.
DOI: 10.1007/978-3-319-31750-2_41

or cold start problem[1]. Common recommendation strategies can be categorized into three types: collaborative filtering (CF), content-based filtering (CBF), and hybrid strategies combining CF and CBF.

CF [19], based on a user's preference and behaviors of other users with similar preferences, can accurately recommend videos of interest given sufficient historical records. It is therefore widely used in online video systems. However, as reported in [19], CF is not effective for users with little viewing history. CBF [13] is based on clustering items with similar descriptions and matching them to users' current selection. This strategy can exploit various advanced information retrieval techniques. However, a major weakness of the strategy is over-specification (keep on recommending items of the same type or with similar descriptions). The hybrid system combines the strengths of CF and CBF to overcome the cold-start and over-specification problems, however, its advantage over the other two strategies has so far been marginal [26].

An alternative strategy for overcoming the cold-start problem is to use social information [20]. By exploiting online social networks, videos viewed by a user's friends can be used for recommendation. Although this strategy is quite promising, it is not always effective because social information could be scarce for some users. In this study, instead of exploiting social information associated only with friends, we used social-group-based information for video recommendation. Our collaborator (Tencent Inc.) runs an online platform that provides multiple services, including online games, online videos, and instant messaging (QQ), and facilitates the formation of QQ groups by QQ users. QQ groups allow users to easily communicate within a small circle of typically 50 to 100 users, sharing common interests; for example, there are classmate groups, colleague groups, and interest groups of various types. We find that group affiliation is quite prevalent, and a user is typically affiliated with multiple groups. If we assume that a video viewed by a group mate is of potential interest, then QQ groups can provide considerably more candidate videos for recommendation compared with the number of candidate videos obtained by considering only the circle of friends.

Although QQ groups substantially increase the pool of candidate videos available for recommendation, the relevance of these videos depends strongly on the type of group the videos originate from. We propose an algorithm for ranking candidate videos from different groups and identifying the top R videos for recommendation[2]. This is done in two steps. First, videos from a single group are ranked. If this is the only group the user is affiliated with, then this ranking provides the order for recommendation. Otherwise, the ranked videos from multiple groups are (weighted) rank aggregated to obtain the final order for recommendation, where the weight of each group is learned by a supervised learning method based on several calculated group features.

---

[1] Cold start can be seen as the extreme case of data sparsity. This study refers cold start to situations where available collaborative information is inadequate for effective recommendation.

[2] R is the number of videos recommended to a user, which depends on the system setting.

Our objective was to recommend videos with a high hit rate, high diversity because both these metrics are crucial to user satisfaction. To the best of our knowledgement, this is the first study to develop a video recommendation method based on different group affiliations and merely implicit feedback data[3], and the method was tested in an online video system. Our contributions are summarized as follows.

- We determined and analyzed the difference between the number of candidate videos obtained from social groups and that acquired from only a friend circle.
- We proposed a video recommendation method based on social group information; the method can rank candidate videos from a single group as well as multiple groups that a user is affiliated with.
- We evaluated the proposed social-group-based video recommendation algorithm by implementing it on the Tencent video system, and showed that it improves both click-through rate and video diversity.

The rest of this paper is organized as follows. We first discuss related work in Sect. 2 for identifying research gaps. We then describe (in Sect. 3) the studied system and demonstrate how the social-group-based approach can considerably increase the number of candidate videos, as the motivation of the social-group-based strategy. The ranking methodology and algorithms are detailed in Sect. 4. In Sect. 5, experimental results are presented. Finally, Sect. 6 concludes the paper.

2 Related Works

Collaborative filtering (CF) is a well-developed framework utilizing viewing history of different users to provide personalized recommendation. A detailed survey of collaborative filtering is provided by [19], and various enhancements have also been developed [12,24]. The main drawback of collaborative filtering is the cold start problem, which has been discussed and studied in [3,27].

As a promising approach to overcome the cold start problem, social recommendation which capitalizes on social information has become more popular. Existing studies of social recommendation can be grouped into three types: (1) Extra social information is used to improve an existing recommendation system [10,11,14]; (2) Social information is used to create or enable a recommender system [6]; (3) Studying user trust and item reputation [23]. Despite the popularity of social recommender systems, [20] points out that there are still some negative experiences in applying social recommender systems: (a) Social relations are too noisy and may have negative impact on recommender systems; (b) For cold start users, they may also have few social relations; (c) Different types of social relations may affect social recommender systems differently, and the success of one type of social relation may not be applicable to others.

---

[3] Implicit feedback means only user viewing records (view or not) without explicit numerical rating data.

While those social recommender systems utilize social information of friends, others turn to communities of users for recommendation. For example, [17] utilizes community detection approaches to find communities from different dimensions of social networks, and then performs collaborative filtering within community members. [25] develops a circle-based recommender system that infers category-specific social trust circles from available rating data combined with social network data, and it has also proposed several variants to weight friends based on their inferred expertise levels.

Different from those community based approaches, which generate groups by virtue of community detection on top of interest similarity or the social relationship network, our work is based on explicitly defined groups formed autonomously by members because of common interest and other reasons. Furthermore, in our setting users can be affiliated with multiple groups which leads to a new challenge of how to exploit these groups together for recommendation.

3   Analysis of Candidate Video Pool

Tencent Video [22] is one of the largest online VoD (Video-on-Demand) service providers in China, supporting more than 50 million active users on a daily basis. During peak hours, more than 2 million concurrent users are served. Tencent Video's video catalog includes movie, TV episodes, MVs (music video), news, UGC (user generated content) and many other types.

The online social network used by us is Tencent QQ [15]. QQ is one of the most popular Instant Messaging service in China, through which one can make friends, chat with friends and join QQ groups. In June 2015, there are roughly 843 million active QQ accounts, with a peak of 233 million online QQ users [16]. Moreover, most users join multiple QQ groups (more than ten). For most QQ groups, there are more than 50 members.

To help discuss our ideas precisely, we first define some notations. All users, videos and groups are assigned unique IDs. For user i, his viewing record list for the last 30 days is represented by set $\mathcal{V}_i$, i.e., if a video j is viewed by a user i, then $j \in \mathcal{V}_i$. The set $\mathcal{G}_i$ represents the groups that user i joins. For a group k, all its members are represented by set $\mathcal{U}_k$. If user u is a group mate of user i, there exists a group k, such that $u, i \in \mathcal{U}_k$. The set $\mathcal{G}$ is all the groups, and the total number of groups is G. For a group k, the video pool, i.e., the set of all videos that have been viewed by any user within the group, is represented by $\mathcal{P}_k = \cup_{i \in \mathcal{U}_k} \mathcal{V}_i$. The group video candidate pool for user i is $\mathcal{P}_{\mathcal{G}_i} = \cup_{k \in \mathcal{G}_i} \mathcal{P}_k$. The set of friends for user i is $\mathcal{F}_i$, i.e., if user u is one of user i's friends, then $u \in \mathcal{F}_i$. The friend video candidate pool for user i is represented by $\mathcal{P}_{\mathcal{F}_i} = \cup_{u \in \mathcal{F}_i} \mathcal{V}_u$.

Currently, recommender systems based on collaborative filtering are already deployed in Tencent Video to provide personalized recommendation. One primary challenge faced by the system is new-user cold start issue and data sparsity issue. Take the recommender system for movies as an example. According to system measurements, around 25 % of the daily users have not watched any movie in Tencent Video in the last 30 days, and thus can be treated as new users for a

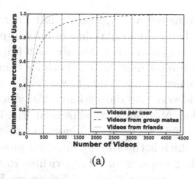
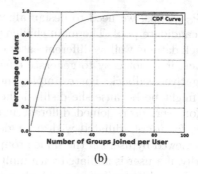

(a)   (b)

Fig. 1. (a). Cummulative distribution of number of videos from individual users, their friends and group mates; (b). Cummulative distribution of the number of affiliated groups per user.

Table 1. Statistics of friends, group mates and video candidate pools

Per user	Mean value	Per user	Mean value
Number of friends	35.9	Size of friend video candidate pool	48.8
Number of group mates	213.2	Size of group video candidate pool	142

one-month window. In addition, the user viewing behavior is fairly sparse on the individual level, indicating data sparsity, as illustrated by the distribution curve of number of videos per user in Fig. 1(a). In other words, the non-zero entries in the user-item consumption matrix take less than 0.01 % of the whole matrix.

We can also analyze the candidate video pool for a given user in Tencent Video and Tencent QQ together. For a particular user, his viewing records, his friends' viewing records and his group mates' viewing records are jointly measured over 30 days. The results are summarized in Table 1 and Fig. 1(a). The results show that for a user, there are more group mates than friends and more video records from group mates accordingly, which implies we may discover more interesting videos via groups. This motivates us to design the social-group-based recommendation strategy.

4   Ranking Algorithms

4.1   Objective and Challenges

The problem we want to solve in this paper is to recommend a set of relevant and diverse[4] videos $\mathcal{R}_i$ (with size R) for user i from the group video candidate

---

[4] In this paper, relevant recommendation means the results conform to users' interest or the recommended items can attract users, which is similar to the accuracy measure. Diversity refers to sales diversity, which measures how unequally different items are chosen by users or recommended to users when a particular recommender system is used.

pool $\mathcal{P}_{\mathcal{G}_i}$. Thus, we need to design algorithms to rank videos in user i's group video candidate pool. The main challenges lie in the availability of only implicit feedback data as well as different cases of group affiliation. Merely with implicit feedback, it is not easy to generate effective recommendation, because there is no negative feedback. For example, the reason a user didn't watch a certain video might be because she disliked the video or she didn't know the video at all. Moreover, users joined different number of groups, as shown in Fig. 1(b). Firstly, for users affiliated with a single group, we need an algorithm to rank videos viewed by users in the same group, which is the *single affiliation problem*. Secondly, if a user is affiliated with multiple groups, we need algorithms to rank videos from different groups, which refers to the *multiple affiliation problem*. In the real system, most users have joined multiple groups, however, we begin with the ranking algorithm in the single affiliation case for ease of clarifying the proposed algorithms.

4.2   Video Ranking for Single Affiliation Problem

In this case, we need to resolve the intra-group video ranking problem. Different from previous studies of group profiling [21] and recommendation with rank aggregation [2] that utilize explicit user preference, we only have implicit feedback data. Assume user i only joins group k, then $\mathcal{P}_{\mathcal{G}_i} = \mathcal{P}_k$.

Generally speaking, videos that are more representative of the group should be ranked higher. As argued by [7], a representative item is supposed to be *frequent* and *discriminative*, in other words, it should be frequent so as to be a "pattern", and it can be used to distinguish one from others. Intuitively, the more members of a group viewed a video, the more the video is likely to attract other users in the group. However, frequent items are not necessarily discriminative. For example, for a group comprised of sports fans, some very hot videos, such as breaking news, are less discriminative than a certain sports video, although those hot videos were viewed by more members of this group.

To capture videos that are both frequent and discriminative, we firstly define scores to quantify these two characteristics of a video respectively. We use a local popularity score to denote how frequently a video appears in the group video candidate pool, i.e., how many group members have viewed the video:

$$\eta_{k,j} = \sum_{i \in \mathcal{U}_k} I(j \in \mathcal{V}_i), \tag{1}$$

where $I(j \in \mathcal{V}_i)$ is an indicator with value 1 if video j was viewed by user i, otherwise 0.

To measure how discriminative a video is for a group, we compare the total number of groups with the number of groups whose members have viewed video j, and define the discrimination score as

$$\beta_j = \log_2 \frac{G}{\sum_{k' \in \mathcal{G}} I(j \in \mathcal{P}_{k'})}, \tag{2}$$

where G is the total number of groups and $I(j \in \mathcal{P}_{k'})$ is an indicator with value 1 if video j is viewed by any user from group k', otherwise 0. Therefore, videos liked by a less number of groups will get higher discrimination scores.

Then, for each video j, we assess its representative level for group k by combining its local popularity score and discrimination score to generate a score $W_{k,j}$:

$$W_{k,j} = \eta_{k,j} * \beta_j, \tag{3}$$

where $\eta_{k,j}$ is the local popularity score, and β_j is the discrimination score. The values of $\eta_{k,j}$ and β_j are affected by the number of members in the target group and the total number of groups in the system respectively. For a certain video system, the values of $\eta_{k,j}$ and β_j can be scaled (such as max-min scaling) before being multiplied. In sum, we prefer videos that are locally popular in the target group rather than videos favored by most groups.

Videos in group k's video pool $\mathcal{P}_k$ will be ranked according to their scores $W_{k,j}$ in the decreasing order. The ranked video list for group k is denoted by l_k. If user i is only affiliated with group k, then the top R videos from l_k, after removing videos viewed by user i, will be recommended to user i.

4.3 Video Ranking for Multiple Affiliation Problem

In this case, we can firstly apply the intra-group video ranking algorithm described in Sect. 4.2 to each of the affiliated groups. Then, we should consider how to merge the ranked video lists from those groups, which is a rank aggregation problem [2]. However, groups are of different values in video recommendation. For example, a highly interactive interest group may be more valuable than a colleague group because like-minded users are more likely to enjoy common videos. Thus, we need a group scoring algorithm to discriminate those groups.

Group Scoring. To assess a group, we firstly define features that can distinguish groups, and then use these features to calculate the group score.

Group Features. Basically, we focus on two kinds of group features: social features and interest features. The social features we exploited comprise of social activeness and social conformity, which mainly take the social influence into account. For example, good friends may share common interests in viewing videos. We detect the social influence from the density of friendship inside the group and the strength of interaction among group members. For interest features, we consider interest activeness and interest conformity. Preferred groups are those whose members are fond of viewing (especially representative) videos and like-minded with each other.

According to the advice-taking theory used in recommender systems [1], tie strength, trustworthiness, and homophily are three significant factors to affect the likelihood of users seeking and accepting someone's advice for decision making. In online social network, tie strength can be measured by the frequency and duration of interaction, while trustworthiness corresponds to social relationship, and homophily means interest similarity.

With the knowledge of the group information and historical viewing records, we calculate the feature scores for group k as below:

- *Social activeness:* $S_k^a = \frac{\sum_{i \in \mathcal{U}_k} M_{k,i}}{|\mathcal{U}_k|}$, where $M_{k,i}$ is the number of group messages sent by user i in group k in last 30 days, and $|\mathcal{U}_k|$ is the number of members in group k. Social activeness measures the intra-group interaction strength, which reflects the group-level tie strength.
- *Social conformity:* $S_k^c = \frac{\sum_{u,i \in \mathcal{U}_k, u \neq i} I_{u,i}}{|\mathcal{U}_k|(|\mathcal{U}_k|-1)/2}$, where $I_{u,i}$ is an indicator to show whether u and i are friends in QQ. Social conformity is the density of the friendship network inside this group, representing the trustworthiness on a group basis.
- *Interest activeness:* $I_k^a = \sum_{j \in \mathcal{P}_k} W_{k,j}$, where $W_{k,j}$ is the score of video j in group k calculated in Sect. 4.2. This feature measures how representative the group video candidate pool is in total.
- *Interest conformity:* $I_k^c = \frac{\sum_{u,i \in \mathcal{U}_k, u \neq i} \cos(u,i)}{|\mathcal{U}_k|(|\mathcal{U}_k|-1)/2}$, where $\cos(u,i) = \frac{|\mathcal{V}_u \cap \mathcal{V}_i|}{|\mathcal{V}_u|^{\frac{1}{2}}|\mathcal{V}_i|^{\frac{1}{2}}} \in$ $(0,1)$. As shown in [2], the more alike the users in the group are, the more effective the group recommendations will be. This is a group-level video interest similarity measure corresponding to homophily. The larger this value, the higher degree of common interest among this group.

Calculation of Group Scores. Those feature scores provide a four-dimension comparison among different groups. However, in order to merge video lists from multiple groups, we need to combine these features to generate a single score for each group.

It is not easy to generate a group score with the feature scores. We should assign each feature a reasonable weight to combine them. We adopt the logistic function introduced in logistic regression to generate the group score, which is a value between 0 and 1 indicating the likelihood of a group to be effective for recommendation[5].

$$J_k = \frac{1}{1 + e^{-(\theta_0 + \theta_1 S_k^a + \theta_2 S_k^c + \theta_3 I_k^a + \theta_4 I_k^c)}}, \tag{4}$$

where $\theta_0, \ldots, \theta_4$ are weights that can be tuned to combine feature scores.

We can use the supervised learning approach to learn the five feature weights. To prepare the training set, we choose in total m users randomly from all the online users. For each user i, we randomly select a group $k_i = \xi(\mathcal{G}_i)$, and use the proposed intra-group video ranking algorithm to recommend videos to the user. If any recommended videos is then selected and viewed by user i, then group k_i is effective to recommend videos for user i, denoted by $y_{k_i} = 1$, otherwise $y_{k_i} = 0$. For each instance in the training set, such as using group k_i to recommend videos

---

[5] The reason we chose logistic regression is that it returns well calibrated predictions by default as it directly optimizes log-loss.

for user i, the empirical error is calculated by the log-loss cost function $f_i(J_{k_i})$, which is defined as

$$f_i(J_{k_i}) = \begin{cases} -\log(J_{k_i}), & \text{if } y_{k_i} = 1 \\ -\log(1 - J_{k_i}). & \text{if } y_{k_i} = 0 \end{cases} \tag{5}$$

We can obtain optimal weights $\theta_0, \ldots, \theta_4$ by minimizing the regularized cost function shown below (using gradient descent approach):

$$\min_{\theta_0, \theta_1, \theta_2, \theta_3, \theta_4} \frac{1}{m} \sum_{i=1}^{m} f_i(J_{k_i}) + \frac{\lambda}{2m} \sum_{j=1}^{4} \theta_j^2, \tag{6}$$

$$= -\frac{1}{m} \sum_{i=1}^{m} [y_{k_i} \log(J_{k_i}) + (1 - y_{k_i}) \log(1 - J_{k_i})] + \frac{\lambda}{2m} \sum_{j=1}^{4} \theta_j^2. \tag{7}$$

Video Aggregation. With the ranked video lists from multiple groups as well as the score of each group, we should address a weighted ranking aggregation problem. Borda Fuse [2] is a widely known approach proposed to merge ranking lists. In Borda Fuse, each ranked video list is like a voter, and each voter ranks a partial set of c video candidates. For each voter, the top ranked video is assigned c scores, the second ranked video is assigned $c - 1$ points, and the like. If some videos left unranked by the voter, i.e., not in this ranked video list, the remaining scores are divided evenly among the unranked videos. In our case, we use a weighted Borda Fuse method to obtain the merged result R_i. For user i, the score of video j after merging multiple video lists is

$$W_{i,j} = \sum_{k \in \mathcal{G}_i} J_k * (D_i - p_{l_k}(j) + 1), \tag{8}$$

where J_k is the score of group k, and D_i is the number of distinct videos in user i's group candidate video pool $\mathcal{P}_{\mathcal{G}_i}$. If $j \in l_k$, the variable $p_{l_k}(j) \in [1, D_i]$ is the position of video j in list l_k, otherwise, to allocate the remaining scores among the unranked videos,

$$p_{l_k}(j) = \frac{|l_k| + D_i + 1}{2}, \tag{9}$$

where $|l_k|$ is the length of list l_k. After ranking all the videos according to $W_{i,j}$ and removing videos viewed by user i, we can obtain the final recommendation list $\mathcal{R}_i$.

One thing to note is that the social-group-based algorithm in this paper does not utilize some model-based methods like those in CF[6], for example, the proposed algorithm does not factorize the group information directly. Thus, the proposed algorithm is more like a memory-based one. Despite the higher accuracy in general achieved by model-based methods, memory-based methods show the superiority in terms of simplicity, interpretability, and the ability of incremental updating [19], which makes them still prevalent in real systems.

---

[6] CF can be divided into two classes: memory-based CF and model-based CF.

5   Experimental Results

Instead of using historial data to conduct offline evaluations, we implement the social-group-based algorithm in the Tencent Video system to test it online. In our experiments, we focus on movie recommendation. In the current VoD service of Tencent Video, there are more than 5 million daily views of movies.

To learn the feature weights in the group scoring algorithm, we randomly select 10 % of daily users along with random group selection, and use the intra-group video ranking algorithm to recommend movies. Then we collect the users' (implicit) feedback to conduct feature weighting training. To evaluate the performance of feature weighting, we collect feedback data in two days, using the former day's feedback data for training and the latter day's for testing. The receiver operating characteristic (ROC) curve of the feature weighting is shown in Fig. 2.

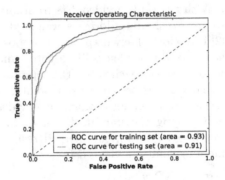

Fig. 2. The receiver operating characteristic curve of the feature weighting learning.

For online testing of the proposed social-group-based algorithm, we also implement two state-of-art approaches as benchmark for comparison.

- Implicit feedback-based collaborative filtering [9]: A matrix factorization model tailored for implicit feedback is utilized, where implicit feedback data is treated as indication of positive and negative preference associated with various confidence levels.
- Ontology-content-based filtering [18]: Each item profile is represented with a set of concepts taken from a video-related ontology. And each user's content-based profile, generated according to the user's implicit feedback, consists of a weighted list of ontology concepts representing his/her interests. A cosine similarity measure is adopted to match users and items.

We conduct abundant A/B testing [8] online to evaluate the recommendation performance of the three algorithms. In the A/B test, users whose past viewing behaviors follow the distribution in Fig. 1(a) are diverted into several distinct sets evenly and randomly, where each set adopts a distinct setting for one targeted character and all other characters are fixed. These sets are then compared against one another over a set of predefined metrics. In our experiments, the targeted character is the adoption of different algorithms. To evaluate the performances of these algorithms in terms of relevance and diversity, we use two metrics in the experiments, namely, click-through rate (CTR) [5] and Gini coefficient [4].

- $CTR = \frac{\# \ of \ click}{\# \ of \ impression}$, where '$\# \ of \ impression$' is the total number of recommendation, and '$\# \ of \ click$' is the number of recommendation whose recommended video lists are clicked after they are shown to users.
- $Gini = \frac{2 \sum_{j=1}^{n} j*d_j}{n \sum_{j=1}^{n} d_j} - \frac{n+1}{n}$, where n is the number of distinct videos recommended, and d_j is j-th lowest frequency of occurrence in all recommended video lists. The smaller the Gini coefficient, the more diverse the recommendation results are.

We measured the performance over a period of 21 days. We normalize the second largest daily CTR to be 1 and the other two CTRs to be the ratio of it. The results of the normalized CTR and Gini coefficient are shown in Fig. 3, where each value is averaged over the same day of three weeks[7].

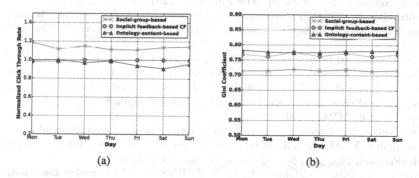

(a)   (b)

Fig. 3. Per-day average CTR and Gini coefficient for different algorithms over a period of 3 weeks.

The results show that the social-group-based algorithm achieved the highest CTR and smallest Gini coefficient, which means it can generate the most relevant recommendation in terms of hit rate and provide the most diverse results compared with the other two approaches. And this also indicates that the proposed algorithm can perform well in the cold-starting online video system.

6   Conclusion

In this paper, we propose a social-group-based video recommendation framework by virtue of explicitly formed groups (QQ groups). We elaborate on three algorithms of the framework, namely, intra-group video ranking, group scoring, and

---

[7] Using a huge amount of users in a long period of experiments for the online tests could yield statistically significant results.

video aggregation. To validate the effectiveness of our approach, we deployed it in the online video system and compare it with two state-of-the-art algorithms. The evaluation results show that our design can produce recommendation results with both high relevance and diversity. In the future work, we will implement more algorithms for comparison, such as hybrid CF and CBF algorithm, social-friend-based algorithm.

Acknowledgement. The authors wish to acknowledge the support from HK RGC grant 14201814. This work was partially supported by the Natural Science Foundation of China under Grant No. 61402297.

References

1. Arazy, O., Kumar, N., Shapira, B.: Improving social recommender systems. IT Prof. **11**(4), 38–44 (2009)
2. Baltrunas, L., Makcinskas, T., Ricci, F.: Group recommendations with rank aggregation and collaborative filtering. In: Proceedings of the Fourth ACM Conference on Recommender systems, pp. 119–126. ACM (2010)
3. Bobadilla, J., Ortega, F., Hernando, A., Bernal, J.: A collaborative filtering approach to mitigate the new user cold start problem. Knowl. Based Syst. **26**, 225–238 (2012)
4. Gini Coefficient: http://en.wikipedia.org/wiki/gini
5. Davidson, J., Liebald, B., Liu, J., Nandy, P., Van Vleet, T., Gargi, U., Gupta, S., He, Y., Lambert, M., Livingston, B., et al.: The youtube video recommendation system. In: Proceedings of the Fourth ACM Conference on Recommender Systems, pp. 293–296. ACM (2010)
6. Dell'Amico, M., Capra, L.: Sofia: social filtering for robust recommendations. In: Karabulut, Y., Mitchell, J., Herrmann, P., Jensen, C.D. (eds.) Trust Management II. IFIP, vol. 263, pp. 135–150. Springer, New York (2008)
7. Doersch, C., Singh, S., Gupta, A., Sivic, J., Efros, A.: What makes Paris look like Paris? ACM Trans. Graph. **31**(4), 101 (2012)
8. Search evaluation at Google: http://googleblog.blogspot.hk/2008/09/search-evaluation-at-google.html
9. Hu, Y., Koren, Y., Volinsky, C.: Collaborative filtering for implicit feedback datasets. In: Eighth IEEE International Conference on Data Mining, ICDM 2008, pp. 263–272. IEEE (2008)
10. Jamali, M., Ester, M.: Trustwalker: a random walk model for combining trust-based and item-based recommendation. In: Proceedings of the 15th ACM SIGKDD International Conference on Knowledge Discovery and Data Mining, pp. 397–406. ACM (2009)
11. Jamali, M., Ester, M.: A matrix factorization technique with trust propagation for recommendation in social networks. In: Proceedings of the Fourth ACM Conference on Recommender Systems, pp. 135–142. ACM (2010)
12. Koren, Y.: Factorization meets the neighborhood: a multifaceted collaborative filtering model. In: Proceedings of the 14th ACM SIGKDD International Conference on Knowledge Discovery and Data Mining, pp. 426–434. ACM (2008)
13. Lops, P., De Gemmis, M., Semeraro, G.: Content-based recommender systems: state of the art and trends. In: Ricci, F., Rokach, L., Shapira, B., Kantor, P.B. (eds.) Recommender Systems Handbook, pp. 73–105. Springer, New York (2011)

14. Ma, N., Lim, E.P., Nguyen, V.A., Sun, A., Liu, H.: Trust relationship prediction using online product review data. In: Proceedings of the 1st ACM International Workshop on Complex Networks Meet Information and Knowledge Management, pp. 47–54. ACM (2009)
15. QQ: http://www.qq.com/
16. Tencent   Report   (2015).   http://www.tencent.com/en-us/content/at/2015/attachments/20150812.pdf
17. Sahebi, S., Cohen, W.W.: Community-based recommendations: a solution to the cold start problem. In: Workshop on Recommender Systems and the Social Web, RSWEB (2011)
18. Shoval, P., Maidel, V., Shapira, B.: An ontology-content-based filtering method (2008)
19. Su, X., Khoshgoftaar, T.M.: A survey of collaborative filtering techniques. Adv. Artif. Intell. **2009** (2009). Article no. 4
20. Tang, J., Hu, X., Liu, H.: Social recommendation: a review. Soc. Netw. Anal. Min. **3**(4), 1113–1133 (2013)
21. Tang, L., Wang, X., Liu, H.: Understanding emerging social structures - a group profiling approach. School of Computing, Informatics, and Decision Systems Engineering, Arizona State University, Tech. report TR-10-002 (2010)
22. Tencent Video: http://v.qq.com/
23. Wen, Z., Lin, C.Y.: On the quality of inferring interests from social neighbors. In: Proceedings of the 16th ACM SIGKDD International Conference on Knowledge Discovery and Data Mining, pp. 373–382. ACM (2010)
24. Xu, B., Bu, J., Chen, C., Cai, D.: An exploration of improving collaborative recommender systems via user-item subgroups. In: Proceedings of the 21st International Conference on World Wide Web, pp. 21–30. ACM (2012)
25. Yang, X., Steck, H., Liu, Y.: Circle-based recommendation in online social networks. In: Proceedings of the 18th ACM SIGKDD International Conference on Knowledge Discovery and Data Mining, pp. 1267–1275. ACM (2012)
26. Yao, L., Sheng, Q.Z., Segev, A., Yu, J.: Recommending web services via combining collaborative filtering with content-based features. In: IEEE 20th International Conference on Web Services (ICWS 2013), pp. 42–49. IEEE (2013)
27. Zhou, K., Yang, S.H., Zha, H.: Functional matrix factorizations for cold-start recommendation. In: Proceedings of the 34th International ACM SIGIR Conference on Research and Development in Information Retrieval, pp. 315–324. ACM (2011)

FeRoSA: A Faceted Recommendation System for Scientific Articles

Tanmoy Chakraborty(✉), Amrith Krishna, Mayank Singh, Niloy Ganguly,
Pawan Goyal, and Animesh Mukherjee

Department of CSE, Indian Institute of Technology, Kharagpur 721302, India
{its_tanmoy,amrith.krishna,mayank.singh,niloy,
pawang,animeshm}@cse.iitkgp.ernet.in

Abstract. The overwhelming number of scientific articles over the years calls for smart automatic tools to facilitate the process of literature review. Here, we propose for the first time a framework of *faceted recommendation* for scientific articles (abbreviated as FeRoSA) which apart from ensuring quality retrieval of scientific articles for a query paper, also efficiently arranges the recommended papers into different facets (categories). Providing users with an interface which enables the filtering of recommendations across multiple facets can increase users' control over how the recommendation system behaves. FeRoSA is precisely built on a random walk based framework on an induced subnetwork consisting of nodes related to the query paper in terms of either citations or content similarity. Rigorous analysis based an experts' judgment shows that FeRoSA outperforms two baseline systems in terms of faceted recommendations (overall precision of 0.65). Further, we show that the faceted results of FeRoSA can be appropriately combined to design a better flat recommendation system as well. An experimental version of FeRoSA is publicly available at www.ferosa.org (receiving as many as 170 hits within the first 15 days of launch).

1 Introduction

One of the most common ways of doing any literature survey is perhaps the following – start from a known article and then traverse along those articles which have either cited the known article or have been cited by the known article. In particular, when a researcher reads the known article, she starts ruminating and asking recurrent questions pertaining to it that can further lead her to browse the other articles. These questions are most often synthesized based on the *knowledge context* of the users. For instance, an expert user, while reading a paper, might want to find papers presenting "alternative approach" of the query paper; while on the other hand, a naïve user might be interested to understand the "background" of the query paper. A smart recommendation engine should be able to organize the recommended papers into multiple such facets/tags. This would not only reduce the tedious effort of searching related articles, but also should answer a more fundamental question: what is the *role* of a recommended

© Springer International Publishing Switzerland 2016
J. Bailey et al. (Eds.): PAKDD 2016, Part II, LNAI 9652, pp. 528–541, 2016.
DOI: 10.1007/978-3-319-31750-2_42

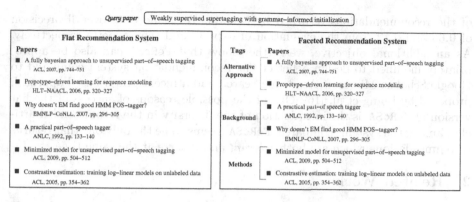

Fig. 1. An illustrative example of a flat (left) and a faceted (right) recommendation system. The figure in the right side shows three facets (see Sect. 3) through which the recommended papers are related to the query paper.

paper in relation to the query paper. However, the traditional paper recommendation systems primarily aim at improving the *relevance* of the recommendations and therefore tend to overlook the above fundamental aspect.

In this paper, we attempt to build a "Faceted Recommendation System for Scientific Articles" (FeRoSA) that given a query paper, in addition to recommending the relevant scientific papers, organizes the recommendations into facets, thereby, suitably catering to the appropriate knowledge context of the end user. Our methodology is based on a principled framework of *random walk with restarts* that attempts to simulate the traversal mechanism of a user initiating from the known article. The model takes into consideration both the citation links as well as the content information to systematically produce the most relevant results. FeRoSA groups the recommendations into four naturally observed facets, namely, Background, Alternative Approaches, Methods and Comparison. This grouping has been formulated from the most intuitive forms of the knowledge context of the end users, which directly map to the different broad sections of any paper; however, the current system can very easily adapt to any other suitable form of grouping. A representative example of a flat and a faceted paper recommendation system is shown in Fig. 1.

To the best of our knowledge, ours is the first *faceted* recommendation system for scientific articles. Moreover, the evaluation of such kind of systems is non-trivial and requires expert judgment. Therefore, we develop novel evaluation schemes. Due to the lack of standard baseline for faceted recommendation, we design two baseline systems and compare the performance of FeRoSA with these systems on the AAN (ACL Anthology Network) dataset [22]. The evaluation is conducted in three steps. First, the experts having significant domain knowledge are asked to develop the ground-truth dataset which is limited in size, based upon which initial evaluation of the faceted system is conducted. Second, a set of researchers having partial knowledge of the domain ("semi-experts") are asked for a mass-scale evaluation of the faceted system. Finally, we shortlist a few papers and request one of the authors of each paper to judge the quality of the recommendations returned by FeRoSA. We achieve an overall precision of 0.65 and 0.72 from the evaluation of experts and semi-experts respectively. As an additional objective, we further show that FeRoSA can also be appropriately modified to design a better flat recommendation system compared to Google Scholar, Microsoft Academic Search and a recent graph-based approach proposed by Liang et al. [16]. One of the possible reasons of success of the flat version of FeRoSA is the introduction of the diversity in the recommended articles; among the systems compared, FeRoSA seems to be the only one that returns recommendations which are both *relevant* and *diverse* at the same time.

2   Related Work

Techniques applied for the traditional recommendation systems can be broadly classified into three categories: (i) *Collaborative filtering (CF)* [24], (ii) *Content-based methods* [13] and (iii) *Hybrid and other approaches* [1]. Several systems have also been developed particularly for scientific paper recommendation. Sugiyama and Kan [25] designed scholarly paper recommendation with citation and reference information. Lee et al. [15] proposed a personalized academic research paper recommendation system. Gipp et al. [8] developed "Scienstein" that can improve the approach of the typically used keyword-based search. Collaborative filtering in the domain of research paper recommendation has been criticized for various reasons [2].

In the context of designing search systems, there have been few attempts to incorporate facets [20]. Tunkelang [26] presented a novel approach that addresses the vocabulary problem for faceted data. Hearst [10] proposed a design guideline for faceted search interfaces. Bast and Weber [3] demonstrated the Semantic GrowBag approach to automatically organize facets for community-specific document collections. Recently, Diederich et al. [7] developed "FacetedDBLP" that allows to search computer science publications starting from some keyword and returns the result set along with a set of facets, e.g., distinguishing publication years, authors, or conferences. Sacco and Tzitzikas [23] presented theory and research results in dynamic taxonomy and faceted search systems. Vallet et al. [27] examined the use of multi-faceted recommendations to aid users while carrying out exploratory video retrieval tasks. In his master's thesis, Celma [4] studied music recommendation using a multi-faceted approach.

However, to the best of our knowledge, ours is the first attempt to introduce faceted paper recommendation system which unlike other real-time systems, is not only able to retrieve the relevant set of papers against a query paper, but also draws a semantic relation between the recommended papers and the query paper.

3   Dataset

We collected the AAN dataset[1] [22] which is an assemblage of all papers included in ACL[2] publication venues. In the full dataset, most of the papers had raw

---

[1] http://clair.eecs.umich.edu/aan/xml/.

[2] https://www.aclweb.org/.

text. The texts were pre-processed where sentences, paragraphs and sections were properly separated using different markers. A significant part of the corpus had word splits and word joins. These were rectified using a dictionary based approach. The filtered dataset contains 9,843 papers (average 6.21 references per paper pointing to the papers within the dataset) and 7,892 unique authors.

We categorize the citation links based on their occurrence in various sections of the paper. Therefore, in addition to the citing-cited paper pair for each citation, we also need to know the context and the section heading where the citation has occurred, in order to assign the facet. We use *Parscit* [6] to identify the citation contexts from the dataset and then extract the section headings for the pair of papers within the network.

Extraction of Section Heading. To extract the section heading, a list of 25,483 unique headings is collected and manually annotated into five different categories: Introduction, Related Work, Method, Results and Conclusion. The categories are further mapped into four facets, namely Background (Introduction), Alternative Approaches (Related Work), Method (Method) and Comparison (Results and Conclusion), as also suggested by Zhigang et al. [11]. A brief description of the facets/tags is as follows:

- **Background (BG):** These are the citations which are prerequisite for understanding the basic notions of the citing paper. These citations generally point either to some seminal papers in that particular area, or to some papers which describe certain concepts that are relevant in understanding the framework of the citing paper. For instance, [9] is a suitable background paper for this study.
- **Alternative Approaches (AA):** If there are citations to the approaches, which can be seen as alternative to the method proposed in the citing paper, then such citations are categorized as AA. These references are often found in system-oriented research papers where new methods/frameworks are proposed. For instance, for this paper, [16] may be treated as AA.
- **Methods (MD):** If the citing paper borrows any such tools, techniques, datasets, measures or other concepts from the paper, or if both the papers have some overlap in usage of any of the entities mentioned above, then such a citation is treated as MD. For instance, [22] is a potential MD for this paper.
- **Comparison (CM):** As mentioned in [16], a relation is said to be comparable if the citing paper has been compared to a cited paper in terms of differences or resemblances. Most of the times, these types of references tend to occur in the evaluation section of the citing paper. For instance, [16] is related to this paper by the CM tag. Essentially, one can argue that all the AA-tagged papers can be treated as CM papers. However, the AA-tagged papers may be irreproducible or difficult to be reimplemented, and thus may not be used for comparison. We only consider the cited paper as CM if it is used by the citing paper for comparison.

A facet is assigned to each pair of citing-cited paper, depending on the section information. Note that if a cited paper occurs multiple times in different sections of a citing paper, multiple facets would get assigned to this paper pair.

Out of total 61,051 citation contexts extracted, the proportion in each facet is as follows: 23,022 (BG), 10,797 (AA), 8,828 (MD) and 18,404 (CM). To validate our approach of mapping section information to facets, we took experts' opinions[3] and obtained an average precision of 0.66. In parallel, we also performed an automated way of annotating references with the facets by Stanford MaxEnt classifier [17] with the features mentioned in [12], and obtained average precision of 0.68 (after 10-fold cross validation)[4]. We observed that the results obtained from the annotations using the section information and that from the supervised classification model were comparable. Moreover, the former is straightforward to compute and can be easily incorporated into a real application. Therefore, we proceed with the annotations obtained directly from the section information.

4   Recommendation Method

In this section, we describe in detail the working principle of our proposed recommendation system. Figure 2 shows a schematic diagram of the proposed work-flow.

The Citation Network. We build the citation network which is a directed graph $G = (V, E)$ with edge labels. The labeling is a mapping from the edge set E to the set of facets based on the data obtained from the citation contexts. An edge may be tagged with multiple facets, if a paper cites another paper in multiple sections.

The Induced Subgraph. We construct an induced subgraph of the network for each query paper. An initial pool of vertices is obtained by following two criteria: (i) we consider all the papers which are at 1-hop or 2-hop distance from the query paper in the citation network irrespective of the label and directionality of edges; (ii) we also consider those papers that have a cosine similarity of at least 0.49 with the query paper (top 100 papers if the number of papers exceeds 100). Then we construct an induced subgraph of nodes present in the initial pool for each facet individually. For instance, for AA we only consider those citation edges in the induced subgraph which are labeled as AA. Note that in this process, few nodes might get disconnected or remain isolated. We connect these nodes with the query node through teleportation probability as discussed below.

Random Walk on the Induced Subgraphs. One of the simplest ways to simulate the process of literature review based on knowledge context would be a suitable form of random walk that can mimic the article surfing behavior. Here, in order to obtain the importance of the nodes with respect to the query paper,

---

[3] The expert opinion was taken from the annotators, who were later involved in evaluating the systems as discussed in Sect. 5. For a direct reference of a paper, we asked experts whether the reference indicates BG, AA, MD or CM and then compared their opinion with our section annotation (in four categories).

[4] In the interest of space, the detailed experiments and results of the supervised classification are not presented in this paper.

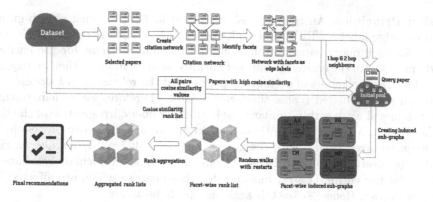

Fig. 2. (Color online) The work-flow diagram of FeRoSA.

we perform random walk with restarts (RWR)[19] on the induced subgraph with query paper being the starting node. RWR is defined in Eq. 1: consider a random walker that starts the walk from node i. The walker iteratively moves to its neighborhood with a probability proportional to the edge weights. At each step of the random walk, it has some probability c to return to the starting node i. The relevance score of node j with respect to node i is defined by the steady-state probability $r_{i,j}$ that the walker will finally stay at node j:

$$\overrightarrow{r}_i = (1 - c)\hat{A}\overrightarrow{r_i} + c\overrightarrow{e}_i \tag{1}$$

where $\overrightarrow{r}_i = [r_{i,j}]$ is an $n \times 1$ ranking vector; $r_{i,j}$ is the relevance score of node j with respect to node i; c is the restart probability, $0 \le c \le 1$ (we consider $c = 0.4$ [19]); $\hat{A}$ is the normalized weighted matrix associated with the weighted adjacency matrix $A = [a_{ij}]$; $\overrightarrow{e}_i$ is the restart vector, with all its elements 0 except the ith element.

Apart from the citation links in the induced subgraph, we also consider the isolated nodes by assigning a *teleportation probability* (i.e., a probability of randomly jumping to any one of the isolated nodes) as 0.3, thus eliminating the chance of the isolated nodes remaining unreachable by the random walker.

Rank Aggregation. We use the above framework to obtain a rank list of nodes present in the induced subgraph for each facet separately. Additionally, we consider content similarity by measuring the cosine-similarity between the query paper and each of the papers present in the induced subgraph. Next, we utilize a rank aggregation method to combine these two types of rankings. In our work, we use *RankAggreg* [21], where the rank aggregation is considered as an optimization problem to find an ordered list δ that minimizes the total distance between each of the provided lists L_i and δ. Note that for each facet T, we aggregate the ranking obtained for T and the cosine-similarity based ranking to obtain the final rank list. In addition, we also perform a total rank aggregation in order to design a flat version of FeRoSA (f-FeRoSA) by combining all the facet-wise rankings and the cosine similarity based ranking together (see Sect. 5.3).

Design Principles. All the sub-tasks involved in FeRoSA, such as sub-graph creation, edge labeling, RWR and rank aggregation, can be performed independently for each query paper. To find out the recommendations for the entries, we do not require the entire (global) network to be in-memory; rather we require only the 2-hop neighbors of each query paper, and the relevant top k documents based on pre-calculated cosine similarity values. Therefore, we anticipate that the system will work within similar time bound per query, irrespective of the size of the dataset. Moreover, we noticed that on an average the entire process for a single query takes 382 ms, with the largest one being 497 ms. When new entries need to be inserted into the existing system, we need not recalculate recommendations for the entire dataset, but only for those entries which are affected by the new entries. Hence, FeRoSA is scalable and light-weight.

5   Experimental Results

In this section, we present the performance of FeRoSA. We design a new evaluation framework, consisting of three independent steps: (i) experts' judgment on a limited set of papers, (ii) semi-experts' judgment for mass-scale evaluation, and (iii) the judgment by the original authors of the paper. Finally, we show that FeRoSA can also be used to design a better flat recommendation system.

5.1   Evaluation Metrics

We use *Overall Precision* (OP) and *Overall Impression* (OI) for comparative evaluation. OP measures the ratio between the number of relevant recommendations (according to the experts' judgments) and the total number of recommendations provided for a query paper by each competing system. The OP of each system is then measured by averaging OPs for all the query papers. OI measures that among all the query papers, in how many cases a particular system is rated to have an overall better performance. We measure this value for a system on the basis of precision majority. Similarly for faceted evaluation, we measure the OP of the recommended papers under each individual facet.

5.2   Evaluation of Faceted Recommendation

In this section, we first describe the process of ground-truth generation, followed by a brief description of the baseline algorithms, and then elaborate the comparative evaluation of all the systems for faceted recommendation.

Ground-Truth Generation. Because of the unavailability of a benchmark dataset for the evaluation of scientific article recommendation especially for the faceted recommendation, we conducted an expert judgment to generate a set of faceted and flat recommendations (used later in Sect. 5.3) as our ground-truth. First, we shortlisted a set of 30 query papers that cover the fields of expertise of 10 experts. For each query paper, we presented 30 recommendations that we pulled from four separate systems: FeRoSA, Google Scholar (GS)[5], Microsoft Academic Search (MAS)[6] and a graph based paper recommendation system proposed by Liang et al. [16] (LLQ henceforth). Note that the three latter systems which are quite popular for paper recommendation are further used in Sect. 5.3 as competing systems to FeRoSA for flat recommendation. The experts were provided with web based interfaces[7], in which they were shown 30 recommendations for each query paper (the name of the systems remained anonymous). Each expert had to mark whether each recommended paper was relevant to the query paper, and if so, the possible facet(s).

Baseline Systems. Due to the lack of faceted recommendation system for scientific literature, we design two competitive baseline systems to compare with FeRoSA.

- **VanillaPR:** For each query paper, we form a single induced subgraph $G'(V', E')$ which is exactly the same as that built for FeRoSA but ignores the facet labeling. Once the graph is formed, we perform RWR from the query paper. We then retrieve the nodes having the highest values from RWR. Finally, to label the retrieved papers with facets, we train a supervised model using the ground-truth data we collected in the experts' judgment. We use the following three types of features to train the supervised model. For each pair of the query and the recommended paper, we use total 12 (4 Boolean + 8 real valued) features: (i) section of the recommended paper, if it appears in 1-hop of the query paper (4 Boolean features, one for each section), (ii) within V' for the query paper, fractional number of times a particular recommended paper appears in a given section (4 real-valued features, one for each section), and (iii) for a given recommended paper, fraction of times it is cited in a given section by any paper in the whole dataset (4 real-valued features, one for each section). We then learn the weights for features with a rankSVM model [14]. We report the average precision of the system after performing a three fold cross-validation over the ground-truth data.
- **FeRoSA-CS:** Our second baseline recommends papers by relying only on RWR, performed on subgraphs of papers within 2-hop distance from the query paper, without considering the cosine-similarity (CS) based papers. This in turn answers the necessity of considering cosine-similarity based papers while constructing the initial pool.

Comparative Analysis. We conduct an empirical study on the results obtained from FeRoSA. Figure 3(a) represents a Venn diagram of all the recommended papers under different facets, i.e., in what facets have a particular paper been recommended for different queries. For example, 3.54% of the recommended papers appear only as BG to any of the query paper. We observe that 58.48% of

---

[5] http://scholar.google.co.in.

[6] http://academic.research.microsoft.com/.

[7] http://www.ferosa.org/evaluation/.

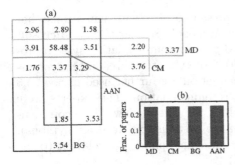

Fig. 3. (Color online) (a) Venn diagram of the recommended papers in four facets; (b) distribution of papers which are tagged by all four facets.

Table 1. Faceted evaluation of all the competing faceted recommendation systems. For VanillaPR, the results are reported by performing 3-fold cross validation.

Facets	VanillaPR	FeRoSA-CS	FeRoSA
BG	0.65	0.51	**0.79**
AA	0.48	0.34	**0.56**
MD	**0.62**	0.39	**0.62**
CM	0.44	0.38	**0.62**
Average	0.55	0.40	**0.65**

the recommended papers appear under all the four facets for various queries. For these papers, we further show in Fig. 3(b) their distribution in different facets, which seems to be fairly uniform.

We report in Table 1 the OP of all the systems for different facets. We observe that FeRoSA attains the highest OP (0.65) amongst all other systems, which is 18 % and 62.5 % higher than VanillaPR and FeRoSA-CS respectively. The maximum OP of FeRoSA is obtained for BG (0.79), which is followed by MD (0.62), CM (0.62), and AA (0.56). The pattern is also similar for FeRoSA-CS. For the case of MD, however, we observe similar performance for VanillaPR and FeRoSA.

For further analysis, we present the confusion matrix for the facets in Table 2 along with the false positive rate (FPR) for each system. The FPR is quite low for the facets (i.e., the specificity values of the facets are quite high). To understand the reason behind the misclassification, we unfold the tagged citation network once again and observe that it arises due to the frequent occurrences of a single edge being tagged by multiple facets. For instance, Table 2 shows that AA is mostly misclassified to MD for FeRoSA. In the tagged citation network, we observe the same phenomenon that among the multi-faceted edges with AA tag, around 50 % of edges are tagged by both AA and MD (similarly for CM and MD with 36.6 % of occurrences together).

Table 2. Confusion matrix for the faceted evaluation (false positive rate: FPR). For VanillaPR, we report the confusion matrix for that run where maximum OP for all the facets is achieved.

Facets	AA	BG	CM	MD	FPR
AA	29	11	15	8	0.08
BG	13	87	11	17	0.09
CM	7	9	19	11	0.06
MD	11	6	7	32	0.05

(a) VanillaPR

Facets	AA	BG	CM	MD	FPR
AA	23	14	9	17	0.09
BG	18	69	13	28	0.14
CM	13	8	18	7	0.06
MD	11	14	5	26	0.05

(b) FeRoSA-CS

Facets	AA	BG	CM	MD	FPR
AA	33	11	6	13	0.07
BG	14	99	5	10	0.07
CM	2	6	28	10	0.04
MD	3	7	12	34	0.04

(c) FeRoSA

Mass-Scale Evaluation. To broaden the evaluation of FeRoSA, we perform a mass-scale evaluation, aiming for more coverage on the system output and targeting a wider set of evaluators. All the selected evaluators had a good knowledge of the NLP domain. This time we reverse engineer the process by selecting few papers from the ground-truth data, each of which appears in the recommendation of multiple query papers. To start with, we shortlisted a collection of 31 such recommended papers. For each recommended paper, we enlisted the set of query papers (and the facets) in which the recommended paper has appeared[8]. The evaluators then evaluated the relevance of the recommended paper, as well as the relevance of the facet with respect to each query paper in which the given recommended paper has appeared. Total 26 experts participated in this evaluation task. For each recommended paper, we calculate OP per facet for all its corresponding query papers and show the results in Table 3(a). Similarly, we calculate facet-wise OP for all the query papers and report it in Table 3(b). We observe that for both the cases, FeRoSA significantly outperforms other baselines.

Table 3. Overall precision per facet for (a) recommended paper to query paper and (b) query paper to recommended paper.

<table>
<tr><th colspan="4">(a)</th><th colspan="4">(b)</th></tr>
<tr><th>Facets</th><th>VanillaPR</th><th>FeRoSA–CS</th><th>FeRoSA</th><th>Facets</th><th>VanillaPR</th><th>FeRoSA–CS</th><th>FeRoSA</th></tr>
<tr><td>BG</td><td>0.57</td><td>0.70</td><td>**0.73**</td><td>BG</td><td>0.66</td><td>0.82</td><td>**0.85**</td></tr>
<tr><td>AA</td><td>0.43</td><td>0.41</td><td>**0.53**</td><td>AA</td><td>0.45</td><td>0.48</td><td>**0.54**</td></tr>
<tr><td>CM</td><td>0.37</td><td>0.59</td><td>**0.64**</td><td>CM</td><td>0.42</td><td>0.54</td><td>**0.73**</td></tr>
<tr><td>MD</td><td>0.58</td><td>0.55</td><td>**0.69**</td><td>MD</td><td>0.51</td><td>0.68</td><td>**0.77**</td></tr>
<tr><td>Avg.</td><td>0.49</td><td>0.56</td><td>**0.64**</td><td>Avg.</td><td>0.51</td><td>0.63</td><td>**0.72**</td></tr>
</table>

As a related objective we investigate whether our system performs better for the highly-cited query papers, or whether the same accuracy is achieved for all citation ranges of the query papers. Generally, a standard recommendation system should perform equally well for all ranges of query papers [15]. Here we divide the entire range of incoming citations of the query paper into three buckets and measure the facet-wise OP of all the competing systems for each bucket separately. In Table 4, we see that FeRoSA performs significantly better than the other baseline systems even for low-cited query papers.

Evaluation by the Authors. There is no better alternative than the authors themselves when it comes to evaluating the recommendation for a particular paper. We were curious to know whether FeRoSA could impress the authors with its recommendations. We therefore designed a judgment experiment by selecting a set of 30 authors and sent each of them, a judgment form, where we specified one of his/her papers as query paper, and one (top) recommendation from FeRoSA for each facet. The author had to make a binary judgment about the relevance of recommendation to the query as well as the relevance of the facet for the recommendation separately. A sample response of an anonymous author can be found at http://ferosa.org/#authorresponse. Twelve authors responded to the survey. We obtain an average precision of 0.50 (BG: 0.49, AA: 0.42,

---

[8] This indeed reduced the evaluators' effort of reading multiple papers.

Table 4. Performance of three competing faceted systems for different query papers divided into three citation ranges (Low: 0 to 6, Medium: 7 to 28, High: 29 to 343).

Facets	VanillaPR			FeRoSA-CS			FeRoSA		
	Low	Medium	High	Low	Medium	High	Low	Medium	High
BG	0.53	0.73	0.71	0.44	0.54	0.55	0.65	0.84	0.87
AA	0.41	0.52	0.49	0.28	0.35	0.41	0.53	0.56	0.61
MD	0.57	0.59	0.71	0.40	0.34	0.44	0.65	0.55	0.67
CM	0.29	0.55	0.48	0.33	0.39	0.41	0.56	0.62	0.69
Avg	0.45	0.59	0.59	0.36	0.40	0.45	**0.59**	**0.64**	**0.71**

MD: 0.52, CM: 0.59). In 75 % cases the recommended papers are marked as relevant. Four authors marked three out of four faceted recommendations as relevant. Overall, the authors appreciated the attempt of designing a faceted recommendation system for scientific articles.

5.3 Evaluation of Flat Recommendation

We further posit that FeRoSA can also be used as a flat recommendation system if the rank lists obtained from the different facets and the cosine-similarity based ranking can be appropriately combined. Therefore, we use the rank-aggregation method discussed in Sect. 4 in order to obtain a flat recommendation list.

In this section, we discuss the performance of the flat version of FeRoSA (f-FeRoSA henceforth) and compare it with three state-of-the-art flat baseline systems: Google Scholar (GS), Microsoft Academic Search (MAS) and a graph based paper recommendation system, LLQ [16]. We consider LLQ as a baseline system because similar to our approach, it also classifies citation relations into three categories, namely Based-on, Comparable and General using the approach proposed in [18], and these categories are further used to compute a final combined score. Note that while GS and MAS are mostly known for searching scientific papers, an inherent nature of ranking of the retrieved results has lead us in using them as potential baseline systems.

We perform a broad analysis of the performance of all the competing methods. Table 5(a) reports the values of individual metrics mentioned earlier, averaged over all the judgments conducted by the experts. For top three recommendations per system, f-FeRoSA achieves OP of 0.79 which is 29 %, 75 % and 62 % higher than GS, MAS and LLQ respectively. One can also notice that for 43 % of the cases, f-FeRoSA fares better than all other systems in terms of OI. Clearly, f-FeRoSA is preferred nearly twice more than GS, which is the second best performing system. This indeed shows that f-FeRoSA outperforms the state-of-the-art recommendation systems by a reasonable margin. We also see in Table 5(b) that f-FeRoSA is quite consistent in recommending highly relevant papers within top rank list.

Table 5. (a) Flat evaluation of the competing systems based on relevance and diversity; (b) overall precision of `f-FeRoSA` at different number of recommendations.

(a)

System	Relevance		Diversity			(b)	
	OI@3	OP@3	ISP	σ_1	σ_2	OP	f-FeRoSA
GS	0.27	0.61	0.043	24.92	66.29	OP@3	0.79
MAS	0.17	0.45	0.043	16.51	53.85	OP@5	0.78
LLQ	0.13	0.41	0.054	21.43	63.82	OP@10	0.71
f-FeRoSA	0.43	0.79	0.003	54.77	108.93		

As mentioned earlier, the reason behind the success of `f-FeRoSA` can be the introduction of *diversity*, i.e., inclusion of highly relevant papers from each facet into the aggregated list. To substantiate this argument quantitatively, we further take two graph-based measures from [5] and evaluate the competing flat systems based on diversity: (i) *Inverse Shortest Path* (ISP) $= \frac{\sum_{u,v \in S} 1/d_{uv}}{|S| \times (|S|-1)}$, (ii) *Expansion Ratio* $(\sigma_l) = \frac{\bigcup_{u \in S} N_u^l}{|S|}$, where S: set of recommended papers, d_{uv}: shortest path between u and v in the citation network, and N_u^l: neighbors within l-hops of u in the citation network (we take $l = 1, 2$). The less (more) the ISP (σ_l), the more the diversity. Results in Table 5(a) corroborate our argument that `f-FeRoSA` is indeed the most diverse system among others.

6   Discussions and Future Work

In this paper, we proposed a faceted recommendation system, `FeRoSA` for scientific articles that not only recommends relevant articles for a particular query paper, but also relates the recommended articles with the query paper through four pre-defined facets. As a by-product of this study, we also obtain an annotated citation network where each citation link between a citing paper and a cited paper gets labeled, and a ground-truth dataset for evaluating scientific recommendation systems. `FeRoSA` is designed to be light-weight, so that it can easily be deployed as an online system. We evaluated our system in three stages based on human judgment and observed significant performance improvement. Although `FeRoSA` is designed for faceted recommendation, we further showed that it significantly outperforms the baselines in flat recommendation.

The scalability of our proposed model can be guaranteed after its extensive usage and testing on other datasets. We are also interested in the design aspects related to the ergonomics of the user interface so that it can significantly reduce user's cognitive overload, while providing high user satisfaction at the same time. We anticipate that the framework used in `FeRoSA` can be adopted to design faceted recommendations for items such as movies, books, videos etc. The annotated citation network and the human evaluation results are available at www.ferosa.org/data.

References

1. Adomavicius, G., Tuzhilin, A.: Toward the next generation of recommender systems: a survey of the state-of-the-art and possible extensions. IEEE TKDE **17**(6), 734–749 (2005)
2. Agarwal, N., Haque, E., Liu, H., Parsons, L.: Research paper recommender systems: a subspace clustering approach. In: Fan, W., Wu, Z., Yang, J. (eds.) WAIM 2005. LNCS, vol. 3739, pp. 475–491. Springer, Heidelberg (2005)
3. Bast, H., Weber, I.: Type less, find more: fast autocompletion search with a succinct index. In: SIGIR, Seattle, WA, pp. 364–371. ACM (2006)
4. Celma, Ò.: Music Recommendation: A multi-faceted approach. Master's thesis (2006)
5. Chakraborty, T., Modani, N., Narayanam, R., Nagar, S.: Discern: a diversified citation recommendation system for scientific queries. In: ICDE, Seoul, pp. 555–566 (2015)
6. Councill, I.G., Giles, C.L., Kan, M.Y.: Parscit: an open-source CRF reference string parsing package. In: LREC, Marrakech, Morocco (2008)
7. Diederich, J., Balke, W.T., Thaden, U.: Demonstrating the semantic growbag: automatically creating topic facets for facetedDBLP. In: JCDL, pp. 505–505. ACM, New York (2007)
8. Gipp, B., Beel, J., Hentschel, C.: Scienstein: a research paper recommender system. In: ICETiC, Virudhunagar, India, pp. 309–315 (2009)
9. He, Q., Pei, J., Kifer, D., Mitra, P., Giles, L.: Context-aware citation recommendation. In: WWW, pp. 421–430. ACM, New York (2010)
10. Hearst, M.A.: Design recommendations for hierarchical faceted search interfaces. In: Proceedings of SIGIR 2006, Workshop on Faceted Search, Seattle, Washington, pp. 26–30 (2006)
11. Hu, Z., Chen, C., Liu, Z.: Where are citations located in the body of scientific articles? A study of the distributions of citation locations. J. Informetrics **7**(4), 887–896 (2013)
12. Jochim, C., Schütze, H.: Towards a generic and flexible citation classifier based on a faceted classification scheme. In: COLING, Mumbai, India, pp. 1343–1358 (2012)
13. Koren, Y.: Factorization meets the neighborhood: a multifaceted collaborative filtering model. In: SIGKDD, New York, USA, pp. 426–434 (2008)
14. Lee, C.P., Lin, C.J.: Large-scale linear rankSVM. Neural Comput. **26**(4), 781–817 (2014)
15. Lee, J., Lee, K., Kim, J.G.: Personalized academic research paper recommendation system. CoRR abs/1304.5457 (2013)
16. Liang, Y., Li, Q., Qian, T.: Finding relevant papers based on citation relations. In: Wang, H., Li, S., Oyama, S., Hu, X., Qian, T. (eds.) WAIM 2011. LNCS, vol. 6897, pp. 403–414. Springer, Heidelberg (2011)
17. Manning, C., Klein, D.: Optimization, maxent models, and conditional estimation without magic. In: NAACL, p. 8. ACL, Stroudsburg (2003)
18. Nanba, H., Okumura, M.: Towards multi-paper summarization using reference information. In: IJCAI, pp. 926–931. Morgan Kaufmann, Stockholm (1999)
19. Pan, J.Y., Yang, H.J., Faloutsos, C., Duygulu, P.: Automatic multimedia cross-modal correlation discovery. In: SIGKDD, Seattle, WA, pp. 653–658 (2004)
20. Peng, T., Seng-cho, T.: itrustu: a blog recommender system based on multi-faceted trust and collaborative filtering. In: SAC, Hawaii, USA, pp. 1278–1285. ACM (2009)

21. Pihur, V., Datta, S., Datta, S.: Rankaggreg, an R package for weighted rank aggregation. BMC Bioinform. **10**(1), 62 (2009)
22. Radev, D., Muthukrishnan, P., Qazvinian, V., Abu-Jbara, A.: The ACL anthology network corpus. In: LREC, pp. 1–26 (2013)
23. Sacco, G.M., Tzitzikas, Y.: Dynamic Taxonomies and Faceted Search: Theory, Practice, and Experience, 1st edn. Springer, Heidelberg (2009)
24. Sarwar, B., Karypis, G., Konstan, J., Riedl, J.: Item-based collaborative filtering recommendation algorithms. In: WWW, pp. 285–295. ACM, New York (2001)
25. Sugiyama, K., Kan, M.Y.: Scholarly paper recommendation via user's recent research interests. In: JCDL, pp. 29–38. ACM, New York (2010)
26. Tunkelang, D.: Dynamic category sets: an approach for faceted search. In: ACM SIGIR. ACM, New York (2006)
27. Vallet, D., Halvey, M., Hannah, D., Jose, J.M.: A multi faceted recommendation approach for explorative video retrieval tasks. In: IUI, New York, USA, pp. 389–392 (2010)

Dual Similarity Regularization
for Recommendation

Jing Zheng[1], Jian Liu[1], Chuan Shi[1(✉)], Fuzhen Zhuang[2],
Jingzhi Li[3], and Bin Wu[1]

[1] Beijing Key Lab of Intelligent Telecommunications Software and Multimedia,
Beijing University of Posts and Telecommunications, Beijing 100876, China
`oliviacheng@126.com`, `fullback@yeah.net`, {`shichuan,wubin`}`@bupt.edu.cn`
[2] Key Laboratory of Intelligent Information Processing, Institute of Computing
Technology, Chinese Academy of Sciences, Beijing 100190, China
`zhuangfz@ics.ict.ac.cn`
[3] Department of Mathematics, Southern University of Science and Technology,
Shenzhen 518055, China
`lijz@sustc.edu.cn`

Abstract. Recently, social recommendation becomes a hot research
direction, which leverages social relations among users to alleviate data
sparsity and cold-start problems in recommender systems. The social
recommendation methods usually employ simple similarity information
of users as social regularization on users. Unfortunately, the widely used
social regularization may suffer from several aspects: (1) the similarity
information of users only stems from users' social relations; (2) it only
has constraint on users; (3) it may not work well for users with low similarity. In order to overcome the shortcomings of social regularization,
we propose a new dual similarity regularization to impose the constraint
on users and items with high and low similarities simultaneously. With
the dual similarity regularization, we design an optimization function to
integrate the similarity information of users and items, and a gradient
descend solution is derived to optimize the objective function. Experiments on two real datasets validate the effectiveness of the proposed
solution.

Keywords: Social recommendation · Regularization · Heterogeneous
information network

1 Introduction

Recommender system, as an effective way to tackle information overload problems, has attracted much attention from multiple disciplines. Many techniques
have been proposed to build recommender systems. As a popular technique, the
low rank matrix factorization has shown its effectiveness and efficiency, which fac-
torizes user-item rating matrix into two low rank user-specific and item-specific
matrices, then utilizes the factorized matrices to make further predictions [10].

© Springer International Publishing Switzerland 2016
J. Bailey et al. (Eds.): PAKDD 2016, Part II, LNAI 9652, pp. 542–554, 2016.
DOI: 10.1007/978-3-319-31750-2_43

With the boom of social media, social recommendation has become a hot research topic, which utilizes the social relations among users for better recommendation. Some researchers utilized trust information among users [5,6], and some began to use friend relationship among users [7,12] or other types of information [1,2]. Most of these social recommendation methods employ social regularization to confine similar users under the low rank matrix factorization framework. Specifically, we can obtain the similarity of users from their social relations as a constraint term to confine the latent factors of similar users to be closer. It is reasonable, since similar users should have similar latent features.

However, the social regularization used in social recommendation has several shortcomings. (1) The similarity information of users is only generated from social relations of users. But we can obtain users' similarity from many ways, such as users' contents. (2) The social regularization only has constraint on users. In fact, we can obtain the similarity of items to impose constraint on the latent factors of items. (3) The social regularization may be less effective for dissimilar users, which may lead to dissimilar users having similar factors. The analysis and experiments in Sect. 2 validate this point.

In order to address the limitations of traditional social recommendation, we propose a Dual Similarity Regularization based recommendation method (called DSR) in this paper. Inspired by the success of Heterogeneous Information Network (HIN) in many applications, we organize objects and relations in a recommender system as a HIN, which can integrate all kinds of information, including interactions between users and items, social relations among users and attribute information of users and items. Based on the HIN, we can generate rich similarity information on users and items by setting proper meta paths. Furthermore, we propose a new similarity regularization which can impose the constraint on users and items with high and low similarity. With the similarity regularization, DSR adopts a new optimization objective to integrate those similarity information of users and items. Then we derive its solution to learn the weights of different similarities. The experiments on real datasets show that DSR always performs best compared to social recommendation and HIN-based recommendation methods. Moreover, DSR also achieves the best performance for cold-start users and items due to the dual similarity regularization on users and items.

The rest of the paper is organized as follows. We analyze the limitations of social recommendation in Sect. 2 and introduce the rich similarity information of users and items generated from HIN in Sect. 3. Then we propose the similarity regularization and the DSR model in Sect. 4. We do experiments in Sect. 5, describe related work in Sect. 6 and finally draw the conclusion in Sect. 7.

2   Limitations of Social Recommendation

Recently, with the increasing popularity of social media, there is a surge of social recommendations which leverage rich social relations among users to improve recommendation performance. Ma et al. [7] first proposed the social regularization to extend low-rank matrix factorization, and then it is widely used in a lot of work [4,13]. A basic social recommendation method is illustrated as follows:

$$\min_{U,V} \mathcal{J} = \frac{1}{2} \sum_{i=1}^{m} \sum_{j=1}^{n} I_{ij}(R_{ij} - U_i V_j^T)^2$$

$$+ \frac{\alpha}{2} \sum_{i=1}^{m} \sum_{j=1}^{m} S_U(i,j)\|U_i - U_j\|^2 + \frac{\lambda_1}{2}(\|U\|^2 + \|V\|^2), \qquad (1)$$

where $m \times n$ rating matrix R depicts users' ratings on n items, R_{ij} is the score user i gives to item j. I_{ij} is an indicator function which equals to 1 if user i rated item j and equals to 0 otherwise. $U \in \mathbb{R}^{m \times d}$ and $V \in \mathbb{R}^{n \times d}$, where $d << min(m,n)$ is the dimension number of latent factor. U_i is the latent vector of user i derived from the ith row of matrix U while V_j is the latent vector of item j derived from the jth row of V. S_U is the similarity matrix of users and $S_U(i,j)$ denotes the similarity of user i and user j. $\| \cdot \|^2$ is the Frobenius norm. Particularly, the second term is the social regularization which is defined as follows:

$$SocReg = \frac{1}{2} \sum_{i=1}^{m} \sum_{j=1}^{m} S_U(i,j)\|U_i - U_j\|^2. \qquad (2)$$

As a constraint term in Eq. (1), $SocReg$ forces the latent factors of two users to be close when they are very similar. However, it may have two drawbacks.

- The similarity information may be simple. In social recommendation, the similarity information of users is usually generated from rating information or social relations and only one type of similarity information is employed. However, we can obtain much rich similarity information of users and items from various ways, such as rich attribute information and interactions.
- The constraint term may not work well when two users are not very similar. The minimization of optimization objective should force the latent factors of two similar users to be close. But we note that when two users are not similar (i.e., $S_U(i,j)$ is small), $SocReg$ may still force the latent factors of these two users to be close. In fact, these two users is dissimilar which means their latent factors should have a large distance.

In order to uncover the limitations of social regularization, we apply the model detailed in Eq. (1) to conduct four experiments each with different levels of similarity information (*None, Low, High, All*). *None* denotes that we utilize no similarity information in the model (i.e., *alpha* = 0 in the model), *Low* denotes that we utilize bottom 20 % users' similarity information generated in the model, *High* is that of top 20 %, *All* denotes we utilize all users' similarity information. The Douban dataset detailed in Table 1 is employed in the experiments and we report MAE and RMSE (defined in Sect. 5.2) in Fig. 1. The results of *Low, High* and *All* are better than that of *None*, which implies social regularization really works in the model. However, in terms of performance improvement compared to *None, Low* does not improve as much as *High* and *All* do. The above analysis reveals that the social regularization may not work well in recommender models when users are with low similarity.

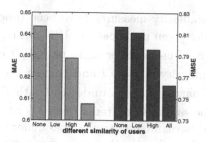

Fig. 1. Limitations of social regularization.

3 Rich Similarity Generated from HIN

Traditional social recommendations only consider the constraint of users with their social relations. However, rich similarity information on users and items can be generated in a heterogeneous information network. A **heterogeneous information network** [11] is a special information network with multiple types of entities and relations. Figure 2(a) shows a typical HIN extracted from a movie recommender system. The HIN contains multiple types of objects, e.g., users (U), movies (M), groups (G), and actors (A).

Two types of objects in a HIN can be connected via various **meta path** [11], which is a composite relation connecting these two types of objects. A meta path $\mathcal{P}$ is a path defined on a schema $\mathcal{S} = (\mathcal{A}, \mathcal{R})$, and is denoted in the form of $A_1 \xrightarrow{R_1} A_2 \xrightarrow{R_2} \cdots \xrightarrow{R_l} A_{l+1}$ (abbreviated as $A_1 A_2 \cdots A_{l+1}$), which defines a composite relation $R = R_1 \circ R_2 \circ \cdots \circ R_l$ between type A_1 and A_{l+1}, where $\circ$ denotes the composition operator on relations. As an example in Fig. 2(a), users can be connected via "User-User" (UU), "User-Movie-User" (UMU) and so on. Different meta paths denote different semantic relations, e.g., the UU path means that users have social relations while the UMU path means that users have watched the same movies. Therefore we can evaluate the similarity of users (or movies) based on different meta paths. For example, for users, we can consider UU, UGU, UMU, etc. Similarly, meaningful meta paths connecting movies include MAM, MDM, etc.

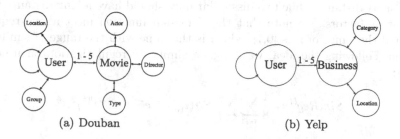

(a) Douban   (b) Yelp

Fig. 2. Network schema of HIN examples.

Several path-based similarity measures have been proposed to evaluate the similarity of objects under given meta path in HIN [9,11]. We assume that $S_U^{(p)}$ denotes similarity matrix of users under meta path $\mathcal{P}_U^{(p)}$ connecting users, and $S_U^{(p)}(i,j)$ denotes the similarity of users i and j under the path $\mathcal{P}_U^{(p)}$. Similarly, $S_I^{(q)}$ denotes similarity matrix of items under the path $\mathcal{P}_I^{(q)}$ connecting items, and $S_I^{(q)}(i,j)$ denotes the similarity of items i and j.

Since users (or items) have different similarities under various meta paths, we combine their similarities on all paths through assigning weights on these paths. For users and items, we define S_U and S_I to represent the similarity matrix of users and items on all meta paths, respectively.

$$S_U = \sum_{p=1}^{|\mathcal{P}_U|} w_U^{(p)} S_U^{(p)}, \tag{3}$$

$$S_I = \sum_{q=1}^{|\mathcal{P}_I|} w_I^{(q)} S_I^{(q)}, \tag{4}$$

where $w_U^{(p)}$ denotes the weight of meta path $\mathcal{P}_U^{(p)}$ among all meta paths $\mathcal{P}_U$ connecting users, and $w_I^{(q)}$ denotes the weight of meta path $\mathcal{P}_I^{(q)}$ among all meta paths $\mathcal{P}_I$ connecting items.

4   Matrix Factorization with Similarity Regularization

In this section, we propose our dual similarity regularization based matrix factorization method **DSR** and infer its learning algorithm.

4.1   Similarity Regularization

Due to the limitations of social regularization, we design a new similarity regularization to constrain users and items simultaneously with much similarity information of users and items. The basic idea of similarity regularization is that the distance of latent factors of two users (or items) should be negatively correlated to their similarity, which means two similar users (or items) should have a short distance while two dissimilar ones should have a long distance with their latent factors. We note that the Gaussian function meet above requirement and the range of it is [0,1], which is the same with the range of similarity function. Following the idea, we design a similarity regularization on users as follows:

$$SimReg^{\mathcal{U}} = \frac{1}{8} \sum_{i=1}^{m} \sum_{j=1}^{m} (S_U(i,j) - e^{-\gamma \|U_i - U_j\|^2})^2, \tag{5}$$

where γ controls the radial intensity of Gaussian function and the coefficient $\frac{1}{8}$ is convenient for deriving the learning algorithm. This similarity regularization can enforce constraint on both similar and dissimilar users. In addition, the similarity matrix S_U can be generated from social relations or the above HIN. Similarly, we can also design the similarity regularization on items as follows:

$$SimReg^{\mathcal{I}} = \frac{1}{8}\sum_{i=1}^{n}\sum_{j=1}^{n}(S_I(i,j) - e^{-\gamma\|V_i-V_j\|^2})^2. \qquad (6)$$

The Proposed DSR Model. We propose the **D**ual **S**imilarity regularization for **R**ecommendation (called DSR) through adding the similarity regularization on users and items into low-rank matrix factorization framework. Specifically, the optimization model is proposed as follows:

$$\min_{U,V,\boldsymbol{w}_U,\boldsymbol{w}_I} \mathcal{J} = \frac{1}{2}\sum_{i=1}^{m}\sum_{j=1}^{n}I_{ij}(R_{ij} - U_iV_j^T)^2 \qquad (7)$$

$$+\frac{\lambda_1}{2}(\|U\|^2 + \|V\|^2) + \frac{\lambda_2}{2}(\|\boldsymbol{w}_U\|^2 + \|\boldsymbol{w}_I\|^2)$$

$$+\alpha SimReg^{\mathcal{U}} + \beta SimReg^{\mathcal{I}}$$

$$s.t. \quad \sum_{p=1}^{|\mathcal{P}_U|}\boldsymbol{w}_U^{(p)} = 1, \boldsymbol{w}_U^{(p)} \geq 0$$

$$\sum_{q=1}^{|\mathcal{P}_I|}\boldsymbol{w}_I^{(q)} = 1, \boldsymbol{w}_I^{(q)} \geq 0,$$

where α and β control the ratio of similarity regularization term on users and items, respectively.

4.2   The Learning Algorithm

The learning algorithm of DSR can be divided into two steps. (1) Optimize the latent factor matrices of users and items (i.e., U, V) with the fixed weight vectors $\boldsymbol{w}_U = [\boldsymbol{w}_U^{(1)}, \boldsymbol{w}_U^{(2)}, \ldots, \boldsymbol{w}_U^{(|\mathcal{P}_U|)}]^T$ and $\boldsymbol{w}_I = [\boldsymbol{w}_I^{(1)}, \boldsymbol{w}_I^{(2)}, \ldots, \boldsymbol{w}_I^{(|\mathcal{P}_I|)}]^T$. (2) Optimize the weight vectors $\boldsymbol{w}_U$ and $\boldsymbol{w}_I$ with the fixed latent factor matrices U and V. Through iteratively optimizing these two steps, we can obtain the optimal U, V, $\boldsymbol{w}_U$, and $\boldsymbol{w}_I$.

Optimize U and V. With the fixed $\boldsymbol{w}_U$ and $\boldsymbol{w}_I$, we can optimize U and V by performing stochastic gradient descent.

$$\frac{\partial \mathcal{J}}{\partial U_i} = \sum_{j=1}^{n}I_{ij}(U_iV_j^T - R_{ij})V_j \qquad (8)$$

$$+\alpha\sum_{j=1}^{m}\gamma[(S_U(i,j) - e^{-\gamma\|U_i-U_j\|^2})e^{-\gamma\|U_i-U_j\|^2}(U_i - U_j)] + \lambda_1U_i,$$

$$\frac{\partial \mathcal{J}}{\partial V_j} = \sum_{i=1}^{m} I_{ij}(U_i V_j^T - R_{ij})U_i \tag{9}$$

$$+ \beta \sum_{i=1}^{n} \gamma[(S_I(i,j) - e^{-\gamma\|V_i-V_j\|^2})e^{-\gamma\|V_i-V_j\|^2}(V_i - V_j)] + \lambda_1 V_j.$$

Optimize w_U and w_I. With the fixed U and V, the minimization of $\mathcal{J}$ with respect to w_U and w_I is a well-studied quadratic optimization problem with non-negative bound. We can use the standard trust region reflective algorithm to update w_U and w_I at each iteration. We can simplify the optimization function of w_U as the following standard quadratic formula:

$$\min_{w_U} \frac{1}{2} w_U^T H_U w_U + f_U^T w_U \tag{10}$$

$$s.t. \sum_{p=1}^{|\mathcal{P}_U|} w_U^{(p)} = 1, w_U^{(p)} \geq 0.$$

Here H_U is a $|\mathcal{P}_U| \times |\mathcal{P}_U|$ symmetric matrix as follows:

$$H_U(i,j) = \begin{cases} \frac{\alpha}{4}(\sum\sum S_U^{(i)} \odot S_U^{(j)}) & i \neq j, 1 \leq i,j \leq |\mathcal{P}_U| \\ \frac{\alpha}{4}(\sum\sum S_U^{(i)} \odot S_U^{(j)}) + \lambda_2 & i = j, 1 \leq i,j \leq |\mathcal{P}_U|, \end{cases}$$

$\odot$ denotes the dot product. f_U is a column vector with length $|\mathcal{P}_U|$, which is calculated as follows:

$$f_U(p) = -\frac{\alpha}{4} \sum_{i=1}^{m} \sum_{j=1}^{m} S_U^{(p)}(i,j)e^{-\gamma\|U_i-U_j\|^2}.$$

Similarly, we can also infer the optimization function of w_I.

5   Experiments

In this section, we conduct experiments to validate the effectiveness of DSR and further explore the cold-start problem.

5.1   Dataset

We use a real dataset from Douban[1], a well known social media network in China, which includes 3,022 users and 6,971 movies with 195,493 ratings ranging from 1 to 5. And another real dataset is employed from Yelp[2], a famous user review website in America, which includes 14,085 users and 14,037 movies with 194,255 ratings ranging from 1 to 5. The description of two datasets can be seen in Table 1 and their network schemas are shown in Fig. 2. The Douban dataset has sparse social relationship with dense rating information while the Yelp dataset has dense social relationships with sparse rating information.

---

[1] http://movie.douban.com/.
[2] http://www.yelp.com/.

Table 1. Statistics of Douban and Yelp dataset

Datasets	Relations of (A − B)	Number of A/B/A − B	Ave. degrees of A/B
Douban	User-Movie	3022/6971/195493	64.69/28.04
	User-User	779/779/1366	1.75/1.75
	User-Group	2212/2269/7054	3.11/3.11
	User-Location	2491/244/2491	1.00/10.21
	Movie-Director	3014/789/3314	1.09/4.20
	Movie-Actor	5438/3004/15585	2.87/5.19
	Movie-Type	6787/36/15598	2.29/433.28
Yelp	User-Business	14085/14037/194255	4.6/20.7
	User-User	9581/9581/150532	10.0/10.0
	Business-Category	14037/575/39406	2.8/73.9
	Business-Location	14037/62/14037	1.0/236.1

5.2 Comparison Methods and Metrics

In order to validate the effectiveness of DSR, we compare it with following representative methods. Besides the classical social recommendation method SoMF, the experiments also include two recent HIN based methods HeteCF and HeteMF. In addition, we include the revised version of SoMF with similarity regularization (i.e., SoMF$_{SR}$) to validate the effectiveness of similarity regularization.

- **UserMean**. It employs a user's mean rating to predict the missing ratings directly.
- **ItemMean**. It employs an item's mean rating to predict the missing ratings directly.
- **PMF** [8]. Salakhutdinov and Minh proposed the basic low-rank matrix factorization method for recommendation.
- **SoMF** [7]. Ma et al. proposed the social recommendation method with social regularization on users.
- **HeteCF** [4]. Luo et al. proposed the social collaborative filtering algorithm using heterogeneous relations.
- **HeteMF** [13]. Yu et al. proposed the HIN based recommendation method through combining user ratings and items' similarity matrices.
- **SoMF$_{SR}$**. It adapts SoMF through only replacing the social regularization with the similarity regularization $SimReg^{\mathcal{U}}$.

For Douban dataset, we utilize 7 meta paths for user (i.e., UU, UGU, ULU, UMU, UMDMU, UMTMU, UMAMU) and 5 meta paths for item (i.e., MTM, MDM, MAM, MUM, MUUM). For Yelp dataset, we utilize 2 meta paths for user (i.e., UB, UU) and 2 meta paths for item (i.e., BC, BL). HeteSim [9] is employed to evaluate the object similarity based on above meta paths. These similarity matrices are fairly utilized for HeteCF, HeteMF, and DSR. We set $\gamma = 1$, $\alpha = 10$, and $\beta = 10$ through parameter experiments on Douban dataset.

In the experiments on Yelp dataset, we set the parameters $\gamma = 1$, $\alpha = 10$, $\beta = 10$. Meanwhile, optimal parameters are set for other models in the experiments.

We use Mean Absolute Error (MAE) and Root Mean Square Error (RMSE) to evaluate the performance of rating prediction:

$$MAE = \frac{\sum_{(u,i)\in R} |R_{u,i} - \hat{R}_{u,i}|}{|R|}, \tag{11}$$

$$RMSE = \sqrt{\frac{\sum_{(u,i)\in R} (R_{u,i} - \hat{R}_{u,i})^2}{|R|}}, \tag{12}$$

where R denotes the whole rating set, $R_{u,i}$ denotes the rating user u gave to item i, and $\hat{R}_{u,i}$ denotes the rating user u gave to item i as predicted by a certain method. A smaller MAE or RMSE means a better performance.

5.3   Effectiveness Experiments

For Douban dataset, we use different ratios (80 %, 60 %, 40 %) of data as training sets and the rest of the dataset for testing. Considering the sparse density of Yelp dataset, we use 90 %, 80 %, 70 % of data as training sets and the rest of the dataset for testing for Yelp dataset. The random selection is carried out 10 times independently and we report the average results in Table 2.

It is clear that three HIN based methods (DSR, HeteCF, and HeteMF) all achieve significant performance improvements compared to PMF, UserMean, ItemMean and SoMF. It implies that integrating heterogeneous information is a promising way to improve recommendation performance. Particularly, DSR always has the best performance on all conditions compared to other methods. It indicates that the dual similarity regularization on users and items may be more effective than traditional social regularization. It can be further confirmed by the better performance of SoMF$_{SR}$ over SoMF. Although the superiority of SoMF$_{SR}$ over SoMF is not significant, the improvement is achieved on the very weak social relations in Douban dataset. In addition, we can also find that DSR has better performance improvement for less training data. It reveals that DSR has the potential to alleviate the cold-start problem.

5.4   Study on Cold-Start Problem

To validate the superiority of DSR on cold-start problem, we run PMF, SoMF, HeteCF, HeteMF, DSR on Douban dataset with 40 % training ratio. Four levels of users are seted: three types of cold-start users with various numbers of rated movies (e.g., [0,8] denotes users rated no more than 8 movies and "All" means all users in Fig. 3). We conduct similar experiments on cold-start items and users & items (users and items are both cold-start). The experiments are shown in Fig. 3. Once again, we find that 3 HIN-based methods all are effective for cold-start

Table 2. Effectiveness experimental results on Douban and Yelp (The improvement is based on PMF)

Dataset	Training	Metrics	PMF	UserMean	ItemMean	SoMF	HeteCF	HeteMF	SoMF$_{SR}$	DSR
Douban	80 %	MAE	0.6444	0.6954	0.6284	0.6396	0.6101	0.5941	0.6336	**0.5856**
		Improve		−7.92 %	2.47 %	0.73 %	5.32 %	7.79 %	1.68 %	9.12 %
		RMSE	0.8151	0.8658	0.7928	0.8098	0.7657	0.7520	0.8000	**0.7379**
		Improve		−6.23 %	2.73 %	0.64 %	6.05 %	7.73 %	1.85 %	9.46 %
	60 %	MAE	0.6780	0.6967	0.6370	0.6696	0.6317	0.6056	0.6648	**0.5946**
		Improve		−2.76 %	6.05 %	1.25 %	6.84 %	10.68 %	1.96 %	12.31 %
		RMSE	0.8569	0.8687	0.8135	0.8445	0.7901	0.7665	0.8358	**0.7483**
		Improve		−1.37 %	5.07 %	1.45 %	7.80 %	10.56 %	2.46 %	12.68 %
	40 %	MAE	0.7364	0.7009	0.6629	0.7245	0.6762	0.6255	0.7141	**0.6092**
		Improve		4.83 %	9.99 %	1.63 %	8.18 %	15.07 %	3.03 %	17.28 %
		RMSE	0.9221	0.8747	0.8747	0.9058	0.8404	0.7891	0.8950	**0.7629**
		Improve		5.14 %	5.13 %	1.76 %	8.86 %	14.42 %	2.94 %	17.27 %
Yelp	90 %	MAE	0.8475	0.9543	0.8822	0.8460	0.8461	0.8960	0.8459	**0.8158**
		Improve		−12.60 %	−4.09 %	0.18 %	0.17 %	−5.72 %	0.18 %	3.74 %
		RMSE	1.0796	1.3138	1.2106	1.0772	1.0773	1.1272	1.0772	**1.0369**
		Improve		−21.69 %	−12.13 %	0.22 %	0.21 %	−4.41 %	0.22 %	3.95 %
	80 %	MAE	0.8528	0.9621	0.8931	0.8527	0.8528	0.8907	0.8526	**0.8206**
		Improve		−12.82 %	−4.72 %	0.01 %	0.00 %	−4.44 %	0.01 %	3.78 %
		RMSE	1.0850	1.3255	1.2304	1.0849	1.0850	1.1195	1.0848	**1.0413**
		Improve		−22.17 %	−13.40 %	0.01 %	0.00 %	−3.18 %	0.02 %	4.03 %
	70 %	MAE	0.8576	0.9706	0.9062	0.8575	0.8576	0.8976	0.8575	**0.8250**
		Improve		−13.17 %	−5.67 %	0.01 %	0.00 %	−4.66 %	0.01 %	3.80 %
		RMSE	1.0894	1.3395	1.2547	1.0936	1.0894	1.1313	1.0894	**1.0461**
		Improve		−22.96 %	−15.17 %	−0.39 %	0.00 %	−3.85 %	0.00 %	3.97 %

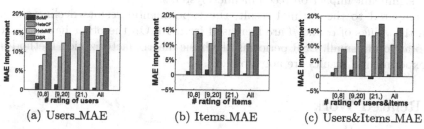

(a) Users_MAE   (b) Items_MAE   (c) Users&Items_MAE

Fig. 3. MAE improvement against PMF on various cold-start levels.

users and items. Moreover, DSR always has the highest MAE improvement on almost all conditions, due to dual similarity regularization on users and items. It's reasonable since the DSR method takes much constraint information of users and items into account which would play a crucial role when there's little available information of users or items.

5.5 Parameter Study on α and β

The DSR model is based on the low-rank matrix factorization framework and the similar regularization on users and items is applied to constrain the model

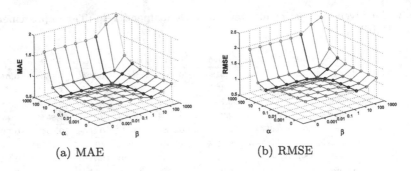

(a) MAE   (b) RMSE

Fig. 4. Parameter study on MAE and RMSE

learning process. The relevant parameters of the basic matrix factorization have been studied in other matrix factorization methods. In this section we only study α and β which are the parameters of dual similarity regularization on Douban dataset.

Figure 4 shows that the impacts of α and β on MAE and RMSE are quite similar. When the values of α and β are both around 10, the experiment has the best performance. When the values of α and β are quite large or small, the results are not ideal. When α and β set the proper value (in our experiments they are both 10), regularization and rating information take effect on the learning process simultaneously so that the experiments could get better performance. It indicates that integrating the similarity information of users and items in a HIN has a significant impact on recommender systems.

Compared to the optimal result, the experimental results decline sharply when the values of α and β are increased from 10. On the other hand, when α and β are quite small, DSR performs like basic matrix factorization method but the experimental results are not too bad.

6   Related Work

With the prevalence of social media, social recommendation has attracted many researchers. Ma et al. [6] fused user-item matrix with users' social trust networks. In [7], the social regularization ensures that the latent feature vectors of two friends with similar tastes to be closer. Yang et al. [12] inferred category-specific social trust circles from available rating data combined with friend relations.

To further improve recommendation performance, more and more researchers have been aware of the importance of heterogeneous information network (HIN), in which objects are of different types and links among objects represent different relations. Zhang et al. [14] investigated the problem of recommendation over heterogeneous network and proposed a random walk model to estimate the importance of each object in the heterogeneous network. Considering heterogeneous network constructed by different interactions of users, Jamali and Lakshmanan [3] proposed HETEROMF to integrate a general latent factor and context-dependent latent factors. Yu et al. [13] proposed Hete-MF through combining rating information and items' similarities derived from meta paths in HIN. More recently, Luo et al. [4] proposed a collaborative filtering-based social recommendation method, called Hete-CF, using heterogeneous relations.

7 Conclusions

In the paper, we analyzed the limitations of social regularization and designed a similarity regularization whose basic idea is to enforce the constraint on both similar and dissimilar objects. Then, we employ the similarity regularization on low-rank matrix factorization framework and proposed the DSR method. Experiments validate the effectiveness of DSR, especially on alleviating the cold-start problem.

Acknowledgments. This work is supported in part by National Key Basic Research and Department (973) Program of China (No. 2013CB329606), and the National Natural Science Foundation of China (No. 71231002, 61375058, 11571161), and the CCF-Tencent Open Fund, the Co-construction Project of Beijing Municipal Commission of Education, and Shenzhen Sci.-Tech Fund No. JCYJ20140509143748226.

References

1. BellogíN, R., Cantador, I., Castells, P.: A comparative study of heterogeneous item recommendations in social systems. Inf. Sci. **221**, 142–169 (2013)
2. Cantador, I., Bellogin, A., Vallet, D.: Content-based recommendation in social tagging systems. In: RecSys, pp. 237–240 (2010)
3. Jamali, M., Lakshmanan, L.V.: Heteromf: recommendation in heterogeneous information networks using context dependent factor models. In: WWW, pp. 643–653 (2013)
4. Luo, C., Pang, W., Wang, Z.: Hete-cf: Social-based collaborative filtering recommendation using heterogeneous relations. In: ICDM, pp. 917–922 (2014)
5. Ma, H., King, I., Lyu, M.R.: Learning to recommend with social trust ensemble. In: SIGIR, pp. 203–210 (2011)
6. Ma, H., Yang, H., Lyu, M.R., King, I.: Sorec: Social recommendation using probabilistic matrix factorization. In: CIKM, pp. 931–940 (2008)
7. Ma, H., Zhou, D., Liu, C., Lyu, M.R., King, I.: Recommender systems with social regularization. In: WSDM, pp. 287–296 (2011)
8. Salakhutdinov, R., Mnih, A.: Probabilistic matrix factorization. In: NIPS, pp. 1257–1264 (2012)
9. Shi, C., Kong, X., Huang, Y., Yu, P.S., Wu, B.: Hetesim: a general framework for relevance measure in heterogeneous networks. IEEE Trans. Knowl. Data Eng. **26**(10), 2479–2492 (2014)
10. Srebro, N., Jaakkola, T.: Weighted low-rank approximations. In: ICML, pp. 720–727 (2003)
11. Sun, Y., Han, J., Yan, X., Yu, P., Wu, T.: Pathsim: meta path-based top-k similarity search in heterogeneous information networks. In: VLDB, pp. 992–1003 (2011)

12. Yang, X., Steck, H., Liu, Y.: Circle-based recommendation in online social networks. In: KDD, pp. 1267–1275 (2012)
13. Yu, X., Ren, X., Gu, Q., Sun, Y., Han, J.: Collaborative filtering with entity similarity regularization in heterogeneous information networks. In: IJCAI-HINA Workshop (2013)
14. Zhang, J., Tang, J., Liang, B., Yang, Z., Wang, S., Zuo, J., Li, J.: Recommendation over a heterogeneous social network. In: The Ninth International Conference on Web-Age Information Management, WAIM 2008, pp. 309–316. IEEE (2008)

Collaborative Deep Ranking: A Hybrid Pair-Wise Recommendation Algorithm with Implicit Feedback

Haochao Ying[1]([⊠]), Liang Chen[2], Yuwen Xiong[1], and Jian Wu[1]

[1] College of Computer Science and Technology,
Zhejiang University, Hangzhou, China
{haochaoying,orpine,wujian2000}@zju.edu.cn
[2] School of Computer Science and Information Technology, RMIT,
Melbourne, Australia
liang.chen@rmit.edu.au

Abstract. Collaborative Filtering with Implicit Feedbacks (e.g., browsing or clicking records), named as CF-IF, is demonstrated to be an effective way in recommender systems. Existing works of CF-IF can be mainly classified into two categories, i.e., point-wise regression based and pair-wise ranking based, where the latter one relaxes assumption and usually obtains better performance in empirical studies. In real applications, implicit feedback is often very sparse, causing CF-IF based methods to degrade significantly in recommendation performance. In this case, side information (e.g., item content) is usually introduced and utilized to address the data sparsity problem. Nevertheless, the latent feature representation learned from side information by topic model may not be very effective when the data is too sparse. To address this problem, we propose collaborative deep ranking (CDR), a hybrid pair-wise approach with implicit feedback, which leverages deep feature representation of item content into Bayesian framework of pair-wise ranking model in this paper. The experimental analysis on a real-world dataset shows CDR outperforms three state-of-art methods in terms of recall metric under different sparsity level.

1 Introduction

With the growing community value of personalized services, recommendation techniques have been playing an significant role in online applications [15]. To provide personalized services, users' preference from their past feedback of items is critical. Generally, users' feedback can be classified into two categories: explicit and implicit. Explicit feedback (e.g. the graded ratings 1–5 in Netflix) expresses the users' true preference, which has been well studied in many literatures. However, users may be compelled to convey their rating values in some scenarios. Moreover, users just express their behaviors implicitly in many more situations, such as browsing or not browsing, clicking or not clicking in Web sites. The meaning of unobserved items are ambiguous because users may not like these items

J. Bailey et al. (Eds.): PAKDD 2016, Part II, LNAI 9652, pp. 555–567, 2016.
DOI: 10.1007/978-3-319-31750-2_44

or may be unaware of these items. Therefore, the scenario of recommendation with implicit feedback is more challenging.

Previous works based on implicit feedback include point-wise regression and pair-wise ranking preference algorithm [1, 9]. The point-wise regression algorithm supposes that users don't like all unobserved items and optimizes the absolute rating scores, while pair-wise ranking algorithm assumes that users' preference of observed items is stronger than unobserved items and then directly convert the prediction to rank. The latter algorithm actually relaxes assumption and usually obtains better performance in empirical studies. However, a user typically observes limited number of items and doesn't interact with thousands of items. Therefore, the data sparsity is a big problem for pair-wise ranking algorithm. With the increasing availability of auxiliary information about items (e.g., movie plots and item description), referred to as side information, it is wise to explore the possibility of using such information to improve the performance of pair-wise ranking algorithm through alleviating data sparsity [4].

Collaborative topic ranking (CTRank) is a recently proposed hybrid method, which seamlessly combines latent dirchlet allocation and pair-wise ranking model. Although this model learn the feature of side information associated with items, LDA is often not effective enough to learn the latent representation especially when side information is sparse [13]. Alternatively, deep learning, as a set of representation-learning methods, models multiple levels of representation of raw input by composing simple but non-linear modules that each transform the representation at one level into a representation at a higher, slightly more abstract level [3]. It has been proved that deep learning methods are expert in automatically mining and representing intricate structures in high-dimensional data.

In this paper, we propose a hybrid pair-wise recommendation approach with implicit feedback, named collaborative deep ranking (CDR), which integrates abstract representation of side information about items into Bayesian framework of pair-wise ranking model. Specifically, Stacked Denoising Autoencoders (SDAE) is exploited to extract the feature representation of item content. Cooperating with this, pair-wise ranking component can tackle sparsity problem to some extent and improve the recommendation accuracy. Note that, although CDR employs SDAE for feature representation, CDR as a generic framework also can collaborate with other deep learning methods, such as convolutional neural networks and recurrent neural networks.

The main contribution is summarized as follows:

1. We propose a hierarchical Bayesian framework, named as CDR, which combines deep feature presentation of item content and user implicit preference for sparsity reduction.
2. We conduct experiments on a real-world dataset to evaluate the effectiveness of CDR. Experimental result shows CDR outperforms three state-of-art methods in terms of recall metric under different sparsity level.

The remainder of this paper is organized as follows. Section 2 gives an overview of the related work. Section 3 demonstrates details of our proposed model. Section 4 shows the experimental results and Sect. 5 concludes the paper.

2   Related Work

In many practical recommendation scenarios, users rarely express their explicit behaviors, while implicit ones are more common (e.g. clicking and browsing history). This class of collaborative filtering with only positive examples is also called One-Class Collaborative Filtering (OCCF) [5]. There are mainly two types of existing approaches for OCCF: point-wise and pair-wise [14].

Point-wise algorithm directly optimizes the absolute value of binary rating. Hu et al. [1] estimate user-item pair preference whether the user would like or dislike the item and then assign a confidence level for this. After defining preference and confidence level, they join them into traditional probabilistic matrix factorization. However, the performance of CF-based models degrades significantly when facing data sparsity problem. Many models explore side information about items and users to alleviate this problem. Wang et al. [11] propose collaborative topic regression (CTR) for recommending scientific articles. In CTR, item content based on probabilistic topic model incorporates into traditional collaborative filtering. Based on this work, Purushotham [7] further study the influence of users' social network and propose CTR with SMF model. Recently, Wang et al. [13] employs deep learning model to automatically learn effective representation of content of items. From their experiments, we can observe that deep learning models are more appealing than traditional topic model in feature representation.

Different from point-wise model with intermediate step of predicting rating for recommendation, pair-wise algorithm generates a preference ranking of items for each user. Rendle et al. [8,9,15] assume that user prefers observed items than unobserved items and propose a generic Bayesian Personalized Ranking (BPR) framework optimization criterion. They also demonstrate matrix factorization and adaptive kNN learned by BPR are superior to the same model with respect to other criteria under AUC evaluation. Pan et al. [6] relax individual and independence assumption in BPR by adding the interaction of users and propose group Bayesian personalized ranking (GBPR). For sparsity reduction, several models extend pair-wise ranking techniques via taking extra side information into consideration. Grimberghe et al. [2] combine social graph and BPR matrix factorization for social network data. Yao et al. [14] propose a hierarchical Bayesian framework, which integrates latent dirchelet allocation into BPR matrix factorization. However, the topic model is not effective enough when side information is sparse. Therefore, in this paper, we integrate deep representation learning of content of items into pair-wise ranking model and propose a generalized hierarchical Bayesian model, called CDR.

3   Collaborative Deep Ranking

In this section, we present details of our proposed algorithm, CDR, which integrates pair-wise ranking models and side information about the items.

Notation and Problem Definition. Let $\mathcal{U}$ denote the set of users and $\mathcal{I}$ denote the set of items. The size of $\mathcal{U}$ and $\mathcal{I}$ are n and m, respectively. This paper focuses on implicit feedback recommendation scenarios, which means the implicit interaction matrix $R \in \mathcal{U} \times \mathcal{I}$ is available. Specifically, the elements $r_{ij} = 1$ in matrix R denotes user i prefers item j, while $r_{ij} = 0$ implies that user i is not interesting in item j or might not observe item j yet. For a given user i, pair-wise algorithms [9] suppose that user i prefers item j over item k if and only if $j \in \mathcal{I}^+$ and $k \in \mathcal{I}\backslash\mathcal{I}^+$, where $\mathcal{I}^+ = \{j : r_{ij} = 1\}$.

Except the observed binary matrix R, side information about items could be collected in many scenarios (e.g. item profile in Amazon). Given an $m \times s$ matrix X_c presents the side information about all items, the j-th row denotes the bag-of-words vector of item j based on vocabulary of size S (i.e. $X_{c,j*}$). Let u_i, v_j denote the latent factor with low dimension K of user i and item j, respectively. Our objective is to learn the latent factor $U = (u_i)_{i=1}^n$ and $V = (v_j)_{j=1}^m$ from implicit interaction and item information matrix for recommending an personalized ranking list for users.

Stacked Denoising Autoencoders. Generally, a good representation of side information about items can improve performance of Recommender System. Denosing Autoencoders (DAE) [10] learns an compressed representation from corrupted input to recover the clean input through a feedforward neural network. SDAE stacks DAE to form a deep network by feeding the output code of DAE found on the layer below as input to the current layer and the highest level output representation is used as item feature. An SDAE network is to minimize the regularized optimization problem as below:

$$\min_{\{W_l\},\{b_l\}} \|X_c - X_L\|_F^2 + \lambda_w \sum_l (\|W\|_F^2 + \|b\|_F^2), \tag{1}$$

where W_l and b_l is the weight matrix and bias vector of layer l, L is the number of layers, and λ_w is the regularization hyperparameter.

Supposing that the corrupted input X_0 and the clean input X_c are observed variables, SDAE can be generalized as a probabilistic model [12]. The generative process is as follows:

1. For each layer l of the SDAE network,
   (a) For each column n of the weight matrix W_l, draw $W_{l,*n} \sim \mathcal{N}(0, \lambda_w^{-1}I_{K_l})$.
   (b) Draw the bias vector $b_l \sim \mathcal{N}(0, \lambda_w^{-1}I_{K_l})$
   (c) For each row j of X_l, draw $X_{l,j*} \sim \mathcal{N}(\sigma(X_{l-1,j*}W_l + b_l), \lambda_s^{-1}I_{K_l})$
2. For each item j, draw a clean input,

$$X_{c,j*} \sim \mathcal{N}(X_{L,j*}, \lambda_n^{-1}I_m)$$

where I_{K_l} is a K-dimensional identity matrix of layer l, λ_w, λ_s, λ_n is the hyperparameters and $\sigma(.)$ is the sigmoid function Through maximizing a posteriori estimation, the model will degenerate to be the original SDAE if λ_s goes to infinity (i.e. $X_{l-1,j*} = \sigma(X_{l-1,j*} * W_l + b_l)$). After this process, $X_{\frac{L}{2}}$ could effectively present the latent feature representation of side information about all items.

Collaborative Deep Ranking. CDR exploits pair-wise preferences and content-based items feature together for collaborative filtering. Figure 1 shows the

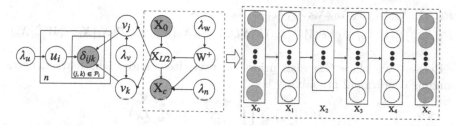

Fig. 1. The graphic model of CDR. SDAE with $L = 4$ is presented inside the dashed rectangle. Note that W^+ denotes the set of weight matrices and bias vectors of all layers (Color figure online).

graphic model of CDR. Obviously, there are two generative processes in our model. First, the original SDAE process (in the red dashed rectangle) extracts feature representation from side information about items and then integrates them into latent factor of items in pair-wise ranking model. Second, the pair-wise ranking model captures special relationship $\delta_{ijk} = r_{ij} - r_{ik}$, which delegates the preference of user i on item j and k. Unlike the point-wise approach directly predicting the value r_{ij}, pair-wise approach instead classifies the difference of $r_{ij} - r_{ik}$ [9]. The generative process of CDR is as follows:

1. For each layer l of the SDAE network,
   (a) For each column q, draw the weight matrix and bias vector W_l^+, draw $W_{l,*q}^+ \sim \mathcal{N}(0, \lambda_w^{-1} I_{K_l})$.
   (b) For each row j of X_l, draw $X_{l,j*} \sim \mathcal{N}(\sigma(X_{l-1,j*}W_l + b_l), \lambda_s^{-1} I_{K_l})$
2. For each item j,
   (a) Draw a clean input $X_{c,j*} \sim \mathcal{N}(X_{L,j*}, \lambda_n^{-1} I_m)$
   (b) Draw a latent item offset vector $\epsilon_j \sim \mathcal{N}(0, \lambda_v^{-1} I_K)$ and then set the latent item vector to be:
   $$v_j = \epsilon_j + X_{\frac{L}{2},j*}^T$$
3. For each user i,
   (a) Draw user factor vector $u_i \sim \mathcal{N}(0, \lambda_u^{-1} I_K)$
   (b) For each pair-wise preference $(j, k) \in \mathcal{P}_i$, where $\mathcal{P}_i = \{(j,k) : r_{ij} - r_{ik} > 0\}$, draw the estimator,
   $$\delta_{ijk} \sim \mathcal{N}(u_i^T v_j - u_i^T v_k, c_{ijk}^{-1})$$

Where confidence parameter c_{ijk} denotes how much user i prefers item j than item k. For simplicity, we set $c_{ijk} = 1$ in the experiments. Note that the linkage between SDAE and pair-wise ranking model is the middle layer $X_{\frac{L}{2},j*}$. In the extreme case, if $\varepsilon_j = 0$, the latent factor of items completely generates from content information, which will ignore the information contained in user preference matrix.

For learning model parameters, maximum a posterior probability (MAP) estimator can be utilized. Through Bayesian inference, we have

$$P(U, V, X_l, W^+|\delta, X_0, X_c, \lambda_u, \lambda_v, \lambda_s, \lambda_w, \lambda_n) \propto$$
$$P(U|\lambda_u)P(V|\lambda_v, X_{\frac{L}{2}})P(W^+|\lambda_w)P(\delta|U, V)P(X_c|X_L, \lambda_n)P(X_l|X_{l-1}, W_l^+, \lambda_s)$$
$$(2)$$

Because we place Gaussian priors on user, item and wight matrix, corresponding conditional probability is

$$P(U|\lambda_u) = \prod_{i=1}^{n} \mathcal{N}(u_i|0, \lambda_u^{-1}I_K)$$
$$P(V|\lambda_v, X_{\frac{L}{2}}) = \prod_{j=1}^{m} \mathcal{N}(v_j|X_{\frac{L}{2}}, \lambda_v^{-1}I_K) \tag{3}$$
$$P(W^+|\lambda_w) = \prod_{l=1}^{L} \mathcal{N}(0, \lambda_w^{-1}I_{K_l})$$

Similarly, we have below corresponding conditional probability based on assumption of Gaussian distribution:

$$P(X_c|X_L, \lambda_n) = \prod_{j=1}^{m} \mathcal{N}(X_{c,j*}|X_{L,j*}, \lambda_n^{-1}I_m)$$
$$P(X_l|X_{l-1}, W_l^+, \lambda_s) = \prod_{j=1}^{m} \mathcal{N}(X_{l,j*}|\sigma(X_{l-1,j*}W_l + b_l), \lambda_s^{-1}I_{K_l}) \tag{4}$$
$$P(\delta|U, V) = \prod_{ijk} \mathcal{N}(\delta_{ijk}|u_i^T v_j - u_i^T v_k, c_{ijk}^{-1})$$

Similar to the generalized SDAE, λ_s goes to infinity and maximization of posterior probability is equivalent to maximizing the joint log-likelihood of U, V, X_l, X_c, W^+, and δ given λ_u, λ_v, λ_n,

$$\mathcal{L} = -\sum_{ijk} \frac{c_{ijk}}{2}(\delta_{ijk} - (u_i^T v_j - u_i^T v_k))^2 - \frac{\lambda_w}{2}\sum_j (W_l^2 + b_l^2)$$
$$-\frac{\lambda_u}{2}\sum_i u_i^T u_i - \frac{\lambda_v}{2}\sum_j (v_j - X_{\frac{L}{2},j*}^T)^2 - \frac{\lambda_n}{2}\sum_j (X_{L,j*} - X_{c,j*})^2 \tag{5}$$

In the generative process, we assume that the preference δ_{ijk} follows Gaussian Distribution. However, similar loss function can be obtained with different assumption on δ_{ijk} (e.g. Bernoulli distribution in [14]). Note that the model CTRank, proposed in [14], is analogous to our model, which also combines pair-wise ranking and side information about items. The big difference is that CTRank extracts topic proportions from content of items to conduct the learning of latent factors for ranking, while CDR exploits deep network to mine effective feature representation of items. Note that prior distribution of LDA-based models is difficult to define. What's worse, topic proportions can not effectively represent the latent feature of items when side information is very sparse. As showed in the experiments, CDR gets better performance.

The first term in Eq. 5 extracts user preference from implicit matrix $\mathcal{R}$ to construct pair-wise ranking loss, while the fourth term integrates content of items. Therefore, if two item j and k have similar side information (i.e. similar $X_{\frac{L}{2}}$), the distance between v_j and v_k will be reduced. As we have mentioned,

$X_{\frac{L}{2}}$ serves as a bridge between pair-wise ranking and SDAE model. When λ_v/λ_n goes to positive infinity, the disappeared reconstruction error will lead to invalid feature representation $X_{\frac{L}{2}}$, meanwhile $X_{\frac{L}{2}}$ dominate the learning process of V. On the other hand, when λ_v/λ_n approaches to zeros, CDL will decouple into two models and the learned V is not influenced by side information about items. Both extreme cases demonstrate bad performance in the experiments.

Parameter Learning. Similar to [1,13,14], we optimize this function using coordinate ascent by alternatively optimizing latent factors u_i, v_j and weight matrix & bias vector W^+. Given a current estimate of W^+, we update u_i, v_j and v_k based on the following stochastic Newton-Raphson rules:

$$u_i = u_i - \alpha \frac{\lambda_u u_i - c_{ijk}\mathcal{E}_{ijk}(v_j - v_k)}{\lambda_u + c_{ijk}(v_j - v_k)^T(v_j - v_k)}$$

$$v_j = v_j - \alpha \frac{\lambda_v(v_j - X^T_{\frac{L}{2},j*}) - c_{ijk}\mathcal{E}_{ijk}u_i}{\lambda_v + c_{ijk}u_i^T u_i} \tag{6}$$

$$v_k = v_k - \alpha \frac{\lambda_v(v_k - X^T_{\frac{L}{2},k*}) + c_{ijk}\mathcal{E}_{ijk}u_i}{\lambda_v + c_{ijk}u_i^T u_i}$$

where α is learning rate and $\mathcal{E}_{ijk} = \delta_{ijk} - (u_i^T v_j - u_i^T v_k)$. Note that when updating, bootstrap sampling is applied to sample observed item j and unobserved item k of user i [9].

Given U and V, wight matrix W_l and bias vector b_l for each layer update by back-propagation learning algorithm. The gradient of $\mathcal{L}$ with respect to W_l and b_l is as follows:

$$\nabla_{W_l}\mathcal{L} = -\lambda_w W_l - \lambda_v \sum_j \nabla_{W_l} X^T_{\frac{L}{2},j*}(X_{\frac{L}{2},j*} - v_j) - \lambda_n \sum_j \nabla_{W_l} X_{L,j*}(X_{L,j*} - X_{c,j*})$$

$$\nabla_{b_l}\mathcal{L} = -\lambda_w b_l - \lambda_v \sum_j \nabla_{b_l} X^T_{\frac{L}{2},j*}(X_{\frac{L}{2},j*} - v_j) - \lambda_n \sum_j \nabla_{b_l} X_{L,j*}(X_{L,j*} - X_{c,j*}) \tag{7}$$

Prediction. After learning the optimal parameters U,V,W^+, we predict R_{ij} from its expectation:

$$E[R_{ij}|U,V,W^+,...] \approx u_i^T(X_{\frac{L}{2}} + \epsilon_j) = u_i v_j,$$

and then a ranked list of items is generated for each user based on these prediction values.

Complexity Analysis. According to updating rules, the complexity of computing U is approximately $O(nrK)$ where r is the average number of items a user interacts. The complexity of computing the output of encoder is controlled by the computation of first layer. Therefore, the complexity of updating V is $O(nrK + sK_1)$, where K_1 is the dimension of first layer. The complexity of updating all the wights and bias is $O(msK_1)$. Hence, the total complexity is $O(2nrK + sK_1 + msK_1)$.

4   Experiments

In this section, we compare performance of our approach with some state-of-art algorithms. All experiments are conducted on a server with 2 Intel E5-2620 CPUs and 1 GTX Titan GPU.

Datasets. To effectively illustrate the performance of CDR, we use the same dataset in [11,13,14]. The dataset is collected from CiteULike[1], which provides service for managing and discovering articles for users. In this dataset, if a user has collected an article in his library, we consider that the user implicitly prefers the article, rating as '1' otherwise '0'. The preliminary statistics shows that the dataset contains 5,551 users and 16,980 articles with 204,986 user-item preference pairs. Note that the sparsity is 99.78 % and each user has at least 10 articles in their preference library. To obtain the side information about articles, the title and abstract of articles are exploited. After removing stop words, we extract 8000 distinct words through sorting their TF-IDF values. As a result, the size of X_c is 16980 × 8000 as clean input of SDAE.

Evaluation. Similar to [7,11], we employ the metric recall to quantize the performance of recommendation, since the metric precision is not suited to implicit feedback datasets. Because the meaning of zero entry in the user-item matrix is ambiguous, which represent either user don't like item or is unaware of item. Instead, the positive rating (e.g. $r_{ij} = 1$) only hints the user i likes the item j, we focus on recall metric. Specifically, after predicting the ratings in the test dataset, we sort them and recommend top M items for each user. The recall@M is defined as follows:

$$recall@M = \frac{\text{number of items the user likes in Top M}}{\text{total number of items the user likes}}$$

Average recall from all users points out the performance of method.

Baselines and Experiments Setting. In order to evaluate effectiveness of our approach, CDR, we compare it with three state-of-art hybrid recommendation algorithm for implicit feedback as follows:

- **CTR.** Collaborative Topic Regression is a point-wise algorithm, which combines probability matrix factorization and latent dirchelet allocation [11].
- **CTRank.** Collaborative Topic Ranking, a pair-wise algorithm, which integrates side information of items into Bayesian personalized ranking [14]. With different assumption of preference, Two algorithms, CTRank-log and CTRank-squared, have been proposed. We choose CTRank-squared as our compared approach, because it has higher performance than CTRank-log.
- **CDL.** Collaborative deep learning is a point-wise hierarchical Bayesian model, which first tightly couples deep representation feature of content information and collaborative filtering [13].
- **CDR.** Collaborative Deep Ranking is our proposed model described in Sect. 3.

---

[1] http://www.citeulike.org/.

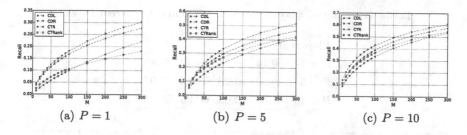

Fig. 2. Performance comparison of CDR, CDL, CTRank, CTR under different P.

Similar to [13], we randomly choose P items from each user to consist of train set and take all the rest as test set in the experiments. In particular, we vary train set sparsity is $0.006\,\%$, $0.03\,\%$, $0.06\,\%$ (i.e. $P = 1, 5, 10$). Each approach is performed 5 times with different random seeds for each sparsity and the average performance is reported. The grid search is applied to find optimal hyperparameters for each approach. For CTR, $\lambda_u = 0.1$, $\lambda_v = 10$, $a = 1$, $b = 0.01$, $K = 50$, and $\alpha = 1$ can reach good performance (note that α is the dirchelet prior). For CTRank, we find $\lambda_u = 0.025$, positive $\lambda_v = 0.25$, negative $\lambda_v = 0.025$, $K = 200$, and $c = 1$ can achieve best results. For CDL, we set the same parameters of $a = 1$, $b = 0.01$, $K = 50$, $\lambda_w = 0.0001$, and a 2-layer SDAE architecture '8000-200-50-200-8000' for different P. Otherwise, $\lambda_u = 1$, $\lambda_v = 10$, and $\lambda_n = 1000$ when $P = 1$ and 5, while $\lambda_u = 0.1$, $\lambda_v = 1$, and $\lambda_n = 100$ when $P = 10$.

In the pretrain of CDR, SDAE employs a mixture of edge detectors and grating filters (i.e. masking noise) with a noise level of $30\,\%$ to obtain the corrupted input X_0 from the clean input X_c. Meanwhile, dropout rate is set to 0.1 for achieving adaptive regularization when the number of layers is more than 2. The number of hidden units K_l is set to 1000 ($l \neq \frac{L}{2}$), while the number of middle layer is 200. That is, the dimension of feature representation and latent factor u_i, v_j is 200. Note that K_0 and K_L are equal to the size of vocabulary. After grid searching, we find that the hyperparameters $\lambda_u = 0.01$, $\lambda_v = 0.1$, $\lambda_n = 5$, and $\lambda_w = 0.0001$ can achieve good performance when $P = 5$ and $P = 10$, while we set larger hyperparameters (i.e. $\lambda_u = 1$, $\lambda_v = 10$, $\lambda_n = 1000$, and $\lambda_w = 0.0001$) to prevent overfitting when $P = 1$.

Comparison. Figure 2 provides comparison results of CDR, CDL, CTRank and CTR under different sparsity. A 3-layers CDR '8000-1000-200-1000-8000' is used. It can be observed that our proposed approach outperforms other three methods at all sparsity levels. As a whole, baseline CDL performs better than CTR and CTRank. That is, deep learning approach (e.g. SDAE) can admire better feature quality of side information about items than topic model (e.g. LDA). The reason may be that deep learning approach captures distributed features, while the features (i.e., topics) in LDA is independent. Otherwise, CTRank outperforms CTR in most case (both models use LDA model) and CDR outperforms CDL (both model employ deep learning architecture). Therefore, pair-wise algorithm

Table 1. Impact of #layers at recall @300 under different P(%)

#layers	1	2	3	4
$P = 1$	9.74	13.43	30.32	30.89
$P = 5$	49.62	49.26	51.84	47.07
$P = 10$	61.09	59.03	60.96	59.41

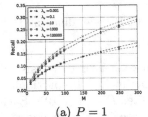 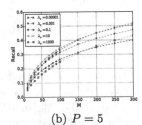 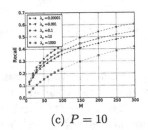

(a) $P = 1$ (b) $P = 5$ (c) $P = 10$

Fig. 3. The impact of λ_v/λ_n under different P.

with directly optimizing ranking has advantage over point-wise method which optimizes rating. Concretely, when recommending 300 articles CDR relatively improves 36.58 %, 66.96 %, 9.22 % than CTR, CTRank, CDL respectively at $P = 1$ while the value is 13.44 %, 11.34 %, 5.25 % at $P = 10$. Thus, when data is sparse, the relative improvement is more significant. That is, CDR can alleviate data sparsity to some extent.

Impact of #layers. Table 1 presents the recall@300 under different P with various #layers. As we can observe, when the number of layers is 1 and 2, the recall is quite low at $P = 1$, while the performance significantly enhance with the number of layers growing. That is, the performance of recommendation depends on the quality of feature representation of side information about items when the data is extremely sparsity. As reducing the sparsity degree, the effective of pair-wise ranking component begins to present and CDR starts to overfit when $\#layers = 4$ and $P = 5$. We also can find that the recall value is similar when $P = 10$ in different number of layers. That means with increasing of train set size, pair-wise ranking model can guide the further learning of features in CDR model.

Impact of λ_v/λ_n. Figure 3 shows the impact of λ_v/λ_n under different P, via changing λ_v and fixing other hyperparameters. We can observe that as increasing or reducing the λ_v from the optimal λ_v, the performance degrades gradually. This result is consistent to the explanation in Sect. 3. When λ_v/λ_n is large, the side information about items dominates the learning process of V and the performance purely depend on $X_{\frac{L}{2}}$. When λ_v/λ_n is small, the performance purely generates by the pair-wise ranking component. The experimental result indicates that appropriately combining pair-wise ranking and content of items can achieve better performance than in above two extreme case.

Table 2. An example of validity of CDR.

Top 3 topics	1. users-user-semantic-similarity-collaborative-filtering-items-recommendations-recommender-implicit	
	2. social-individuals-tags-tagging-tag-navigation-networking-ties-emergent-popularity	
	3. web-search-pages searching-page-engine-engines-google-searches-pagerank	
Top 10 articles	1. getting our head in the clouds toward evaluation studies of tagclouds	False
	2. usage patterns of collaborative tagging systems	True
	3.tagbased social interest discovery	False
	4. recommending scientific articles using citeulike	True
	5. collaborative filtering recommender systems	True
	6. open user profiles for adaptive news systems help or harm	True
	7. can all tags be used for search	True
	8. optimizing web search using social annotations	True
	9. evaluating collaborative filtering recommender systems	True
	10. can social bookmarking improve web search	True

Latent Factor Interpretability. To demonstrate the validity of CDR deeply, Table 2 show one example users of top 3 topic of his all articles and the top 10 recommended articles under the setting $P = 10$. From the top 2 topics, we can speculate the user focus on tag recommendation research, while the user also study web search based on the third topic. CDR successfully captures all three topics and reach 80 % recall when recommending top 10 articles. It is worth mentioning, the rank of recommended articles of the first topic is higher than the third topic.

5   Conclusion

In this paper, we propose a hybrid recommendation approach (CDR) with implicit feedback. Specifically, CDR employs SDAE to extract deep feature representation from side information and then integrates into pair-wise ranking model for alleviating sparsity reduction. Our study presents that CDR outperforms other three state-of-art algorithms at all sparsity level. In the future, we plan to use other deep learning methods to replace SDAE for boosting further performance in our hierarchical Bayesian framework. For example, the convolutional neural network which considers the context and order of words may improve the performance. Beyond that, we also consider how to incorporate other side information into our framework, such as users social network and items relationship.

Acknowledgement. This research was partially supported by the Natural Science Foundation of China under grant of No. 61379119, Science and Technology Program of Zhejiang Province under grant of No. 2013C01073, the Open Project of Qihoo360 under grant of No. 15-124002-002.

References

1. Hu, Y., Koren, Y., Volinsky, C.: Collaborative filtering for implicit feedback datasets. In: Eighth IEEE International Conference on Data Mining, pp. 263–272. IEEE (2008)
2. Krohn-Grimberghe, A., Drumond, L., Freudenthaler, C., Schmidt-Thieme, L.: Multi-relational matrix factorization using Bayesian personalized ranking for social network data. In: Proceedings of the Fifth ACM International Conference on Web Search and Data Mining, pp. 173–182. ACM (2012)
3. LeCun, Y., Bengio, Y., Hinton, G.: Deep learning. Nature **521**(7553), 436–444 (2015)
4. Ning, X., Karypis, G.: Sparse linear methods with side information for top-n recommendations. In: Proceedings of the Sixth ACM Conference on Recommender Systems, pp. 155–162. ACM (2012)
5. Pan, R., Zhou, Y., Cao, B., Liu, N.N., Lukose, R., Scholz, M., Yang, Q.: One-class collaborative filtering. In: Eighth IEEE International Conference on Data Mining, pp. 502–511. IEEE (2008)
6. Pan, W., Chen, L.: GBPR: group preference based bayesian personalized ranking for one-class collaborative filtering. In: Proceedings of the Twenty-Third International Joint Conference on Artificial Intelligence, pp. 2691–2697. AAAI Press (2013)
7. Purushotham, S., Liu, Y., Kuo, C.C.J.: Collaborative topic regression with social matrix factorization for recommendation systems. In: Proceedings of the 29th International Conference on Machine Learning (2012)
8. Rendle, S., Freudenthaler, C.: Improving pairwise learning for item recommendation from implicit feedback. In: Proceedings of the 7th ACM International Conference on Web Search and Data Mining, pp. 273–282. ACM (2014)
9. Rendle, S., Freudenthaler, C., Gantner, Z., Schmidt-Thieme, L.: BPR: Bayesian personalized ranking from implicit feedback. In: Proceedings of the Twenty-Fifth Conference on Uncertainty in Artificial Intelligence, pp. 452–461. AUAI Press (2009)
10. Vincent, P., Larochelle, H., Lajoie, I., Bengio, Y., Manzagol, P.A.: Stacked denoising autoencoders: learning useful representations in a deep network with a local denoising criterion. J. Mach. Learn. Res. **11**, 3371–3408 (2010)
11. Wang, C., Blei, D.M.: Collaborative topic modeling for recommending scientific articles. In: Proceedings of the 17th ACM SIGKDD International Conference on Knowledge Discovery and Data Mining, pp. 448–456. ACM (2011)
12. Wang, H., Shi, X., Yeung, D.Y.: Relational stacked denoising autoencoder for tag recommendation. In: Twenty-Ninth AAAI Conference on Artificial Intelligence (2015)
13. Wang, H., Wang, N., Yeung, D.Y.: Collaborative deep learning for recommender systems. In: Twenty-First ACM SIGKDD Conference on Knowledge Discovery and Data Mining (KDD) (2015)

14. Yao, W., He, J., Wang, H., Zhang, Y., Cao, J.: Collaborative topic ranking: leveraging item meta-data for sparsity reduction. In: Twenty-Ninth AAAI Conference on Artificial Intelligence(2015)
15. Zhong, H., Pan, W., Xu, C., Yin, Z., Ming, Z.: Adaptive pairwise preference learning for collaborative recommendation with implicit feedbacks. In: Proceedings of the 23rd ACM International Conference on Conference on Information and Knowledge Management, pp. 1999–2002. ACM(2014)

Author Index

Adnan, Md. Nasim I-304
Aghaee, Amin I-253
Agrawal, Rakesh II-376
Alm, Cecilia Ovesdotter I-477
Al-Maskari, Sanad I-578
Araujo, Miguel I-461
Ashely, David I-152

Baek, Jae Yeon I-52
Baghshah, Mahdieh Soleymani I-253
Banerjee, Arunava II-3
Barlow, Michael II-168
Bélisle, Eve I-578
Bergeron, Frédéric II-42
Berkovsky, Shlomo I-527
Blömer, Johannes II-296
Boström, Henrik I-77
Bouchard, Bruno II-42
Bouchard, Kevin II-42
Bujna, Kathrin II-296

Calvelli, Cara I-477
Cao, Jianping I-127
Cao, Lele II-257
Cao, Xiaohuan I-449
Cats, Oded I-552
Cerqueira, Vitor I-552
Chakraborty, Tanmoy II-528
Chandra, Anca I-514
Chang, Chia-Hui II-478
Chawla, Sanjay I-139, I-409
Chen, Enhong II-130
Chen, Fang I-527, I-540, I-565
Chen, Jian I-3
Chen, Liang I-436, II-515, II-555
Chen, Ling I-203
Chen, Ming-Chuan II-478
Chen, Shijiang II-104
Chen, Tian-Sheng II-478
Chen, Wei II-67, II-245
Chen, Wenjie I-191
Cheng, Reynold II-363
Cheung, Yiu-ming I-14
Chinnakotla, Manoj II-454

Chiu, Dah Ming II-515
Christen, Peter II-283, II-338, II-415

Das, Arpita II-454
Deng, Jeremiah D. I-165
Ding, Guiguang II-104
Ding, Jhung-Li II-478
Dobbie, Gillian II-233
Du, Changde I-239
Du, Changying I-239
duVerle, David I-277

Endo, Yuki II-54
Erfani, Sarah M. II-183

Faloutsos, Christos I-461, II-376
Fard, Mahtab J. I-139
Fatima, Asra I-489
Feng, Xiaodong II-79
Fisher, Jeffrey II-338
Fong, Simon I-565
Fu, Bin I-191

Gaboury, Sébastien II-42
Ganguly, Niloy II-528
Gao, Jingyi II-283
Gao, Lu II-271
Garimella, Kiran I-409
Ge, Jiaqi II-17
Ghadiri, Mehrdad I-253
Ghafoori, Zahra II-183
Ghanavati, Mojgan I-565
Giacometti, Arnaud II-196
Gionis, Aristides I-409
Giroux, Sylvain II-42
Goyal, Pawan II-528
Guan, Naiyang I-591
Guan, Xin II-503
Guan, Yu II-503
Guo, Guangming II-130
Guo, Ji-Min II-443
Guo, Jinma I-227
Guo, Xuan I-477
Guo, Yuchen II-104

Guo, Yunhui   I-502
Gupta, Manish   I-489
Gupta, Prashant   I-489
Gupta, Sunil Kumar   I-102, I-152, II-388

Ha, Yajun   II-233
Haake, Anne   I-477
Han, Shuchu   II-309
Haraguchi, Makoto   I-423
He, Qing   I-239
Hooi, Bryan   II-376
Hsieh, Hsun-Ping   I-177
Hu, Qinghua   I-65
Huang, Kuan-Hao   II-143
Huang, Tian   II-233
Huang, Yourong   I-191

Ibrahim, A.   I-265
Iqbal, Muhammad   II-117
Islam, Md. Zahidul   I-304

Jadav, Divyesh   I-514
Jiang, Yong   II-443
Johansson, Ulf   I-77
Joy, Tinu Theckel   I-102

Kantarcioglu, Murat   II-350
Karunasekera, Shanika   II-183
Karypis, George   I-89
Kawanobe, Akihisa   II-54
Khiari, Jihed   I-552
Koprinska, Irena   I-527
Kramer, Stefan   I-328
Krishna, Amrith   II-528
Kulkarni, Ashish   I-290
Kumar, Vishwajeet   I-290

Le Digabel, Sébastien   I-578
Le, Trung   I-27
Leckie, Christopher A.   II-183
Lee, Sunhwan   I-514
Lei, Kai   II-67, II-245
Li, Bin   I-3, I-203, I-527
Li, Bingyang   II-245
Li, Chang-Tsun   II-503
Li, Cheng   I-152
Li, Cheng-Te   I-177
Li, Chenxiao   II-104
Li, Dan   I-52

Li, Hongyan   II-67
Li, Jianmin   I-227
Li, Jingzhi   I-449, II-542
Li, Rui   I-477
Li, Shun   II-67
Li, Xiaoli   I-395
Li, Xinyang   II-233
Li, Xue   I-578
Liang, Biwei   II-67
Liang, Tingting   I-436
Liao, Qing   I-591
Liao, Yongxin   I-3
Lin, Hailun   II-325
Lin, Hanhe   I-165
Lin, Hsuan-Tien   I-115, II-143
Lin, Shou-De   II-491
Lin, Zheng   II-325
Linusson, Henrik   I-77
Liu, Jian   II-542
Liu, Mengyun   II-233
Liu, Qi   II-130
Liu, Wei   I-540
Liu, Yingling   II-130
Lo, Hung-Yi   II-491
Löfström, Tuve   I-77
Long, Guodong   I-40
Long, Guoping   I-239
Lu, Jiaheng   II-156
Lu, Yang   I-14
Luo, Ali   I-239
Luo, Jiebo   II-92
Luo, Ling   I-527
Luo, Wei   I-152

Mahmood, Abdun Naser   II-168
Mai, Khai   II-431
Mai, Sang   II-431
Mao, Yishu   II-233
Mendes-Moreira, João   I-215
Moreira-Matias, Luis   I-552
Mukherjee, Animesh   II-528

Nakajima, Yuta   II-221
Nawahda, Amin   I-578
Nguyen, Anh   II-431
Nguyen, Hoang   I-540
Nguyen, Khanh   I-27
Nguyen, Minh Luan   I-369
Nguyen, Thanh Dai   II-388

Nguyen, Vu   I-27
Nishida, Kyosuke   II-54
Nishimura, Naoki   II-221

Ojha, Himanshu   I-290
Okubo, Yoshiaki   I-423

Palshikar, Girish Keshav   II-208
Papalexakis, Evangelos   II-376
Park, Laurence A.F.   I-382
Petitjean, François   I-341
Pham, Trang   II-30
Phung, Dinh   I-27, I-152, II-30
Pinto, Fábio   I-215
Polyzou, Agoritsa   I-89

Qiao, Fengcai   I-127
Qin, Hong   II-309
Qiu, Minghui   II-130

Rajasegarar, Sutharshan   II-183
Ramakrishnan, Ganesh   I-290
Rana, Santu   I-102, I-152, II-388
Ranbaduge, Thilina   II-415
Reddy, Chandan K.   I-139
Ribeiro, Pedro   I-461
Rivera, Paul   I-540
Roos, Teemu   I-316

Sahu, Kuleshwar   II-208
Sastry, P.S.   I-265
Sastry, Shivakumar   I-265
Shah, Zubair   II-168
Shen, Bilong   I-354
Shi, Chuan   I-449, II-542
Shi, Hong   I-65
Shi, Pengcheng   I-477
Shrivastava, Manish   II-454
Singh, Mayank   II-528
Singh, Pankaj   I-290
Sk, Minhazul Islam   II-3
Soares, Carlos   I-215
Song, Hyun Ah   II-376
Soulet, Arnaud   II-196
Spanos, Costas J.   I-52
Srivastava, Rajiv   II-208
Stone, Glenn   I-382
Sun, Fuchun   II-257
Sun, Haiqi   II-363
Sun, Yuqing   II-363

Takata, Noboru   II-221
Takeishi, Naoya   II-221
Tang, Yuan Yan   I-14
Tang, Zhiwei   II-79
Terada, Aika   I-277
Than, Khoat   II-431
Ting, Chia-Hsin   II-491
Toda, Hiroyuki   II-54
Tomita, Etsuji   I-423
Tran, Truyen   II-30
Tsang, Dominic   I-409
Tsuda, Koji   I-277
Tyukin, Andrey   I-328

Vaddavalli, Pravin K.   I-489
Van Linh, Ngo   II-431
Vatsalan, Dinusha   II-415
Venkatesh, Svetha   I-102, I-152, II-30, II-388
Verykios, Vassilios   II-415

Wan, Yao   I-436
Wang, Dong   I-191
Wang, Feiyue   I-127
Wang, Hui   I-127
Wang, Lulu   I-395
Wang, Qing   II-283, II-338
Wang, Senzhang   I-127
Wang, Tengjiao   II-67, II-245
Wang, Wei   II-156
Wang, Weiping   II-325
Wang, Xiaodong   II-466
Wang, Xiaorong   II-245
Wang, Xin   I-502
Wang, Yang   I-565
Wang, Yuanzhuo   II-325
Wang, Zhikuan   II-401
Webb, Geoffrey I.   I-341
Wen, Zhaoduo   I-354
Wicker, Jörg   I-328
Wong, Raymond K.   I-565
Woodford, Brendon J.   I-165
Wu, Bin   I-449, II-542
Wu, Chengkun   I-591
Wu, Jian   I-436, II-555
Wu, Jun   II-92
Wu, Le   II-130
Wu, Qingyao   I-3
Wu, Sen   II-79
Wu, Wei   I-203

Wu, Yafei   II-233
Wu, Yingjie   II-466

Xia, Shu-Tao   II-443
Xia, Yuni   II-17
Xiang, Zhiyang   I-191
Xiao, Chunjing   I-395
Xiao, Zhu   I-191
Xie, Zongxia   I-65
Xiong, Yuwen   II-555
Xu, Congfu   I-502
Xu, Xin   II-156
Xue, Bing   II-117

Yairi, Takehisa   II-221
Yan, Rui   I-177
Yang, Chunfeng   II-515
Yang, Haolin   II-257
Ying, Haochao   II-555
You, Sheng-Chi   I-115
Yu, Hong   II-271
Yu, Man   I-65
Yu, Philip S.   I-127
Yu, Qi   I-436, I-477
Yuan, Jianbo   II-92
Yuan, Shenxi   I-3
Yue, Yinliang   II-325

Zaidi, Nayyar A.   I-341
Zhai, Hongjie   I-423
Zhang, Chengqi   I-40, I-203

Zhang, Mengjie   II-117
Zhang, Peng   I-40, II-325
Zhang, Qian   I-591
Zhang, Qin   I-40
Zhang, Qinzhe   I-40
Zhang, Shifeng   I-227
Zhang, Xianchao   II-271
Zhang, Xiaopeng   II-515
Zhang, Yue   II-401
Zhang, Yuxiang   I-395
Zhang, Yuxiao   II-245
Zhao, Deli   II-257
Zhao, Ying   I-354
Zhe, Shandian   I-239
Zheng, Hai-Tao   II-443
Zheng, Jing   II-542
Zheng, Weimin   I-354
Zheng, Yuyan   I-449
Zhong, Jiang   I-578
Zhou, Dongliang   I-354
Zhou, Wenjun   II-79
Zhou, Yan   II-350
Zhou, Yipeng   II-515
Zhou, Yuxun   I-52
Zhu, Daxin   II-466
Zhu, Feida   II-130
Zhu, Yongxin   II-233
Zhu, Youwen   II-401
Zhuang, Fuzhen   II-542
Zou, Yuan   I-316

Printed in the United States
By Bookmasters